Clinical Regenerative Medicine in Urology

Bup Wan Kim
Editor

Clinical Regenerative Medicine in Urology

Springer

Editor
Bup Wan Kim
Department of Urology
Kyungpook National University Hospital
Daegu
South Korea

ISBN 978-981-10-2722-2 ISBN 978-981-10-2723-9 (eBook)
DOI 10.1007/978-981-10-2723-9

Library of Congress Control Number: 2017954376

Printed on acid-free paper

This Springer imprint is published by Springer Nature
The registered company is Springer Nature Singapore Pte Ltd.
The registered company address is: 152 Beach Road, #21-01/04 Gateway East, Singapore 189721, Singapore

Preface

Regenerative medicine has emerged as an innovative scientific field that focuses on developing new approaches to repairing cells, tissues, and organs. Over the years, many therapeutic strategies have been developed to address clinical shortcomings by enhancing cellular functions and building viable tissues and organs in the field of medicine. This book gathers recent research and clinical efforts on regenerative medicine applications in urology. Specifically, this multidisciplinary book contains reviews on the development of clinical approaches, which combine stem/progenitor cells with biomaterials and scaffolds and growth factors and other bioactive agents in the field of urology. *Clinical Regenerative Medicine in Urology* aims to provide clinicians and researchers a broad perspective on the development of regenerative medicine technologies with the intention of disseminating regenerative medicine principles and various therapeutic approaches that target clinical applications in urology. These include upper and lower urinary tract dysfunctions, urinary incontinence, neurogenic bladder, and erectile dysfunction.

I would like to thank the authors who contributed to the making of this book as well as the publishing team at Springer. I hope that this book will encourage urologists to investigate the new paradigms in regenerative medicine research and explore a means to apply the translational concepts to their clinical practices.

Daegu, South Korea | Bup Wan Kim

Contents

Part I

Introduction

Current Developments and Future Perspectives of Tissue Engineering and Regenerative Medicine

1

Ji Hyun Kim and James J. Yoo

1.1 Introduction

Tissue engineering and regenerative medicine (TERM) is an interdisciplinary field encompassing many disciplines, including engineering, medicine, and science. The field has gained an enormous attention due to its potential to replace damaged human tissues and organs and restore normal function [1–3]. Although the early concept of tissue engineering was developed based on cell culture techniques, recent advances in the field combine multiple innovative technologies to accelerate the translation of clinical therapies. In fact, a number of TERM technologies have advanced to human clinical trials and commercialized [4, 5]. In this chapter, we review the current developments and recent progresses made in the field of TERM and discuss the most relevant challenges and future perspectives in the translation of TERM research.

J.H. Kim (✉) • J.J. Yoo
Wake Forest Institute for Regenerative Medicine, Wake Forest University School of Medicine, Winston-Salem, NC, USA
e-mail: jihkim@wakehealth.edu; jyoo@wakehealth.edu

1.2 Current Developments in TERM

Aligned with the goals of TERM, research activities have advanced in many disciplines toward developing new therapies that overcome the current limitations associated with conventional medical and surgical therapies. Over the years, numerous strategies have been applied for regenerating or replacing damaged tissues and organs, which led to the development of new therapies and products for patients [4, 6–9]. Accordingly, several urological tissue technologies have been developed for clinical use, including bladder, urethra, and ureter [9–11]. In this section, we review the current status of TERM therapies and discuss new and innovative technologies that hold promise in the future.

1.2.1 Cell Source

Cells are one of the main basic components of TERM. Cells can be injected or infused into damaged tissues alone or by combining with cell carrier materials. Cells can also be used to engineer tissue constructs by attaching on a scaffolding system fabricated with biomaterials. These cells can mature into tissues and organs by direct proliferation and differentiation into

B.W. Kim (ed.), *Clinical Regenerative Medicine in Urology*, DOI 10.1007/978-981-10-2723-9_1

tissue-specific cells. Implanted cells can also stimulate tissue formations indirectly by releasing biological factors. In TERM therapies, allogenic and autologous cells have been used frequently. Autologous cells are preferred over allogenic cells because these cells are isolated from the host and therefore minimize rejection and eliminate side effects caused by immunosuppressive medications [4]. Several types of autologous cells, such as smooth muscle cells, urothelial cells, and muscle-derived cells, have been used in the preclinical and clinical settings with successful outcomes in the field of urology [12–14]. Autologous cells are usually isolated from diseased or damaged tissues of the host; however, obtaining sufficient amount of normal cell population may be difficult in selected cases. Additionally, fully differentiated cells may show limited proliferative capacity [15]. Therefore, alterative cell sources may be needed to obtain sufficient numbers for clinical therapies.

Stem cells are an alternative cell source that can overcome the limitations of autologous somatic cells [16–20]. Stem cells have the ability of self-renewal and can be differentiated into one or more specialized cell types. Stem cells are traditionally classified according to their origin: embryonic stem cells, fetal stem cells, or adult stem cells. Stem cells can also be classified by their plasticity: totipotent, pluripotent, or multipotent. Embryonic stem cells have a strong capacity for self-renewal with the potential to differentiate into any type of cells, so they are an effective stem cell source for regenerating tissues and organs. However, the use of these cells is very controversial at the present time. Embryonic stem cells are obtained from the inner cell mass of embryos. During the isolation process, the embryo is destroyed, and somatic nuclear transfer and cloning processes are required [21, 22]. Aside from the ethical issues, embryonic stem cells are known to form teratomas in vivo [22]. Thus, many countries prohibit the use of embryonic stem cells in research.

Adult stem cells can be obtained in most tissues and organs in the body, including the bone marrow, fat tissue, blood, heart, nerve tissue, muscle, skin, and urine [18, 23]. Although adult stem cells have lower self-renewal and differentiation capacity than embryonic stem cells, they are relatively free from ethical and safety issues and are easily translated to patients. Adult mesenchymal stem cells, which are derived from bone marrow or adipose tissues, have been used in numerous studies [24]. Bone marrow-derived stem/stromal cells can differentiate into various cell types, including osteocytes, chondrocytes, and myocytes, and can develop into many tissue types including the bone, cartilage, muscle, vessels, and liver. Since Zuk and colleagues have first reported on how to obtain multipotent stem cells from human adipose tissues in 2001 [25], the use of adipose-derived stem cells has been enormously increased. Adipose-derived stem cells are multipotent cells, like bone marrow-derived stem cells. However, these cells are easier to obtain with high yields than bone marrow-derived stem cells [26, 27]. With these favorable characteristics, adult mesenchymal stem cells have been frequently used in bladder and urethral reconstructive applications [9, 28].

Another attractive source of stem cells in TERM is amniotic fluid and placental stem cells. These cells are a mixture of multipotent stem cells that express embryonic and adult stem cell markers [29]. They can differentiate into a variety of cell types, including chondrogenic, osteogenic, hepatic, myogenic, neuronal, and hepatic lineages. They have over 250 population doublings and do not form tumors in vivo [29–31]. With these advantages, amniotic fluid stem cells have shown therapeutic effects in various tissue regeneration studies. For instance, the injection of a combination of early differentiated amniotic fluid stem cells into damaged urethral sphincter in mice showed new muscle fiber formation and improved urodynamic function [32]. Thus, amniotic fluid and placental stem cells are considered a promising cell source for cell-based TERM research. Their applications are expected to increase in preclinical and clinical settings [31].

Recently, induced pluripotent stem cells (iPSC) have gained an enormous attention as novel cell sources for TERM applications. These cells are generated by dedifferentiation of adult somatic cells through genetic reprogramming

[33]. iPSC were first discovered by Takahashi and Yamanaka in 2006 through the reprogramming of adult mouse fibroblasts by overexpressing four transcription factors, Oct-4, Sox-2, c-Myc, and Klf4 [34]. Pluripotent stem cells can be obtained without the use of embryos. Recently, a few studies have shown the possibility of reprogramming human cells, as they show similar morphology, gene expression, and surface markers as human embryonic stem cells [35–37]. These results have opened a new era in TERM, as they can be utilized as personalized cell sources, not only for regenerating damaged tissues or organs of the patients but also for establishing patient-specific platforms in drug discovery. In addition, a recent study reported that human iPSC-derived smooth muscle precursor cells improve urethral sphincter function in a rat model. As such these cell sources have received much attention in urology [38]. Issues related to incomplete reprograming and teratoma formation still remain to be solved, but research of iPSC has seen a rapid and exponential growth [39–42].

Stem cells have been also found in urine [28]. Urine-derived stem cells have high self-renewal and expansion capacity and low telomerase activity and secrete paracrine factors. These cells have multipotent differentiation potential, including osteogenic, chondrogenic, myogenic, adipogenic, neurogenic, and endothelial differentiation. Urine-derived stem cells have become a promising cell source in urology, because the origin of these cells is from the urinary tract system, and they can be differentiated into bladder cells [23, 28, 43].

1.2.2 Biomaterials

Biomaterials are an essential element in TERM research. Biomaterials can be used alone as scaffolds or used with cells as tissue constructs. Biomaterials serve as a bridge to fill defect sites and promote new tissue formations by inducing the body's ability to regenerate. Biomaterials can also deliver cells into the body to facilitate rapid regeneration. Biomaterials provide environment for cell attachment, growth, and differentiation. Therefore, an ideal biomaterial should be biocompatible and support tissue developments by appropriate control of cell attachment, proliferation, migration, and differentiation [44]. When biomaterials are implanted in the body, they should readily integrate with host tissues without inducing unfavorable inflammation. Biomaterial scaffolds possess porous structure to facilitate cell migration and induce vascularization, as well as provide structural support. Biodegradable biomaterials specifically should provide mechanical support in early tissue development and should degrade at a controlled rate over time, thereby promoting new tissue formation.

Numerous biomaterials with specific properties have been developed and applied in TERM research. Selection of biomaterials is extremely important for the successful outcome of research studies. Biomaterials are traditionally classified according to their origin: synthetic and natural polymer [44, 45]. Biodegradable synthetic polymers such as poly-lactic acid (PLA), poly-glycolic acid (PGA), poly-lactic-co-glycolic acid (PLGA), and poly-caprolactone (PCL) have been widely used in TERM studies due to several advantages. These polymers are easy to obtain and handle as compared to natural polymers, although synthetic polymers are generally not as biocompatible as natural polymers. Biodegradable synthetic polymers can be manufactured in a large scale with reproducible physical and chemical properties, and their degradation rates are controllable. Natural polymers that are frequently used in TERM include collagen, gelatin, alginate, fibrin, silk, chitosan, and hyaluronic acids. Although naturally derived materials are difficult to control their degradation rates, they have advantages of possessing biologic recognition and biocompatibility.

Recently, injectable biodegradable hydrogels, such as collagen, gelatin, alginate, fibrin, hyaluronic acid, and polyethylene glycol (PEG), have been extensively used in TERM research. Hydrogels can cross-link, absorb water up to 1000 times their dry weight, and provide microenvironments similar to that of native tissues [46]. They can be injected in combination with cells and bioactive molecules, such as growth factors or genes. Despite these advantages in TERM research, the use of hydrogels is limited due to the weak mechanical properties and rapid degradation [47]. As such, various types of synthetic hydrogels and hybrid/composite biomaterials have been investigated in order to improve these properties [46, 48].

Current approaches in biomaterials research have focused on developing smart biomaterials that provide bioactive functions. Smart biomaterials have been investigated with the purpose of accelerating or enhancing tissue regeneration by controlling biological activities of cells and releasing regulatory factors [49, 50]. Consequently, a strategy of in situ tissue regeneration has gained an increasing interest in the field. This strategy uses the body's own regenerative capacity by activating and recruiting host stem/progenitor cells into the scaffold implant. For facilitating migration, recruitment, homing, growth, and differentiation of host cells, functional scaffolding systems containing bioactive molecules, including growth factors, peptides, and drugs, have been used. By controlling spatiotemporal release of the bioactive molecules contained in the scaffolding system, tissues can be regenerated by using only the host cells [51–53]. This strategy eliminates ex vivo cell manipulations, which are labor intensive and time-consuming and require enormous resources.

Decellularized tissues and organs have been used as scaffold biomaterials. Decellularization is the process that removes cellular components from donor tissues and organs, while retaining extracellular matrices. Decellularized scaffolds preserve the architectural structures and shape of organs and tissues. Due to lack of cellular components, immunorejection can be minimized. Therefore, use of decellularized scaffolds has become an alternative strategy to allogeneic organ transplantation [54]. In urology, decellularized urinary bladder matrix and small intestinal submucosa have been investigated extensively for tissue reconstruction. These approaches have been successfully translated into the clinic for bladder augmentations [8, 12, 55–57]. Moreover, recellularization of decellularized lung, kidney, and heart scaffolds has been tried preclinically to achieve functional tissues for whole-organ engineering [58–60].

1.2.3 Vascularization

Vascularization is a key factor in the successful outcomes of tissue regeneration. Tissue survival primarily depends on oxygen and nutrient supply from host blood vessels in vivo. Diffusion distance of oxygen and nutrients is generally 200 μm, so it is difficult for implanted cells that are more than 200 μm away from host vessels to survive [61–63]. Therefore, vascularization of volumetric, three-dimensional (3D) tissue constructs is necessary to the cells for a long-term survival. As such, several vascularization strategies have been investigated in engineered tissues such as the urethra [64], ureter [9], bone [65], cardiac tissue [66], skeletal muscle [67], skin [68], and lung [69], to overcome the current challenges associated with diffusion limitation.

A common approach for vascularization in TERM is to stimulate vascular ingrowth into the tissue constructs from adjacent host tissues. Angiogenic/vasculogenic growth factors, including vascular endothelial growth factor (VEGF), basic fibroblast growth factor (bFGF), and platelet-derived growth factor (PDGF), have been used to promote angiogenesis [63, 70]. VEGF has been widely used in tissue engineering applications in urology; for instance, VEGF-binding biomaterials were used to promote angiogenesis that resulted in tissue regeneration such as urethral and bladder tissues [71, 72]. Stem/progenitor cells have been implanted for inducing vascularization in the target tissues to be regenerated [73, 74]. These cells can release angiogenic factors or partly differentiate into vascular endothelial cells following implantation. Stem cells, vascular cells, and growth factors can be combined in various ways to achieve synergistic effects [70, 75, 76].

These strategies, however, would not be sufficient to form volumetric vascularized tissues of clinically relevant size. 3D scaffolds implanted in vivo take time to be fully vascularized, despite the use of angiogenic factors. This is due to the transient activity of factors that induce angiogenesis, and the rate of neovascularization is only ~5 μm/h [77]; thus, the angiogenic activity cannot be maintained long-term [63]. Consequently, the cells residing in the core region of the implanted scaffolds are unable to receive necessary oxygen and nutrients in a timely manner. To overcome the limitations associated with delayed vascularization within the 3D scaffolds, a strategy of pre-vascularization has been extensively investigated. This strategy involves in vitro as well as in situ formation of vascular structures within the scaffolds prior to implantation [78]. The rationale for this strategy is that preformed vascular structure would be readily integrated with the host vascular system following implantation, resulting in a rapid perfusion of blood into the implanted scaffolds. Recent advances in scaffold fabrication techniques, such as microfabrication and 3D printing, have enabled the development of 3D scaffolds that possess complex and perfusable vascular structures [79–81].

Another vascularization strategy is to facilitate vascularization by controlling the properties of microstructure of biomaterials. Highly controlled porous scaffolds facilitate the formation of vascular networks as well as enhancing supply of oxygen and nutrients [82]. The use of decellularized tissue scaffolds presents with an advantage due to the preservation of inherent vascular structures [69, 83]. While establishing adequate vascularization of 3D volumetric scaffolds still remains a challenge, continued development of innovative strategies using advanced techniques may lead to a solution that could accelerate the translation of various TERM therapies to patients.

1.2.4 Bioreactors

Bioreactors have been used in TERM to improve structural and functional maturation of engineered tissues in vitro. Bioreactors provide biomechanical forces such as compression, shear stress, and pulsatile flow under real-time monitored, highly controlled conditions [84, 85]. Bioreactors have been applied in a variety of tissues, including cardiovascular, musculoskeletal, skin, and bladder tissues [85–89]. In urology, bioreactors have been used to culture engineered tissues under the condition mimicking the filling pressures of bladder. For instance, human urothelial cells seeded on extracellular matrix scaffolds were exposed to cyclical filling pressures in the urinary bladder bioreactor system. This resulted in a significant increase of urothelial cell growth as compared with the static culture conditions [90]. In another example, 3D engineered muscle constructs were subjected to uniaxial cyclic bioreactors, which resulted in improved organization of unidirectional myotubes [91]. Bioreactors also have been used in cardiovascular tissue engineering to perfuse sufficient oxygen and nutrients to engineered tissue constructs [92]. In addition, preconditioning of cellular vascular grafts by a pulsatile bioreactor could maintain cellular viability and improve mechanical properties [93]. Recent advances in computer engineering, mechanical engineering, and materials science facilitated in designing of more sophisticated bioreactors that simulate physiological environments of tissues or organs [94]. With the advances in bioreactor technology, it is expected that its utility will be expanded to large, complex engineered tissue applications such as the kidney, liver, and heart.

1.2.5 New and Innovative Technologies for TERM Applications

Recent research works in the field of TERM have shown rapid technological evolutions that target clinical applications via multidisciplinary research. These technological evolutions reboost research areas that had previously been stagnated by innovative solutions and convergence technologies. Some of these include scaffold fabrication methods, monitoring and evaluating tools, modeling, and preservation. Utilization of these technologies will further enable the field to develop TERM therapies for complex tissue and organ systems.

3D printing, an additive manufacturing technology, is one of the most explored scaffold fabrication technologies currently in TERM [95]. It has become a powerful tool to fabricate clinically relevant 3D scaffolds and tissue constructs in a precise manner with high reproducibility. With this technology, precise fabrication is possible for complex tissue and organs of the human body as well as scaffolding tailored for the geometry of patients' lesions. This technology also has been used to fabricate surgical models such as kidney, liver, bone, and cardiovascular tissues [96, 97]. In urology, 3D printing has been used to fabricate bladder scaffolds that are later covered with the patient's own cells [98, 99]. Recently, 3D printing technology has rapidly improved so that various types of cells as well as biomaterials can be printed. 3D bioprinting technology has allowed for the fabrication of living tissues and organs having complex structures [99, 100]. Various 3D bioprinting methods, such as inkjet printing, microextrusion printing, and laser printing, have been explored [99]. These bioprinting modalities have been modified and combined in various ways to produce tissue constructs effectively. With the remarkable improvement of 3D bioprinting technology, 3D human-scale living tissues have been fabricated [101]. While 3D bioprinting technology holds promise as a tool to generate complex tissue constructs, several issues remain to be solved, including the development of clinically usable software and development of diverse bioinks as well as improvement of bioprinters that could be reliably used in the clinic. In the future, one can envision treating a patient with a tissue deformation. Medical data such as CT or MRI can be converted into a 3D CAD model to generate a patient-specific tissue construct using a bioprinter for immediate surgical reconstruction in the operating room.

Another research area that is gaining increasing attention is developing tools for monitoring and evaluation of TERM products. Histological analysis using sample tissues is the most common method to assess tissue implants. Animals are often sacrificed for tissue collections, and no further follow-up will be possible. To overcome this limitation, noninvasive high-resolution image acquisition techniques have been developed at the cell, molecular or tissue level [102]. These techniques allow researchers to collect accurate morphological and functional information without compromising an ongoing study. To this end, various cell tacking technologies and related tissue monitoring methods have been introduced to track implanted or injected cells [102]. Noninvasive magnetic resonance imaging (MRI) has been used to assess tissue functions associated with the movement of the body, hemodynamics, and metabolism [102, 103]. For instance, to monitor and evaluate stem cell homing and tissue regeneration effects in vivo, magnetic nanoparticle-absorbed mesenchymal stem cells were injected, and the cells were tracked with MRI [104]. Optical imaging techniques are also used in noninvasive real-time measurements of the tissue. With those tools, a series of tissue formation processes such as gene expression, differentiation, apoptosis, and vascularization from implantation to remodeling can be investigated [102, 103]. These monitoring methods progress in conjunction with the advancement of relevant nanoengineering, molecular biology, and computer science technologies. In the future, noninvasive image acquisition methods are expected to contribute to the analysis of tissue function in smaller scale, making it more reproducible, less invasive, faster, more precise, and economical.

Modeling technology is becoming increasingly important in TERM. 3D computer simulations of cell behaviors predict not only physical structures but also dynamic interactions between cells and environments. "BioSPICE" is open-source software to predict spatiotemporal behaviors of cells, and "cellular automata" simulate cellular dynamics, including migration and differentiation in two-dimensional (2D) space [105, 106]. These technologies might be essential in the design of improved tissue engineering products, as tissue development depends on cellular phenotypes and behavior patterns. In the near future, multi-scale modeling from the submolecular to the tissue level could be applied to provide a framework for bioreactor design and bioinspired tissue engineering constructs for TERM.

Cells and tissues' preservation technology is essential in clinical applications and commercialization of TERM products. The most common technology used in TERM is cryopreservation. Successful cryopreservation protocols have been developed for different cell types with parameters that control cryopreservation speed and the type of cryoprotectant chemicals [107, 108]. While the goal of cryopreservation is to preserve cells for tissue fabrication of engineered tissue products, it is becoming increasingly popular for individuals to preserve reproductive tissues for potential infertility or chemotherapy. To this end, appropriate preservation facilities are required to preserve cells and tissues without loss of their fertility over a long period of time [109]. More recently, a testicular tissue preservation system has been established for utilizing spermatogonial stem cells, which follows good tissue practice (GTP) protocol of US Food Drug Administration [110]. Establishing preservation protocols for various cell types is crucial for the fabrication and commercialization of cellular composite tissue constructs.

1.3 Challenges and Future Perspectives

Over the past decades, a considerable research progress has been made to bring TERM technologies toward clinical translation. Consequently, several TERM products have been successfully translated to human clinical trials, some of which achieved favorable outcomes involving functional improvement. For instance, tissue-engineered tubularized urethras using autologous smooth muscle and epithelial cells were implanted into urethral defect sites, and that led to functional tissue restoration for up to 6 years in 5 patients [111]. In another example, various types of cells have been implanted in patients for corneal regeneration [112] and for treating myocardial infarction [59, 113, 114]. With the successful clinical translation of initial TERM therapies, it is expected that TERM research will continue to focus on developing engineered tissue constructs and therapies that can be translated to patients. However, several challenges still remain, and these must be addressed in order to bring more complex tissue therapies to the clinic [3, 4, 115].

One practical challenge that hampers clinical translation is building a clinically relevant-sized composite tissue constructs [98, 116]. In the clinical situation, composite tissue damage is more common rather than single tissue damage. For instance, a patient with severe facial trauma frequently has composite tissue injuries involving the skin, muscle, nerve, and bone, which should all be reconstructed for functional restoration. Considering the clinical relevance, current research should focus on developing implantable composite tissue constructs for patients. Accordingly, the composite tissue constructs should be able to combine at least two different tissue types in a single contiguous construct and mimic the structural and mechanical characteristics of each tissue type. However, fabricating composite tissue constructs remains a technological challenge. As previously noted, 3D bioprinting and decellularization/recellularization technologies have shown the potential of fabricating complex, composite, human-scale tissues [60, 101, 117]. In addition to these technologies, other fabrication methods that can produce composite tissue constructs need to be developed.

The issue associated with identifying ideal cell sources should also be considered. Tissues and organs in the body are composed of multiple cell types. As such, several types of cells have been used to engineer composite tissue constructs to mimic the cellular composition, tissue anatomy, and the function of native tissues [98, 116]. However, placing all cell types into a space-confined construct may not be as feasible. In addition, ex vivo cell preparation requires enormous labor force, time, facility, and resources. To facilitate accelerated translation of engineered tissues into the clinic, ex vivo cell manipulation should be simplified and improved. Thus, the selection and placement of cells for target-tissue regeneration should be carefully addressed. Furthermore, in situ tissue regeneration through host cell recruitment may gain increased attention in the future, as this strategy would not need ex vivo cell manipulations [51–53]. Further development of this strategy will accelerate clinical translation of many cell-based therapies.

Integration of implanted engineered constructs with host tissue is another critical consideration for successful translational outcomes. Specifically, integration of engineered tissues with host nervous system is critically important in achieving target tissue function. In engineered bladder tissue, for instance, appropriate innervation is directly related to voiding function [10]. In engineered muscle tissues, failure of innervation results in abnormal contraction and skeletal muscle atrophy following implantation in vivo [118]. With the significance of innervation on the functional restoration, biological factors, including growth factors, neurotransmitters, and neuronal cells, have been applied to the engineered tissues [10, 119, 120]. However, the use of those factors should be further investigated, in terms of clinical relevance and effectiveness. Meanwhile, enormous efforts have been made to create vascularized engineered tissue constructs for implant survival. However, vascularization of clinically relevant-sized constructs and functional integration with host vasculature still remain a challenge. Therefore, developing methods to achieve rapid vascular and neuronal integration of the engineered tissues is critical.

To date, the available commercial products in TERM are mostly for skin and cartilage regeneration. Additional efforts are needed to diversify and commercialize TERM products. However, a solid understanding on the manufacturing of TERM products has been limited, and a road map has yet to be developed. Therefore, developing an infrastructure for TERM manufacturing system that target clinical applications and commercialization is necessary. In addition, systems for quality control should also be established through standardization of manufacturing processes. Manufactured products should be economically competitive. For instance, manufacturing of TERM products using cells requires multiple processing steps, including tissue biopsy, cell isolation and expansion, scaffold design and fabrication, cell seeding, construct preconditioning, and preservation and storage. This series of processes need to be performed in a GMP (good manufacturing practice) facility. Monitoring and evaluating cells and scaffolds should also be developed for each processing step. Additionally, optimized and standardized protocols compatible with ethical and safety regulations should be established and adhered to. Regular and thorough discussions between researchers, physicians, engineers, manufacturers, enterprisers, and legislators will contribute to the successful development of TERM product road map, definition of required infrastructure/protocols for each step, and a system of regulation [121].

Conclusion

The field of TERM has achieved a remarkable progress over the years. TERM is a translational academic field for developing clinical applications into the clinic, which may provide a valuable opportunity to overcome the limitations of conventional therapies. Multiple translational products have been applied in patients, and these technologies are expected to be available to a wider population in the coming years. While recent technological developments have addressed some of the limitations that hamper the translational progress, a majority of engineering challenges still remain a problem. Continued development of innovative and technological advances must be made in order to translate complex and composite tissue and organ therapies to the clinic.

Acknowledgments The authors thank Margaret van-Schaayk for editorial assistance with this manuscript.

References

1. Kemp P. History of regenerative medicine: looking backwards to move forwards. Regen Med. 2006;1(5):653–69.
2. Vacanti CA. The history of tissue engineering. J Cell Mol Med. 2006;10(3):569–76.
3. Mao AS, Mooney DJ. Regenerative medicine: current therapies and future directions. Proc Natl Acad Sci U S A. 2015;112(47):14452–9.
4. Fisher MB, Mauck RL. Tissue engineering and regenerative medicine: recent innovations and the

transition to translation. Tissue Eng Part B Rev. 2013;19(1):1–13.
5. Gemmiti C. Tissue engineering/regenerative medicine ventures should invest early in market research to understand the future market's needs. Tissue Eng Part B-Re. 2013;19(2):97–8.
6. Vacanti JP. Tissue engineering: from bench to bedside via commercialization. Surgery. 2008;143(2):181–3.
7. Martin I, Smith T, Wendt D. Bioreactor-based roadmap for the translation of tissue engineering strategies into clinical products. Trends Biotechnol. 2009;27(9):495–502.
8. Caione P, Boldrini R, Salerno A, Nappo SG. Bladder augmentation using acellular collagen biomatrix: a pilot experience in exstrophic patients. Pediatr Surg Int. 2012;28(4):421–8.
9. de Jonge PK, Simaioforidis V, Geutjes PJ, Oosterwijk E, Feitz WF. Recent advances in ureteral tissue engineering. Curr Urol Rep. 2015;16(1):465.
10. Horst M, Madduri S, Gobet R, Sulser T, Milleret V, Hall H, et al. Engineering functional bladder tissues. J Tissue Eng Regen Med. 2013;7(7):515–22.
11. de Kemp V, de Graaf P, Fledderus JO, Ruud Bosch JL, de Kort LM. Tissue engineering for human urethral reconstruction: systematic review of recent literature. PLoS One. 2015;10(2):e0118653.
12. Atala A, Bauer SB, Soker S, Yoo JJ, Retik AB. Tissue-engineered autologous bladders for patients needing cystoplasty. Lancet. 2006;367(9518):1241–6.
13. Carr LK, Robert M, Kultgen PL, Herschorn S, Birch C, Murphy M, et al. Autologous muscle derived cell therapy for stress urinary incontinence: a prospective, dose ranging study. J Urol. 2013;189(2):595–601.
14. Orabi H, AbouShwareb T, Zhang Y, Yoo JJ, Atala A. Cell-seeded tubularized scaffolds for reconstruction of long urethral defects: a preclinical study. Eur Urol. 2013;63(3):531–8.
15. Schnabel M, Marlovits S, Eckhoff G, Fichtel I, Gotzen L, Vecsei V, et al. Dedifferentiation-associated changes in morphology and gene expression in primary human articular chondrocytes in cell culture. Osteoarthr Cartil. 2002;10(1):62–70.
16. Wagers AJ. The stem cell niche in regenerative medicine. Cell Stem Cell. 2012;10(4):362–9.
17. Bianco P, Cao X, Frenette PS, Mao JJ, Robey PG, Simmons PJ, et al. The meaning, the sense and the significance: translating the science of mesenchymal stem cells into medicine. Nat Med. 2013;19(1):35–42.
18. Mahla RS. Stem cells applications in regenerative medicine and disease therapeutics. Int J Cell Biol. 2016;2016:6940283.
19. Kim JH, Song YS. Current status of stem cell therapy in urology. Korean J Urol. 2015;56(6):409–11.
20. Fu Q, Cao YL. Tissue engineering and stem cell application of urethroplasty: from bench to bedside. Urology. 2012;79(2):246–53.
21. King NM, Perrin J. Ethical issues in stem cell research and therapy. Stem Cell Res Ther. 2014;5(4):85.
22. Tachibana M, Amato P, Sparman M, Gutierrez NM, Tippner-Hedges R, Ma H, et al. Human embryonic stem cells derived by somatic cell nuclear transfer. Cell. 2013;153(6):1228–38.
23. Liu G, Pareta RA, Wu R, Shi Y, Zhou X, Liu H, et al. Skeletal myogenic differentiation of urine-derived stem cells and angiogenesis using microbeads loaded with growth factors. Biomaterials. 2013;34(4):1311–26.
24. Kobolak J, Dinnyes A, Memic A, Khademhosseini A, Mobasheri A. Mesenchymal stem cells: identification, phenotypic characterization, biological properties and potential for regenerative medicine through biomaterial micro-engineering of their niche. Methods. 2016;99:62–8.
25. Zuk PA, Zhu M, Mizuno H, Huang J, Futrell JW, Katz AJ, et al. Multilineage cells from human adipose tissue: implications for cell-based therapies. Tissue Eng. 2001;7(2):211–28.
26. Aust L, Devlin B, Foster SJ, Halvorsen YD, Hicok K, du Laney T, et al. Yield of human adipose-derived adult stem cells from liposuction aspirates. Cytotherapy. 2004;6(1):7–14.
27. Rhie JW. Adipose-derived stem cells: characterization and clinical application. J Korean Med Assoc. 2012;55(8):757–69.
28. Qin D, Long T, Deng J, Zhang Y. Urine-derived stem cells for potential use in bladder repair. Stem Cell Res Ther. 2014;5(3):69.
29. Dziadosz M, Basch RS, Young BK. Human amniotic fluid: a source of stem cells for possible therapeutic use. Am J Obstet Gynecol. 2016;214(3):321–7.
30. De Coppi P, Bartsch G Jr, Siddiqui MM, Xu T, Santos CC, Perin L, et al. Isolation of amniotic stem cell lines with potential for therapy. Nat Biotechnol. 2007;25(1):100–6.
31. Murphy SV, Atala A. Amniotic fluid and placental membranes: unexpected sources of highly multipotent cells. Semin Reprod Med. 2013;31(1):62–8.
32. Chun SY, Kwon JB, Chae SY, Lee JK, Bae JS, Kim BS, et al. Combined injection of three different lineages of early-differentiating human amniotic fluid-derived cells restores urethral sphincter function in urinary incontinence. BJU Int. 2014;114(5):770–83.
33. Wilson KD, Wu JC. Induced pluripotent stem cells. JAMA. 2015;313(16):1613–4.
34. Takahashi K, Yamanaka S. Induction of pluripotent stem cells from mouse embryonic and adult fibroblast cultures by defined factors. Cell. 2006;126(4):663–76.
35. Shi Y. Induced pluripotent stem cells, new tools for drug discovery and new hope for stem cell therapies. Curr Mol Pharmacol. 2009;2(1):15–8.
36. Burridge PW, Li YF, Matsa E, Wu H, Ong SG, Sharma A, et al. Human induced pluripotent stem cell-derived cardiomyocytes recapitulate the predilection of breast cancer patients to doxorubicin-induced cardiotoxicity. Nat Med. 2016;22(5):547–56.
37. Yu J, Vodyanik MA, Smuga-Otto K, Antosiewicz-Bourget J, Frane JL, Tian S, et al. Induced pluripotent stem cell lines derived from human somatic cells. Science. 2007;318(5858):1917–20.

38. Wang Z, Wen Y, Li YH, Wei Y, Green M, Wani P, et al. Smooth muscle precursor cells derived from human pluripotent stem cells for treatment of stress urinary incontinence. Stem Cells Dev. 2016;25(6):453–61.
39. Lister R, Pelizzola M, Kida YS, Hawkins RD, Nery JR, Hon G, et al. Hotspots of aberrant epigenomic reprogramming in human induced pluripotent stem cells. Nature. 2011;471(7336):68–73.
40. Das AK, Pal R. Induced pluripotent stem cells (iPSCs): the emergence of a new champion in stem cell technology-driven biomedical applications. J Tissue Eng Regen Med. 2010;4:413–21.
41. O'Malley J, Woltjen K, Kaji K. New strategies to generate induced pluripotent stem cells. Curr Opin Biotechnol. 2009;20(5):516–21.
42. Vierbuchen T, Ostermeier A, Pang ZP, Kokubu Y, Sudhof TC, Wernig M. Direct conversion of fibroblasts to functional neurons by defined factors. Nature. 2010;463:1035–41.
43. Zhang Y, McNeill E, Tian H, Soker S, Andersson KE, Yoo JJ, et al. Urine derived cells are a potential source for urological tissue reconstruction. J Urol. 2008;180(5):2226–33.
44. Keane TJ, Badylak SF. Biomaterials for tissue engineering applications. Semin Pediatr Surg. 2014;23(3):112–8.
45. Dawson E, Mapili G, Erickson K, Taqvi S, Roy K. Biomaterials for stem cell differentiation. Adv Drug Deliv Rev. 2008;60(2):215–28.
46. Hunt JA, Chen R, van Veen T, Bryan N. Hydrogels for tissue engineering and regenerative medicine. J Mater Chem B. 2014;2(33):5319–38.
47. El-Sherbiny IM, Yacoub MH. Hydrogel scaffolds for tissue engineering: Progress and challenges. Global cardiology science & practice. 2013;2013(3):316–42.
48. Koutsopoulos S. Self-assembling peptide nanofiber hydrogels in tissue engineering and regenerative medicine: progress, design guidelines, and applications. J Biomed Mater Res A. 2016;104(4):1002–16.
49. Mieszawska AJ, Kaplan DL. Smart biomaterials – regulating cell behavior through signaling molecules. BMC Biol. 2010;8:59.
50. Zambuzzi WF, Coelho PG, Alves GG, Granjeiro JM. Intracellular signal transduction as a factor in the development of "smart" biomaterials for bone tissue engineering. Biotechnol Bioeng. 2011;108(6):1246–50.
51. Lee SJ, Van Dyke M, Atala A, Yoo JJ. Host cell mobilization for in situ tissue regeneration. Rejuvenation Res. 2008;11(4):747–56.
52. Kim JH, Jung Y, Kim BS, Kim SH. Stem cell recruitment and angiogenesis of neuropeptide substance P coupled with self-assembling peptide nanofiber in a mouse hind limb ischemia model. Biomaterials. 2013;34(6):1657–68.
53. Ko IK, Lee SJ, Atala A, Yoo JJ. In situ tissue regeneration through host stem cell recruitment. Exp Mol Med. 2013;45:e57.
54. Badylak SF, Taylor D, Uygun K. Whole-organ tissue engineering: decellularization and recellularization of three-dimensional matrix scaffolds. Annu Rev Biomed Eng. 2011;13:27–53.
55. Schaefer M, Kaiser A, Stehr M, Beyer HJ. Bladder augmentation with small intestinal submucosa leads to unsatisfactory long-term results. J Pediatr Urol. 2013;9(6 Pt A):878–83.
56. Zhang F, Liao L. Tissue engineered cystoplasty augmentation for treatment of neurogenic bladder using small intestinal submucosa: an exploratory study. J Urol. 2014;192(2):544–50.
57. Joseph DB, Borer JG, De Filippo RE, Hodges SJ, McLorie GA. Autologous cell seeded biodegradable scaffold for augmentation cystoplasty: phase II study in children and adolescents with spina bifida. J Urol. 2014;191(5):1389–95.
58. Uygun BE, Soto-Gutierrez A, Yagi H, Izamis ML, Guzzardi MA, Shulman C, et al. Organ reengineering through development of a transplantable recellularized liver graft using decellularized liver matrix. Nat Med. 2010;16(7):814–20.
59. Guyette JP, Gilpin SE, Charest JM, Tapias LF, Ren X, Ott HC. Perfusion decellularization of whole organs. Nat Protoc. 2014;9(6):1451–68.
60. Kitahara H, Yagi H, Tajima K, Okamoto K, Yoshitake A, Aeba R, et al. Heterotopic transplantation of a decellularized and recellularized whole porcine heart. Interact Cardiovasc Thorac Surg. 2016;22(5):571–9.
61. Folkman J, Hochberg M. Self-regulation of growth in three dimensions. J Exp Med. 1973;138(4):745–53.
62. Helmlinger G, Yuan F, Dellian M, Jain RK. Interstitial pH and pO2 gradients in solid tumors in vivo: high-resolution measurements reveal a lack of correlation. Nat Med. 1997;3(2):177–82.
63. Novosel EC, Kleinhans C, Kluger PJ. Vascularization is the key challenge in tissue engineering. Adv Drug Deliv Rev. 2011;63(4–5):300–11.
64. Imbeault A, Bernard G, Rousseau A, Morissette A, Chabaud S, Bouhout S, et al. An endothelialized urothelial cell-seeded tubular graft for urethral replacement. Can Urol Assoc J. 2013;7(1–2):E4–9.
65. Mercado-Pagan AE, Stahl AM, Shanjani Y, Yang Y. Vascularization in bone tissue engineering constructs. Ann Biomed Eng. 2015;43(3):718–29.
66. Sun X, Altalhi W, Nunes SS. Vascularization strategies of engineered tissues and their application in cardiac regeneration. Adv Drug Deliv Rev. 2016;96:183–94.
67. Qazi TH, Mooney DJ, Pumberger M, Geissler S, Duda GN. Biomaterials based strategies for skeletal muscle tissue engineering: existing technologies and future trends. Biomaterials. 2015;53:502–21.
68. Frueh FS, Menger MD, Lindenblatt N, Giovanoli P, Laschke MW. Current and emerging vascularization strategies in skin tissue engineering. Crit Rev Biotechnol. 2017;37:613–25.
69. Stabler CT, Caires LC Jr, Mondrinos MJ, Marcinkiewicz C, Lazarovici P, Wolfson MR, et al. Enhanced re-endothelialization of decellularized rat lungs. Tissue Eng Part C Methods. 2016;22(5):439–50.

70. Rouwkema J, Khademhosseini A. Vascularization and angiogenesis in tissue engineering: beyond creating static networks. Trends Biotechnol. 2016;34(9):733–45.
71. Jia W, Tang H, Wu J, Hou X, Chen B, Chen W, et al. Urethral tissue regeneration using collagen scaffold modified with collagen binding VEGF in a beagle model. Biomaterials. 2015;69:45–55.
72. Jiang X, Xiong Q, Xu G, Lin H, Fang X, Cui D, et al. VEGF-loaded nanoparticle-modified BAMAs enhance angiogenesis and inhibit graft shrinkage in tissue-engineered bladder. Ann Biomed Eng. 2015;43(10):2577–86.
73. Shepherd BR, Enis DR, Wang F, Suarez Y, Pober JS, Schechner JS. Vascularization and engraftment of a human skin substitute using circulating progenitor cell-derived endothelial cells. FASEB J. 2006;20(10):1739–41.
74. Levenberg S, Rouwkema J, Macdonald M, Garfein ES, Kohane DS, Darland DC, et al. Engineering vascularized skeletal muscle tissue. Nat Biotechnol. 2005;23(7):879–84.
75. Kaully T, Kaufman-Francis K, Lesman A, Levenberg S. Vascularization-the conduit to viable engineered tissues. Tissue Eng Part B Rev. 2009;15(2):159–69.
76. Kim JH, Jung Y, Kim SH, Sun K, Choi J, Kim HC, et al. The enhancement of mature vessel formation and cardiac function in infarcted hearts using dual growth factor delivery with self-assembling peptides. Biomaterials. 2011;32(26):6080–8.
77. Utzinger U, Baggett B, Weiss JA, Hoying JB, Edgar LT. Large-scale time series microscopy of neovessel growth during angiogenesis. Angiogenesis. 2015;18(3):219–32.
78. Laschke MW, Menger MD. Prevascularization in tissue engineering: current concepts and future directions. Biotechnol Adv. 2016;34(2):112–21.
79. Baranski JD, Chaturvedi RR, Stevens KR, Eyckmans J, Carvalho B, Solorzano RD, et al. Geometric control of vascular networks to enhance engineered tissue integration and function. Proc Natl Acad Sci U S A. 2013;110(19):7586–91.
80. Kolesky DB, Homan KA, Skylar-Scott MA, Lewis JA. Three-dimensional bioprinting of thick vascularized tissues. Proc Natl Acad Sci U S A. 2016;113(12):3179–84.
81. Lee VK, Kim DY, Ngo H, Lee Y, Seo L, Yoo SS, et al. Creating perfused functional vascular channels using 3D bio-printing technology. Biomaterials. 2014;35(28):8092–102.
82. Tuzlakoglu K, Reis RL. Biodegradable polymeric fiber structures in tissue engineering. Tissue Eng Part B Rev. 2009;15(1):17–27.
83. Quint C, Kondo Y, Manson RJ, Lawson JH, Dardik A, Niklason LE. Decellularized tissue-engineered blood vessel as an arterial conduit. Proc Natl Acad Sci U S A. 2011;108(22):9214–9.
84. Plunkett N, O'Brien FJ. Bioreactors in tissue engineering. Technol Health Care. 2011;19(1):55–69.
85. Hansmann J, Groeber F, Kahlig A, Kleinhans C, Walles H. Bioreactors in tissue engineering – principles, applications and commercial constraints. Biotechnol J. 2013;8(3):298–307.
86. Gardel LS, Serra LA, Reis RL, Gomes ME. Use of perfusion bioreactors and large animal models for long bone tissue engineering. Tissue Eng Part B-Re. 2014;20(2):126–46.
87. Tocchio A, Tamplenizza M, Martello F, Gerges I, Rossi E, Argentiere S, et al. Versatile fabrication of vascularizable scaffolds for large tissue engineering in bioreactor. Biomaterials. 2015;45:124–31.
88. Amrollahi P, Tayebi L. Bioreactors for heart valve tissue engineering: a review. J Chem Technol Biot. 2016;91(4):847–56.
89. Murphy SV, Atala A. Organ engineering–combining stem cells, biomaterials, and bioreactors to produce bioengineered organs for transplantation. BioEssays. 2013;35(3):163–72.
90. Davis NF, Mooney R, Piterina AV, Callanan A, McGuire BB, Flood HD, et al. Construction and evaluation of urinary bladder bioreactor for urologic tissue-engineering purposes. Urology. 2011;78(4):954–60.
91. Heher P, Maleiner B, Pruller J, Teuschl AH, Kollmitzer J, Monforte X, et al. A novel bioreactor for the generation of highly aligned 3D skeletal muscle-like constructs through orientation of fibrin via application of static strain. Acta Biomater. 2015;24:251–65.
92. Massai D, Cerino G, Gallo D, Pennella F, Deriu MA, Rodriguez A, et al. Bioreactors as engineering support to treat cardiac muscle and vascular disease. J Healthcare Eng. 2013;4(3):329–70.
93. Ahn H, Ju YM, Takahashi H, Williams DF, Yoo JJ, Lee SJ, et al. Engineered small diameter vascular grafts by combining cell sheet engineering and electrospinning technology. Acta Biomater. 2015;16:14–22.
94. Depprich R, Handschel J, Wiesmann HP, Jasche-Meyer J, Meyer U. Use of bioreactors in maxillofacial tissue engineering. Br J Oral Maxillofac Surg. 2008;46(5):349–54.
95. Billiet T, Vandenhaute M, Schelfhout J, Van Vlierberghe S, Dubruel P. A review of trends and limitations in hydrogel-rapid prototyping for tissue engineering. Biomaterials. 2012;33(26):6020–41.
96. Shafiee A, Atala A. Printing Technologies for Medical Applications. Trends Mol Med. 2016;22(3):254–65.
97. Soliman Y, Feibus AH, Baum N. 3D printing and its urologic applications. Rev Urol. 2015;17(1):20–4.
98. Atala A, Kasper FK, Mikos AG. Engineering complex tissues. Sci Transl Med. 2012;4(160):160rv12.
99. Murphy SV, Atala A. 3D bioprinting of tissues and organs. Nat Biotechnol. 2014;32(8):773–85.
100. Ozbolat IT, Peng W, Ozbolat V. Application areas of 3D bioprinting. Drug Discov Today. 2016;21(8):1257–71.
101. Kang HW, Lee SJ, Ko IK, Kengla C, Yoo JJ, Atala A. A 3D bioprinting system to produce human-scale tissue constructs with structural integrity. Nat Biotechnol. 2016;34(3):312–9.

102. Nam SY, Ricles LM, Suggs LJ, Emelianov SY. Imaging strategies for tissue engineering applications. Tissue Eng Part B Rev. 2015;21(1):88–102.
103. Appel AA, Anastasio MA, Larson JC, Brey EM. Imaging challenges in biomaterials and tissue engineering. Biomaterials. 2013;34(28):6615–30.
104. Huang X, Zhang F, Wang Y, Sun X, Choi KY, Liu D, et al. Design considerations of iron-based nanoclusters for noninvasive tracking of mesenchymal stem cell homing. ACS Nano. 2014;8(5):4403–14.
105. Hatzikirou H, Deutsch A. Cellular automata as microscopic models of cell migration in heterogeneous environments. Curr Top Dev Biol. 2008;81:401–34.
106. Kumar SP, Feidler JC. BioSPICE: a computational infrastructure for integrative biology. OMICS. 2003;7(3):225.
107. Hunt CJ. Cryopreservation of human stem cells for clinical application: a review. Transfusion Med Hemother. 2011;38(2):107–23.
108. Wang Z, Qin TW. Review: vitreous cryopreservation of tissue-engineered compositions for tissue repair. J Med Biol Eng. 2013;33(2):125–31.
109. Donnez J, Dolmans MM, Pellicer A, Diaz-Garcia C, Sanchez Serrano M, Schmidt KT, et al. Restoration of ovarian activity and pregnancy after transplantation of cryopreserved ovarian tissue: a review of 60 cases of reimplantation. Fertil Steril. 2013;99(6):1503–13.
110. Sadri-Ardekani H, McLean TW, Kogan S, Sirintrapun J, Crowell K, Yousif MQ, et al. Experimental testicular tissue banking to generate spermatogenesis in the future: a multidisciplinary team approach. Methods. 2016;99:120–7.
111. Raya-Rivera A, Esquiliano DR, Yoo JJ, Lopez-Bayghen E, Soker S, Atala A. Tissue-engineered autologous urethras for patients who need reconstruction: an observational study. Lancet. 2011;377(9772):1175–82.
112. Rama P, Matuska S, Paganoni G, Spinelli A, De Luca M, Pellegrini G. Limbal stem-cell therapy and long-term corneal regeneration. N Engl J Med. 2010;363(2):147–55.
113. Makkar RR, Smith RR, Cheng K, Malliaras K, Thomson LE, Berman D, et al. Intracoronary cardiosphere-derived cells for heart regeneration after myocardial infarction (CADUCEUS): a prospective, randomised phase 1 trial. Lancet. 2012;379(9819):895–904.
114. Bolli R, Chugh AR, D'Amario D, Loughran JH, Stoddard MF, Ikram S, et al. Cardiac stem cells in patients with ischaemic cardiomyopathy (SCIPIO): initial results of a randomised phase 1 trial. Lancet. 2011;378(9806):1847–57.
115. Dimmeler S, Ding S, Rando TA, Trounson A. Translational strategies and challenges in regenerative medicine. Nat Med. 2014;20(8):814–21.
116. Badylak SF, Weiss DJ, Caplan A, Macchiarini P. Engineered whole organs and complex tissues. Lancet. 2012;379(9819):943–52.
117. Sicari BM, Rubin JP, Dearth CL, Wolf MT, Ambrosio F, Boninger M, et al. An acellular biologic scaffold promotes skeletal muscle formation in mice and humans with volumetric muscle loss. Sci Transl Med. 2014;6(234):234ra58.
118. Kang SB, Olson JL, Atala A, Yoo JJ. Functional recovery of completely denervated muscle: implications for innervation of tissue-engineered muscle. Tissue Eng Part A. 2012;18(17–18):1912–20.
119. Ko IK, Lee BK, Lee SJ, Andersson KE, Atala A, Yoo JJ. The effect of in vitro formation of acetylcholine receptor (AChR) clusters in engineered muscle fibers on subsequent innervation of constructs in vivo. Biomaterials. 2013;34(13):3246–55.
120. Morimoto Y, Kato-Negishi M, Onoe H, Takeuchi S. Three-dimensional neuron-muscle constructs with neuromuscular junctions. Biomaterials. 2013;34(37):9413–9.
121. Hunsberger J, Harrysson O, Shirwaiker R, Starly B, Wysk R, Cohen P, et al. Manufacturing road map for tissue engineering and regenerative medicine technologies. Stem Cells Transl Med. 2015;4(2):130–5.

Part II

Fundamentals of Regenerative Medicine

2 Biomaterials and Tissue Engineering

Sang Jin Lee, James J. Yoo, and Anthony Atala

2.1 Introduction

Tissue engineering, a major technique in regenerative medicine, follows the principles of cell biology, materials science, and engineering to develop biological substitutes that mimic anatomical and functional features of native tissues in order to restore and maintain normal function of injured or diseased tissues [1–4]. Currently, tissue engineering strategies can be divided into two major categories: cell-based and scaffold-based approaches. The classic tissue engineering approach attempts to create an engineered tissue construct by combining cells with a natural and/or synthetic scaffold material under suitable culture conditions. The scaffold-based tissue engineering approach is more dependent on the body's natural ability to regenerate for proper orientation and direction of new tissue in growth. These scaffolds can be prepared by manufacturing artificial microenvironment derived from natural or synthetic materials or by removing cellular components from tissues using mechanical and chemical manipulation to produce collagen-based tissue matrices [5–9]. These scaffold materials slowly degrade following implantation and are subsequently replaced by the extracellular matrix (ECM) proteins secreted by the in-growing cells. Additionally, cells can be used for therapy via injection either with carriers, such as hydrogels, or alone [10].

A small piece of donor tissue is dissociated into individual cells when cells are used for tissue engineering. These cells may be implanted directly into the host, or they may be expanded in culture and attached to a scaffold for reimplantation into the host. The source of donor tissue can be heterologous (such as bovine), allogeneic (same species, different individual), or autologous (from the host). Ideally, both structural and functional tissue replacement will occur with minimal complications. Autologous cells are preferential to heterologous or allogeneic cell sources because the biopsy tissue is obtained from the host, the cells are dissociated and expanded in culture, and the expanded cells are implanted into the original host. While an inflammatory response is possible, the use of autologous cells avoids rejection and eliminates the need for immunosuppressive medications, which have deleterious side effects.

Currently, most tissue engineering strategies are dependent upon obtaining autologous cells from the diseased organ of the host. Patients with extensive end-stage organ failure, however, are poor candidates for obtaining autologous cells since a tissue biopsy may not yield enough healthy cells for expansion and transplantation. Moreover, primary autologous human cells cannot be expanded from certain organs, such as the

S.J. Lee (✉) • J.J. Yoo • A. Atala
Wake Forest Institute for Regenerative Medicine, Wake Forest School of Medicine, Medical Center Boulevard, Winston-Salem, NC 27157, USA
e-mail: sjlee@wakehealth.edu

B.W. Kim (ed.), *Clinical Regenerative Medicine in Urology*, DOI 10.1007/978-981-10-2723-9_2

pancreas. Stem cells are envisioned as being an alternative source of cells from which the desired tissue can be derived. Recovery of stem cells is possible from discarded human embryos, from fetal-related tissue (amniotic fluid or placenta), or from adult sources (bone marrow, fat, or skin) [11]. Therapeutic cloning has also played significant a role in the development of the field of regenerative medicine.

This chapter describes biomaterials used in tissue engineering, including regeneration of specific structures (mostly soft tissues and organs) in the body. Therapies at the cellular, tissue, and organ levels are described, as well as the specific challenges and applications associated with each.

2.2 Types of Biomaterials

Basically, three major classes of biomaterials have been utilized for tissue engineering applications: natural polymers (e.g., collagen and alginate), acellular tissue matrices (e.g., decellularized tissues or organs), and synthetic biodegradable polymers [e.g., poly(glycolic acid) (PGA), poly(lactic acid) (PLA), and their copolymers]. These biomaterials have been tested with respect to their biological, biochemical, and biomechanical properties [12]. Natural polymers and acellular tissue matrices have the potential advantage of biological recognition; however, synthetic polymers can be produced reproducibly on a large scale with controlled properties of their mechanical strength, degradation rate, and microstructure (Table 2.1).

2.2.1 Natural Polymers

Natural polymers are defined as those that are produced by the cells of a living organism. Particularly, structural proteins such as collagen, laminin, elastin, and fibronectin have been used as scaffold materials for tissue engineering applications and as vehicles for cell delivery. Among these, collagen has been found a common use as a scaffold and carrier for cells in various tissue

Table 2.1 Advantages and disadvantages of biomaterials in tissue engineering

Biomaterials	Advantages	Disadvantages
Biodegradable synthetic polymers	Easy to process	Less biological properties than natural polymers
	Controlled biodegradation rate	Degrade to form by-products
	Controlled surface characteristics	Hydrophobic nature
	Tunable mechanical properties	
	Pathogen-free	
Natural polymers	Biological properties	Difficult to control biodegradation rate
	Mimic natural ECM structure and composition	Poor mechanical stability
	Hydrophilic nature	Temperature sensitive
		Transfer of pathogens possible
Acellular tissue matrices	Appropriate biomechanical properties to apply	Immune responses
	Similar structure and composition in native tissues	Difficult to re-cellularize
		Transfer of pathogens possible
Composite materials	Mostly natural polymers and synthetic polymers	Limited fabrication processing
	Synergic effects can be expected	In some cases, poor interaction of interface
	Composition can be controlled	

engineering applications [13–15], because it is the most abundant protein in mammalian tissues as the major component of the ECM materials. Collagen, which has a triple helix structure, mainly provides the structural integrity of vertebrates and organisms. Collagen can be extracted and purified from these animal tissues by enzymatic digestion and salt/acid extraction methods. The sources of collagens are mostly animal tissues such as bovine skin and tendon, porcine skin and tendon, and rat tail. Since the use of animal-derived collagens has been concerned due to the possible transmission of infectious agents, the recombinant human collagens have been produced [16]. Although collagen has excellent biological functions [17, 18], its low biomechanical properties and fast degradation by enzymes as well as antigenicity potential have been limited for clinical applications.

Silk fibroin is an insoluble fibrous protein obtained by silk worms (*Bombyx mori*), which contains approximately 90% amino acids, including glycine, alanine, and serine [19–21]. Purified silk fibroin can be extracted from degummed silk (boiling off), which refers to the removal of serine. During the past decades, silk fibroin has been commonly used as a suture material (Surusil®, Suru; Sofsilk™, Covidien). To date, various studies reported the use of silk materials in nanofibers, films, scaffolds, hydrogels, and microspheres for tissue engineering applications with various cell types and drug delivery applications [21–23]. The unique physicochemical properties of silk include a hydrophobic structure, strong intramolecular and intermolecular interactions, and crystal polymorphism. Therefore, silk fibroin enables tunable and slow biodegradation kinetics without toxic chemical cross-linkers. The hydrolysis of silk fibroin is mainly attributed to the proteolytic enzymes, including part of the foreign-body response cascade [24].

Natural carbohydrates have been utilized as hydrogels for tissue engineering applications and drug carrier materials for drug delivery system [25]. For instance, the hydrophilic polysaccharide hyaluronic acid (HA), which is composed of repeating disaccharide units of β-D-glucuronic acid and *N*-acetyl-β-D-glucosamine, is broadly distributed in the ECM and shows a critical role in vertebrate tissue morphogenesis [26]. Due to its excellent biocompatibility and hydrodynamic features, HA has been approved for use in human patients as viscous fluid and sheet formulations for knee pain control and surgical adhesives, respectively. To date, many clinical trials have confirmed the safety and effectiveness of HA for human clinical trials [27–31]. The bioactivity of HA, like that of other carbohydrates, can be also improved by physical and/or chemical modification to modulate the cell migration, spreading, and multiplication.

Other carbohydrate polymers such as alginate, dextran, and chitosan have been used in various biomedical applications. Alginate (alginic acid), composed of linear block copolymers of [1–4]-linked β-D-mannuronic acid (M) and α-L-guluronic acid (G) covalently linked together in different blocks, has been used extensively in a hydrogel form for cell encapsulation and drug delivery [32] as well as in tissue engineering applications [33]. Alginate is ionically cross-linked by the addition of divalent cations (e.g., Ca^{2+}) in aqueous solution. The production of purified alginates with consistently high medical-grade quality has been considered. The purity of the input algal source is critical, along with the development of validation strategies for alginate extraction and purification processes. Alginate hydrogel undergoes a slow and unpredictable dissolution process in vivo, mainly due to the sensitivity of the hydrogels toward calcium chelating compounds [34]. Moreover, a key engineering challenge in designing immunoisolation of the alginate-based microcapsules for cell encapsulation is that of maintaining unhampered exchange of nutrients, oxygen, and therapeutic molecules released by the encapsulated cells while avoiding swelling and subsequent rupture of the alginate microcapsules [35]. Although many advances in alginate-based biomaterials have been informed, numerous progresses will be required for clinical applications.

Chitosan is also a linear polysaccharide of [1–4]-linked D-glucosamine and *N*-acetyl-D-glucosamine residues, which is a polycationic material produced by the deacetylation of chitin. Chitosan is insoluble in aqueous solution (above pH 7) because of its crystalline structure, while it is fully soluble in dilute acids (below pH 5). It has been used in a number of gene and drug delivery applications due to the abundant positive charged functional groups. Also, its applications in tissue engineering and regenerative medicine have been broadly investigated [36–38].

2.2.2 Acellular Tissue Matrices

ECM contributes to biomechanical/structural integrity and signaling and regulatory functions in the development, maintenance, and regeneration of tissues. ECM components participate in the tissue-specific control of gene expression through a variety of transduction mechanisms in synergic effects with soluble biochemical signals provided by growth factors, cytokines, and hormones [39–41]. Moreover, ECM is itself a dynamic structure that is actively remodeled by direct interactions with the cells [42]. Development of novel scaffolding systems that more closely recapitulate the biological and biomechanical functions of native ECM remains an important area of tissue engineering approaches [43]. Understanding ECM's complex functions in mature or regenerating tissues continues to be an important task to design a synthetic microenvironment.

In the current limitations for de novo construction of a true ECM mimic from purified ECM components, decellularized tissue matrices have been considered as an ideal scaffolding material due to their biological, structural, and biomechanical similarity to native tissues and their possession of tissue-specific ECM proteins which remain after the decellularization process [44]. As a result, these matrices have been used for a number of tissue engineering applications and have yielded an understanding of the physicochemical properties of ECM as well as a tissue-specific scaffold for engineering functional tissues or organs. For example, by harnessing the cell-matrix interactions that are crucial for cellular behaviors, including adhesion, proliferation, and differentiation [40], numerous studies have encouraged the tissue formation by combining cells with tissue-specific ECM scaffolds from decellularized heart [45], urinary bladder [46], liver [47], lung [48], kidney [49], and small intestinal submucosa (SIS) [50]. Each of these decellularized tissue matrices maintained the biomechanical properties and structural integrity of the original tissue or organ with minimal disruption to the ECMs. More importantly, these ECM proteins play a vital role in tissue maintenance and regeneration of a specific tissue type, and they can modulate cell adhesion and migration, growth factor storage and release, and progenitor cell activation and differentiation [51, 52].

One of the most common acellular tissue matrices is small intestinal submucosa (SIS), which is prepared by mechanical removal of tunica mucosa and muscularis, and serosal layer [53]. Acellular SIS contains various bioactive molecules such as fibronectin, glycosaminoglycans (GAGs), and growth factors [54]. Recently, commercially available SIS-based matrices have been successfully implanted in several reconstructive surgical procedures, which include hernia repair [55, 56], abdominal wall closure [57], urinary incontinence [58, 59], bladder augmentation [60], pelvic organ prolapse repair [61], and so on. As SIS provides high biological variability, standardized protocol for preparation, and manufacturing procedure with quality control, SIS-based tissue products can be a reliable therapeutic option for clinical applications.

Skeletal muscle tissue matrices have been decellularized, characterized in vitro, and validated in animal models [62–64]. We believe that the intact ECM proteins play an important role in tissue maintenance and reconstruction of skeletal muscle tissue; furthermore, they modulate cell adhesion and migration, growth factor storage and release, and satellite cell activation and differentiation [51, 52]. Moreover, immune responses to allogeneic or xenogeneic cell sources and tissues involved the recognition of major histocompatibility complex (MHC) molecules expressed on the surface of the cells by the immune cells of the host. The decellularization process removes all cellular components, including DNA, and leaves a structure comprised of ECM materials. Thus, minimal immune rejection response may occur with the transplantation of these allogeneic or xenogeneic nature scaffolds [64].

The use of native heart ECM is of special interest for cardiac-related cells due to the presence of intrinsic regulatory factors for cardiac function [65–67]. To engineer a tissue construct, cells are dependent on the ECM providing tissue-specific molecular, structural, and mechanical signals that regulate cell behaviors. Thus, scaffold materials for engineered cardiac tissue should mimic the native heart ECM in order to provide a natural microenvironment for the cardiac cells. Recently, incorporation of heart ECM into a bioink formulation for bioprinting cardiac tissue has been suggested [68]. In an attempt to produce such a unique ECM component, decellularization technique of native tissues has been a very attractive method, while natural polymers and synthetic polymers are unable to replicate the precise spatial organization of complex structures like cardiac tissue [69]. Generally, decellularization strategy is to remove cellular components from tissues via mechanical or chemical manipulation to produce ECM-rich matrices.

The high degree of evolutionary preservation of ECM components allows the use of acellular matrices derived from xenogeneic sources. Various decellularized tissue matrices have been utilized successfully for tissue engineering in animal models, and several xenogeneic products have received regulatory approval for clinical use [70]. These decellularized xenogeneic medical products are being introduced into the market. Despite many advantages, there are concerns about the use of decellularized xenogeneic tissue matrices. These include the potential for immunogenicity, the possible presence of infectious agents, variability among preparations, and the inability to completely specify and characterize the bioactive molecules of the material. Many of the soluble bioactive factors in the tissue matrices can be lost during the decellularization process.

2.2.3 Synthetic Biodegradable Polymers

Synthetic polymers are high molecular weight macromolecules consisted of covalently bound repeating units (monomers). These repeating units can react with each other to form high molecular weight homopolymers by mainly addition polymerization, step-growth polymerization, and ring-opening polymerization. In tissue engineering, biodegradable synthetic polymers offer a number of benefits as scaffolds or drug delivery vehicles [71, 72]. These polymers can be synthesized with reproducible quality controls and high purity and fabricated into various shapes with anticipated bulk and surface features. Specific advantages include the ability to tailor the mechanical properties and degradation kinetics of these polymeric materials to suit various tissue engineering applications. Linear aliphatic poly(α-hydroxy acids), such as PGA, PLA, and their copolymer [poly(lactide-co-glycolide) (PLGA)], are the most commonly used biodegradable polymers in tissue engineering (Fig. 2.1) [72]. These biodegradable polymers have gained popularity due to their processing reliability, tunable mechanical properties, and controlled biodegradability, and they are Food and Drug Administration (FDA)-approved for human use in a variety of medical applications, including suture material, implant, and drug delivery systems.

The ester bonds in these polymers degrade by nonenzymatic hydrolysis, and their nontoxic degradation products are eliminated from the body in the form of carbon dioxide and water. The degradation rate of these polymers can be controlled by alteration of their crystallinity, molecular weight, and the copolymer ratio of monomers. Furthermore, the degradation rates that can be achieved range from several weeks to several months. Because these polymers are thermoplastics, they can be configured into a three-dimensional (3D) structure with a desired microarchitecture, shape, and dimension. However, synthetic polymers generally lack intrinsic biological properties, and their degradation products may cause adverse effects or alter local microenvironment in vivo. In addition, the

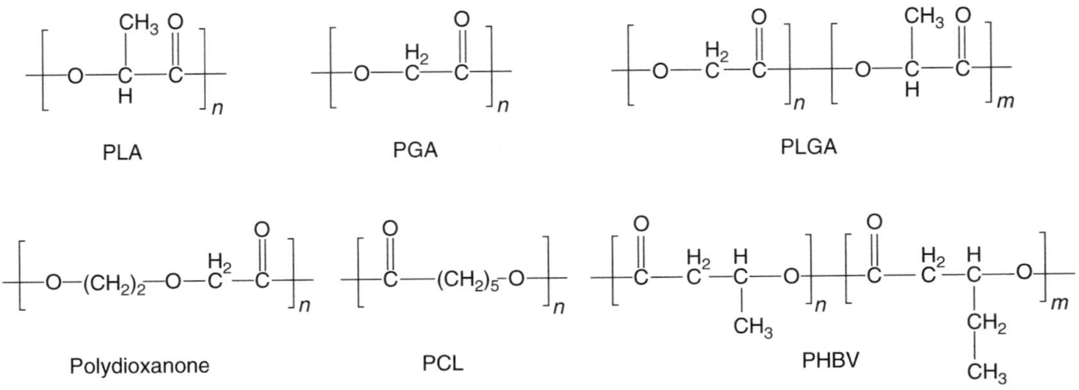

Fig. 2.1 Chemical structure of the commonly used biodegradable synthetic polymers in tissue engineering

surface hydrophobicity of synthetic polymers may mediate protein denaturation in the vicinity of the implant and induce fibrous encapsulation [73, 74].

Several research teams have explored the synthesis of novel polymeric biomaterials that unite the advantages of "smart" polymeric systems with the biological activities of proteins at the chemical level. The concept of these smart polymeric systems was initially derived from the development of polymeric materials that show large conformational changes in response to microenvironmental stimuli, such as temperature, ionic strength, pH, or light [75]. The responses of the polymers may include precipitation or gelation, reversible adsorption on a surface, collapse of a hydrogel or surface grafting, and alternation between hydrophilic and hydrophobic states [76]. In most of cases, the change in the state of the polymer is reversible. Beyond the physical properties of these polymers, imparting smart biomaterials with the specific properties of signaling molecules, such as ECM components and growth factors, can be achieved. Hence, smart polymeric biomaterials aim to actively participate in the biological system in the body for accelerating tissue maturation and, eventually, regeneration.

To maximize the role of biomaterials in tissue regeneration, a combination of two or more materials with unique properties is often selected for use as a scaffold. These hybrid biomaterial systems that possess biological and physiochemical functionality have been used. The main strategy of this hybrid biomaterial system is to achieve synergy between the desirable biomechanical properties of one component with the biological compatibility and physiological relevance of the other component [77]. For example, natural polymers possess microstructures that contain chemically and structurally defined components. Combination of these materials with synthetic polymers would offer a convenient bridge between chemical and biosynthetic approaches, often within the context of thermodynamically driven assembly processes [78].

2.3 Basic Requirements of Biomaterials in Tissue Engineering

Biomaterials replicate the biologic and biomechanical features of the native ECMs found in tissues in the body by serving as a synthetic microenvironment. Biomaterials provide a 3D space for the cells to form new tissues with tissue-specific structure and function; moreover, they could allow for the delivery of cells and bioactive molecules (such as cell adhesive proteins and growth factors), to the targeted sites in the body [74]. As the majority of mammalian cell types are anchorage-dependent, biomaterials provide a cell adhesion substrate that can deliver cells to specific sites with high loading efficiency and homogeneity. Biomaterials can also provide biomechanical support against in vivo forces to maintain the predefined 3D tissue structure during the neo-tissue formation. Also, bioactive molecules, including adhesive proteins and growth factors, can be loaded along with cells to regulate the cellular behaviors.

In order to minimize the host responses, including inflammation, biomaterials should be biodegradable and bioresorbable. Incompatible materials produce an inflammatory or foreign-body reaction that eventually leads to rejection and/or tissue necrosis. Degradation products, if produced, should be removed from the body through the metabolic pathway at a designed rate that keeps the concentration of these degradation products in the body at a tolerable level [79]. Biomaterials should also provide a tissue-specific microenvironment in which appropriate regulation of cell behaviors (adhesion, proliferation, migration, and differentiation) can occur so that functional neo-tissue can be formed. Cell behaviors in the newly formed tissue have been shown to be regulated by multiple interactions of the cells with their microenvironment, including interactions with cell adhesion ligands [80] and with soluble growth factors [81]. Since biomaterials provide temporary mechanical support while the cells undergo spatial tissue reorganization, the properly chosen biomaterials should allow the engineered tissue to maintain sufficient biomechanical integrity to support itself in early developmental stage. In late development, biomaterials begin degradation; hence tissue growth is unhindered [74].

2.3.1 Scaffolding System: Template

Biomaterials for cell-based tissue engineering approaches should be biodegradable and absorbable without eliciting inflammatory responses that interfere with cellular function and neo-tissue formation; thus it has been considered to their biological properties, biodegradability, and structural and mechanical properties. In tissue engineering, biomaterials should [1] facilitate the localization and delivery of somatic cells to specific sites in the body, [2] maintain a 3D architecture that permits the formation of neo-tissues, and [3] guide the development of neo-tissues with tissue-specific functions [7]. In details, biomaterials can provide an adhesion substrate for transplanted cells and serve as a delivery vehicle into the sites of interest. A large surface area-to-volume ratio is desirable in order to allow for accommodation of a high number of cells, which dictates the use of porous scaffolds. It is often desirable for neo-tissues to have a predefined structure, which can be achieved by an appropriate selection of biomaterials with controlled degradable rates. In addition, biomaterials can provide a temporary mechanical support that is sufficient to withstand in vivo forces and maintain a spatial structure until neo-tissue is able to support itself. This is essential for hard, weight-bearing tissue regeneration like the bone and cartilage. The stability of biomaterials depends primarily on physical properties and chemical degradation [8]. The mechanical properties of biomaterials can be adjusted to suit the needs of the target tissue or organ along with the fabrication technologies.

A porous architecture with interconnectivity that allows the cellular ingrowth and maximizes the oxygen and nutrient diffusion is considered as an essential scaffold fabrication requirement. Configuration of such a structure may require tissue-specific considerations, which include variations in pore morphology and surface characteristics. In addition, biomaterial should be biodegradable or amenable to long-term integration with the host tissue. Biodegradability allows for the gradual and orderly replacement of the scaffold with functional tissue and prevents the development of adverse chronic responses to the artificial tissue structure [9]. Therefore, biomaterials should be selected based on an understanding of target tissues or organs biologically, anatomically, pathologically, and biomechanically.

2.3.2 Biodegradability/Bioabsorbability

The mechanism of biodegradation of synthetic polymeric biomaterials is mainly processed through passive hydrolysis, with the monomeric acids as the final metabolized product [73]. The degradation rates are closely related to molecular weight, degree of crystallization, hydrophilicity/hydrophobicity, and surface-to-volume ratio. For instance, PGA is generally available in a highly crystalline form, and it possesses a relatively high melting point (~230 °C) and low solubility in organic solvents. However, its hydrophilic nature allows for rapid absorption of water leading to a fast degradation within a few weeks. In contrast, PLA is available in semicrystalline or amorphous forms, depending on the types of the D and L optical isomers of lactic acid present in the molecule. D-PLA and L-PLA are both highly crystalline, whereas the racemic copolymer D,L-PLA is completely amorphous. The presence of a methyl group makes PLA more hydrophobic than PGA, resulting in lower water absorption rates and correspondingly slower degradation, in the order of months [10]. Moreover, the factors affecting biodegradation include chemical and structural configuration, fabrication parameters, sterilization methods, and implantation sites. For example, sterilization using γ-irradiation can decrease the polymer's Mw by 30–40%. The irradiated polymers continue to decrease in Mw during storage at room temperature. The decline in Mw affects the mechanical properties and the biodegradation rate of the polymers [8, 11].

The degradation of polymeric biomaterials results in erosion (loss of mass), when the degradation products diffuse or dissolve into the body fluid. Polymer erosion is generally classified by bulk erosion and surface erosion [82]. The polymer chain scission by the bulk erosion is wholly occurred, while the surface erosion is only occurred in the surface of the polymeric scaffolds. In the bulk erosion, the molecular weight and mechanical properties of the polymeric scaffolds decrease in time; however, the external dimension of the scaffolds remains essentially unchanged until the scaffold is fully degraded. In the surface erosion, the external dimension and mass of the scaffolds decrease in time.

Unlike the synthetic polymers, the biodegradation of the natural polymers is mainly dependent on the enzymatic reaction. This is the microenvironmental responsiveness of the natural polymers through degradation and remodeling by cell-secreted enzymes [83]. The biodegradation rate of these natural polymers can be controlled by various cross-linking processes. For instance, a collagen scaffold that is designed for a long-term in vivo use would require a cross-linking process to improve its resistance to the proteolytic degradation and hydrolysis and to enhance its mechanical strength [12]. Glutaraldehyde is a commonly used cross-linker for tissue matrices because it has amine-reactive homobifunctional groups [84]. Carbodiimide chemistry has been used extensively to cross-link collagen scaffolds and to immobilize other proteins to the scaffolds. Particularly, 1-ethyl-3-(3-dimethylaminopropyl) carbodiimide hydrochloride (EDC) has been used as a zero-length cross-linker that directly binds between collagen molecules [85].

Biomaterials used for cell-based tissue engineering serve as an artificial ECM and bring forth biological, biochemical, and biomechanical functionalities of native ECM found in tissues and organs. Therefore, cell-biomaterial interactions, such as cellular adhesion on the matrix surface, cell maturation, ECM production, and proliferation, are critical for the success of this strategy. ECM is known to affect cellular differentiation and function; thus, biomaterials can be designed to deliver cells and bioactive molecules (e.g., cytokines and growth factors) to a designated region for neo-tissue formation with enhanced functionality [4–6].

2.4 Advanced Biomaterial Systems in Tissue Engineering

Future advances in biomaterials for tissue engineering rely on the development of the biomimetic materials that can utilize in vivo microenvironment for efficient tissue regeneration and better integration with host tissue

(Fig. 2.2). Desirable biomimetic materials could be designed as a smart scaffolding system that mimics the natural biological system and integrates the necessary structural and biological properties. Thorough understanding of biomaterials science combined with extensive knowledge of the clinical challenges and cell biology is vital for development of novel advanced biomaterials to be used for tissue engineering applications.

2.4.1 Surface Modification

One commonly used technique is the surface modification of polymeric biomaterials. Chemical and physical modification of the biomaterials surface can increase their biological properties (Fig. 2.3). Chemical procedures, such as oxidation, hydrolysis, and quaternization, change the surface chemistry and functionality of biomaterials. Polymerization or grafting of water-soluble polymers to a biomaterial surface is usually performed to improve the protein adsorption and further cell adhesion [86, 87]. In addition, bioactive molecules, such as enzymes, drugs, proteins, peptide sequences, and antibodies, have been immobilized onto biomaterial surfaces to achieve added functionality [88]. These approaches provide an opportunity to tailor a biomaterial surface so that it has specific affinities, site recognition properties, and controlled mobility, while the bulk properties of biomaterials are still maintained.

2.4.2 Drug Delivery System

Biomaterial functionalization has also allowed enhanced cellular interactions via delivery of bioactive molecules from an implanted scaffold [89]. Bioactive molecules, such as cytokines and growth factors, are powerful regulators of biological functions, which include the cell migration, proliferation, and differentiation. Incorporation of bioactive molecules into biomaterial scaffolds is a technique to improve the outcome of cell-based tissue engineering. To employ this technique, an understanding of the biological activities of molecules is necessary. For example, the biological activity of growth factors is dependent not only on their presence in solution but also on their interactions with the surrounding ECM. Some growth factors are most effective

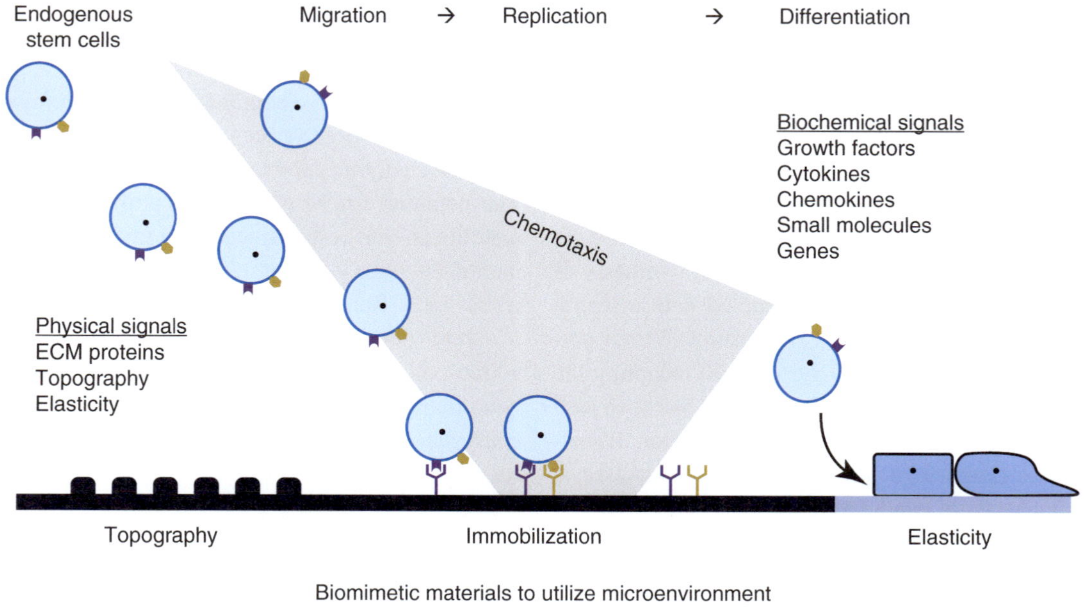

Fig. 2.2 Schematic illustration of interactions between endogenous stem cells and biomimetic materials. Stem cell fate in a particular microenvironment is regulated by intricate reciprocal molecular interactions with its surroundings

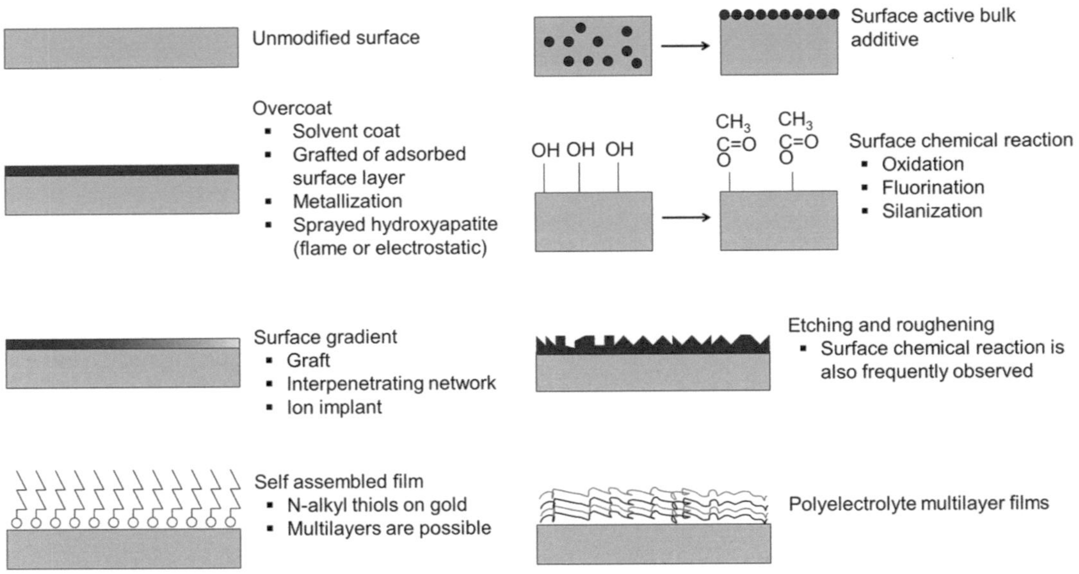

Fig. 2.3 Schematic diagram of surface modification methods of Polymeric biomaterials

when released over a prolonged period of time, whereas others are more effective when delivered in a bolus. Some factors are active while tethered to a material, whereas others are only active when they have been released from the biomaterial and are internalized into a cell. These considerations must be taken into account when designing a delivery system [90].

2.4.3 ECM-Mimetic Biomaterials

Development of advanced biomaterials has been extended to the design, discovery, and evaluation of biologically active materials. This required the relatively straightforward chemical synthesis of new polymeric materials, coupled with a search for novel activities and examination of their performance in biological systems. By adapting the combinatorial library approach that has been well established for synthetic peptides and small molecule drugs with novel high-throughput assays, thousands of candidate chemicals can be synthesized and tested. As one example, screening process of a combinatorial library derived from commercially available monomers in the acrylate family revealed novel synthetic polymers that influenced the attachment, growth, and differentiation of human embryonic stem cells (ESCs) in unexpected ways [91]. More revolutionary developments in biomaterials science will continue to arise as tissue engineering is combined with technology-based approaches. Basic understanding of the 3D structure of existing biological molecules is being applied to a "bottom-up" approach to generate new, self-assembling supramolecular architectures [92].

Self-assembling peptides offer promise because of the large variety of sequences that can be made easily by automated chemical synthesis. The potential for bioactivity, the ability to form nanofibers, and responsiveness to microenvironmental cues are inherent in some of these biomaterials [93]. Recent advances include the design of short peptides (e.g., heptamers) based on coiled-coil motifs that reversibly assemble into nanofilaments and nanoropes, without excessive aggregation [94]. These novel peptide amphiphiles can be induced to self-assemble by changes in concentration, pH, or the level of divalent cations [95]. Branched structures can be designed to present bioactive sequences such as arginine-glycine-aspartic acid (RGD) to cells by nanofibrous hydrogels or as coatings on typical tissue engineering scaffolds [96]. In addition, assembly can occur under conditions that permit the entrapment of cells in the resulting nanofibrous matrix [97]. The entrapped cells retain motility and the ability to proliferate. Peptide-based nanofibers can be designed to present bioactive sequences to cells at very high density, substantially exceeding that of corresponding peptide epitopes in biological ECM.

A pentapeptide epitope of laminin, isoleucine-lysine-valine-alanine-valine (IKVAV), known to promote neurite extension, was incorporated into peptide amphiphiles capable of self-assembly into nanofibers that form a hydrogel [98]. When neural progenitor cells (NPCs) capable of differentiating into neurons or glial cells were encapsulated during assembly of the nanofibers, they survived over a few weeks in culture. Moreover, even without the addition of neurotrophic factors, they displayed neuronal differentiation as exemplified by the extension of large neurites that were obvious after 1 day in culture and by their expression of β-tubulin III. The production of neuron-like cells from neural progenitors, whether dissociated or grown as clustered neurospheres, was more rapid and robust in the IKVAV-containing peptide amphiphile (PA) gels than on laminin-coated substrates or with soluble IKVAV. By contrast, the production of cells expressing glial fibrillary acidic protein (GFAP), a marker of astrocytic differentiation, was suppressed significantly in the IKVAV-PA hydrogels even when compared to growth on laminin, which favors neuronal differentiation. The ability to direct stem or progenitor cell differentiation via a chemically synthesized biomaterial, without the need to incorporate growth factors, offers many potential advantages in tissue engineering and regenerative medicine.

Proteins have been conjugated either randomly or in a site-specific manner, through engineering of protein to introduce a reactive amino acid at a particular site [99]. If a conjugation site is introduced near the ligand-binding domain of protein, it has been shown that induction of a change in conformational state of the novel polymers can serve to regulate the protein's activity [100–103]. This may allow selective capture and recovery of specific cell types, delivery of cells to a desired location, and modulation of enzymes that influence the tissue remodeling process. The discovery of new classes of biomaterials may provide yet another opportunity to overcome the current limitations in tissue engineering.

2.4.4 Biophysical Signals

Cells are exposed to a tissue-specific microenvironment in vivo, and they respond to numerous chemical and physical stimuli. As a result, cellular functions like cell adhesion and proliferation, protein synthesis, and cytoskeletal architecture are critically influenced by the microenvironment of the cells. To maintain the phenotype expression and differentiation of tissue-specific cells, numerous cues have been applied to biomaterial scaffolds to mimic natural ECM microenvironments. Cells can be directly influenced by topographical cues of their surrounding microenvironment. The topography of the biomaterial surface can influence the cell adhesion and proliferation, which further affects cellular functions. It has been determined that natural ECM has a nanoscaled fibrous structure that supports the cellular interactions and stimulates the cellular functions. This substrate nanoscale topography provides a tool for investigating complex cell functions, such as adhesion, migration, cytoskeleton reorganization, and cell polarization [104, 105].

Physical contact of cells on the biomaterial surface is essential to anchorage dependence and function. Current understanding reveals that stem cell fates can be significantly affected by the substrate elasticity from the surrounding microenvironment, while stem cell differentiation induced by soluble biological signals has been widely developed. For example, neural cell differentiation can be induced on soft matrices (high elasticity), which mimic the mechanical properties of the brain tissue, while myogenic and osteogenic differentiation require stiffer matrices (low elasticity) that mimic muscle and collagenous bone tissues. Soft matrices that mimic brain are neurogenic, stiffer matrices that mimic muscle are myogenic, and comparatively rigid matrices that mimic collagenous bone prove osteogenic [106]. These studies have demonstrated that specific stem cell differentiation can be achieved using substrates or matrices with particular surface elasticity/stiffness. Modification of the cross-linking degree of polymeric hydrogels is the basis of substrate-controlled elasticity.

2.5 Scaffold Fabrication Technologies

Biomaterials used for tissue engineering applications require different spatial configurations, depending on the desired outcome. In tissue engineering, biomaterials need to be fabricated to facilitate regeneration of the target tissues or organs. Thus, starting with simple foams and fibers, the porosity and pore interconnectivity of biomaterial scaffolds has quickly become a central theme, with pore dimensions and physical properties being vital in promoting cell seeding, migration, proliferation, and de novo production of ECM molecules. Many fabrication technologies have been employed to configure biomaterials in this way for tissue-engineered scaffolds. In such instances, biomaterials provide the structural support for cells to proliferate and maintain their function. To permit cell accommodation, biomaterials must possess adequate surface area for cell attachment and distribution. This is usually achieved by creating interconnecting pores that allow for maximum diffusion of oxygen and nutrients. Thus, numerous fabrication technologies have been developed to create porous scaffolds, which include solvent casting/particulate leaching, freeze-drying, fiber bonding, electrospinning, phase separation, gas foaming, high pressure processing, melt molding, membrane lamination, hydrocarbon templating, cell printing, and solid free-form fabrication (SFF) methods.

2.5.1 Hydrogels

Hydrogels have been used in injectable approaches for cell therapy and tissue engineering, which offer many advantages. These include the repair of tissue defects or injection of cells for achieving cellular function [33]. However, while injection therapy is simple and does not require invasive surgical procedures, its application is limited to areas of the body where injection is feasible. Injectable biomaterials usually take the form of liquids or gels, which can be configured from synthetic and natural polymers. A "hydrogel" is defined as a 3D network swollen by water which is usually the major component of the hydrogel system. Hydrogels are formed by cross-linking water-soluble polymers to form an insoluble, hydrophilic polymer network. These polymer networks are capable of absorbing water from a small fraction of their dry weight up to thousands of times their dry weight [107]. Representative hydrogels include poly(ethylene glycol) (PEG), poly(ethylene oxide) (PEO), poly(vinyl alcohol) (PVA), poly(acrylic acid) (PAA), Pluronics (PEO-PPO-PEO copolymers), agarose, alginate, chitosan, collagen, fibrin, gelatin, and hyaluronic acid (HA).

In the past decade, progress in creating 3D cellular microenvironments using hydrogels has appeared to have similar elasticity to that of native tissues. The structure and composition of hydrogels can be tailored to bear the suitable chemical, biological, and physical cues that encourage the development of tissue constructs [108]. For example, matrix metalloproteinase (MMP)-sensitive PEG-based hydrogels have been developed to direct differentiation of pluripotent cardioprogenitor cells using P19 embryonal carcinoma cells [109]. In order to systematically probe 3D microenvironmental effects on P19 embryonal carcinoma cell differentiation, matrix elasticity, MMP-sensitivity, and the concentration of a matrix-bound RGDSP peptide were modulated. Results indicated that soft PEG-based hydrogels mimicking the elasticity of embryonic cardiac tissue could direct cardiac commitment of pluripotent P19 cells.

The most common cell encapsulation system involves alginate microcapsules, which can enclose living cells or other molecules. Cell encapsulation techniques such as this allow allogeneic and xenogeneic cell sources to be considered for clinical translation, and this is important as autologous cells can often be limited in supply. In order to use allogeneic or xenogeneic cell sources, the immune responses of the host need to be addressed. An immunoisolation technique using cell encapsulation permits the use of these cell sources without the use of immunosuppressants. Cell microencapsulation can be used for cell therapy as well as for delivering biological products. This technique consists of enclosing biologically active substances within a polymeric hydrogel surrounded by a semipermeable membrane that is designed to prevent immune recognition and subsequent rejection. Ideally, the encapsulating membrane allows the bidirectional diffusion of nutrients, oxygen, and waste as well as the secretion of therapeutic products while preventing immune cells and antibodies, which would normally destroy the enclosed cells, from entering the capsule [110]. Pancreatic islets, hepatocytes, renal cells, parathyroid cells, and adrenal chromaffin cells have been immunoisolated for cell-based therapeutic delivery systems in this manner. Several polymeric encapsulation systems have been developed and are currently being tested in clinical trials [111, 112].

2.5.2 Electrospinning

While many other scaffold fabrication methods are used in tissue engineering applications, few can provide scaffolds with critical similarities to natural ECM that electrospinning of nanofibers can provide. Other fabrication processes allow for the inclusion of natural polymers such as collagen into the scaffold, but some, such as melt molding, require high temperatures which may denature proteins. Others, such as solvent casting and particulate leaching, provide scaffolds with very high porosity but potentially low interconnectivity and may not allow for easy shaping [113]. Electrospinning has become a popular alternative fabrication method, as it can be applied to many disciplines and is a relatively simple and inexpensive scaffold fabrication process. It relies on a high-voltage direct or alternating current source to charge a polymer solution or melt contained in a syringe [114]. Production of nanofibers by electrospinning provides excellent pore interconnectivity and precise control of porosity through adjustment of fiber size [115]. This interconnectivity can allow for integration of cells into the scaffold if the pore size is large enough [116] and also allows for dissolution of soluble factors and nutrients throughout the scaffold. These properties closely mimic the natural properties of ECM, and the parameters of the electrospinning process can be controlled so as to adjust to features of targeted tissue types. For instance, arteries are composed of a layer called the intima, which consists of an endothelial cell (EC) layer lining the lumen of the artery, which sits on a basal lamina. Surrounding this layer is the tunica media, which is a layer of connective tissue and smooth muscle cells (SMCs). To mimic this vessel structure, our lab showed that it is possible to create a bilayered vascular scaffold which supports EC growth on an electrospun intimal layer and SMC growth on an electrospun media layer (Fig. 2.4) [116].

Electrospun fibrous scaffolds can also be functionalized by adding biochemical and biomechanical cues to enhance cellular interactions; however, specific interactions between cells and electrospun scaffolds functionalized through surface modifications and bioactive factor incorporation are still poorly understood. A better understanding of the specific cues that enhance cell adhesion, proliferation, and guidance of cells seeded on a scaffold, as well as the cues that could affect host cell infiltration, differentiation, and vascularization both in vitro and in vivo, is crucial for the advancement of tissue engineering applications of electrospinning.

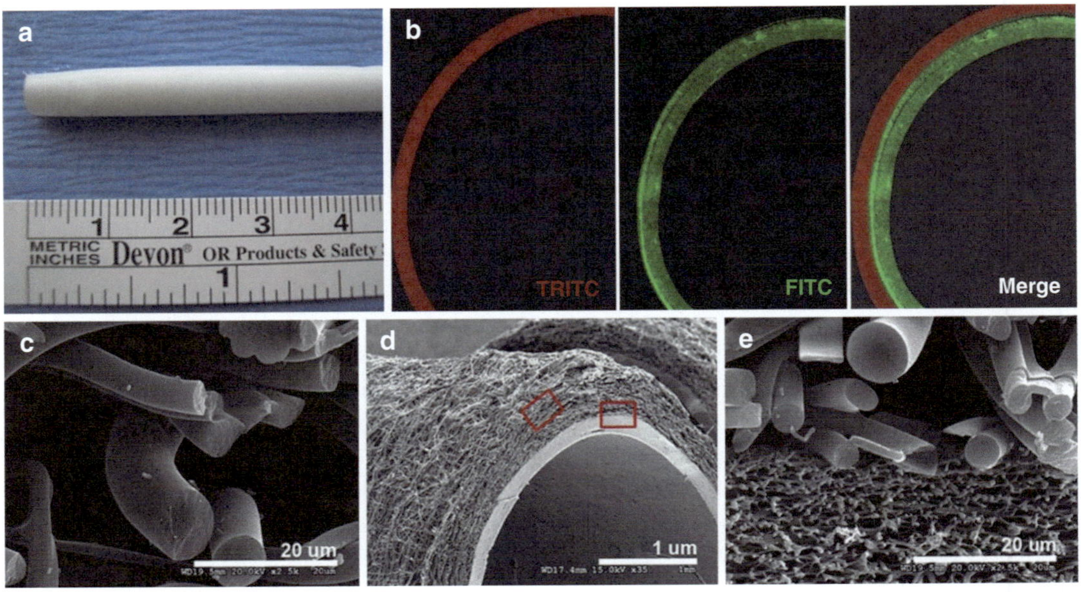

Fig. 2.4 Bilayered electrospun PCL/collagen vascular scaffolds; (**a**) gross appearance, (**b**) fluorescent images of bilayered scaffolds incorporated fluorescent dyes. SEM images of cross-sectional (**c**) outer layer, (**d**) entire, and (**e**) interface between outer and inner layers of bilayered scaffolds. Reproduced with permission from Ju et al. [116]

2.5.3 Computer-Aided Scaffold Fabrication

For clinical translation, a fabrication method should provide a scaffold with appropriate shape and size to reconstruct damaged tissues in patients. Computer-aided fabrication or additive manufacturing of tissue-engineered constructs is a rapidly emerging technology that promises to accelerate the clinical translation of many tissue engineering applications. This technology is based on the ability to build tissue-engineered constructs with highly reproducible architecture and to provide compositional variation across the entire scaffold through tightly controlled, computer-driven fabrication. To accomplish this, a conventional inkjet printer can be used [117–120]. This thermal inkjet printer produces small air bubbles consisting of cells and a hydrogel by heating, and then bubbles are collapsed to provide the pressure pulse to eject a very tiny drop of "bioink" out of the nozzle. This printing technique uses aqueous processes that have been shown to induce little damage to cells. In the extrusion printing method, a pneumatic dispensing system has been utilized to build various 3D architectures by depositing a continuous stream of cell-laden hydrogel [121]. These approaches take advantage of rapid prototyping assisted by computer-assisted design (CAD) procedures to build a tissue construct with geometric control of its internal structure and external shape. We have recently showed that viable 3D heterogeneous construct with multiple cell types could be created by printing the cells with hydrogel layer-by-layer [120]. This cell-printed constructs were able to survive and mature into tissues with adequate vascularization when implanted in vivo (Fig. 2.5) [122].

In 2016, our laboratory has reported a novel integrated tissue and organ printing (ITOP) system that makes it possible to fabricate a high strength tissue construct and precisely place various cell types within a single architecture [122, 123]. This system consists of multiple nozzles that allow delivery of multiple types of biomaterials and cells. The goal of ITOP is to concurrently deliver a supportive 3D template (mostly biodegradable thermoplastic polymers) and 3D–patterned deposition of cell-laden hydrogel

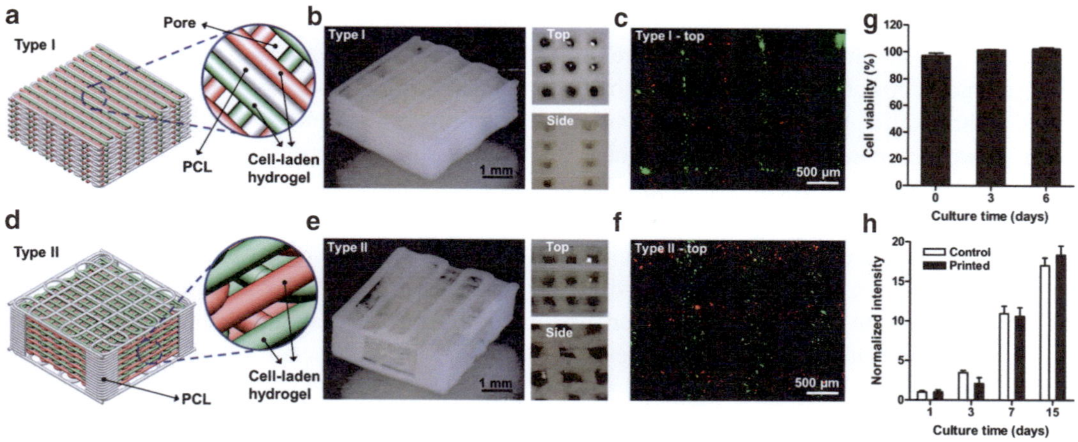

Fig. 2.5 3D multiple cell positioning: (**a–c**) type I pattern and (**d–f**) type II pattern. (**a, d**) Two types of 3D patterning, including cell-A (*red*), cell-B (*blue*), and PCL (*green*), were fabricated. (**b, e**) Photographs and (**c, f**) fluorescent image of the 3D printed patterns. (**g**) Cell viability (%) was over 95% on day 0 and then maintained on days 3 and 6 ($n = 3$). (**h**) Cell proliferation results showed that the number of cells continuously increased over a 15-day period, and no significant differences between the control and the printed constructs were observed ($n = 5$). Reproduced with permission from Kang et al. [122]

biomaterials in a precise manner. This approach is borne out of the challenges of tissue engineering 3D tissue constructs.

2.6 Tissue Engineering Applications

Several therapeutic approaches on the cellular level have been investigated for direct treatment of medical conditions or for use of implanted engineered tissue constructs. A variety of tissue engineering products have been tested by injection, gene delivery, implantation, encapsulation, or combined systems. Applications in tissue engineering of the several therapies are described.

2.6.1 Bulking Agents for Urinary Incontinence and Vesicoureteral Reflux

Injectable bulking agents have been used endoscopically in the treatment of both urinary incontinence and vesicoureteral reflux. The advantages in treating urinary incontinence and vesicoureteral reflux with this minimally invasive approach include the simplicity of a quick outpatient procedure and the low morbidity associated with it. Investigators are seeking alternative biomaterials that would be safe for human use. The ideal substance for the endoscopic treatment of reflux and incontinence should be injectable, nonantigenic, nonmigratory, volume stable, and safe for human use. Toward this goal, long-term studies were conducted to determine the effect of injectable chondrocytes in vivo. It was initially determined that alginate, a liquid solution suspended with chondrocytes, could serve as a synthetic substrate for the injectable delivery and maintenance of cartilage architecture in vivo. This system was adapted for the treatment of vesicoureteral reflux in a mouse model [124] and a porcine model [125]. The use of autologous cartilage for the treatment of vesicoureteral reflux in humans would satisfy all the requirements for an injectable substance.

Two multicenter clinical trials were conducted using this technology. Patients with vesicoureteral reflux were treated at ten centers throughout the USA. The patients had a similar success rate as with other injectable substances, in terms of cure. Chondrocyte formation was not noted in patients who had treatment failure. It is supposed that the patients who were cured have a biocompatible region of engineered autologous tissue present, rather than a foreign material [126]. Patients with urinary incontinence were also treated endoscopically with injected chondrocytes at three different medical centers. Phase 1 trials showed an approximate success rate of 80% at follow-up 3 and 12 months postoperatively [127]. Therefore, human application of cell-based tissue engineering for urologic applications occurred successfully with the injection of chondrocytes for the correction of vesicoureteral reflux in children and for urinary incontinence in adults [126, 127].

2.6.2 Injection Therapy Using Muscle Progenitor Cells

The potential use of injectable cultured primary muscle progenitor cells (MPCs) for the treatment of stress urinary incontinence has been investigated [128]. Fluorescence-labeled MPCs were injected directly into the proximal urethra and lateral bladder walls of nude mice with a microsyringe in an open surgical procedure. Tissue harvested up to 35 days after injection contained the labeled MPCs, as well as evidence of differentiation of the labeled MPCs into newly formed myofibers. It was reported that a significant portion of the injected MPC population persisted in vivo. Similar techniques of sphincteric-derived MPCs have been used for the treatment of urinary incontinence in a clinical trial [129]. The fact that MPCs can be labeled, survive after injection, and begin the process of myogenic differentiation further supports the feasibility of using cultured cells of muscular origin as an injectable bioimplant.

The use of injectable MPCs has also been investigated for use in the treatment of urinary incontinence due to irreversible urethral sphincter injury or maldevelopment. MPCs are the quiescent satellite cells found in each myofiber that proliferate to form myoblasts and, eventually, myotubes and new muscle tissue. Previously, intrinsic MPCs have been shown to play an active role in the regeneration of injured striated urethral sphincter. In a subsequent study, autologous MPCs were injected into a rat model [130] and a canine model [10, 131] of urethral sphincter injury, and both replacement of mature myotubes, as well as restoration of functional motor units, were noted in the regenerating sphincteric muscle tissue. This was the first demonstration of the replacement of both sphincter muscle tissue and its innervation by the injection of MPCs. As a result, MPCs may be a minimally invasive solution for urinary incontinence in patients with irreversible urinary sphincter muscle insufficiency.

2.6.3 Endocrine Replacement

Patients, who have a testicular dysfunction and hypogonadal disorders, are dependent on androgen replacement therapy to restore and maintain physiological levels of serum testosterone and its metabolites, dihydrotestosterone, and estradiol [132]. Currently available androgen replacement modalities, such as testosterone tablets and capsules, depot injections, and skin patches, may be associated with fluctuating serum levels and complications such as fluid and nitrogen retention, erythropoiesis, hypertension, and bone density changes [133]. Since Leydig cells of the testes are the major source of testosterone in men, implantation of heterologous Leydig cells or gonadal tissue fragments has previously been proposed as a method for chronic testosterone replacement [134]. However, these approaches were limited by the failure of the tissues and cells to produce testosterone.

Encapsulation of cells in biocompatible and semipermeable polymeric membranes has been an effective method to protect against host immune responses while maintaining viability of the cells and allowing the secretion of desired therapeutic agents [135, 136]. Encapsulation of Leydig cells in alginate-poly(L-lysine) was used as a novel method for testosterone delivery in vivo [132]. Elevated stable serum testosterone levels were noted in castrated adult rats over the course of the study, which suggests microencapsulated Leydig cells may be a potential therapeutic modality for testosterone supplementation.

2.6.4 Urethral Tissue

Various approaches have been proposed for the regeneration of urethral tissue. PGA nonwoven meshes without cells [137, 138] or with cells [139] were used to regenerate urethra tissues in various animal models. Acellular tissue matrices such as decellularized bladder submucosa [5] and urethral submucosa [140] have also been tested in various animal models for urethral reconstruction. Bladder submucosa matrix (BSM) proved to be a suitable graft for repair of urethral defects in rabbits [5]. The grafts demonstrated a normal urothelial luminal lining and organized muscle bundles on the outer portion of the graft. These results were confirmed clinically in a series of patients with a history of failed hypospadias reconstruction. The urethral defects in these patients were repaired with human BSMs [141]. An engineered urethral tissue was created by anastomosing the matrix to the urethral plate in an onlay fashion. The size of the engineered urethra tissue ranged from 5 to 15 cm. At the 3 years follow-up, three of the four patients had a successful outcome in regard to cosmetic appearance and function. One patient, who had a 15 cm neo-urethra tissue formed, developed a subglanular fistula. Similar results were obtained in pediatric and adult patients with primary urethral stricture disease using the same BSMs [142].

Another study in 30 patients with recurrent stricture disease showed that a healthy urethral bed (two or fewer prior urethral surgeries) was needed for successful urethral reconstruction using the BSM grafts [143]. More than 200 pediatric and adult patients with urethral disease have been successfully treated in an onlay manner with BSMs. One advantage of this acellular scaffold over the traditional nongenital tissue grafts that have been used for urethroplasty is that the acellular scaffold is "off the shelf." Eliminating the necessity of additional surgical procedures for graft harvesting may decrease operative time as well as the potential morbidity due to the harvest procedure.

These techniques, which employed acellular tissue scaffolds that had not been reseeded with cells, were applied experimentally and clinically in a successful manner for onlay urethral repairs. Further study, however, indicated that when tubularized urethral repairs with unseeded scaffolds were attempted experimentally, adequate urethral tissue regeneration was not achieved, and complications, such as graft contracture and stricture formation, occurred [144]. To determine if seeding the scaffold with cells from the urinary tract could improve the results of tubularized urethral repairs, autologous rabbit bladder epithelial cells and SMCs were grown and seeded onto tubular scaffolds. Entire urethra segments were then resected, and urethroplasties were performed with tubularized scaffolds either seeded with cells or without cells. The tubularized scaffolds seeded with autologous cells formed neo-tissue which was histologically similar to native urethra tissue. The tubularized scaffolds without cells lead to poor tissue development, fibrosis, and stricture formation. These findings were confirmed clinically when a trial using tubularized non-seeded scaffolds for urethral stricture repair was performed in eight evaluable patients. Two patients with short inflammatory strictures maintained urethral patency. Stricture recurrence developed in the other six patients within 3 months of surgery [145].

In 2011, Raya-Rivera et al. were able to show that synthetic polymers can also be used in urethral reconstruction when they are tubularized and seeded with autologous cells [146]. This group used PGA-PLGA scaffolds seeded with autologous cells derived from bladder biopsies taken from each patient. These cell-seeded scaffolds were then used to repair urethral defects in five boys. Upon follow-up evaluation, it was found that most of the boys had excellent urinary flow rates postoperatively, and voiding cystourethrograms indicated that these patients maintained wide urethral calibers. Urethral biopsies revealed that the grafts had developed a normal appearing architecture consisting of urothelial and muscular tissues (Fig. 2.6).

2.6.5 Bladder

The challenges in the bladder augmentation are often associated with renal function and quality of

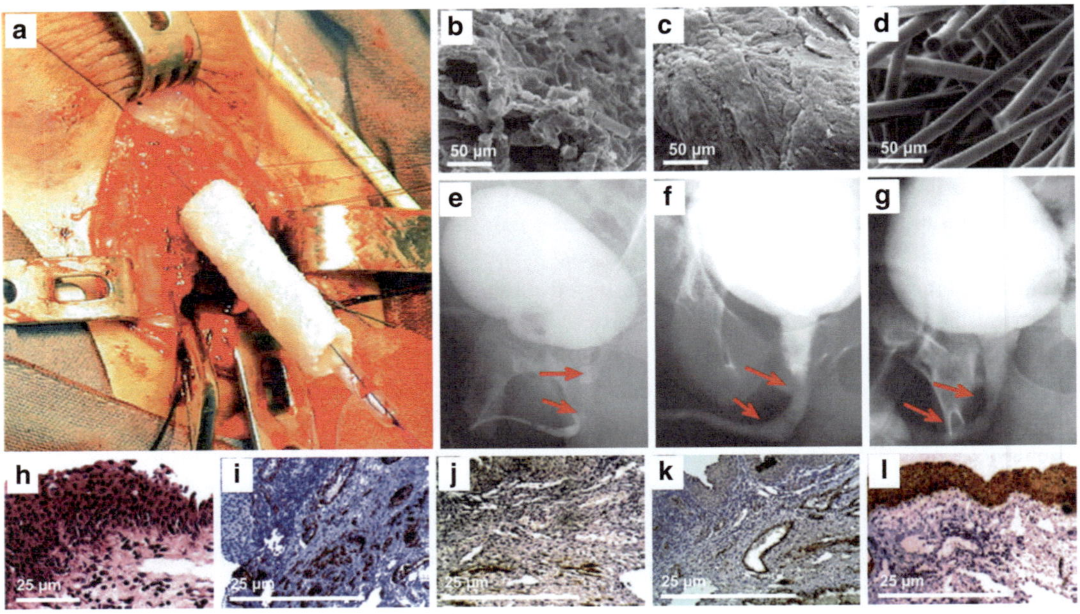

Fig. 2.6 Neo-urethra implantation and clinical outcomes; (**a**) a cell-seeded graft sutured to the normal urethral margins. Scanning electron microscopy of seeded scaffolds 6 days after seeding; (**b**) urothelial cells, (**c**) smooth muscle cells, and (**d**) scaffold without cells. Voiding cystourethrograms of a patient (**e**) before surgery (*arrows* show the abnormal margins), (**f**) 12 months after surgery (*arrows* show margins of tissue-engineered urethras), and (**g**) at last follow-up. Histological and immunohistochemical analyses; (**h**) H&E, (**i**) α-actin, (**j**) desmin, (**k**) mvosln, and (**l**) AE1/AE3. Reproduced with permission from Raya-Rivera et al. [146]

life [147]. Gastrointestinal segments are most commonly used for this purpose in the clinic, but they are designed to absorb specific solutes, whereas bladder tissue is designed for the excretion of solutes. Numerous studies have attempted alternative biomaterials and tissue-engineered constructs for bladder replacement or repair due to the problems encountered with the use of gastrointestinal segments. The ability to use donor tissue efficiently and to provide the right conditions for long-term survival, differentiation, and growth determines the success of cell transplantation strategies for bladder reconstruction. These principles were applied in the creation of tissue-engineered bladders in an animal model that required a subtotal cystectomy with subsequent replacement with a tissue-engineered organ in beagle dogs [148]. Bladder urothelial cells and SMCs were separately expanded from an autologous bladder biopsy and seeded onto a bladder-shaped biodegradable polymeric scaffold. The results from this study revealed the possibility of engineering bladder constructs that are anatomically and functionally normal.

Even though the clinical translation of tissue engineering and regenerative medicine is slowly progressing, the success of tissue engineering for some applications was already demonstrated clinically. A clinical experience involving engineered bladder tissue for cystoplasty reconstruction was conducted starting in 1999. A small pilot study of seven patients was reported, using a collagen scaffold seeded with cells either with or without omentum coverage, or a combined PGA/collagen scaffold seeded with cells and omental coverage. Patient's own bladder cells were isolated and expanded in vitro prior to seeding on a scaffolding system of collagen and PGA. The scaffold was designed to replace the bladder size to improve its compliancy and patient continence (Fig. 2.7a–c). The tissue-engineered constructs were implanted for a period up to 61 months. Our outcomes show the definitive improvement of bladder compliance and capacity while restoring physiological function (Fig. 2.7d, e). In addition, the engineered bladder biopsies showed an adequate structural architecture and

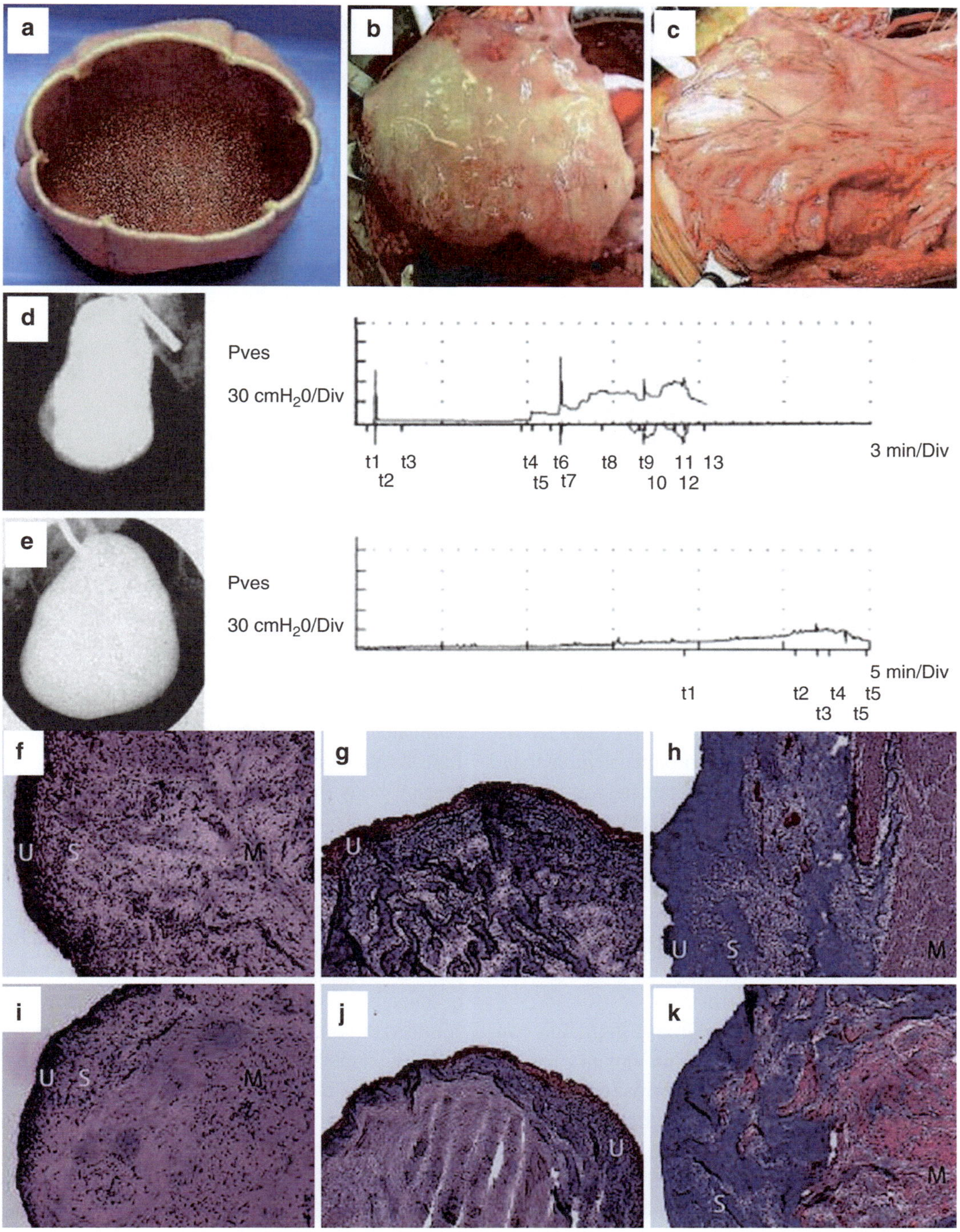

Fig. 2.7 (**a**) Scaffold seeded with cells, (**b**) engineered bladder anastomosed to native bladder with running 4–0 PGA sutures, and (**c**) implant covered with fibrin glue and omentum. (**d**) Preoperative and (**e**) 10 months postoperative cystograms and urodynamic findings in patient with a collagen-PGA scaffold engineered bladder. (**f–h**) cystoscopic biopsies of implanted engineered bladders 31 months after augmentation shows extent of regeneration. Engineered bladder tissue showed tri-layered structure, consisting of lumen lined with urothelial cells, U, surrounded by submucosa, S, and muscle, M. (**f**) H&E, immunocytochemical analysis with (**g**) anti-pancytokeratin AE1/AE3 antibodies, and (**h**) anti-α smooth muscle actin antibodies showed presence of phenotypically normal urothelium and smooth muscle. (**i–k**) native bladder tissue. Original magnification: ×100. Reproduced with permission from Atala et al. [147]

phenotype (Fig. 2.7f–k). Although the experience is promising in terms of showing that engineered tissues can be implanted safely, it is just a start in terms of accomplishing the goal of engineering fully functional bladders. Further experimental and clinical work is being conducted.

2.6.6 Male and Female Reproductive Organs

Reconstructive surgery is required for a wide variety of pathologic penile conditions such as penile carcinoma, trauma, severe erectile dysfunction, and congenital conditions such as ambiguous genitalia, hypospadias, and epispadias. One of the major limitations of phallic reconstructive surgery is the scarcity of sufficient autologous tissue. The major components of the phallus are corporal SMCs and ECs. The creation of autologous functional and structural corporal tissue de novo would be beneficial. Autologous cavernosal SMCs and ECs were harvested, expanded, and seeded on decellularized tissue matrices and implanted in a rabbit model [149, 150]. The engineered corpora demonstrated structural and functional parameters similar to native tissue and male rabbits receiving the bilateral implants were able to successfully impregnate females. This study demonstrated that neo-corpora could be engineered for total pendular penile corporal body replacement [150]. Such technology has considerable potential for patients requiring penile reconstruction.

Several pathologic conditions, including congenital malformations and malignancy of the vagina, can adversely affect normal development or anatomy. Vaginal reconstruction has traditionally been challenging due to the paucity of available native tissue. The feasibility of engineering vaginal tissue in vivo was investigated [151]. Vaginal epithelial cells and SMCs of female rabbits were harvested, grown, and expanded in culture. These cells were seeded onto biodegradable polymer scaffolds, and the cell-seeded constructs were then implanted into nude mice for up to 6 weeks. The presence of vaginal tissue phenotypes was confirmed by histology, immunohistochemistry, and Western blotting. Electrical field stimulation studies in the tissue-engineered constructs showed similar functional properties to those of normal vaginal tissue. When these constructs were used for autologous total vaginal replacement, patent vaginal structures were noted in the tissue-engineered specimens, while the non-cell-seeded structures were noted to be stenotic. In 2014, an engineered vaginal tissue derived from the patient's own cells and implanted showed normal structural and functional variables with a follow-up of up to 8 years in the clinic [152].

Congenital malformations of the uterus may have profound clinical implications. Patients with cloacal exstrophy and intersex disorders may not have sufficient uterine tissue present for future reproduction. Campbell et al. performed another type of study in which the peritoneal cavity of rats and rabbits were used as in vivo bioreactors to produce uterine tissue grafts [153]. In this study, biomaterial templates of the appropriate shape were implanted in the peritoneal cavities of rats or rabbits. After 2–3 weeks, the templates were removed, and the encapsulating myofibroblast-rich tissue that resulted from the foreign-body response to the biomaterial was harvested. This tissue was then used to replace resected segments of the uterus in the same animals in which the tissue was grown. At 12 weeks after grafting, it was reported that this novel uterine graft tissue thickened and developed the morphology of normal uterus. This structure included a lumen lined with endometrium, which was surrounded by several layers of SMCs (myometrium-like) interspersed with collagen. Importantly, these grafted uterine horns supported embryos to the late stages of gestation. The dysfunctional uterine cervix may also benefit from tissue engineering strategies. Spontaneous preterm birth is a frequent complication of pregnancy, and in some cases, abnormalities of the cervix have been implicated in this issue.

House et al. have shown that tissue engineering techniques can be used to create 3D cervical-like tissue constructs [154]. Cervical cells were isolated from two premenopausal women and seeded on porous silk scaffolds. These constructs were cultured under dynamic or static culture conditions. After an 8 weeks culture interval, cervical cells had proliferated in the 3D construct and had synthesized an ECM with biochemical constituents and morphology similar to native tissue, but this matrix was better formed under dynamic culture conditions. This study suggests that it may be possible to engineer cervical tissue for a variety of conditions.

2.6.7 Kidney

The kidney is a complex organ with multiple cell types and a complex functional anatomy that renders it one of the most difficult to reconstruct [155]. Previous efforts in tissue engineering of renal tissues have been directed toward the development of extracorporeal renal support systems made of biological and synthetic components [156–160], and ex vivo renal replacement devices are known to be life-sustaining. Obviously, patients with end-stage kidney disease (ESRD) would benefit if these devices could be implanted long term without the need for an extracorporeal perfusion circuit or immunosuppressive drugs.

We have previously applied the principles of both tissue engineering and therapeutic cloning in an effort to produce genetically identical renal tissue in a large animal model, the cow (*Bos taurus*) [161]. Bovine skin fibroblasts from adult Holstein steers were obtained by ear notch, and single donor cells were isolated and microinjected into the perivitelline space of donor-enucleated oocytes (nuclear transfer). The resulting blastocysts were implanted into progestin-synchronized recipients to allow for further in vivo growth. After 12 weeks, cloned renal cells were harvested, expanded in vitro, and then seeded on scaffolds consisting of three collagen-coated cylindrical polycarbonate membranes. The ends of the three membranes of each scaffold were connected to catheters that terminated into a collecting reservoir. This created a renal neo-organ with a mechanism for collecting the excreted urinary fluid. The scaffolds with the collecting devices were transplanted subcutaneously into the same steer from which the genetic material originated and then retrieved at 12 weeks after implantation. Chemical analysis of the collected urine-like fluid, including urea nitrogen and creatinine levels, electrolyte levels, specific gravity, and glucose concentration, revealed that the implanted renal cells possessed filtration, reabsorption, and secretory capabilities. Histological examination of the retrieved implants revealed extensive vascularization and self-organization of the cells into glomeruli and tubule-like structures. A clear continuity between the glomeruli, the tubules, and the polycarbonate membrane was noted that allowed the passage of urine into the collecting reservoir. Immunohistochemical analysis with renal-specific antibodies revealed the presence of renal proteins. RT-PCR analysis confirmed the transcription of renal-specific RNA in the cloned specimens, and Western blot analysis confirmed the presence of elevated renal-specific protein levels.

Studies have shown that bovine clones harbor the oocyte mitochondrial DNA [162]; therefore, the mitochondrial DNA (mtDNA) of the donor egg was thought to be a potential source of immunologic incompatibility. Differences in mtDNA-encoded proteins expressed by cloned cells could stimulate a T-cell response specific for mtDNA-encoded minor histocompatibility antigens when the cloned cells are implanted back into the original nuclear donor [163]. We used nucleotide sequencing of the mtDNA genomes of the clone and fibroblast nuclear donor to identify potential antigens in the muscle constructs. Only two amino acid substitutions were noted to distinguish the clone and the nuclear donor, and, as a result, a maximum of two minor histocompatibility antigens could be defined. Given the lack of knowledge regarding peptide-binding motifs for bovine MHC class I molecules, reliable methods to predict the impact of these amino acid substitutions on bovine histocompatibility are unavailable. Oocyte-derived mtDNA was also thought to be a potential source of immunologic incompatibility in the cloned renal cells. Maternally transmitted minor histocompatibility antigens in mice have been shown to stimulate both skin allograft rejection in vivo and cytotoxic T lymphocyte expansion in vitro [163]. These antigens could prevent the use of the cloned constructs in patients with chronic rejection of major histocompatibility-matched human renal transplants [164]. We tested for a possible T-cell response to the cloned renal devices using delayed-type hypersensitivity testing in vivo and Elispot analysis of interferon-gamma secreting T cells in vitro. Both analyses revealed that the cloned renal cells lacked evidence of a T-cell response and suggested that rejection will not necessarily occur in the presence of oocyte-derived mtDNA. This finding may represent a step forward in overcoming the histocompatibility problem of stem cell therapy [155]. These studies demonstrated that cells derived from nuclear transfer can be successfully harvested, expanded in culture, and transplanted in vivo with the use of biodegradable scaffolds on which the single suspended cells can organize into tissue structures that are genetically identical to that of the host. This was the first demonstration of the use of therapeutic cloning for regeneration of tissues in vivo.

One such approach for whole organ bioengineering is to combine functional renal cells with a decellularized porcine kidney scaffold [49, 69, 165, 166]. Given the complexity of the kidney structure, which includes over 30 different cell types, vascular network as well as a large array of functional structures, the use of the underlying ECM composing the unique microenvironment is a logical starting point for the bioengineering of a transplantable whole kidney. The efficacy of cellular removal and biocompatibility of the preserved porcine matrices, coupled with scaffold reproducibility, are critical to the success of this approach. To perform the decellularization, we have designed and constructed high-throughput system for decellularization process. This system showed significant cellular removal (<50 ng DNA/mg dry tissue). And, decellularized organs retained intact microarchitecture including the renal vasculature and essential ECM components (Fig. 2.8). Our method ensures clearance of porcine cellular material, which directly impacts

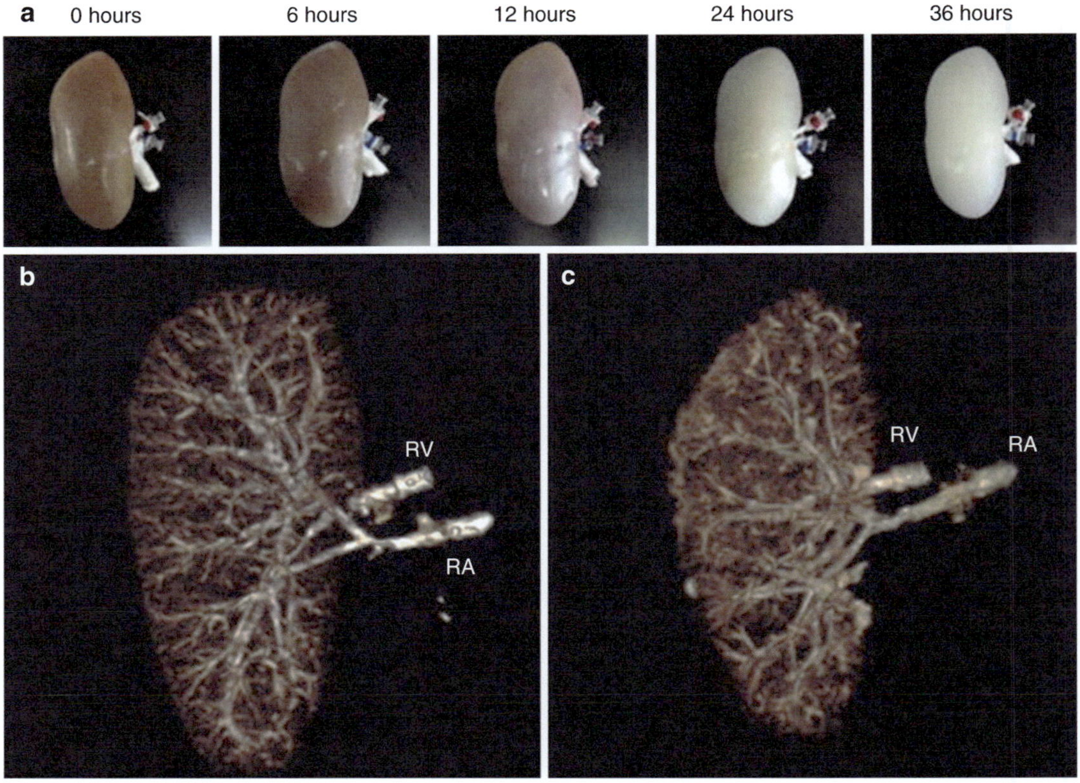

Fig. 2.8 Decellularization of porcine kidney. (**a**) Representative time lapse of decellularization process of porcine kidney. Comparison of whole organ vasculature system between (**b**) native and (**c**) decellularized porcine kidney illustrating the preservation of the arterial–venous renal system by continuation of the renal artery (RA) to the renal vein (RV). Histological analyses of (**d–f**) native and (**g–i**) decellularized porcine kidney; (**d**, **g**) H&E, (**e**, **h**) *Alcian Blue/Sirius Red*, and (**f**, **i**) Masson's trichrome. Reproduced with permission from Sullivan et al. [49]

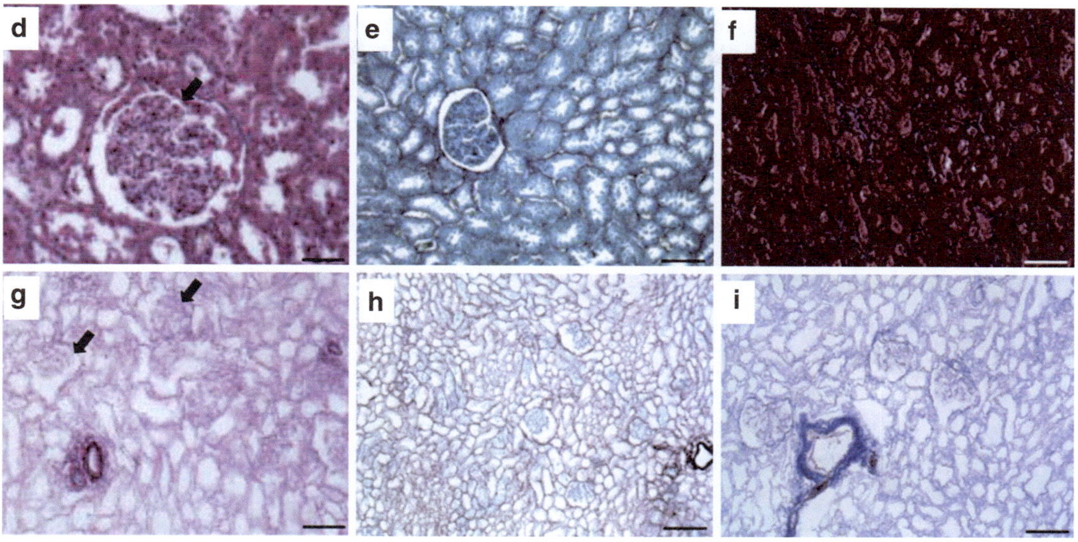

Fig. 2.8 (continued)

immunoreactivity during transplantation, and preserves the ECM and cellular compatibility of these renal scaffolds. Thus, we have developed a rapid decellularization method that can be scaled up for use in other large organs. Our results represent a step toward development of a transplantable organ using tissue engineering techniques. A preclinical large animal study for the application of this technology is currently being conducted.

2.6.8 Blood Vessels

Cardiovascular disease, including coronary artery and peripheral vascular diseases, is the leading cause of mortality in the USA [167]. There remains a major clinical demand to improve technology for vascular grafts. Segments of autologous vessels continue to be the standard for many revascularization procedures. Unfortunately, autografts are limited in supply and dimensions, while allografts and xenografts are limited by strong immunogenic response [168–170]. Synthetic vascular grafts have been explored for over half a century. Substances such as expanded polytetrafluoroethylene (ePTFE) or Dacron (polyethylene terephthalate fiber) work well as large diameter (>5 mm) bypass conduits, but such synthetics have not proven generally satisfactory for small diameter grafts, due to the high frequency of thrombosis, stenosis, occlusion, and infection [171, 172].

Tissue engineering offers an attractive approach to vascular grafting, particularly for small diameter (<5 mm) vessels. The basic approach is to create vessels by combining autologous cells with a natural and/or synthetic scaffold under suitable culture conditions, resulting in a tubular construct [116] that can be implanted in vivo. Coating the vascular lumen of synthetic grafts with ECs has been shown to overcome acute thrombosis [169, 170]. The presence of a permanent synthetic graft remains an issue, as it may lead to chronic inflammatory responses and other problems. A preferred solution would be the use of biodegradable scaffolds that, over time, can be replaced by cells to generate an essentially normal bioengineered blood vessel. Indeed, vascular scaffolds combined with viable cells have been shown to allow grafts to remodel when introduced in vivo.

Previous work by our group demonstrated that a decellularized porcine vessel scaffold coated with autologous ECs could replace a small diameter blood vessel in a sheep carotid arterial interposition model [173, 174]. The implanted decellularized vascular grafts could remain patent for at least 130 days. While the use of cells bound to a naturally derived vascular tissue matrix demonstrates the principle of vascular tissue engineering, this approach has several drawbacks that would be expected to hinder clinical translation. These include the limited supply of vessels of the required dimensions, inconsistency in donor tissue composition, and potential contamination by pathogens. Therefore, it is essential to develop a synthetic vascular scaffolding system that would allow for the consistent production of small diameter vessel substitutes of any desired dimensions.

A number of approaches have been attempted with the goal of producing an ideal vascular scaffold. Among these, electrospinning technology, which uses high-voltage electrostatic fields to generate nanofibers, appears particularly attractive because it provides a biomimetic environment that may be designed to resemble that of ECM architecture of native vasculature. Electrospinning permits fabrication of fibrous matrices of nano- to microscale. In addition, it is possible to control the composition, structure, dimension, and biomechanical properties of fabricated scaffolds. Numerous reports have documented that synthetic biodegradable polymer-based materials can be electrospun to generate candidate vascular scaffolds with in vitro characteristics of strength and biocompatibility that appear consistent with clinical requirements [77, 116, 175–177]. Many efforts attempted to develop a durable biomaterial capable of resisting physiological forces and maintaining structural integrity until mature vascular tissue forms in vivo. It is also important that engineered vessels should be compliant, resistant to kinking and compression, and possess sufficient tensile and shear strength to resist fraying at cut edges and tearing out of sutures [178].

With these considerations in mind, we recently developed a vascular scaffold that is a composite of poly(ε-caprolactone) (PCL) and type I collagen [77, 116]. We observed that this scaffold can withstand physiologically relevant vascular conditions over a period of 1 month in vitro and after implantation in vivo in a rabbit aortoiliac bypass model [179]. The introduction of collagen to PCL appeared to increase the structural stability under high physiological pressure as well as cell compatibility. Interestingly, the hydrated PCL/collagen scaffolds showed tensile properties similar to those of native blood vessels due to the stiffness of collagen. Most recently, we have developed a bilayered scaffolding system that provides different pore sizes to facilitate adequate cellular interactions. Thus, our system allows for EC adhesion onto the luminal surface and homogenous infiltration of SMCs into the outer layer (Fig. 2.4). Moreover, this scaffold provides sufficient mechanical properties that withstand physiologically relevant vascular conditions. Our findings suggest that bilayered scaffolds may facilitate endothelialization and smooth muscle maturation for improved vessel tissue function. A preclinical large animal study for the application of this fully cellularized vascular scaffold is currently being conducted. These studies will evaluate the utility of vascular tissue engineering to provide platform technologies for rehabilitation of patients recovering from severe and devastating cardiovascular diseases. The long-term goals are to provide alternatives to vascular grafting using tissue-engineered blood vessels derived from the patient's own cells. We believe that completion of this study has the potential to lead directly to clinical translation and address an important unmet medical need.

2.6.9 Skeletal Muscle and Innervation

Skeletal muscle defects due to trauma or tumor ablation usually require reconstructive procedures in order to restore normal tissue function. Currently, muscle pedicle flap from adjacent regions is the primary method practiced. This option is challenged by host muscle tissue availability and donor site morbidity such as functional loss and volume deficiency [180, 181]. Recent advances in cell therapy using myoblasts have provided an alternate therapeutic opportunity to regenerate muscle tissue for functional augmentation [182, 183]. Injection of cultured myoblasts has shown some efficacy [184–186]; however, this approach may not be suitable for treating large volumetric muscle loss injury [185]. Hence, creation of an implantable functional muscle tissue that could restore muscle tissue defects may be a possible solution [187, 188].

An essential step in bioengineering functional skeletal muscle tissue construct is to mimic the structure of native tissue which is comprised of highly oriented myofibers formed from numerous fused mononucleated muscle cells [189]. It is well known that structure and organization of myofibers dictate tissue function. Thus, muscle cell alignment that permits organized myotube formation is a crucial step in the musculoskeletal myogenesis [190]. The ability to efficiently organize muscle cells to form aligned myotubes in vitro would greatly benefit efforts in skeletal muscle tissue engineering. To date, significant deficits in reconstructing complex structure with spatial organization have been problematic and have limited construction to primarily hollow and simple tissue and organ structures. 3D printing technology has been shown to be a viable option for creating a tissue-engineered construct designed to regenerate or replace a damaged tissue/organ [191, 192]. This 3D printing method can precisely place cells with hydrogel-based materials in a layer-by-layer fashion, allowing it to replicate the 3D complex structure of tissues or organs. Our group has used this technology to fabricate various tissue-like structures such as cartilage [193], bone [194], and vascularized tissue [120]. We have previously demonstrated our ability to print a skeletal muscle bundle using the 3D bioprinting technology [122]. A 3D muscle construct was fabricated using muscle cells with hydrogel material. After 3D bioprinting, the printed muscle structures contained muscle fiber-like bundles. Notably, the printed cells started stretching the shape along the longitudinal axis of the printed fiber structure. Our result demonstrates that the 3D bioprinting can mimic the tissue organization in the human body for reconstruction; moreover, this may result in improved function and maturation of engineered skeletal muscle tissue constructs. Findings from our study show that these tissue constructs were able to maintain structural and functional characteristics in vitro. In a pilot animal study, we have showed that the implanted 3D bioprinted muscle construct was robust enough to maintain its structural characteristics and induce vascularization and nerve integration in vivo [122].

Another critical component that is required to successfully engineer a functional muscle tissue in vivo is effective innervation of the skeletal muscle [195]. The established contacts of tissue-engineered muscle constructs with host nervous tissue is vitally important following implantation, as improper or incomplete innervation leads to atrophy of muscle tissue and loss of contractile function [196]. When muscle cells/fibers are implanted in the body, host nerves contact with the muscle fibers to form neuromuscular junctions (NMJs). Denervated muscle, which is analogous to in vitro bioengineered muscle constructs, has been shown to be reinnervated upon direct transplantation of host nerve [197]. The process of innervations into denervated muscle is slow, however, and substantial time is required before muscle tissue is functional. Therefore, methods to accelerate innervations are needed. Previous reports indicate that the expression of acetylcholine receptors (AChRs) and their clustering on muscle fibers are critical factors required to induce neural contacts on newly formed muscle fibers in a natural biological system [198–200]. For instance, in NMJ development in the mouse, AChRs are pre-patterned at sites that are used by the outgrowing motor neurons to form synaptic contacts [201]. On each muscle fiber, one synaptic site (often contacted by several different nerves) becomes stabilized, and the mature endplate pattern is established at a later embryonic stage. During postnatal development, further structural and functional changes are observed. Multiple innervations of muscle fibers are reduced until motor neurons form single synaptic contacts, and embryonic AChRs are replaced by aggregates of adult AChRs that acquire a characteristically "pretzel-like" pattern. During the process of maturation of NMJs between muscle fibers and motor neurons, the formation of embryonic AChR clusters on muscle fibers clearly appears to be a prerequisite step that controls the overall process leading to mature innervations of muscle fibers [199]. Although this well-organized process of pre-forming AChR clusters on embryonic muscle fibers is present in normal vertebrate systems, tissue-engineered muscle constructs are unable to provide such a sophisticated process and remain a major concern for the use of bioengineered muscle tissue in vivo.

To address this issue, we investigated whether prefabrication of AChR clusters on bioengineered, bulk muscle fibers using agrin, a neural-released trophic factor, would accelerate innervation when implanted in vivo. Agrin has a critical role in inducing AChR expression and their clustering on muscle cells [202–205]. Neural agrin stimulates the rapid phosphorylation of a muscle-specific kinase (MuSK), which has been shown to be necessary for the formation of the NMJ and leads to the redistribution of previously unlocalized AChR proteins, to synaptic sites [206]. The function of agrin was demonstrated in an agrin-deficient mutant mouse model, which failed to form functional endplates [207]. This deficit was overcome by gene transfer experiments and/or injection of recombinant agrin into rat soleus muscle in vivo [208]. Thus, we tested whether agrin molecules would facilitate the formation of AChR clusters on the bioengineered muscle fibers in fibrin hydrogel constructs [209]. Our results indicate that agrin treatment increased AChR cluster formation on the differentiated muscle cells, and these prefabricated AChR clusters facilitated accelerated contacts with dorsal root ganglion (DRG) nerve in vitro or with the host nerve in bioengineered muscle fibers treated with agrin in vivo (Fig. 2.9) [209]. Moreover, the implanted agrin-treated muscle constructs displayed a mature myofiber structure with myosin heavy chain (MHC) expression in the vicinity of the host nerve, indicating enhanced interaction with the host nerve. It is our belief that the effective use of a NMJ-inducing factor, agrin, could induce proper maturation of bioengineered muscle constructs for accelerating further host nerve integration (innervation).

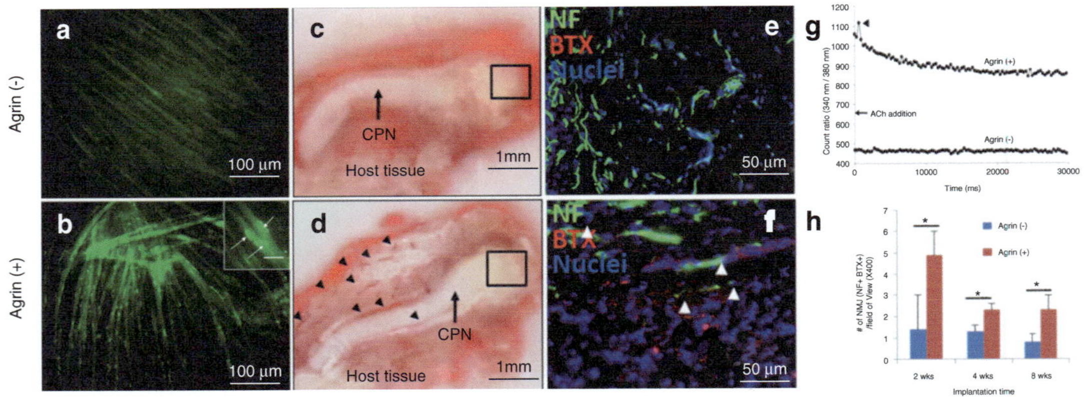

Fig. 2.9 Prefabrication of AChR by agrin treatment accelerates innervations in vitro and in vivo. (**a, b**) In vitro myotube formation (**a**) without and (**b**) with agrin treatment; agrin treatment induced enhanced AChR expression on myotubes [α-BTX$^+$ indicated by arrows on the surface of myotubes (box in **b**)] while there are no characteristic evidence in untreated myotubes (**a**). (**c, d**) Prefabricated C2C12 myotubes in fibrin gel were implanted subcutaneously embedded with common peroneal nerve (CPN); gross appearance of the harvested C2C12/fibrin gel with CPN at 2 weeks after implantation; numerous large blood vessels were observed in the implant treated with agrin (**d**), whereas there was no remarkable blood vessels recruitment surrounding the implant without agrin treatment (**c**). (**e, f**) Immunofluorescent images of the harvested C2C12/fibrin gels; higher numbers of innervated structures [neurofilament (NF)$^+$/α-BTX$^+$ double staining (*arrowheads*)] were observed in agrin-treated group (**f**) than in no treated group (**e**) and quantification of their number (**h**). Student *t*-test, $*P < 0.05$, $n = 3$, from three individual animals. (**g**) In vitro functional properties of AChR expression on myotubes by calcium imaging; agrin-treated myotubes displayed depolarization-induced increases in steady-state intracellular calcium levels in the kinetic curve with the sharp increase (*arrow head*) in agrin-treated myotubes. Reproduced with permission from Ko et al. [209]

2.7 Vascularization of Engineered Tissues

A major challenge in tissue engineering is the lack of proper vascularization in the engineered tissue constructs when implanted. To develop complex tissues or organs, fully vascularized tissue constructs should be provided because it is critical to long-term cell survival and function of cell constructs not only at the margin but also at the center of the tissue grafts [210]. In fact, the growth of a new microvascular system persists as one of the major limitations to the successful introduction of tissue engineering products to clinical practice [211]. Accordingly, the focus of much research in tissue engineering has evolved to include a better understanding of angiogenesis. Numerous efforts have been made to overcome this limitation, and attempts to enhance angiogenesis within the host tissue have been pursued using several approaches.

2.7.1 Angiogenic Factor Delivery

It has been demonstrated that the delivery of growth factors and cytokines that play central regulatory roles in the process of angiogenesis, which is thought to induce ingrowth of capillaries and blood vessels into an engineered implant, yields diminishing hypoxia-related cell damage. The delivery of such angiogenic factors has been achieved either by incorporating the desired factors into the scaffold biomaterials to be used or by genetic modification of the cells to be used in the engineering process, which forces the cells to express factors such as vascular endothelial growth factor (VEGF), one of the most potent angiogenic factors [212, 213].

A study evaluated controlled release of VEGF by incorporating VEGF directly into PLGA scaffolds or by incorporating VEGF encapsulated in PLGA microspheres into scaffolds [214]. VEGF incorporated into scaffolds resulted in rapid release of the cytokine, whereas the pre-encapsulated group showed a delayed release. In addition, both systems showed negligible release of VEGF into the systemic circulation, yet their use led to enhanced local angiogenesis in vivo for up to 21 days. These studies demonstrated the delivery of VEGF in a controlled and localized fashion in vivo. This angiogenic factor delivery system was applied to bone regeneration, and its potent ability to enhance angiogenesis within implanted scaffolds was followed by enhanced bone regeneration and outlines a novel approach for engineering tissues in hypovascular environments [215].

As another example, a dual delivery system combining VEGF and insulin-like growth factor 1 (IGF-1) was developed to enhance transplantation and dispersion of cultured myogenic cells [216]. Localized delivery of VEGF and IGF-1 from alginate scaffolds into injured muscle enhanced local angiogenesis, altered muscle fiber type, and enhanced muscle regeneration in a model of severe muscle damage. Transplanting cultured myoblasts on scaffolds that delivered VEGF/IGF-1 dramatically enhanced their direct participation in muscle regeneration and further enhanced angiogenesis and the return to normal tissue perfusion levels, as compared to growth factor delivery alone. This approach holds promise in the transplantation of many cell types used to promote the regenerative response of multiple tissues.

2.7.2 Incorporation of Endothelial Cells

Human vascular ECs and skeletal myoblasts transfected with adenovirus encoding the gene for VEGF were tested for regenerating a vascularized engineering muscle construct [212]. The transfected cells were injected subcutaneously in athymic mice and noted to form a muscle tissue with neovascularization by histological and immunohistochemical analyses with maintenance of their muscle volume, while engineered muscle of non-transfected cells had a significantly smaller mass of cells with loss of muscle volume over time, less neovascularization, and no surviving ECs. Results indicate that a combination of VEGF and ECs may be useful for inducing neovascularization and volume preservation in engineered tissue.

The development of a method of formation and stabilization of endothelialized vessel networks in vitro in engineered skeletal muscle tissue [217] has shown promising results. A 3D multicultural system consisting of myoblasts, embryonic fibroblasts, and ECs co-seeded on highly porous, biodegradable polymeric scaffolds. These results showed that pre-vascularization of the implants improved angiogenesis and cell survival within the scaffolds. Moreover, this research group emphasized that cocultures with ECs and SMCs may also be important for inducing differentiation of engineered tissues. A synthetic, biomimetic hydrogel was developed to allow the rapid formation of a stable and mature vascular network [218]. Hydrogels were fabricated with integrin-binding sites and protease-sensitive substrates to mimic the natural provisional ECM, and ECs cultured in these hydrogels were organized into stable, intricate networks of capillary-like structures. The resulting structures were further stabilized by recruitment of mesenchymal progenitor cells that differentiated into a SMC lineage and deposited collagen IV and laminin.

2.8 Conclusions and Future Directions

Various tissue-engineered constructs are at different stages of development, with some in current clinical use. The tissue-engineered scaffold provides mechanical support, shape, and cell-scale architecture for neo-tissue construction in vitro or in vivo as seeded cells expand and organize. Most degradable biomaterials used to date comprise a class of synthetic polyesters and/or natural polymers. A multitude of fabrication techniques have been devised and afford an abundance of potential shapes, sizes, porosities, and architectures [219, 220]. In addition, scaffolds should provide more than temporary architectural structure to a developing tissue construct. As cells and molecular biology meet with materials science and biomedical engineering, new applications in regenerative medicine will benefit from interactive biomaterials that serve to orchestrate cell attachment and growth, as well as tissue morphogenesis. However, many of the same tools developed for evaluating the biocompatibility of traditional biodegradable polymers are still used to investigate the fundamental interactions between new classes of biomaterials and their host [221, 222].

It is evident from the previous examples that the diversity in biomaterials for tissue engineering applications is immense. Numerous approaches to mimicking the structure and mechanical properties, and more importantly, the function of the ECM have been devised. It is imperative that these important technologies continue to be investigated for their ability to interact in biological systems. An essential tool set has previously been described that will enable comprehensive evaluation of novel biomaterials in their host environment. Successful regulatory approval of new categories of regenerative technologies entering human clinical trials will continue to be based on these fundamental principles of biocompatibility and biological interaction. Recent progress suggests that tissue-engineered constructs may have an expanded clinical applicability in the future and may represent a viable therapeutic option for those who require tissue replacement or repair.

References

1. Atala A. Engineering organs. Curr Opin Biotechnol. 2009;20(5):575–92.
2. Atala A. Regenerative medicine and tissue engineering in urology. Urol Clin North Am. 2009;36(2): 199–209. viii-ix
3. Atala A. Tissue engineering of reproductive tissues and organs. Fertil Steril. 2012;98(1):21–9.
4. Atala A, Kasper FK, Mikos AG. Engineering complex tissues. Sci Transl Med. 2012;4(160):160rv12.
5. Chen F, Yoo JJ, Atala A. Acellular collagen matrix as a possible "off the shelf" biomaterial for urethral repair. Urology. 1999;54(3):407–10.
6. Chun SY, Lim GJ, Kwon TG, Kwak EK, Kim BW, Atala A, et al. Identification and characterization of bioactive factors in bladder submucosa matrix. Biomaterials. 2007;28(29):4251–6.
7. Yoo JJ, Meng J, Oberpenning F, Atala A. Bladder augmentation using allogenic bladder submucosa seeded with cells. Urology. 1998;51(2):221–5.
8. Eberli D, Susaeta R, Yoo JJ, Atala A. Tunica repair with acellular bladder matrix maintains corporal tissue function. Int J Impot Res. 2007;19(6):602–9.

9. Kim BS, Atala A, Yoo JJ. A collagen matrix derived from bladder can be used to engineer smooth muscle tissue. World J Urol. 2008;26(4):307–14.
10. Eberli D, Aboushwareb T, Soker S, Yoo JJ, Atala A. Muscle precursor cells for the restoration of irreversibly damaged sphincter function. Cell Transplant. 2012;21(9):2089–98.
11. Joo S, Ko IK, Atala A, Yoo JJ, Lee SJ. Amniotic fluid-derived stem cells in regenerative medicine research. Arch Pharm Res. 2012;35(2):271–80.
12. Pariente JL, Kim BS, Atala A. In vitro biocompatibility evaluation of naturally derived and synthetic biomaterials using normal human bladder smooth muscle cells. J Urol. 2002;167(4):1867–71.
13. Meaney Murray M, Rice K, Wright RJ, Spector M. The effect of selected growth factors on human anterior cruciate ligament cell interactions with a three-dimensional collagen-GAG scaffold. J Orthop Res. 2003;21(2):238–44.
14. Patino MG, Neiders ME, Andreana S, Noble B, Cohen RE. Collagen as an implantable material in medicine and dentistry. J Oral Implantol. 2002;28(5):220–5.
15. van Susante JLC, Pieper J, Buma P, van Kuppevelt TH, van Beuningen H, van Der Kraan PM, et al. Linkage of chondroitin-sulfate to type I collagen scaffolds stimulates the bioactivity of seeded chondrocytes in vitro. Biomaterials. 2001;22(17):2359–69.
16. Yang C, Hillas PJ, Baez JA, Nokelainen M, Balan J, Tang J, et al. The application of recombinant human collagen in tissue engineering. BioDrugs. 2004;18(2):103–19.
17. Friess W. Collagen--biomaterial for drug delivery. Eur J Pharm Biopharm. 1998;45(2):113–36.
18. Lee CH, Singla A, Lee Y. Biomedical applications of collagen. Int J Pharm. 2001;221(1–2):1–22.
19. Kasoju N, Bora U. Silk fibroin in tissue engineering. Adv Healthc Mater. 2012;1(4):393–412.
20. Talukdar S, Nguyen QT, Chen AC, Sah RL, Kundu SC. Effect of initial cell seeding density on 3D-engineered silk fibroin scaffolds for articular cartilage tissue engineering. Biomaterials. 2011;32(34):8927–37.
21. Altman GH, Diaz F, Jakuba C, Calabro T, Horan RL, Chen J, et al. Silk-based biomaterials. Biomaterials. 2003;24(3):401–16.
22. Li G, Li Y, Chen G, He J, Han Y, Wang X, et al. Silk-based biomaterials in biomedical textiles and fiber-based implants. Adv Healthc Mater. 2015;4(8):1134–51.
23. Yucel T, Lovett ML, Kaplan DL. Silk-based biomaterials for sustained drug delivery. J Control Release. 2014;190:381–97.
24. Wang Y, Rudym DD, Walsh A, Abrahamsen L, Kim HJ, Kim HS, et al. In vivo degradation of three-dimensional silk fibroin scaffolds. Biomaterials. 2008;29(24–25):3415–28.
25. Coviello T, Matricardi P, Marianecci C, Alhaique F. Polysaccharide hydrogels for modified release formulations. J Control Release. 2007;119(1):5–24.
26. Spicer AP, Tien JY. Hyaluronan and morphogenesis. Birth Defects Res C Embryo Today. 2004;72(1):89–108.
27. Beck DE, Cohen Z, Fleshman JW, Kaufman HS, van Goor H, Wolff BG. A prospective, randomized, multicenter, controlled study of the safety of Seprafilm adhesion barrier in abdominopelvic surgery of the intestine. Dis Colon Rectum. 2003;46(10):1310–9.
28. Diamond MP. Reduction of adhesions after uterine myomectomy by Seprafilm membrane (HAL-F): a blinded, prospective, randomized, multicenter clinical study. Seprafilm adhesion study group. Fertil Steril. 1996;66(6):904–10.
29. Kemper F, Gebhardt U, Meng T, Murray C. Tolerability and short-term effectiveness of hylan G-F 20 in 4253 patients with osteoarthritis of the knee in clinical practice. Curr Med Res Opin. 2005;21(8):1261–9.
30. Waddell DD, Bricker DC. Hylan G-F 20 tolerability with repeat treatment in a large orthopedic practice: a retrospective review. J Surg Orthop Adv. 2006;15(1):53–9.
31. Waddell DD, Bricker DC. Clinical experience with the effectiveness and tolerability of hylan G-F 20 in 1047 patients with osteoarthritis of the knee. J Knee Surg. 2006;19(1):19–27.
32. Tonnesen HH, Karlsen J. Alginate in drug delivery systems. Drug Dev Ind Pharm. 2002;28(6):621–30.
33. Gutowska A, Jeong B, Jasionowski M. Injectable gels for tissue engineering. Anat Rec. 2001;263(4):342–9.
34. Boontheekul T, Kong HJ, Mooney DJ. Controlling alginate gel degradation utilizing partial oxidation and bimodal molecular weight distribution. Biomaterials. 2005;26(15):2455–65.
35. Zimmermann H, Zimmermann D, Reuss R, Feilen PJ, Manz B, Katsen A, et al. Towards a medically approved technology for alginate-based microcapsules allowing long-term immunoisolated transplantation. J Mater Sci Mater Med. 2005;16(6):491–501.
36. Khor E, Lim LY. Implantable applications of chitin and chitosan. Biomaterials. 2003;24(13):2339–49.
37. Shi C, Zhu Y, Ran X, Wang M, Su Y, Cheng T. Therapeutic potential of chitosan and its derivatives in regenerative medicine. J Surg Res. 2006;133(2):185–92.
38. Miranda SC, Silva GA, Hell RC, Martins MD, Alves JB, Goes AM. Three-dimensional culture of rat BMMSCs in a porous chitosan-gelatin scaffold: a promising association for bone tissue engineering in oral reconstruction. Arch Oral Biol. 2011;56(1):1–15.
39. Jones PL, Schmidhauser C, Bissell MJ. Regulation of gene expression and cell function by extracellular matrix. Crit Rev Eukaryot Gene Expr. 1993;3(2):137–54.
40. Juliano RL, Haskill S. Signal transduction from the extracellular matrix. J Cell Biol. 1993;120(3):577–85.
41. Reid L, Morrow B, Jubinsky P, Schwartz E, Gatmaitan Z. Regulation of growth and differentiation of epithelial cells by hormones, growth factors,

and substrates of extracellular matrix. Ann N Y Acad Sci. 1981;372:354–70.
42. Birkedal-Hansen H. Proteolytic remodeling of extracellular matrix. Curr Opin Cell Biol. 1995;7(5):728–35.
43. Lutolf MP, Hubbell JA. Synthetic biomaterials as instructive extracellular microenvironments for morphogenesis in tissue engineering. Nat Biotechnol. 2005;23(1):47–55.
44. Hodde J. Naturally occurring scaffolds for soft tissue repair and regeneration. Tissue Eng. 2002;8(2):295–308.
45. Ott HC, Matthiesen TS, Goh SK, Black LD, Kren SM, Netoff TI, et al. Perfusion-decellularized matrix: using nature's platform to engineer a bioartificial heart. Nat Med. 2008;14(2):213–21.
46. Eberli D, Atala A, Yoo JJ. One and four layer acellular bladder matrix for fascial tissue reconstruction. J Mater Sci Mater Med. 2010;22(3):741–51.
47. Baptista PM, Siddiqui MM, Lozier G, Rodriguez SR, Atala A, Soker S. The use of whole organ decellularization for the generation of a vascularized liver organoid. Hepatology. 2010;53(2):604–17.
48. Ott HC, Clippinger B, Conrad C, Schuetz C, Pomerantseva I, Ikonomou L, et al. Regeneration and orthotopic transplantation of a bioartificial lung. Nat Med. 2010;16(8):927–33.
49. Sullivan DC, Mirmalek-Sani SH, Deegan DB, Baptista PM, Aboushwareb T, Atala A, et al. Decellularization methods of porcine kidneys for whole organ engineering using a high-throughput system. Biomaterials. 2012;33(31):7756–64.
50. Sandusky GE, Lantz GC, Badylak SF. Healing comparison of small intestine submucosa and ePTFE grafts in the canine carotid artery. J Surg Res. 1995;58(4):415–20.
51. DeQuach JA, Mezzano V, Miglani A, Lange S, Keller GM, Sheikh F, et al. Simple and high yielding method for preparing tissue specific extracellular matrix coatings for cell culture. PLoS One. 2010;5(9):e13039.
52. Stern MM, Myers RL, Hammam N, Stern KA, Eberli D, Kritchevsky SB, et al. The influence of extracellular matrix derived from skeletal muscle tissue on the proliferation and differentiation of myogenic progenitor cells ex vivo. Biomaterials. 2009;30(12):2393–9.
53. Badylak S, Liang A, Record R, Tullius R, Hodde J. Endothelial cell adherence to small intestinal submucosa: an acellular bioscaffold. Biomaterials. 1999;20(23–24):2257–63.
54. Kim MS, Ahn HH, Shin YN, Cho MH, Khang G, Lee HB. An in vivo study of the host tissue response to subcutaneous implantation of PLGA- and/or porcine small intestinal submucosa-based scaffolds. Biomaterials. 2007;28(34):5137–43.
55. Helton WS, Fisichella PM, Berger R, Horgan S, Espat NJ, Abcarian H. Short-term outcomes with small intestinal submucosa for ventral abdominal hernia. Arch Surg. 2005;140(6):549–60. discussion 60-2
56. Ansaloni L, Catena F, Coccolini F, Gazzotti F, D'Alessandro L, Pinna AD. Inguinal hernia repair with porcine small intestine submucosa: 3-year follow-up results of a randomized controlled trial of Lichtenstein's repair with polypropylene mesh versus Surgisis inguinal hernia matrix. Am J Surg. 2009;198(3):303–12.
57. Karpelowsky JS, Thomas G, Shun A. Definitive abdominal wall closure using a porcine intestinal submucosa biodegradable membrane in pediatric transplantation. Pediatr Transplant. 2009;13(3):285–9.
58. Wiedemann A, Otto M. Small intestinal submucosa for pubourethral sling suspension for the treatment of stress incontinence: first histopathological results in humans. J Urol. 2004;172(1):215–8.
59. Trost L, Elliott D. Small intestinal submucosa urethral wrap at the time of artificial urinary sphincter placement as a salvage treatment option for patients with persistent/recurrent incontinence following multiple prior sphincter failures and erosions. Urology. 2012;79(4):933–8.
60. Caione P, Boldrini R, Salerno A, Nappo SG. Bladder augmentation using acellular collagen biomatrix: a pilot experience in exstrophic patients. Pediatr Surg Int. 2012;28(4):421–8.
61. Chaliha C, Khalid U, Campagna L, Digesu GA, Ajay B, Khullar V. SIS graft for anterior vaginal wall prolapse repair--a case-controlled study. Int Urogynecol J Pelvic Floor Dysfunct. 2006;17(5):492–7.
62. Zhang J, Hu ZQ, Turner NJ, Teng SF, Cheng WY, Zhou HY, et al. Perfusion-decellularized skeletal muscle as a three-dimensional scaffold with a vascular network template. Biomaterials. 2016;89:114–26.
63. Wassenaar JW, Boss GR, Christman KL. Decellularized skeletal muscle as an in vitro model for studying drug-extracellular matrix interactions. Biomaterials. 2015;64:108–14.
64. Fishman JM, Lowdell MW, Urbani L, Ansari T, Burns AJ, Turmaine M, et al. Immunomodulatory effect of a decellularized skeletal muscle scaffold in a discordant xenotransplantation model. Proc Natl Acad Sci U S A. 2013;110(35):14360–5.
65. Duan Y, Liu Z, O'Neill J, Wan LQ, Freytes DO, Vunjak-Novakovic G. Hybrid gel composed of native heart matrix and collagen induces cardiac differentiation of human embryonic stem cells without supplemental growth factors. J Cardiovasc Transl Res. 2011;4(5):605–15.
66. Wassenaar JW, Gaetani R, Garcia JJ, Braden RL, Luo CG, Huang D, et al. Evidence for mechanisms underlying the functional benefits of a myocardial matrix hydrogel for post-MI treatment. J Am Coll Cardiol. 2016;67(9):1074–86.
67. Singelyn JM, Sundaramurthy P, Johnson TD, Schup-Magoffin PJ, Hu DP, Faulk DM, et al. Catheter-deliverable hydrogel derived from decellularized ventricular extracellular matrix increases endogenous cardiomyocytes and preserves cardiac function post-myocardial infarction. J Am Coll Cardiol. 2012;59(8):751–63.

68. Pati F, Jang J, Ha DH, Won Kim S, Rhie JW, Shim JH, et al. Printing three-dimensional tissue analogues with decellularized extracellular matrix bioink. Nat Commun. 2014;5:3935.
69. Arenas-Herrera JE, Ko IK, Atala A, Yoo JJ. Decellularization for whole organ bioengineering. Biomed Mater. 2013;8(1):014106.
70. Gilbert TW, Sellaro TL, Badylak SF. Decellularization of tissues and organs. Biomaterials. 2006;27(19):3675–83.
71. Uhrich KE, Cannizzaro SM, Langer RS, Shakesheff KM. Polymeric systems for controlled drug release. Chem Rev. 1999;99(11):3181–98.
72. Vert M. Aliphatic polyesters: great degradable polymers that cannot do everything. Biomacromolecules. 2005;6(2):538–46.
73. Gunatillake PA, Adhikari R. Biodegradable synthetic polymers for tissue engineering. Eur Cell Mater. 2003;5:1–16.
74. Kim BS, Mooney DJ. Development of biocompatible synthetic extracellular matrices for tissue engineering. Trends Biotechnol. 1998;16(5):224–30.
75. Galaev IY, Mattiasson B. 'Smart' polymers and what they could do in biotechnology and medicine. Trends Biotechnol. 1999;17(8):335–40.
76. Hoffman AS, Stayton PS, Bulmus V, Chen G, Chen J, Cheung C, Founder's Award, Society for Biomaterials, et al. Sixth world biomaterials congress 2000, Kamuela, HI, may 15–20, 2000. Really smart bioconjugates of smart polymers and receptor proteins. J Biomed Mater Res. 2000;52(4):577–86.
77. Lee SJ, Lim GJ, Lee JW, Atala A, Yoo JJ. In vitro evaluation of a poly(lactide-co-glycolide)-collagen composite scaffold for bone regeneration. Biomaterials. 2006;27(18):3466–72.
78. Chaikof EL, Matthew H, Kohn J, Mikos AG, Prestwich GD, Yip CM. Biomaterials and scaffolds in reparative medicine. Ann N Y Acad Sci. 2002;961:96–105.
79. Bergsma JE, Rozema FR, Bos RR, Boering G, de Bruijn WC, Pennings AJ. In vivo degradation and biocompatibility study of in vitro pre-degraded as-polymerized polyactide particles. Biomaterials. 1995;16(4):267–74.
80. Hynes RO. Integrins: versatility, modulation, and signaling in cell adhesion. Cell. 1992;69(1):11–25.
81. Deuel TF. Polypeptide growth factors: roles in normal and abnormal cell growth. Annu Rev Cell Biol. 1987;3:443–92.
82. Gopferich A, Tessmar J. Polyanhydride degradation and erosion. Adv Drug Deliv Rev. 2002;54(7):911–31.
83. Dang JM, Leong KW. Natural polymers for gene delivery and tissue engineering. Adv Drug Deliv Rev. 2006;58(4):487–99.
84. Bedran-Russo AK, Pereira PN, Duarte WR, Drummond JL, Yamauchi M. Application of crosslinkers to dentin collagen enhances the ultimate tensile strength. J Biomed Mater Res B Appl Biomater. 2007;80(1):268–72.
85. Everaerts F, Torrianni M, Hendriks M, Feijen J. Biomechanical properties of carbodiimide crosslinked collagen: influence of the formation of ester crosslinks. J Biomed Mater Res A. 2008;85(2):547–55.
86. Lee SJ, Khang G, Lee YM, Lee HB. The effect of surface wettability on induction and growth of neurites from the PC-12 cell on a polymer surface. J Colloid Interface Sci. 2003;259(2):228–35.
87. Lee JH, Khang G, Lee JW, Lee HB. Interaction of different types of cells on polymer surfaces with wettability gradient. J Colloid Interface Sci. 1998;205(2):323–30.
88. Mansur HS, Orefice RL, Vasconcelos WL, Lobato ZP, Machado LJ. Biomaterial with chemically engineered surface for protein immobilization. J Mater Sci Mater Med. 2005;16(4):333–40.
89. Chai C, Leong KW. Biomaterials approach to expand and direct differentiation of stem cells. Mol Ther. 2007;15(3):467–80.
90. Sakiyama-Elbert SE, Panitch A, Hubbell JA. Development of growth factor fusion proteins for cell-triggered drug delivery. FASEB J. 2001;15(7):1300–2.
91. Anderson DG, Levenberg S, Langer R. Nanoliter-scale synthesis of arrayed biomaterials and application to human embryonic stem cells. Nat Biotechnol. 2004;22(7):863–6.
92. Zhao X, Zhang S. Fabrication of molecular materials using peptide construction motifs. Trends Biotechnol. 2004;22(9):470–6.
93. Fairman R, Akerfeldt KS. Peptides as novel smart materials. Curr Opin Struct Biol. 2005;15(4):453–63.
94. Wagner DE, Phillips CL, Ali WM, Nybakken GE, Crawford ED, Schwab AD, et al. Toward the development of peptide nanofilaments and nanoropes as smart materials. Proc Natl Acad Sci U S A. 2005;102(36):12656–61.
95. Hartgerink JD, Beniash E, Stupp SI. Peptide-amphiphile nanofibers: a versatile scaffold for the preparation of self-assembling materials. Proc Natl Acad Sci U S A. 2002;99(8):5133–8.
96. Guler MO, Hsu L, Soukasene S, Harrington DA, Hulvat JF, Stupp SI. Presentation of RGDS epitopes on self-assembled nanofibers of branched peptide amphiphiles. Biomacromolecules. 2006;7(6):1855–63.
97. Beniash E, Hartgerink JD, Storrie H, Stendahl JC, Stupp SI. Self-assembling peptide amphiphile nanofiber matrices for cell entrapment. Acta Biomater. 2005;1(4):387–97.
98. Silva GA, Czeisler C, Niece KL, Beniash E, Harrington DA, Kessler JA, et al. Selective differentiation of neural progenitor cells by high-epitope density nanofibers. Science. 2004;303(5662):1352–5.
99. Hoffman AS. Bioconjugates of intelligent polymers and recognition proteins for use in diagnostics and affinity separations. Clin Chem. 2000;46(9):1478–86.

100. Stayton PS, Shimoboji T, Long C, Chilkoti A, Chen G, Harris JM, et al. Control of protein-ligand recognition using a stimuli-responsive polymer. Nature. 1995;378(6556):472–4.
101. El-Sayed ME, Hoffman AS, Stayton PS. Smart polymeric carriers for enhanced intracellular delivery of therapeutic macromolecules. Expert Opin Biol Ther. 2005;5(1):23–32.
102. Shimoboji T, Larenas E, Fowler T, Hoffman AS, Stayton PS. Temperature-induced switching of enzyme activity with smart polymer-enzyme conjugates. Bioconjug Chem. 2003;14(3):517–25.
103. Stayton PS, Ding Z, Hoffman AS. Smart polymer-streptavidin conjugates. Methods Mol Biol. 2004;283:37–43.
104. Bettinger CJ, Langer R, Borenstein JT. Engineering substrate topography at the micro- and nanoscale to control cell function. Angew Chem Int Ed Eng. 2009;48(30):5406–15.
105. Ross AM, Jiang Z, Bastmeyer M, Lahann J. Physical aspects of cell culture substrates: topography, roughness, and elasticity. Small. 2012;8(3):336–55.
106. Engler AJ, Sen S, Sweeney HL, Discher DE. Matrix elasticity directs stem cell lineage specification. Cell. 2006;126(4):677–89.
107. Hoffman AS. Hydrogels for biomedical applications. Adv Drug Deliv Rev. 2002;54(1):3–12.
108. Cushing MC, Anseth KS. Materials science. Hydrogel cell cultures. Science. 2007;316(5828):1133–4.
109. Kraehenbuehl TP, Zammaretti P, Van der Vlies AJ, Schoenmakers RG, Lutolf MP, Jaconi ME, et al. Three-dimensional extracellular matrix-directed cardioprogenitor differentiation: systematic modulation of a synthetic cell-responsive PEG-hydrogel. Biomaterials. 2008;29(18):2757–66.
110. Orive G, Gascon AR, Hernandez RM, Igartua M, Luis PJ. Cell microencapsulation technology for biomedical purposes: novel insights and challenges. Trends Pharmacol Sci. 2003;24(5):207–10.
111. King A, Strand B, Rokstad AM, Kulseng B, Andersson A, Skjak-Braek G, et al. Improvement of the biocompatibility of alginate/poly-L-lysine/alginate microcapsules by the use of epimerized alginate as a coating. J Biomed Mater Res A. 2003;64(3):533–9.
112. Klock G, Pfeffermann A, Ryser C, Grohn P, Kuttler B, Hahn HJ, et al. Biocompatibility of mannuronic acid-rich alginates. Biomaterials. 1997;18(10):707–13.
113. Yang S, Leong KF, Du Z, Chua CK. The design of scaffolds for use in tissue engineering. Part I. Traditional factors. Tissue Eng. 2001;7(6):679–89.
114. Greiner A, Wendorff JH. Electrospinning: a fascinating method for the preparation of ultrathin fibers. Angew Chem Int Ed Eng. 2007;46(30):5670–703.
115. Pham QP, Sharma U, Mikos AG. Electrospinning of polymeric nanofibers for tissue engineering applications: a review. Tissue Eng. 2006;12(5):1197–211.
116. Ju YM, Choi JS, Atala A, Yoo JJ, Lee SJ. Bilayered scaffold for engineering cellularized blood vessels. Biomaterials. 2010;31(15):4313–21.
117. Cui X, Boland T. Human microvasculature fabrication using thermal inkjet printing technology. Biomaterials. 2009;30(31):6221–7.
118. Boland T, Xu T, Damon B, Cui X. Application of inkjet printing to tissue engineering. Biotechnol J. 2006;1(9):910–7.
119. Xu T, Jin J, Gregory C, Hickman JJ, Boland T. Inkjet printing of viable mammalian cells. Biomaterials. 2005;26(1):93–9.
120. Xu T, Zhao W, Zhu JM, Albanna MZ, Yoo JJ, Atala A. Complex heterogeneous tissue constructs containing multiple cell types prepared by inkjet printing technology. Biomaterials. 2012;34(1):130–9.
121. Fedorovich NE, Swennen I, Girones J, Moroni L, van Blitterswijk CA, Schacht E, et al. Evaluation of photocrosslinked Lutrol hydrogel for tissue printing applications. Biomacromolecules. 2009;10(7):1689–96.
122. Kang HW, Lee SJ, Ko IK, Kengla C, Yoo JJ, Atala A. A 3D bioprinting system to produce human-scale tissue constructs with structural integrity. Nat Biotechnol. 2016;34(3):312–9.
123. Merceron TK, Burt M, Seol YJ, Kang HW, Lee SJ, Yoo JJ, et al. A 3D bioprinted complex structure for engineering the muscle-tendon unit. Biofabrication. 2015;7(3):035003.
124. Atala A, Cima LG, Kim W, Paige KT, Vacanti JP, Retik AB, et al. Injectable alginate seeded with chondrocytes as a potential treatment for vesicoureteral reflux. J Urol. 1993;150(2 Pt 2):745–7.
125. Atala A, Kim W, Paige KT, Vacanti CA, Retik AB. Endoscopic treatment of vesicoureteral reflux with a chondrocyte-alginate suspension. J Urol. 1994;152(2 Pt 2):641–3. discussion 4
126. Diamond DA, Caldamone AA. Endoscopic correction of vesicoureteral reflux in children using autologous chondrocytes: preliminary results. J Urol. 1999;162(3 Pt 2):1185–8.
127. Bent AE, Tutrone RT, McLennan MT, Lloyd LK, Kennelly MJ, Badlani G. Treatment of intrinsic sphincter deficiency using autologous ear chondrocytes as a bulking agent. Neurourol Urodyn. 2001;20(2):157–65.
128. Chancellor MB, Yokoyama T, Tirney S, Mattes CE, Ozawa H, Yoshimura N, et al. Preliminary results of myoblast injection into the urethra and bladder wall: a possible method for the treatment of stress urinary incontinence and impaired detrusor contractility. Neurourol Urodyn. 2000;19(3):279–87.
129. Strasser H, Marksteiner R, Margreiter E, Pinggera GM, Mitterberger M, Frauscher F, et al. Autologous myoblasts and fibroblasts versus collagen for treatment of stress urinary incontinence in women: a randomised controlled trial. Lancet. 2007;369(9580):2179–86.

130. Yiou R, Yoo JJ, Atala A. Restoration of functional motor units in a rat model of sphincter injury by muscle precursor cell autografts. Transplantation. 2003;76(7):1053–60.
131. Eberli D, Andersson KE, Yoo JJ, Atala A. A canine model of irreversible urethral sphincter insufficiency. BJU Int. 2009;103(2):248–53.
132. Machluf M, Orsola A, Boorjian S, Kershen R, Atala A. Microencapsulation of Leydig cells: a system for testosterone supplementation. Endocrinology. 2003;144(11):4975–9.
133. Bardin CW, Swerdloff RS, Santen RJ. Androgens: risks and benefits. J Clin Endocrinol Metab. 1991;73(1):4–7.
134. Tai J, Johnson HW, Tze WJ. Successful transplantation of Leydig cells in castrated inbred rats. Transplantation. 1989;47(6):1087–9.
135. De Vos P, De Haan B, Van Schilfgaarde R. Effect of the alginate composition on the biocompatibility of alginate-polylysine microcapsules. Biomaterials. 1997;18(3):273–8.
136. Tai IT, Sun AM. Microencapsulation of recombinant cells: a new delivery system for gene therapy. FASEB J. 1993;7(11):1061–9.
137. Bazeed MA, Thuroff JW, Schmidt RA, Tanagho EA. New treatment for urethral strictures. Urology. 1983;21(1):53–7.
138. Olsen L, Bowald S, Busch C, Carlsten J, Eriksson I. Urethral reconstruction with a new synthetic absorbable device. An experimental study. Scand J Urol Nephrol. 1992;26(4):323–6.
139. Atala A, Vacanti JP, Peters CA, Mandell J, Retik AB, Freeman MR. Formation of urothelial structures in vivo from dissociated cells attached to biodegradable polymer scaffolds in vitro. J Urol. 1992;148(2 Pt 2):658–62.
140. Sievert KD, Bakircioglu ME, Nunes L, Tu R, Dahiya R, Tanagho EA. Homologous acellular matrix graft for urethral reconstruction in the rabbit: histological and functional evaluation. J Urol. 2000;163(6):1958–65.
141. Atala A, Guzman L, Retik AB. A novel inert collagen matrix for hypospadias repair. J Urol. 1999;162(3 Pt 2):1148–51.
142. El-Kassaby AW, Retik AB, Yoo JJ, Atala A. Urethral stricture repair with an off-the-shelf collagen matrix. J Urol. 2003;169(1):170–3. discussion 3
143. El-Kassaby A, AbouShwareb T, Atala A. Randomized comparative study between buccal mucosal and acellular bladder matrix grafts in complex anterior urethral strictures. J Urol. 2008;179(4):1432–6.
144. De Filippo RE, Yoo JJ, Atala A. Urethral replacement using cell seeded tubularized collagen matrices. J Urol. 2002;168(4 Pt 2):1789–92. discussion 92-3
145. le Roux PJ. Endoscopic urethroplasty with unseeded small intestinal submucosa collagen matrix grafts: a pilot study. J Urol. 2005;173(1):140–3.
146. Raya-Rivera A, Esquiliano DR, Yoo JJ, Lopez-Bayghen E, Soker S, Atala A. Tissue-engineered autologous urethras for patients who need reconstruction: an observational study. Lancet. 2011;377(9772):1175–82.
147. Atala A, Bauer SB, Soker S, Yoo JJ, Retik AB. Tissue-engineered autologous bladders for patients needing cystoplasty. Lancet. 2006;367(9518):1241–6.
148. Oberpenning F, Meng J, Yoo JJ, Atala A. De novo reconstitution of a functional mammalian urinary bladder by tissue engineering. Nat Biotechnol. 1999;17(2):149–55.
149. Kwon TG, Yoo JJ, Atala A. Autologous penile corpora cavernosa replacement using tissue engineering techniques. J Urol. 2002;168(4 Pt 2):1754–8.
150. Chen KL, Eberli D, Yoo JJ, Atala A. Bioengineered corporal tissue for structural and functional restoration of the penis. Proc Natl Acad Sci U S A. 2010;107(8):3346–50.
151. De Filippo RE, Yoo JJ, Atala A. Engineering of vaginal tissue in vivo. Tissue Eng. 2003;9(2):301–6.
152. Raya-Rivera AM, Esquiliano D, Fierro-Pastrana R, Lopez-Bayghen E, Valencia P, Ordorica-Flores R, et al. Tissue-engineered autologous vaginal organs in patients: a pilot cohort study. Lancet. 2014;384(9940):329–36.
153. Campbell GR, Turnbull G, Xiang L, Haines M, Armstrong S, Rolfe BE, et al. The peritoneal cavity as a bioreactor for tissue engineering visceral organs: bladder, uterus and vas deferens. J Tissue Eng Regen Med. 2008;2(1):50–60.
154. House M, Sanchez CC, Rice WL, Socrate S, Kaplan DL. Cervical tissue engineering using silk scaffolds and human cervical cells. Tissue Eng A. 2010;16(6):2101–12.
155. Auchincloss H, Bonventre JV. Transplanting cloned cells into therapeutic promise. Nat Biotechnol. 2002;20(7):665–6.
156. Amiel GE, Yoo JJ, Atala A. Renal therapy using tissue-engineered constructs and gene delivery. World J Urol. 2000;18(1):71–9.
157. Aebischer P, Ip TK, Panol G, Galletti PM. The bioartificial kidney: progress towards an ultrafiltration device with renal epithelial cells processing. Life Support Syst. 1987;5(2):159–68.
158. Humes HD, Buffington DA, MacKay SM, Funke AJ, Weitzel WF. Replacement of renal function in uremic animals with a tissue-engineered kidney. Nat Biotechnol. 1999;17(5):451–5.
159. Joki T, Machluf M, Atala A, Zhu J, Seyfried NT, Dunn IF, et al. Continuous release of endostatin from microencapsulated engineered cells for tumor therapy. Nat Biotechnol. 2001;19(1):35–9.
160. Lanza RP, Hayes JL, Chick WL. Encapsulated cell technology. Nat Biotechnol. 1996;14(9):1107–11.
161. Lanza RP, Chung HY, Yoo JJ, Wettstein PJ, Blackwell C, Borson N, et al. Generation of histocompatible tissues using nuclear transplantation. Nat Biotechnol. 2002;20(7):689–96.
162. Steinborn R, Schinogl P, Zakhartchenko V, Achmann R, Schernthaner W, Stojkovic M, et al. Mitochondrial DNA heteroplasmy in cloned cattle

produced by fetal and adult cell cloning. Nat Genet. 2000;25(3):255–7.
163. Fischer Lindahl K, Hermel E, Loveland BE, Wang CR. Maternally transmitted antigen of mice: a model transplantation antigen. Annu Rev Immunol. 1991;9:351–72.
164. Yard BA, Kooymans-Couthino M, Reterink T, van den Elsen P, Paape ME, Bruyn JA, et al. Analysis of T cell lines from rejecting renal allografts. Kidney Int Suppl. 1993;39:S133–8.
165. Abolbashari M, Agcaoili SM, Lee MK, Ko IK, Aboushwareb T, Jackson JD, et al. Repopulation of porcine kidney scaffold using porcine primary renal cells. Acta Biomater. 2016;29:52–61.
166. Kim IH, Ko IK, Atala A, Yoo JJ. Whole kidney engineering for clinical translation. Curr Opin Organ Transplant. 2015;20(2):165–70.
167. Ross R. The pathogenesis of atherosclerosis: a perspective for the 1990s. Nature. 1993;362(6423):801–9.
168. Bujan J, Garcia-Honduvilla N, Bellon JM. Engineering conduits to resemble natural vascular tissue. Biotechnol Appl Biochem. 2004;39(Pt 1):17–27.
169. Edelman ER. Vascular tissue engineering: designer arteries. Circ Res. 1999;85(12):1115–7.
170. Gong Z, Niklason LE. Blood vessels engineered from human cells. Trends Cardiovasc Med. 2006;16(5):153–6.
171. Greenwald SE, Berry CL. Improving vascular grafts: the importance of mechanical and haemodynamic properties. J Pathol. 2000;190(3):292–9.
172. Swain TW 3rd, Calligaro KD, Dougherty MD. Management of infected aortic prosthetic grafts. Vasc Endovasc Surg. 2004;38(1):75–82.
173. Amiel GE, Komura M, Shapira O, Yoo JJ, Yazdani S, Berry J, et al. Engineering of blood vessels from acellular collagen matrices coated with human endothelial cells. Tissue Eng. 2006;12(8):2355–65.
174. Kaushal S, Amiel GE, Guleserian KJ, Shapira OM, Perry T, Sutherland FW, et al. Functional small-diameter neovessels created using endothelial progenitor cells expanded ex vivo. Nat Med. 2001;7(9):1035–40.
175. Drilling S, Gaumer J, Lannutti J. Fabrication of burst pressure competent vascular grafts via electrospinning: effects of microstructure. J Biomed Mater Res A. 2009;88(4):923–34.
176. Lee SJ, Yoo JJ, Lim GJ, Atala A, Stitzel J. In vitro evaluation of electrospun nanofiber scaffolds for vascular graft application. J Biomed Mater Res A. 2007;83(4):999–1008.
177. Stitzel J, Liu J, Lee SJ, Komura M, Berry J, Soker S, et al. Controlled fabrication of a biological vascular substitute. Biomaterials. 2006;27(7):1088–94.
178. Baguneid MS, Seifalian AM, Salacinski HJ, Murray D, Hamilton G, Walker MG. Tissue engineering of blood vessels. Br J Surg. 2006;93(3):282–90.
179. Tillman BW, Yazdani SK, Lee SJ, Geary RL, Atala A, Yoo JJ. The in vivo stability of electrospun polycaprolactone-collagen scaffolds in vascular reconstruction. Biomaterials. 2009;30(4):583–8.
180. Bach AD, Beier JP, Stern-Staeter J, Horch RE. Skeletal muscle tissue engineering. J Cell Mol Med. 2004;8(4):413–22.
181. DiEdwardo CA, Petrosko P, Acarturk TO, DiMilla PA, LaFramboise WA, Johnson PC. Muscle tissue engineering. Clin Plast Surg. 1999;26(4):647–56. ix-x
182. Cossu G, Biressi S. Satellite cells, myoblasts and other occasional myogenic progenitors: possible origin, phenotypic features and role in muscle regeneration. Semin Cell Dev Biol. 2005;16(4–5):623–31.
183. Cao B, Deasy BM, Pollett J, Huard J. Cell therapy for muscle regeneration and repair. Phys Med Rehabil Clin N Am. 2005;16(4):889–907.
184. Wernig A, Zweyer M, Irintchev A. Function of skeletal muscle tissue formed after myoblast transplantation into irradiated mouse muscles. J Physiol. 2000;522(Pt 2):333–45.
185. Cesar M, Roussanne-Domergue S, Coulet B, Gay S, Micallef JP, Chammas M, et al. Transplantation of adult myoblasts or adipose tissue precursor cells by high-density injection failed to improve reinnervated skeletal muscles. Muscle Nerve. 2008;37(2):219–30.
186. DeRosimo JF, Washabaugh CH, Ontell MP, Daood MJ, Watchko JF, Watkins SC, et al. Enhancement of adult muscle regeneration by primary myoblast transplantation. Cell Transplant. 2000;9(3):369–77.
187. Payumo FC, Kim HD, Sherling MA, Smith LP, Powell C, Wang X, et al. Tissue engineering skeletal muscle for orthopaedic applications. Clin Orthop Relat Res. 2002;(403 Suppl):S228–42.
188. Powell CA, Smiley BL, Mills J, Vandenburgh HH. Mechanical stimulation improves tissue-engineered human skeletal muscle. Am J Phys Cell Physiol. 2002;283(5):C1557–65.
189. Choi JS, Lee SJ, Christ GJ, Atala A, Yoo JJ. The influence of electrospun aligned poly(epsilon-caprolactone)/collagen nanofiber meshes on the formation of self-aligned skeletal muscle myotubes. Biomaterials. 2008;29(19):2899–906.
190. Wakelam MJ. The fusion of myoblasts. Biochem J. 1985;228(1):1–12.
191. Derby B. Printing and prototyping of tissues and scaffolds. Science. 2012;338(6109):921–6.
192. Jones N. Science in three dimensions: the print revolution. Nature. 2012;487(7405):22–3.
193. Xu T, Binder KW, Albanna MZ, Dice D, Zhao W, Yoo JJ, et al. Hybrid printing of mechanically and biologically improved constructs for cartilage tissue engineering applications. Biofabrication. 2012;5(1):015001.
194. De Coppi P, Bartsch G Jr, Siddiqui MM, Xu T, Santos CC, Perin L, et al. Isolation of amniotic stem cell lines with potential for therapy. Nat Biotechnol. 2007;25(1):100–6.

195. Koning M, Harmsen MC, van Luyn MJ, Werker PM. Current opportunities and challenges in skeletal muscle tissue engineering. J Tissue Eng Regen Med. 2009;3(6):407–15.
196. Kingham PJ, Terenghi G. Bioengineered nerve regeneration and muscle reinnervation. J Anat. 2006;209(4):511–26.
197. Kang SB, Olson JL, Atala A, Yoo JJ. Functional recovery of completely denervated muscle: implications for innervation of tissue-engineered muscle. Tissue Eng A. 2012;18(17–18):1912–20.
198. Bian W, Bursac N. Soluble miniagrin enhances contractile function of engineered skeletal muscle. FASEB J. 2012;26(2):955–65.
199. Kummer TT, Misgeld T, Sanes JR. Assembly of the postsynaptic membrane at the neuromuscular junction: paradigm lost. Curr Opin Neurobiol. 2006;16(1):74–82.
200. Wu H, Xiong WC, Mei L. To build a synapse: signaling pathways in neuromuscular junction assembly. Development. 2010;137(7):1017–33.
201. Witzemann V. Development of the neuromuscular junction. Cell Tissue Res. 2006;326(2):263–71.
202. Bach AD, Beier JP, Stark GB. Expression of Trisk 51, agrin and nicotinic-acetycholine receptor epsilon-subunit during muscle development in a novel three-dimensional muscle-neuronal co-culture system. Cell Tissue Res. 2003;314(2):263–74.
203. Bentzinger CF, Barzaghi P, Lin S, Ruegg MA. Overexpression of mini-agrin in skeletal muscle increases muscle integrity and regenerative capacity in laminin-alpha2-deficient mice. FASEB J. 2005;19(8):934–42.
204. Glass DJ, Bowen DC, Stitt TN, Radziejewski C, Bruno J, Ryan TE, et al. Agrin acts via a MuSK receptor complex. Cell. 1996;85(4):513–23.
205. Lin S, Landmann L, Ruegg MA, Brenner HR. The role of nerve- versus muscle-derived factors in mammalian neuromuscular junction formation. J Neurosci. 2008;28(13):3333–40.
206. DeChiara TM, Bowen DC, Valenzuela DM, Simmons MV, Poueymirou WT, Thomas S, et al. The receptor tyrosine kinase MuSK is required for neuromuscular junction formation in vivo. Cell. 1996;85(4):501–12.
207. Gautam M, Noakes PG, Moscoso L, Rupp F, Scheller RH, Merlie JP, et al. Defective neuromuscular synaptogenesis in agrin-deficient mutant mice. Cell. 1996;85(4):525–35.
208. Bezakova G, Helm JP, Francolini M, Lomo T. Effects of purified recombinant neural and muscle agrin on skeletal muscle fibers in vivo. J Cell Biol. 2001;153(7):1441–52.
209. Ko IK, Lee BK, Lee SJ, Andersson K-E, Atala A, Yoo JJ. The effect of in vitro formation of acetylcholine receptor (AChR) clusters in engineered muscle fibers on subsequent innervation of constructs in vivo. Biomaterials. 2013;34(13):3246–55. doi:10.1016/j.biomaterials.2013.01.029.
210. Folkman J, Hochberg M. Self-regulation of growth in three dimensions. J Exp Med. 1973;138(4):745–53.
211. Nomi M, Atala A, Coppi PD, Soker S. Principals of neovascularization for tissue engineering. Mol Asp Med. 2002;23(6):463–83.
212. De Coppi P, Delo D, Farrugia L, Udompanyanan K, Yoo JJ, Nomi M, et al. Angiogenic gene-modified muscle cells for enhancement of tissue formation. Tissue Eng. 2005;11(7–8):1034–44.
213. Schuch G, Machluf M, Bartsch G Jr, Nomi M, Richard H, Atala A, et al. In vivo administration of vascular endothelial growth factor (VEGF) and its antagonist, soluble neuropilin-1, predicts a role of VEGF in the progression of acute myeloid leukemia in vivo. Blood. 2002;100(13):4622–8.
214. Ennett AB, Kaigler D, Mooney DJ. Temporally regulated delivery of VEGF in vitro and in vivo. J Biomed Mater Res A. 2006;79(1):176–84.
215. Kaigler D, Wang Z, Horger K, Mooney DJ, Krebsbach PH. VEGF scaffolds enhance angiogenesis and bone regeneration in irradiated osseous defects. J Bone Miner Res. 2006;21(5):735–44.
216. Borselli C, Cezar CA, Shvartsman D, Vandenburgh HH, Mooney DJ. The role of multifunctional delivery scaffold in the ability of cultured myoblasts to promote muscle regeneration. Biomaterials. 2011;32(34):8905–14.
217. Levenberg S, Rouwkema J, Macdonald M, Garfein ES, Kohane DS, Darland DC, et al. Engineering vascularized skeletal muscle tissue. Nat Biotechnol. 2005;23(7):879–84.
218. Moon JJ, Saik JE, Poche RA, Leslie-Barbick JE, Lee SH, Smith AA, et al. Biomimetic hydrogels with pro-angiogenic properties. Biomaterials. 2010;31(14):3840–7.
219. Tsang VL, Bhatia SN. Fabrication of three-dimensional tissues. Adv Biochem Eng Biotechnol. 2007;103:189–205.
220. Weigel T, Schinkel G, Lendlein A. Design and preparation of polymeric scaffolds for tissue engineering. Expert Rev Med Devices. 2006;3(6):835–51.
221. Williams DF. A model for biocompatibility and its evaluation. J Biomed Eng. 1989;11(3):185–91.
222. Remes A, Williams DF. Immune response in biocompatibility. Biomaterials. 1992;13(11):731–43.

Cell Technologies

3

So Young Chun

3.1 Available Stem Cell Types for Urology Field

3.1.1 Overview

Stem cells are undifferentiated biological cells that can differentiate into specialized cells and can divide to produce more stem cells. When a stem cell divides, each new cell has the potential either to remain a stem cell or become another type of cell with a more specialized function. They are found in multicellular organisms. In mammals, there are two broad types of stem cells: embryonic stem cells, which are isolated from the inner cell mass of blastocysts, and adult (somatic) stem cells, which are found in various tissues. In adult organisms, stem cells and progenitor cells act as a repair system for the body, replenishing adult tissues. In a developing embryo, stem cells can differentiate into all the specialized cells—ectoderm, endoderm, and mesoderm—but also maintain the normal turnover of regenerative organs, such as blood, skin, or intestinal tissues.

Embryonic stem cells are derived from embryos. Most embryonic stem cells are derived from eggs that have been fertilized in vitro and then donated for research purposes. Embryonic stem cells are described as "pluripotent," meaning that they can generate all the different types of cells. Embryonic stem cells can be obtained from the blastocyst, a very early stage of development that consists of a mostly hollow ball of approximately 150–200 cells. At this stage, there are no organs, just an "inner cell mass" from which embryonic stem cells can be obtained. Human embryonic stem cells are derived primarily from blastocysts that were created by in vitro fertilization (IVF) for assisted reproduction. The fertilized cells that immediately arise in the first few divisions are "totipotent." This means that they can generate a viable embryo. Then these cells transition to become pluripotent.

Adult (somatic) stem cells are found in most major organs and tissues and are currently being isolated from many tissues in the body. The methods of isolation and culture are dependent on the cell source and lineage. Many isolation and purification protocols involve flow cytometry and cell sorting. Positive and negative sorting for cell surface markers can quickly generate enriched populations.

Tissue-specific stem cells, referred to as "adult" or "somatic" stem cells, are already somewhat specialized and can produce some or all of the mature cell types found within the particular tissue or organ. Because of their ability to generate multiple, organ-specific cell types, they are described as "multipotent." Tissue-specific

S.Y. Chun
BioMedical Research Institute,
Kyungpook National Univ. Hospital,
Deagu, 41940, USA
e-mail: soyachun@hotmail.com

B.W. Kim (ed.), *Clinical Regenerative Medicine in Urology*, DOI 10.1007/978-981-10-2723-9_3

stem cells have been found in several organs that need to continuously replenish themselves, such as the blood, skin, and gut. These types of stem cells represent a very small population.

There are three known accessible sources of autologous adult stem cells in humans: bone marrow, which requires extraction by harvesting, that is, drilling into bone; adipose tissue, which requires extraction by liposuction; and blood, which requires extraction through apheresis, wherein blood is drawn from the donor and passed through a machine that extracts the stem cells and returns other portions of the blood to the donor. Stem cells can also be taken from amniotic fluid through amniocentesis, umbilical cord blood just after birth, and voided urine.

Adult stem cells are frequently used in various medical therapies. Stem cells can now be artificially grown and differentiated into specialized cell types with characteristics consistent with cells of various tissues. Research on adult stem cells has generated a great deal of excitement. Scientists have found adult stem cells in many more tissues. This finding has led researchers and clinicians to ask whether adult stem cells could be used for transplants. In fact, stem cells from bone marrow have been used in transplants for more than 40 years. Scientists now have evidence that stem cells exist in the brain and the heart, two locations where adult stem cells were not at first expected to reside. If the differentiation of adult stem cells can be controlled in the laboratory, these cells may become the basis of transplantation-based therapies.

3.1.2 Stem Cell Properties

Stem cells are distinguished from other cell types by two important characteristics: First, they are unspecialized cells capable of renewing themselves through cell division. Second, under certain physiologic or experimental conditions, they can be induced to become tissue-specific cells with special functions. In vivo, stem cell division can either result in two identical daughter cells, or one of the cells can acquire more specialized functions. Depending on stem cell type, location, and local factors, the cell acquires a more specialized function after cell division and becomes a less potent stem cell or becomes a fully differentiated cell.

Two mechanisms exist to ensure that a stem cell population is maintained: asymmetric replication, a stem cell divides into one mother cell that is identical to the original stem cell and another daughter cell that is differentiated; and stochastic differentiation, when one stem cell develops into two differentiated daughter cells, another stem cell undergoes mitosis and produces two stem cells identical to the original.

Stem cells can differentiate into embryonic and extraembryonic cell types. Such cells can construct a complete, viable organism. These cells are produced from the fusion of an egg and sperm cell. Cells produced by the first few divisions of the fertilized egg are also totipotent. Pluripotent stem cells are the descendants of totipotent cells and can differentiate into nearly all cells. Multipotent stem cells can differentiate into a number of cell types, but only those of a closely related family of cells. Oligopotent stem cells can differentiate into only a few cell types. Unipotent cells can produce only one cell type, their own [1].

3.1.3 Similarities and Differences Between Embryonic and Adult Stem Cells

Human embryonic and adult stem cells each have advantages and disadvantages regarding potential use for cell-based regenerative therapies. One major difference between adult and embryonic stem cells is their different abilities in the number and type of differentiated cell types they can become. Embryonic stem cells can become all cell types of the body because they are pluripotent. Adult stem cells are thought to be limited to differentiating into different cell types of their tissue of origin.

Embryonic stem cells can be grown relatively easily in culture. Adult stem cells are rare in mature tissues, so isolating these cells from an adult tissue is challenging, and methods to expand their numbers in cell culture have not yet been worked out. Tissues derived from embryonic and adult stem cells may differ in the likelihood of being rejected after transplantation. It is not yet known for certain whether tissues derived from embryonic stem cells would cause transplant rejection, since relatively few clinical trials have tested the safety of transplanted cells derived.

Adult stem cells are currently believed less likely to initiate rejection after transplantation. This is because a patient's own cells, it could be expanded in culture, differentiated, and then reintroduced into the patient. The use of adult stem cells derived from the patient's own adult stem cells would mean that the cells are less likely to be rejected by the immune system. This represents a significant advantage, as immune rejection can be circumvented only by continuous administration of immunosuppressive drugs.

3.1.3.1 Embryonic Stem Cells, ES Cells

Pluripotency

ES cells are pluripotent stem cells derived from the inner cell mass of a blastocyst, an early-stage preimplantation embryo. Human embryos reach the blastocyst stage 4–5 days post fertilization, at which time they consist of 50–150 cells. ES cells are pluripotent and give rise during development to all derivatives of the three primary germ layers: ectoderm, endoderm, and mesoderm. In other words, they can develop into each of the more than 200 cell types of the adult body when given sufficient and necessary stimulation for a specific cell type. Human ES cells measure approximately 14 μm, and mouse ES cells are closer to 8 μm [2, 3].

Pluripotency distinguishes embryonic stem cells from adult stem cells; while embryonic stem cells can generate all cell types in the body, adult stem cells are multipotent and can produce only a limited number of cell types. If the pluripotent differentiation potential of embryonic stem cells could be harnessed in vitro, it might be a means of deriving cell or tissue types virtually to order. This would provide a radical new treatment approach to a wide variety of conditions where age, disease, or trauma has led to tissue damage or dysfunction.

ES Cell Markers

A human embryonic stem cell is also defined by the expression of several transcription factors and cell surface proteins. The transcription factors Oct4, Nanog, and Sox2 form the core regulatory network that ensures the suppression of genes that lead to differentiation and the maintenance of pluripotency [4]. The cell surface antigens most commonly used to identify hES cells are the glycolipids stage-specific embryonic antigens 3 and 4 and the keratan sulfate antigens Tra-1-60 and Tra-1-81. By using human embryonic stem cells to produce specialized cells like nerve cells or heart cells in the lab, scientists can gain access to adult human cells without taking tissue from patients. They can then study these specialized adult cells in detail to try and catch complications of diseases or to study cell reactions to potentially new drugs. The molecular definition of a stem cell includes many more proteins and continues to be a topic of research [5].

Proliferation and Differentiation

Under defined conditions, embryonic stem cells are capable of propagating themselves indefinitely in an undifferentiated state and have the capacity when provided with the appropriate signals to differentiate, presumably via the formation of precursor cells, to almost all mature cell phenotypes [6]. This allows embryonic stem cells to be employed as useful tools for both research and regenerative medicine, because they can produce limitless numbers of themselves for continued research or clinical use.

Stemness Maintenances

Nearly all research to date has made use of mouse embryonic stem cells or human embryonic stem cells derived from the early inner cell mass. Both have the essential stem cell characteristics, yet they require very different environments in order to maintain an undifferentiated state. Mouse ES cells are grown on a layer of gelatin as an extracellular matrix and require the presence of leukemia inhibitory factor (LIF). Human ES cells are grown on a feeder layer of mouse embryonic fibroblasts (MEFs) and require the presence of basic fibroblast growth factor (bFGF or FGF-2) [7]. Without optimal culture conditions or genetic manipulation, embryonic stem cells will rapidly differentiate.

Potential Clinical use

Current research focuses on differentiating ES into a variety of cell types for eventual use as cell replacement therapies. Some of the cell types that have or are currently being developed include cardiomyocytes, neurons, hepatocytes, bone marrow cells, islet cells, and endothelial cells [8]. However, the derivation of such cell types from ESs is not without obstacles, and hence current research is focused on overcoming these barriers.

Besides in the future becoming an important alternative to organ transplants, ES are also being used in field of toxicology and as cellular screens to uncover new chemical entities that can be developed as small molecule drugs.

Studies have shown that cardiomyocytes derived from ES are validated in vitro models to test drug responses and predict toxicity profiles [8]. ES-derived cardiomyocytes have been shown to respond to pharmacological stimuli and hence can be used to assess cardiotoxicity.

ES-derived hepatocytes are also useful models that could be used in the preclinical stages of drug discovery. However, the development of hepatocytes from ESC has proven to be challenging, and this hinders the ability to test drug metabolism. Therefore, current research is focusing on establishing fully functional ES-derived hepatocytes with stable phase I and II enzyme activity [7].

Researchers have also differentiated ESC into dopamine-producing cells with the hope that these neurons could be used in the treatment of Parkinson's disease [4]. Recently, the development of ESC after somatic cell nuclear transfer of olfactory ensheathing cells to a healthy oocyte has been recommended for neuro-degenerative diseases [5]. The same group also has advocated the use of olfactory ensheathing cells for demyelinating diseases like multiple sclerosis. ESCs have also been differentiated to natural killer cells and bone tissue. Studies involving ES are also underway to provide an alternative treatment for diabetes.

Application to Genetic Disorders

This has been done either by genetically manipulating the cells or more recently by deriving diseased cell lines identified by prenatal genetic diagnosis. This approach may very well prove invaluable at studying disorders such as fragile X syndrome, cystic fibrosis, and other genetic maladies that have no reliable model system.

Yury Verlinsky developed prenatal diagnosis testing methods to determine genetic and chromosomal disorders a month and a half earlier than standard amniocentesis. The techniques are now used by many pregnant women and prospective parents, especially those couples with a history of genetic abnormalities, when the risk of genetically related disorders is higher. In addition, by allowing parents to select an embryo without genetic disorders, they have the potential of saving the lives of siblings that already had similar disorders and diseases using cells from the disease-free offspring [9].

Repair of DNA Damage

ES cells use a different strategy to deal with double-strand breaks (DSBs). Because ES cells give rise to all of the cell types of an organism including the cells of the germ line, mutations arising in ES cells due to faulty DNA repair are a more serious problem than in differentiated somatic cells. Consequently, robust mechanisms are needed in ES cells to repair DNA damages accurately and, if repair fails, to remove those cells with unrepaired DNA damages. Thus, mouse ES cells predominantly use high fidelity homologous recombinational repair (HRR) to repair DSBs. This type of repair depends on the interaction of the two sister chromosomes formed during S phase and present together during the G2 phase of the cell cycle. HRR can accurately repair DSBs in one sister chromosome by using intact information from the other sister chromosome. Cells in the G1 phase of the cell cycle have only one copy of each chromosome. Mouse ES cells lack a G1 checkpoint and do not undergo cell cycle arrest upon acquiring DNA damage. Rather they undergo programmed cell death (apoptosis) in response to DNA damage. Apoptosis can be used as a fail-safe strategy to remove cells with unrepaired DNA damages in order to avoid mutation and progression to cancer. Consistent with this strategy, mouse ES stem cells have a mutation frequency about 100-fold lower than that of isogenic mouse somatic cells [10].

Preparation Method for hES Cells

Human embryonic stem cells can be derived from donated embryos, or additionally they can also be extracted from cloned embryos using a cell from a patient and a donated egg. The inner cell mass, from the blastocyst stage of the embryo, is separated from the trophectoderm, the cells that would differentiate into extraembryonic tissue. Immunosurgery, the process in which antibodies are bound to the trophectoderm and removed by another solution, and mechanical dissection are performed to achieve separation. The resulting inner cell mass cells are plated onto cells that will supply support. The inner cell mass cells attach and expand further to form a human embryonic cell line, which are undifferentiated. These cells are fed daily and are enzymatically or mechanically separated every 4–7 days. For differentiation to occur, the human embryonic stem cell line is removed from the supporting cells to form embryoid bodies, co-cultured with a serum containing necessary signals, or grafted in a three-dimensional scaffold [2].

Difficulties

One potential disadvantage of ESCs is that they can form teratomas, tumors containing cell types from all three germ layers [11]. Studies suggest that teratoma formation may be a result of chromosomal changes that occur in some ES cells during culture. As length of time in culture and the number of passages increase, cells are more likely to display chromosomal instability and mosaicism within the population. ES cell lines must be carefully monitored for changes in phenotype, gene expression, and karyotype. Following differentiation, the cells are subjected to sorting by flow cytometry for further purification. ESCs are predicted to be inherently safer than iPS cells because they are not genetically modified with genes such as c-Myc that are linked to cancer. Nonetheless, ESC expresses very high levels of the iPS-inducing genes, and these genes including Myc are essential for ESC self-renewal and pluripotency.

3.1.3.2 Reprogrammed Stem Cells

One of the hottest topics in stem cell research today is the study of induced pluripotent stem cells (iPSCs). These are adult cells (e.g., skin cells) that are "reprogrammed" to become pluripotent, i.e., behave like an embryonic stem cell. While these iPSCs share many of the same characteristics of embryonic stem cells, including the ability to give rise to all the cell types, the original iPSCs were produced by using viruses to insert extra copies of three to four genes known to be important in embryonic stem cells. It is not yet completely understood how these three to four "reprogramming" genes are able to induce pluripotency. In addition, recent studies have focused on alternative ways of reprogramming cells using methods that are safer for use in clinical settings.

Cell Sources for Reprogramming

The most frequently used sources for reprogramming are blood cells and fibroblasts, and taking cells from urine is less invasive. The latter method does not require a biopsy or blood sampling. Urine-derived stem cells had been differentiated into endothelial, osteogenic, chondrogenic, adipogenic, skeletal myogenic, and neurogenic lineages, without forming teratomas [12]. Their epigenetic memory is suited to reprogramming into iPSCs. However, few cells appear in urine, only low conversion efficiencies had been achieved, and the risk of bacterial contamination is relatively high. Another promising source of cells for reprogramming is mesenchymal stem cells derived from human hair follicles.

Classification by Potency

Induced stem cells (iSC) are stem cells derived from somatic, reproductive, pluripotent, or other cell types by deliberate epigenetic reprogramming. They are classified as totipotent (iTC), pluripotent (iPSC), or progenitor (multipotent, iMSC, also called an induced multipotent progenitor cell, iMPC, or unipotent, iUSC) according to their developmental potential and degree of dedifferentiation. Progenitors are obtained by so-called direct reprogramming or direct differentiation.

iPSC Strategy

Although additional research is needed, iPSCs are already useful tools for drug development and modeling of diseases, and scientists hope to use them in transplantation medicine. Viruses are currently used to introduce the reprogramming factors into adult cells, and this process must be carefully controlled and tested before the technique can lead to useful treatment for humans. In animal studies, the virus used to introduce the stem cell factors sometimes causes cancers. Researchers are currently investigating non-viral delivery strategies. In any case, this breakthrough discovery has created a powerful new way to "dedifferentiate" cells whose developmental fates had been previously assumed to be determined. In addition, tissues derived from iPSCs will be a nearly identical match to the cell donor and thus probably avoid rejection by the immune system. The iPSC strategy creates pluripotent stem cells that, together with studies of other types of pluripotent stem cells, will help researchers learn how to reprogram cells to repair damaged tissues in the human body.

Reprogramming Factors

Human somatic cells are made pluripotent by transducing them with factors that induce pluripotency (OCT 3/4, SOX2, Klf4, c-Myc, NANOG, and LIN28). This results in the production of iPSCs, which can differentiate into any cells of the three embryonic germ layers (mesoderm, endoderm, ectoderm). Reprogramming mechanisms are thus linked, rather than independent and are centered on a small number of genes. iPSC properties are very similar to ESCs. iPSCs have been shown to support the development of all-iPSC mice using a tetraploid (4n) embryo, the most stringent assay for developmental potential.

Rejuvenation Mechanism

An important advantage of iPSC over ESC is that they can be derived from adult cells, rather than from embryos. Therefore, it became possible to obtain iPSC from adult and even elderly patients. Reprogramming somatic cells to iPSC leads to rejuvenation. It was found that reprogramming leads to telomere lengthening and subsequent shortening after their differentiation back into fibroblast-like derivatives. Thus, reprogramming leads to the restoration of embryonic telomere length and hence increases the potential number of cell divisions [13].

Immune Response

However, because of the dissonance between rejuvenated cells and the surrounding niche of the recipient's older cells, the injection of his own iPSC usually leads to an immune response, which can be used for medical purposes or the formation of tumors such as teratoma. The reason has been hypothesized to be that some cells differentiated from ESC and iPSC in vivo continue to synthesize embryonic protein isoforms. So, the immune system might detect and attack cells that are not cooperating properly [14].

Teratoma Formation Mechanism

Teratoma formation by pluripotent stem cells may be caused by low activity of PTEN enzyme, reported to promote the survival of a small population (0, 1–5% of total population) of highly tumorigenic, aggressive, teratoma-initiating embryonic-like carcinoma cells during differentiation. The survival of these teratoma-initiating cells is associated with failed repression of Nanog as well as a propensity for increased glucose and cholesterol metabolism [15]. These teratoma-initiating cells also expressed a lower ratio of p53/p21 when compared to nontumorigenic cells. In connection with the above safety problems, the use iPSC for cell therapy is still limited.

However, they can be used for a variety of other purposes—including the modeling of disease, screening of drugs, and toxicity testing of various drugs.

It is interesting to note that the tissues grown from iPSCs, placed in the "chimeric" embryos in the early stages of mouse development, practically do not cause an immune response (after the embryos have grown into adult mice) and are suitable for autologous transplantation. At the same time, full reprogramming of adult cells in vivo within tissues by transitory induction of the four factors Oct4, Sox2, Klf4, and c-Myc in mice results in teratomas emerging from multiple organs. Furthermore, partial reprogramming of cells toward pluripotency in vivo in mice demonstrates that incomplete reprogramming entails epigenetic changes (failed repression of Polycomb targets and altered DNA methylation) in cells that drive cancer development [16].

3.1.3.3 Reprogramming Methods

Several techniques are widely recognized: transplantation of nuclei taken from somatic cells into an oocyte lacking its own nucleus, fusion of somatic cells with pluripotent stem cells, and transformation of somatic cells into stem cells, using the genetic material encoding reprogramming protein factors, recombinant proteins; microRNA, a synthetic, self-replicating polycistronic RNA and low-molecular weight biologically active substances [17].

Somatic Cell Nuclear Transfer

Induced totipotent cells can be obtained by reprogramming somatic cells with somatic cell nuclear transfer (SCNT). The process involves sucking out the nucleus of a somatic cell and injecting it into an oocyte that has had its nucleus removed [18]. Totipotent cells can be generated by SCNT using dermal fibroblasts nuclei from both a middle-aged 35 years-old male and an elderly, 75 years-old male, suggesting that age-associated changes are not necessarily an impediment to SCNT-based nuclear reprogramming of human cells. Unfortunately, the cells generated by this technology potentially are not completely protected from the immune system of the patient (donor of nuclei), because of mitochondrial DNA in cytoplasm, as a donor of oocytes, instead of the patient's mitochondrial DNA. This reduces their value as a source for autologous stem cell transplantation therapy, until it is not clear how it can induce an immune response of the patient upon treatment.

This technologic may have far-reaching clinical applications for overcoming cytoplasmic defects in human oocytes. For example, the technology could prevent inherited mitochondrial disease from passing to future generations. Mitochondrial genetic material is passed from mother to child. The nucleus from one human egg has been transferred to another, including its mitochondria, creating a cell that could be regarded as having two mothers. The eggs were then fertilized and the resulting embryonic stem cells carried the swapped mitochondrial DNA. As evidence that the technique is safe, the author of this method points to the existence of the healthy monkeys that are now more than 4 years old and are the product of mitochondrial transplants across different genetic backgrounds [19].

Induced Totipotent Cells without SCNT

Recently some researchers succeeded to get the totipotent cells without the aid of SCNT. Totipotent cells were obtained using the epigenetic factors such as oocyte germinal isoform of histone. Reprogramming in vivo, by transitory induction of the four factors Oct4, Sox2, Klf4, and c-Myc in mice, confers totipotency features. Intraperitoneal injection of such in vivo iPS cells generates embryo-like structures that express embryonic and extraembryonic markers [20]. Transplantation of pluripotent/embryonic stem cells into the body of adult mammals usually leads to the formation of teratomas, which can then turn into a malignant tumor teratocarcinoma. However, putting teratocarcinoma cells into the embryo at the blastocyst stage caused them to become incorporated in the cell mass and often produced a normal healthy chimeric animal.

Reprogramming with Reagents

The risk of cancer creates the need to develop methods for safer cell lines suitable for clinical use. An alternative approach is so-called "direct reprogramming"—transdifferentiation of cells without passing through the pluripotent state [21]. The basis for this approach was that 5-azacytidine—a DNA demethylation reagent—can cause the formation of myogenic, chondrogenic, and adipogenic clones in the immortal cell line of mouse embryonic fibroblasts and that the activation of a single gene, later named MyoD1, is sufficient for such reprogramming. Compared with iPSC whose reprogramming requires at least 2 weeks, the formation of induced progenitor cells sometimes occurs within a few days, and the efficiency of reprogramming is usually many times higher. This reprogramming does not always require cell division. The cells resulting from such reprogramming are more suitable for cell therapy because they do not form teratomas.

Single Transcription Factor

Originally only early embryonic cells could be coaxed into changing their identity. Mature cells are resistant to changing their identity once they've committed to a specific kind. However, a brief expression of a single transcription factor, ELT-7 GATA, can convert the identity of fully differentiated, specialized non-endodermal cells of the pharynx into fully differentiated intestinal cells in intact larvae and adult roundworm *Caenorhabditis elegans* with no requirement for a dedifferentiated intermediate.

CRISPR-Mediated Activator

The cell fate can be effectively manipulated by directly activating specific endogenous gene expression with CRISPR-mediated activator. When dCas9 is combined with transcription activators, it can precisely manipulate endogenous gene expression. Using this method, Wei et al. enhanced the expression of endogenous Cdx2 and Gata6 genes by CRISPR-mediated activators, and thus directly converted mouse embryonic stem cells into two extraembryonic lineages, i.e., typical trophoblast stem cells and extraembryonic endoderm cells [22]. An analogous approach was used to induce activation of the endogenous Brn2, Ascl1, and Myt1 genes to convert mouse embryonic fibroblasts to induced neuronal cells. Thus, transcriptional activation and epigenetic remodeling of endogenous master transcription factors are sufficient for conversion between cell types. The rapid and sustained activation of endogenous genes in their native chromatin context by this approach may facilitate reprogramming with transient methods that avoid genomic integration and provides a new strategy for overcoming epigenetic barriers to cell fate specification.

Antibody-Based Reprogramming

The researchers discovered that GCSF-mimicking antibody can activate a growth-stimulating receptor on marrow cells in a way that induces marrow stem cells that normally develop into white blood cells to become neural progenitor cells [23]. The technique enables researchers to search large libraries of antibodies and quickly select the ones with a desired biological effect.

Conditionally Reprogrammed Cells

Schlegel and Liu [24] demonstrated that the combination of feeder cells and a Rho kinase inhibitor induces normal and tumor epithelial cells from many tissues to proliferate indefinitely in vitro. This process occurs without the need for transduction of exogenous viral or cellular genes. These cells have been termed "conditionally reprogrammed cells (CRC)." The induction of CRCs is rapid and results from reprogramming of the entire cell population. CRCs do not express high levels of protein characteristic of iPSCs or embryonic stem cells. This induction of CRCs is reversible and feeders allow the cells to differentiate normally. CRC technology can generate 2×10^6 cells in 5–6 days from needle biopsies and can generate cultures from cryopreserved tissue and from fewer than four viable cells. CRCs retain a normal karyotype and remain nontumorigenic.

Lineage-Specific Enhancers

Differentiated macrophages can self-renew in tissues and expand long term in culture. Under certain conditions macrophages can divide without losing features they have acquired while specializing into immune cells. The macrophages achieve this by activating a gene network similar to the one found in embryonic stem cells. Single-cell analysis revealed that, in vivo, proliferating macrophages can derepress a macrophage-specific enhancer repertoire associated with a gene network controlling self-renewal. This happened when concentrations of two transcription factors named MafB and c-Maf were naturally low or were inhibited for a short time. Genetic manipulations that turned off MafB and c-Maf in the macrophages caused the cells to start a self-renewal program. The similar network also controls embryonic stem cell self-renewal but is associated with distinct embryonic stem cell-specific enhancers.

Indirect Lineage Conversion

Indirect lineage conversion is a reprogramming methodology in which somatic cells transition through a plastic intermediate state of partially reprogrammed cells (pre-iPSC), induced by brief exposure to reprogramming factors, followed by differentiation in a specially developed chemical environment (artificial niche) [25]. This method could be both more efficient and safer, since it does not seem to produce tumors or other undesirable genetic changes and results in much greater yield than other methods.

Outer Membrane Glycoprotein

A common feature of pluripotent stem cells is the specific nature of protein glycosylation of their outer membrane. That distinguishes them from the most nonpluripotent cells. The glycans on the stem cell surface respond rapidly to alterations in cellular state and signaling and are therefore ideal for identifying even minor changes in cell populations. Many stem cell markers are based on cell surface glycan epitopes including the widely used markers SSEA-3, SSEA-4, Tra 1–60, and Tra 1–81 [2]. Suila et al. [26] speculate that human stem cells, extracellular O-GlcNAc and extracellular O-LacNAc, play a crucial role in the fine-tuning of Notch signaling pathway, a highly conserved cell signaling system, that regulates cell fate specification, differentiation, left–right asymmetry, apoptosis, somitogenesis, and angiogenesis and plays a key role in stem cell proliferation.

Changes in outer membrane protein glycosylation are markers of cell states connected in some way with pluripotency and differentiation. The glycosylation change is apparently not just the result of the initialization of gene expression, but performs as an important gene regulator involved in the acquisition and maintenance of the undifferentiated state. For example, activation of glycoprotein ACA, linking glycosylphosphatidylinositol on the surface of the progenitor cells in human peripheral blood, induces increased expression of genes Wnt, Notch-1, BMI1, and HOXB4 through a signaling cascade PI3K/Akt/mTor/PTEN and promotes the formation of a self-renewing population of hematopoietic stem cells. Furthermore, dedifferentiation of progenitor cells induced by ACA-dependent signaling pathway leads to ACA-induced pluripotent stem cells, capable of differentiating in vitro into cells of all three germ layers.

Physical Approach

Cell adhesion protein E-cadherin is indispensable for a robust pluripotent phenotype. During reprogramming for iPS cell generation, N-cadherin can replace function of E-cadherin. These functions of cadherins are not directly related to adhesion because sphere morphology helps maintaining the "stemness" of stem cells. Moreover, sphere formation, due to forced growth of cells on a low attachment surface, sometimes induces reprogramming.

Physical cues, in the form of parallel microgrooves on the surface of cell-adhesive substrates, can replace the effects of small molecule epigenetic modifiers and significantly improve reprogramming efficiency. The mechanism relies on the mechanomodulation of the cells' epigenetic state. Specifically, "decreased histone deacetylase activity and upregulation of the expression of WD repeat domain 5 (WDR5)—a subunit of H3 methyltranferase—by microgrooved surfaces lead to increased histone H3 acetylation and methylation." Nanofibrous scaffolds with aligned fiber orientation produce effects similar to those produced by microgrooves, suggesting that changes in cell morphology may be responsible for modulation of the epigenetic state [27].

Substrate rigidity is also one of the important biophysical cue influencing neural induction and subtype specification. For example, soft substrates promote neuroepithelial conversion while inhibiting neural crest differentiation in a BMP4-dependent manner. Mechanistic studies revealed a multi-targeted mechanotransductive process involving mechanosensitive Smad phosphorylation and nucleocytoplasmic shuttling, regulated by rigidity-dependent Hippo/YAP activities and actomyosin cytoskeleton integrity and contractility [28].

A new method has been developed that turns cells into stem cells faster and more efficiently by "squeezing" them using 3D microenvironment stiffness and density of the surrounding gel. The technique can be applied to a large number of cells to produce stem cells for medical purposes on an industrial scale [29].

3.1.3.4 iPSCs Generate Systems

iPSCs were first generated by the introduction of the transcription factors Oct3/4, Sox2, Klf4, and c-Myc in cells maintained in culture conditions used for ESC [30]. Various introduction systems of these transcription factors have since been used by other investigators.

Viral System

Most iPSCs have been generated using retroviral and lentiviral vectors to introduce transcription factors into cells. However, there are concerns with using these viral vectors. Previous studies of retroviral infection of cells had suggested that retroviruses are silenced in these cells. Silencing is an epigenetic process that suppresses transcription. However, it was demonstrated that silencing was often incomplete and viral genes could still be expressed. A major consideration for using retroviruses to generate iPSCs is that the viruses integrate into the host DNA. Depending on the integration site, integration can have deleterious effects on the cells, altering gene expression and increasing the risk of tumor formation. Adenoviruses [31] and Sendai virus, an RNA virus [32], have been used as alternative vectors to transduce transcription factors because they do not integrate into the genomic DNA.

Sendai virus (SeV) vectors do not integrate into the host genome or alter the genetic information of the host cell [33]. The host cell can be cleared of the vectors and reprogramming factor genes (i.e., Oct4, Sox2, Klf4, and c-Myc) by exploiting the cytoplasmic nature of SeV and the functional temperature sensitivity mutations introduced into the key viral proteins. The reprogramming efficiency may vary among different cell types (~0.01–1%).

Viral-mediated introduction of transcription factors is very inefficient. A more efficient method was found to be the combination of lentiviruses and microRNAs (miRNAs) to reprogram cells [34]. These small RNAs bind to mRNA and either inhibit translation or cause degradation of transcripts. miRNA clusters, including miR-290-295 and miR-302-367, have been shown to enhance reprogramming of somatic cells into iPSCs.

miRNA System

Another approach was the use of miRNA mimics to enhance viral-mediated transduction of transcription factors. miRNA mimics are double-stranded modified RNAs that mimic mature miRNAs. miRNA mimics do not require a vector; they can be transfected directly into cells. The combined use of transcription factors and miRNA mimics produces more homogeneous iPSC clones [35]. An advantage of using miRNA mimics with transcription factors is that the transcription factor c-Myc, an oncogene, is not required.

Chemical Compound System

Several chemical compounds that modulate enzymes controlling epigenetic modifications have been evaluated for increasing the efficiency of transduction by transcription factors. DNA methyltransferase and histone deacetylase (HDAC) inhibitors were shown to potentiate the efficacy of transduction. The HDAC inhibitor valproic acid was the most effective, increasing reprogramming efficiency by 100-fold [36]. As with the miRNA mimics, the use of valproic acid eliminates the need to transduce with the oncogene c-Myc.

Kinase Inhibitor System

A number of other inhibitors of kinases, such as the glycogen synthase kinase-3 (GSK3) inhibitor CHIR99021 and the MEK inhibitor PD0325901, and other enzymes that are in pathways involved in pluripotency have also been shown to enhance the efficiency of reprogramming [37].

Plasmid System

Plasmid expression vectors have been used to introduce transcription factors [38], but the efficiency is low, and occasionally expression plasmids can integrate into genomic DNA.

Episome System

This system contains an optimized mixture of three episomal vectors with an oriP/EBNA-1 (Epstein-Barr nuclear antigen-1) backbone for delivering the reprogramming genes, Oct4, Sox2, Lin28, L-Myc, and Klf4. High transfection efficiency due to oriP/EBNA-1-mediated nuclear import and retention of vector DNA allows iPSC derivation in a single transfection. In addition, silencing of the viral promoter driving EBNA-1 expression and the loss of the episomes at a rate of ~5% per cell cycle due to defects in vector synthesis and partitioning allow the removal of episomal vectors from the iPSCs without any additional manipulation.

Advantages of reprogramming with episome are the following:

1. As a transgene-free and viral-free reprogramming system, it is a safe alternative to other reprogramming methods such as lentiviral delivery for all stages of iPSC research.
2. It allows the reprogramming of a variety of somatic cell types and provides flexibility in somatic cell selection.
3. It is optimized for feeder-free reprogramming, enabling defined and feeder-free reprogramming.
4. High transfection efficiency due to oriP/EBNA-1-mediated nuclear import and retention of vector DNA allows iPSC derivation in a single transfection.
5. The episomal vectors are removed from the iPSCs without any further manipulation due to the silencing of the viral promoter driving EBNA-1 expression and the loss of the episomes due to defects in vector synthesis and partitioning.
6. Do not require the use of small molecules for reprogramming.

Lipofectamine System

Lipofectamine transfection was developed to unleash the power of stem cells by providing a highly efficient (40–70%), cost-effective, deliver DNA into difficult-to-transfect stem cells, alternative method for electroporation. This advanced lipid nanoparticle technology minimizes the stress on cells caused by electroporation, simplifies the reprogramming workflow, and enables advanced gene editing technologies.

3.1.3.5 Mesenchymal Stem Cells, MSCs

MSCs are non-hematopoietic stem cells found in most tissues of the body. The largest populations are in bone marrow and cord blood, with Wharton's jelly, found in the umbilical cord, being a rich source. MSCs are heterogeneous, have multilineage potential, and are capable of differentiating into multiple cell types including adipocytes, chondrocytes, osteoblasts, and cardiomyocytes.

MSCs are the most commonly used cell type in clinical trials for stem cell therapy. Advantages of MSCs include: 1) they can be harvested from several tissues; 2) they have multilineage potential; 3) subpopulations easily isolated by cell sorting can be given; 4) they can attenuate inflammatory responses and secrete many bioactive molecules such as growth factors and cytokines (including angiogenic cytokines); 5) allogeneic (nonself) transplantation usually evokes a minimal immune response; and 6) they may activate local stem cells and repair mechanisms.

Morphology

MSC is derived from human bone marrow showing fibroblast-like morphology. MSCs contain a large, round nucleus with a prominent nucleolus, which is surrounded by finely dispersed chromatin particles, giving the nucleus a clear appearance. The remainder of the cell body contains a small amount of Golgi apparatus, rough endoplasmic reticulum, mitochondria, and polyribosomes. The cells, which are long and thin, are widely dispersed, and the adjacent extracellular matrix is populated by a few reticular fibrils [39].

Cell Sources

MSCs were originally isolated from bone marrow stroma but were subsequently found to be present in most tissues of the body including cord blood, adipose tissue, skin, and periodontal ligaments, which attach teeth to the jaw.

The youngest, most primitive MSCs can be obtained from umbilical cord tissue, namely, Wharton's jelly and the umbilical cord blood. However, MSCs are found in much higher concentration in the Wharton's jelly compared to cord blood. The umbilical cord is easily obtained after a birth and is normally thrown away and poses no risk for collection. The cord MSCs have more primitive properties than other adult MSCs obtained later in life, which might make them a useful source of MSCs for clinical applications.

Another rich source for mesenchymal stem cells is the developing tooth bud of the mandibular third molar. While considered multipotent, they may prove to be pluripotent. They eventually form enamel, dentin, blood vessels, dental pulp, and nervous tissues, a minimum of 24 other different unique end organs. Because of ease in collection at 8–10 years of age before calcification and minimal to no morbidity, they probably constitute a major source for research and multiple therapies.

Amniotic fluid also has been shown to be a rich source of stem cells. As many as 1 in 100 cells collected during amniocentesis have been shown to be a pluripotent mesenchymal stem cell (see detailed information at Sect. 3.1.3.6).

Adipose tissue is becoming an important source of MSCs because adipocytes are easily accessible and present in relatively high numbers in the body. It has been estimated that 1 g of adipose tissue yields 5000 MSCs, whereas bone marrow aspirate contains 100–1000 MSCs per ml [40].

Identification Markers

A cell can be classified as an MSC if it shows plastic adherent properties under normal culture conditions and has a fibroblast-like morphology. In fact, some argue that MSCs and fibroblasts are functionally identical [40, 41]. Furthermore, MSCs can undergo osteogenic, adipogenic, and chondrogenic differentiation ex vivo. The cultured MSCs also express on their surface CD44, CD73, CD90, and CD105 while lacking the expression of CD11b, CD14, CD19, CD34, CD45, CD79a, and HLA-DR surface markers [42].

Differentiation Capacity

The standard test to confirm multipotency is differentiation of the cells into osteoblasts, adipocytes, and chondrocytes as well as myocytes and neurons. MSCs have been seen to even differentiate into neuron-like cells [43], but there is lingering doubt whether the MSC-derived neurons are functional. The degree to which the culture will differentiate varies among individuals and how differentiation is induced, e.g., chemical vs. mechanical; and it is not clear whether this variation is due to a different amount of "true" progenitor cells in the culture or variable differentiation capacities of individuals' progenitors. The capacity of cells to proliferate and differentiate is known to decrease with the age of the donor, as well as the time in culture. Likewise, whether this is due to a decrease in the number of MSCs or a change to the existing MSCs is not known.

Immunomodulatory Effects

Numerous studies have demonstrated that human MSCs avoid allorecognition, interfere with dendritic cell and T-cell function, and generate a local immunosuppressive microenvironment by secreting cytokines [44]. It has also been shown that the immunomodulatory function of human MSC is enhanced when the cells are exposed to an inflammatory environment characterized by the presence of elevated local interferon-gamma levels [45].

Homogenous Isolation

The majority of modern culture techniques still take a colony-forming unit-fibroblast (CFU-F) approach, where raw unpurified bone marrow or Ficoll-purified bone marrow mononuclear cell is plated directly into cell culture plates or flasks. Mesenchymal stem cells, but not red blood cells or hematopoietic progenitors, are adherent to tissue culture plastic within 24–48 h.

Other flow cytometry-based methods allow the sorting of bone marrow cells for specific surface markers, such as STRO-1 [46]. STRO-1+ cells are generally more homogenous and have higher rates of adherence and higher rates of proliferation, but the exact differences between STRO-1+ cells and MSCs are not clear. Methods of immunodepletion using such techniques as MACS have also been used in the negative selection of MSCs.

The supplementation of basal media with fetal bovine serum or human platelet lysate is common in MSC culture. Prior to the use of platelet lysates for MSC culture, the pathogen inactivation process is recommended to prevent pathogen transmission [47].

3.1.3.6 Cord Blood and Amniotic Fluid-Derived Stem Cells

Cord blood and amniotic fluid contain several different types of stem cells, with the predominant cell type being HSC. The various cell types have different potencies, ranging from ESC-like cells to multipotent stem cells.

Cord Blood Stem Cells

Cord blood (and placental blood) has several benefits for clinical therapy. The tissue source is freely and widely available, with no extra risk to the donor and no ethical issues. Banking of cord blood can provide future stem cells for treatment of the donor later in life. These stem cells would have few of the problems of allogeneic stem cells, such as immune incompatibility or the risk of communicable disease transmission. There have been several clinical trials approved for the therapeutic use of cord blood, including the treatment of pediatric stroke and cerebral palsy.

Cord blood contains a large number of stem cells, therefore, substantial numbers of cells can be easily obtained without many passages. This has three advantages: 1) the decreased time in culture and number of cell divisions reduces the potential for chromosomal changes; 2) cord blood stem cells, although they can give rise to all three germ layers, do not seem to be tumorigenic; 3) the ability to rapidly generate large numbers of cells affords the use of high numbers of cells in therapies, which may increase efficacy and reduce the waiting time for treatment.

Amniotic Stem Cells [48]

Amniotic stem cells are the mixture of stem cells that can be obtained from the amniotic fluid as well as the amniotic membrane. They can develop into various tissue types including skin, cartilage, cardiac tissue, nerves, muscle, and bone. The cells also have potential medical applications, especially in organ regeneration (Figs. 3.1 and 3.2).

The stem cells are usually extracted from the amniotic sac by amniocentesis which occurs without harming the embryos. The use of amni-

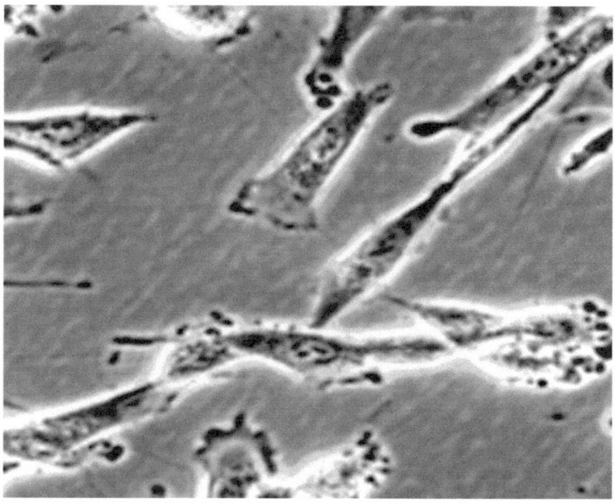

Fig. 3.1 Amniotic stem cells

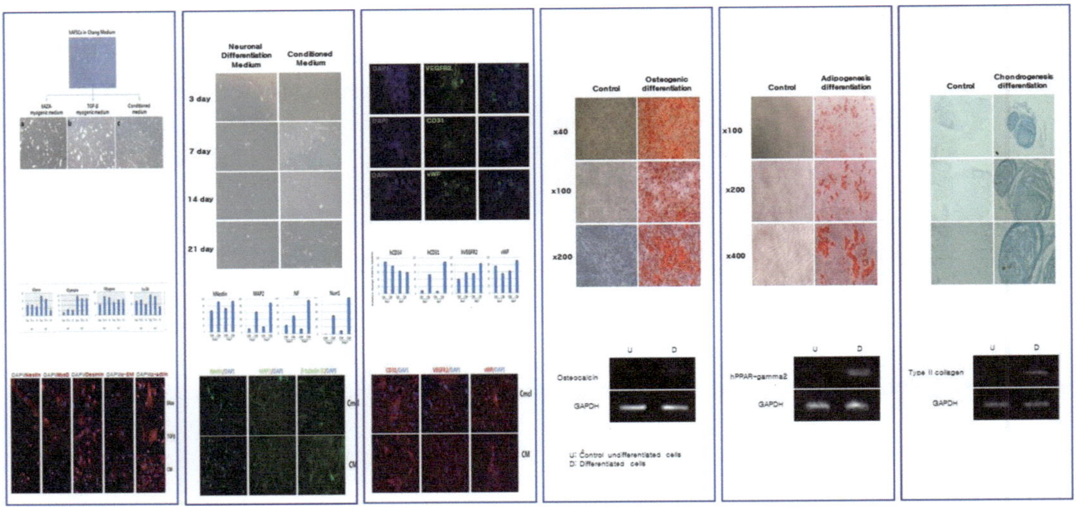

Fig. 3.2 Differentiation potency of amniotic fluid stem cells into muscle, neuron, endothelium, bone, fat, and chondrocyte

otic fluid stem cells is therefore generally considered to lack the ethical problems associated with the use of cells from embryos. The majority of stem cells present in the amniotic fluid share many characteristics, which suggests they may have a common origin.

This cell was confirmed that the amniotic fluid contains a heterogeneous mixture of multipotent cells after it was demonstrated that they were able to differentiate into cells from all three germ layers but they could not form teratomas following implantation into immunodeficient mice. This characteristic differentiates them from embryonic stem cells but indicates similarities with adult stem cells. However, fetal stem cells attained from the amniotic fluid are more stable and more plastic than their adult counterparts making it easier for them to be reprogrammed to a pluripotent state.

A variety of techniques has been developed for the isolation and culturing of amniotic stem cells. One of the more common isolation methods involves the removal of amniotic fluid by amniocentesis. The cells are then extracted from the fluid based on the presence of c-Kit. Several variations of this method exist. There is some debate whether c-Kit is a suitable marker to distinguish amniotic stem cells from other cell types because cells lacking c-Kit also display differentiation potential. Culture conditions may also be adjusted to promote the growth of a particular cell type.

Amniotic fluid highly contained MSCs and several techniques have been described for MSCs isolation. They usually involve the removal of amniotic fluid by amniocentesis, and their distinction from other cells may be based on their morphology or other characteristics. Human leukocyte antigen testing has been utilized to confirm that the MSCs are from the fetus and not from the mother. Comparison of amniotic fluid-derived MSCs to bone marrow-derived ones showed that the former has a higher expansion potential in culture. However, the cultured amniotic fluid-derived MSCs have a similar phenotype to both adult bone marrow-derived MSCs and MSCs originating from second-trimester fetal tissue(Fig. 3.3).

As like mesenchymal stem cells, embryonic-like stem cells also are abundant in the amniotic fluid. Embryonic-like stem cells were originally identified using markers common to embryonic stem cells such as nuclear Oct4, CD34, vimentin, alkaline phosphatase, stem cell factor, and c-Kit. However, these markers were not necessarily concomitantly expressed. In addition, all of these markers can occur on their own or in some combination in other types of cells. The pluripotency of these embryonic-like stem cells remains to be fully established. Although those cells which expressed the markers were able to differentiate into muscle, adipogenic, osteogenic, nephrogenic, neural, and endothelial cells, this did not necessarily occur from a homogenous population of undifferentiated cells. Evidence in favor of their embryonic stem cell nature is the cells' ability to produce clones (Figs. 3.4 and 3.5).

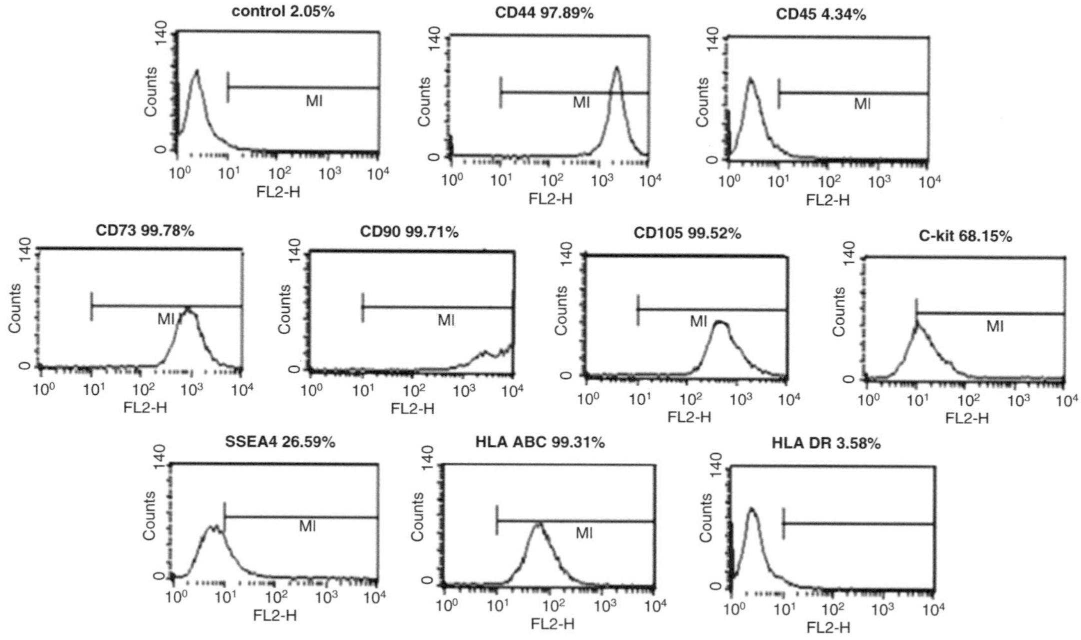

Fig. 3.3 Mesenchymal cell markers expression analysis of amniotic stem cells by FACS

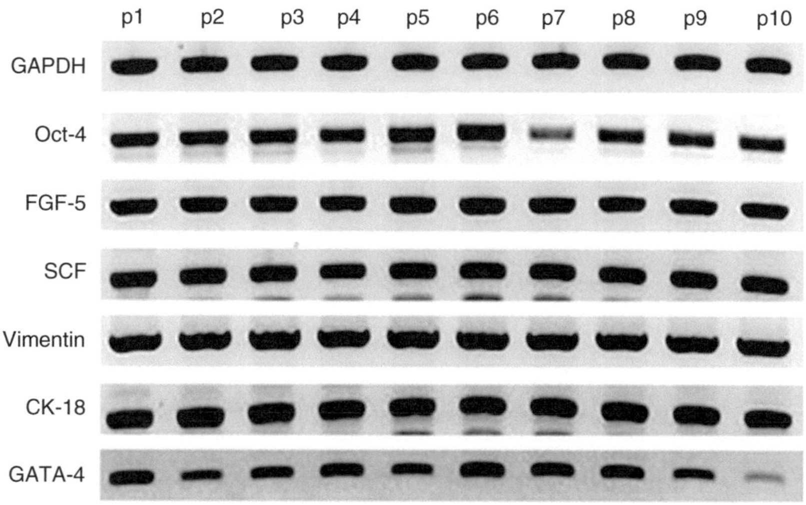

Fig. 3.4 Pluripotency marker expression of amniotic fluid stem cell to the passages

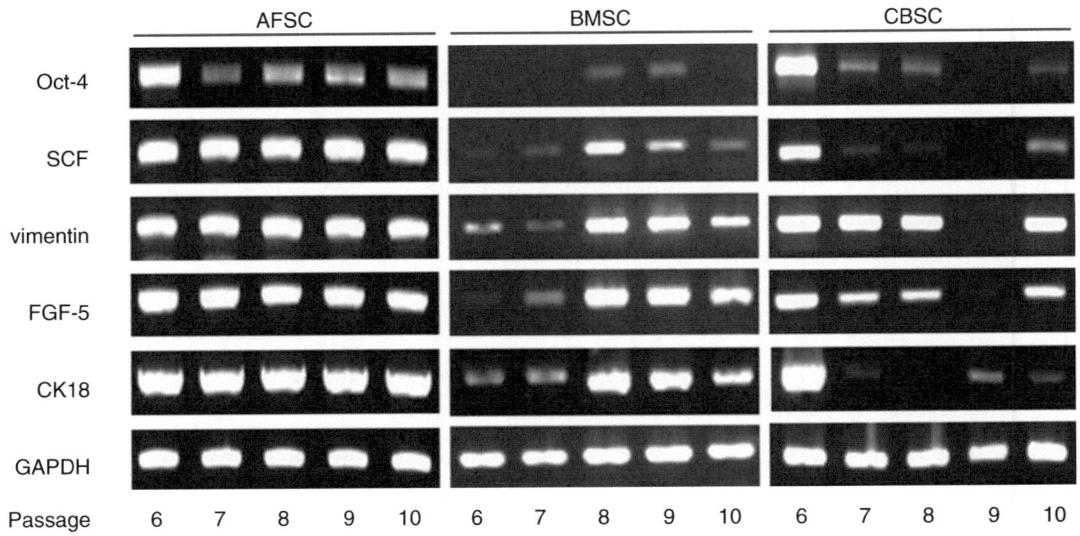

Fig. 3.5 Comparison of pluripotency marker expression among amniotic fluid, bone marrow, and umbilical cord-derived stem cell

3.2 Applications

3.2.1 Stem Cell Culture

3.2.1.1 Human Embryonic Stem Cell Culture Protocol

Isolation of Primary Mouse Embryo Fibroblasts

1. Dissect out the uterus of mouse (pregnant day 13.5, ICR) with sterile forceps and scissors, and place it into a Petri dish with PBS.
2. Isolate the embryos and transfer into a second Petri dish with PBS.
3. Remove the heads, livers, intestines, heart, and all viscera.
4. Transfer the carcasses to a fresh Petri dish, and mince with a blade.
5. Add 5.0 mL of 0.25% trypsin/EDTA per ten fetuses, and triturate through a 10 mL pipette.
6. Transfer the solution to a 10 mL syringe with 18 G needle, slowly push the solution through the needle (two times), and collect in a 50 mL conical tube.
7. Incubate for 15 min at 37°C, triturating three or four times using a 10 mL pipette.
8. Add an equal volume of complete MEF medium.
9. Plate to T175 flask, and add complete MEF medium to make up a final volume of 30 mL.
10. The dissociated MEFs will attach to the flask and begin to divide overnight.
11. Replace the medium the next day with an equal volume of fresh complete MEF medium.
12. 3–4 days after, rinse the cells with Ca^{2+}/Mg^{2+} free PBS.
13. Add 3 mL of 0.05% trypsin/EDTA solution per flask. Inactivate the trypsin solution by adding 5 mL complete MEF medium.
14. Count the number of viable cells using trypan blue exclusion and a hemocytometer. Viability should be between 90 and 95%.
15. Centrifuge the cell suspension at room temperature for 4 min at 200 g.
16. Aspirate medium and resuspend the cells in complete MEF medium (freezing concentration, 2.4×10^7/mL). Add an equal volume of fetal bovine serum with 20% DMSO to get 1x final stocks. Dispense 1 mL per cryovial and freeze.

Preparation of MEF-Conditioned Medium, MEF-CM

1. Plate 4×10^6 mitomycin C-treated MEFs in a T75 flask coated with 0.5% gelatin, in complete MEF medium.

2. The next day, replace the MEF medium with 37.5 ml 20% KSR hESC medium (containing 4 ng/ml bFGF), and incubate for 24 h at 37°C, 5% CO_2.
3. Collect MEF-CM from the flasks after 24 h and 0.22 μM filter sterilize. MEF-CM can be used fresh or can be frozen. Collect MEF-CM for up to 7 days using this procedure.

Microdissection Passaging of hESCs

1. Warm MEF-CM to 37°C in water bath, and add freshly thawed 1 M ß-mercaptoethanol to a final concentration of 0.1 mM.
2. Assess the undifferentiated colonies from differentiated colonies. Use a fine pointed tool (fire drawn Pasteur pipette needles), and score four to five times across a colony and perpendicular to that another four to five times. This creates a grid that cuts the colony into pieces.
3. Use a P1000 pipette tip to lift the colony pieces, and put them into MEF-CM.

3.2.1.2 Human-Induced Pluripotent Stem Cell Generation Protocol

Preparation of Fibroblasts

1. Disinfect skin surface three times with povidone iodine, and administer lidocaine.
2. Push a disposable biopsy punch against the skin, deep enough to reach the fatty layer of subcutaneous tissue.
3. Put the tissue into a 100 mm dish containing FP medium (DMEM containing 10% FBS and 50 units and 50 mg/ml penicillin and streptomycin), and cut it into 1 mm pieces.
4. Transfer them to a 35 mm dish, put the coverslip onto the tissues, and then add 2 ml of FP medium.
5. Incubate at 37°C, 5% CO_2. Leave the dish untouched for a week, and later change the medium twice a week.
6. When cells cover 30–50% of the dish, it's time to transfer the cells to a larger dish.

Lentivirus Production

1. Passaging 293FT cells: Adjust the concentration to 4×10^5 cells per milliliter with 293FT medium without antibiotics. Seed cells at 4×10^6 cells (10 ml) per 100 mm dish, and incubate overnight at 37°C, 5% CO_2.
2. Transfection to 293FT cells: Dilute 9 μg of Virapower packaging mix (pLP1, pLP2, and pLP/VSVG mixture) and 3 μg of pLenti6/UbC encoding the mouse Slc7a1 gene in 1.5 ml of OPTI-MEM I, and mix gently. In a separate tube, dilute 36 μl of Lipofectamine 2000 in 1.5 ml of OPTI-MEM I. Mix gently. After 5 min incubation, combine the diluted DNA with the diluted Lipofectamine 2000. During 20 min incubation, remove the medium from 293FT dishes, and add 9 ml of fresh medium to each dish. Then add 3 ml of the DNA-Lipofectamine 2000 complexes in each dish. Twenty-four hours after transfection, aspirate the medium containing the transfection cocktail, and add 10 ml of fresh FP medium.
3. Collection of virus-containing supernatant: 48 h after transfection, collect the supernatant of the 293FT culture with a 10 ml disposable syringe, and then filtrate it with a 0.45 μm pore size cellulose acetate filter.

Lentiviral Infection

1. Seeding fibroblasts: Adjust the concentration to 8×10^4 cells/ml by adding appropriate volume of FP medium. Seed the cells at 8×10^5 cells (10 ml of cell suspension) per 100 mm dish. Incubate overnight at 37°C, 5% CO_2.
2. Transduction of fibroblasts: Replace medium with 10 ml/dish of the virus-containing supernatant, supplemented with 4 μg/ml Polybrene.
3. Twenty-four hours after transduction, aspirate off the virus-containing medium, wash cells, and add 10 ml of fresh FP medium.

Preparation of SNL Feeder Cells

1. Thawing SNL cells: Suspend the cells with 10 ml of SNL medium, and transfer to a gelatin-coated 100 mm dish. Incubate the cells in a 37°C, 5% CO_2 incubator, until the cells become 80–90% confluent.
2. Mitomycin C-inactivation of SNL cells: Add 0.3 ml of 0.4 mg/ml mitomycin C solution directly to the culture medium of SNL dish,

and incubate 2.25 h at 37°C, 5% CO_2. Then wash the cells twice with 10 ml of PBS. Add 0.5 ml of 0.25% trypsin/1 mM EDTA, and let sit for 1 min at room temperature. Neutralize the trypsin by adding 5 ml of SNL medium, and seed the cells at 1.5×10^6 cells per 100 mm dish.

Preparation of Plat-E Cells

1. Thawing Plat-E cells: Suspend the cells with 10 ml of FP medium, and transfer to a 100 mm dish. Incubate the cells in a 37°C, 5% CO_2 incubator. The next day, replace the medium with 1 μg/ml of puromycin and 10 μg/ml of blasticidin S and incubate until 80–90% confluent.
2. Passage of Plat-E cells: Seed cells on new 100 mm dishes at 1:4–1:6 dilution. Cells should become confluent within 2–3 days.

Generation of iPS Cells

1. Day 1 Plat-E preparation: Count cell number and adjust the concentration to 3.6×10^5 cells/ml with FP medium.
2. Day 2 Retrovirus production (transfection into Plat-E cells): Transfer 0.3 ml of OPTI-MEM I into a 1.5 ml tube. Deliver 27 μl of transfection reagent into the prepared tube, and incubate for 5 min. Add 9 μg of pMXs plasmid DNA (respectively encoding Oct3/4, Sox2, Klf4, and c-Myc) drop by drop, and incubate for 15 min. Add the DNA and incubate overnight.
3. Day 3 Retrovirus production: Aspirate the transfection reagent-containing medium and add 10 ml of fresh FP medium.
4. Preparation of fibroblasts: When human fibroblasts expressing mouse Slc7a1 gene reach to 80–90% confluence, aspirate medium and wash once with 10 ml of PBS. Add 1 ml per dish of 0.05% trypsin/0.53 mM EDTA, add 9 ml of the FP medium per plate, and suspend the cells to a single cell. Count cell numbers to 8×10^4 cells/ml with FP medium.
5. Day 4 Retroviral infection: 48 h post-transfection, collect the medium from each Plat-E dish, filter it through a 0.45 μm pore size cellulose acetate filter, and transfer into a 15 ml conical tube. Add 5 μl of 8 mg/ml Polybrene solution into the filtrated virus-containing medium. Make a mixture of equal parts of the medium containing OCT3/4, SOX2, KLF4, and c-MYC retroviruses. Aspirate the medium from fibroblast dishes, and add 10 ml per dish of the Polybrene/virus-containing medium. Incubate the cells for 4 h.
6. Day 5–10: After 4 h–overnight culture, aspirate the medium from the transduced fibroblasts, and add 10 ml per dish of fresh FP medium.
7. Day 11 Replating transduced fibroblasts onto mitomycin C-treated SNL feeder: Aspirate the culture medium and wash with 10 ml per dish of PBS. Add 1 ml per dish of 0.05% trypsin/0.53 mM EDTA, add 9 ml per dish of FP medium, and suspend the cells. Count cell numbers, and adjust the concentration to 5×10^3 or 5×10^4 cells/ml.
8. Day 12~: Colonies should become visible 2–3 weeks after the retroviral infection. They can be picked up at around day 30.
9. Picking up the iPS colonies: Pick colonies using a P2 or P10 Pipetman L, transfer it into the 96-well plate with 20 μl ES medium, and add 180 μl. Pipet up and down to break up the colony to small clumps, transfer cell suspension into 24-well plates with mitomycin C-treated SNL feeder cells, add 300 μl per well of human ES medium, and incubate until the cells reach 80–90% confluence.
10. Passaging of iPS cells: After washing with PBS, add 0.1 ml per well of CTK solution and incubate at 37°C for 2–5 min. Washing with PBS, add 0.5 ml of human ES medium and suspend the cells to small clumps of 20–30 cells by pipetting. Transfer to a 6-well plates with mitomycin C-treated SNL feeder cells, add 1.5 ml of human ES medium, and incubate until cells become 80–90% confluent in 6-well plates.
11. Expansion of iPS cells: After washing with PBS, add 0.5 ml per well of CTK solution and incubate at 37°C for 2–5 min. Washing with PBS, add 2 ml of human ES medium. Detach iPS cells by using cell scraper and break up the colonies to small clumps by

pipetting. Transfer the cell suspension to a 60 mm dish with SNL feeder. Add 2 ml human ES medium, and incubate until cells reach 80–90% confluence.

3.2.1.3 Mesenchymal Stem Cell Culture Protocol

Mesenchymal Stem Cell Isolation from Human Amniotic Membrane

1. Prepare phosphate-buffered saline solution (PBS), Dulbecco's Modified Eagle's Medium, and high glucose (HG-DMEM), and add 10% fetal bovine serum (FBS), 1% antibiotic-antimycotic solution, and a collagenase II solution at 100 U/ml in HG-DMEM without FBS or antibiotic-antimycotic solution.
2. Sterilize the surgical instruments by autoclaving (standard scissors, simple dissecting forceps, fine point and toothed forceps, and a scalpel handle).
3. Hold the umbilical cord, detach the amniotic membrane (AM) to obtain a fragment of approximately 5 cm^2, and remove residual blood with three washes of 5 ml of PBS. Make a small cut in the AM with the scalpel to generate fragments of approximately 0.5 cm^2.
4. Incubate the AM with 5 ml of 0.125% trypsin/0.5 mM EDTA solution at 37°C for 30 min in a 50 ml conical tube at 37°C in a 5% CO_2. Centrifuge at 250 g and 25°C for 5 min and then remove the trypsin solution.
5. Add 15 ml of collagenase type II solution and incubate for 2 h at 37°C. Immediately after digestion, add 15 ml of PBS and centrifuge at 250 g and 25°C for 10 min. Discard the supernatant and wash the cellular pellet with 15 ml of PBS solution and centrifuge at 250 g and 25°C for 15 min. Discard the supernatant. Resuspend the cellular pellet in 10 ml of HG-DMEM by shaking gently.
6. Seed 1×10^4 cells/cm^2 in culture flask, add HG-DMEM, and incubate the cells at 37°C in a 5% CO_2. Eliminate the nonadherent cell by changing the culture medium after 5–7 days. Change the culture medium every 3 days for an additional 7–10 days to obtain a cellular confluence of 90%.
7. Recover the cells and incubate them for 5 min at 37°C in a 5% CO_2 with 0.125% trypsin/EDTA solution. Then, collect cells and transfer to a 15 ml conical tube and centrifuge at 250 g for 5 min. Discard the supernatant and resuspend the cellular pellet in 10 ml of HG-DMEM. Culture cell suspension in two 75 cm^2 culture flasks (5 ml in each flask), and maintain the culture until 90% confluence. Repeat steps until the 9th passage or subculture.

Mesenchymal Stem Cell Isolation from Adipose Tissue

1. Wash ~250 mL of fat 3–5 times with PBS for 5 min in each wash, add collagenase, and incubate for 1–4 h at 37°C on a shaker.
2. Add 10% FBS to neutralize collagenase, and centrifuge at 800 g for 10 min.
3. Aspirate floating adipocytes, lipids, and liquid, leaving stromal vascular fraction (SVF) pellet, resuspend in 160 mM NH_4Cl, and incubate for 10 min at room temperature. Centrifuge at 400 g for 10 min at room temperatures.
4. Layer cells on Percoll gradient, and centrifuge at 1000 g for 30 min at room temperature. Wash cells twice with PBS and centrifuge at 400 g for 10 min each wash. Resuspend cell pellet in PBS and filter cells through 100 μM nylon mesh. Pass cells through 400 μM mesh. Centrifuge at 400 g for 10 min.
5. Resuspend cell pellet in 40% FBS/DMEM and plate, and incubate at 37°C, 5% CO_2 incubator overnight.

Mesenchymal Stem Cell Isolation from Cord Blood

1. Dilute cord blood with RPMI Medium 1640 with a 3:1 ratio (three-part cord blood to one part RPMI).
2. Using Ficoll-Paque, isolate mononuclear cells (MNCs) by density gradient centrifugation at 400 g for 30 min at room temperature.
3. Transfer MNCs to new centrifuge tube and add PBS with a 1:3 ratio (1 part MNCs to 3 parts PBS). Centrifuge at 400 g for 10 min at room temperature.
4. Remove supernatant and resuspend cells by adding culture medium and plate. Incubate at 37°C, 5% CO_2 incubator overnight.

Mesenchymal Stem Cell Isolation from Amniotic Fluid

1. Collect amniotic fluid from 8 s-trimester pregnancies.
2. Four the AF samples were centrifuge 4 mL of amniotic fluid for 10 min at 300 g.
3. Pellets were resuspended and cultured in DMEM with low glucose supplemented with 15% fetal bovine serum and antibiotics.
4. Culture flasks were incubated at 37°C, 5% CO_2.
5. Medium was changed every 3–4 days until day 14.
6. Treatment with trypsin/EDTA was applied to detach the cells from the bottom of the flasks (passage 0). Cell number and vitality were determined and analyzed by flow cytometry using specific CD73+/CD44+/CD45 cell surface markers.

Mesenchymal Stem Cell Isolation from Mouse Bone Marrow

1. Prepare six 2 weeks-old BALB/c mice; MSCs were collected from femoral and tibial bone marrow of three mice by inserting a 26-gauge syringe at bone cavity, washing it with 10 ml of Knockout DMEM.
2. After centrifuging at 1200 rpm for 10 min, bone marrow cells were resuspended in 1 ml at the corresponding culture media for cell counting and cell feasibility verification
3. Cultures were maintained at 37°C with 5% CO_2.
4. After 72 h of culture, the medium was refreshed in an interval of 3–4 days.

Mesenchymal Stem Cells Isolated from Peripheral Blood

1. The peripheral blood-derived MSCs (PB-MSCs) were obtained from mononuclear cells by density gradient centrifugation, plated at a density of $4 \times 10^5/cm^2$ in 25 cm^2 flasks in growth medium (GM) containing 10% fetal bovine serum (FBS) and 100 units/ml penicillin/streptomycin (pen/strep) in high-glucose Dulbecco's Modified Eagle's Medium and cultured in a humidified atmosphere at 37°C with 5% CO_2.
2. The medium was replaced twice a week and nonadherent cells were discarded.
3. After 2 weeks one colony of adherent fibroblast-like cells was noticed in one of the seeded specimens. When the colony reached the approximate size of 5 cm^2, cells were detached and seeded in a new flask in GM.
4. Cells reached confluence after 10 days. Following first confluence, cells were passaged regularly.

3.2.2 Stem Cell Characteristic Analysis Methods

The characterization of stem cell has two purposes: monitoring the genomic integrity of the cells and tracking the expression of proteins associated with pluripotency. Genomic analysis is necessary to ensure stem cells maintained in culture have not undergone chromosomal changes through chromosomal loss or duplication or changes in their epigenetic profiles. Proteomic analysis ensures that the cells are expressing the factors necessary to maintain pluripotency.

Confirmation of the differentiated state by analyzing key genetic and protein markers ensures identification and propagation of the correct cell type. The general analysis methods for stem cells identification are karyotyping, single-nucleotide polymorphism (SNP) analysis, epigenetic profiling, flow cytometry, immunocytochemistry, RT-qPCR, western blotting, biomarker analysis, and teratoma formation in vivo.

3.2.2.1 Culture-Related Factors

Culture-Related Factors are media composition, cell density, feeder cell type/density, growth factors/additives, feeder-free culture, passage method, number of passages, freezing and thawing protocols, and microbial contamination

Possible Changes

The types of possible changes are chromosomal, phenotype/morphology, differentiation, pluripotency loss, epigenetic changes, tumorigenesis, and loss of self-renewal ability.

A number of techniques are used for genetic characterization of stem cells: karyotyping, fluorescence in situ hybridization (FISH), comparative genomic hybridization, single-nucleotide polymorphism (SNP) analysis, and epigenetic profiling. Some types of stem cells, particularly ESCs and iPSCs, are more prone to being genetically unstable and should be observed frequently for chromosomal changes.

Stem cells express both unique and specific combinations of transcription factors, cell surface proteins, and cytoplasmic proteins. Techniques used for stem cell analysis and characterization include flow cytometry, array-based analysis of the transcriptome, immunocytochemistry, western blots, and biomarker analysis.

Different classes and types of stem cells are characterized by different combinations of markers. In addition to confirming that stem cell lines are stable, markers can be used for screening during the reprogramming of somatic cells into iPSCs or to follow the progression of stem cell differentiation.

Colony-Forming Unit-Fibroblastic (CFU-F)

CFU-F assays were performed by plating MSCs (P3 to P6) in 6-well plates at 10, 50, and 100 cells/well in GM, in two replicas. After 14 days under standard culture conditions, the cells were fixed with ice-cold methanol and stained with 0.3% crystal violet. The number of visible colonies (more than 50 cells) was counted.

Karyotyping

Karyotyping is the examination of cellular chromosome number and morphology. Differences in appearance include size, position of centromeres, and changes in banding patterns. Stem cells in culture can accumulate changes including chromosomal abnormalities. Alterations in the genome can lead to changes in gene expression and cellular functions of stem cells [49] and increase the risk of the stem cells being tumorigenic. Thus, it is crucial to monitor a culture for the any chromosomal abnormalities, particularly in stem cells intended for therapeutic use.

Traditional karyotyping uses dye to stain the chromosomes of a metaphase cell in distinct banding patterns. The most common method is Giemsa staining, known as G-band karyotyping or G-banding; other methods include R-banding (reverse Giemsa staining), C-banding (constitutive heterochromatin staining), Q-banding (quinacrine staining), and T-banding (telomeric staining). Changes in banding patterns are used to identify abnormalities.

Fluorescence In Situ Hybridization (FISH)

In FISH, fluorescently labeled overlapping DNA fragments are hybridized to interphase or metaphase chromosomes. The chromosomes are formalin fixed onto a surface and chemically denatured. Fluorescence microscopy is used to detect and analyze the chromosomes. In fast FISH, chromosomes are denatured with high temperatures instead of chemically.

Spectral karyotyping (SKY) and multiplex fluorescence in situ hybridization (M-FISH) are commonly used forms of FISH for detection of changes. Both are hybridization-based combinatorial techniques using fluorescently labeled DNA probes. Each chromosome is labeled with a different combination of fluorophores specific for that chromosome, giving a unique spectral signature for each chromosome. Chromosomes are identified by their unique signatures, and then the output is pseudo-colored to aid in visualization. The difference between SKY and M-FISH is how the fluorescent signals are collected and processed. These methods can both detect translocations within and between chromosomes more accurately than traditional karyotyping; [50] however, they cannot detect inversions or duplications and deletions of less than approximately 5 Mb.

The main disadvantage of FISH for general screening of stem cells is that probes are required for the length of all the chromosomes. FISH is most commonly used for confirmation of the findings of other techniques using probes for a specific region of a chromosome. Deletions and insertions greater than ~20 kb are detected. Fast Fish is usually used for detecting specific aneuploidies that are commonly found in iPSCs and ESCs.

Comparative Genomic Hybridization

Comparative genomic hybridization can both determine chromosomal ploidy and detect copy number variations (CNVs), which are variable numbers of copies of segments of DNA. In this technique, DNA samples from the test stem cell population and from a reference population are each labeled with different probes. When this technique was first developed, reference and test DNA were denatured then hybridized to metaphase chromosomes. Using microscopy and fluorescence measurements, the chromosomes were scanned for any differences in fluorescent signals between the test and reference samples [51]. This technique detects and locates CNVs without requiring any knowledge of the particular sequence.

The disadvantage of using whole metaphase chromosomes is that the technique is not much more sensitive than G-banding. The adoption of techniques using microarrays with genomic clones has increased detection sensitivity and efficiency. Comparative genomic hybridization does not detect balanced translocations, inversions, or small gains or losses. Both mosaicism in a stem cell population and the presence of repetitive regions can complicate interpretation of results when using arrays. Due to the variability in some chromosomal regions among individuals, the preferred reference is either an early passage of the stem cells or another DNA sample extracted from the source of the stem cells.

SNP Analysis

Single-nucleotide polymorphisms (SNPs), single base-pair mutations within a region of DNA, are the most common type of genetic variation. SNPs may accumulate in a stem cell population over time and can lead to phenotypic changes that influence survival or growth. These genetic changes may result in unwanted outcomes such as loss of pluripotency or gain of tumorigenicity. SNP genotyping with high-density oligonucleotide arrays can specifically identify a stem cell line's origin and monitor its genomic integrity.

SNP arrays are used for detecting loss of heterozygosity (LOH). Normally for a chromosome pair, since the chromosomes come from two parents, there are differences between the sequences of the two chromosomes including many SNPs. When one copy of a region in a pair of chromosomes is lost, this results in LOH. Cultured hESC have been demonstrated to lose heterozygosity with passage. LOH can be associated with an increase in tumorigenic potential.

Aneuploidy, unbalanced translocations, deletions, and duplications can be detected with SNP arrays. Inversions are not detected and low levels of mosaicism may confound interpretation. As with comparative genomic hybridization, due to variability between individuals, choice of reference material is important.

Epigenetic Profiling

A stem cell's pluripotency is dictated to a large extent by its epigenetic profile. DNA methylation and histone modification regulate the accessibility of DNA to the transcriptional machinery and thus regulate gene expression. Analysis of DNA methylation patterns is performed by treating DNA with bisulfite under controlled conditions such that all cytosines are converted to uracils while leaving methylated cytosines unchanged. After conversion, the DNA can be analyzed for global methylation patterns using a chip array, PCR, or DNA sequencing. Chromatin immunoprecipitation (ChIP) analyzes patterns of histone modifications by cross-linking the DNA to histones. Cross-linked chromatin is then sheared and purified using antibodies against a specific protein or histone modification. Analysis of the cross-linked regions can be performed using real-time qPCR, microarrays, or DNA sequencing.

Quantitative assessment of chromatin structure in cultured cells can also be performed using specific epigenetics tools. Chromatin is digested in the presence or absence of nuclease, and then the genomic DNA is purified and quantified. Chromatin structure is assessed via real-time PCR, by comparing results against an epigenetically silenced gene to discriminate open, actively transcribed chromatin regions from closed, transcriptionally silent regions. This quantifies the impact of epigenetic events such as DNA methylation and histone modification on gene expression regulation through chromatin state changes.

Epigenetic profiling has shown that iPSCs can retain some epigenetic memory. As cells differentiate during development, there are epigenetic changes accompanying the changes in gene expression patterns. When a somatic cell is reprogrammed to become an iPSC, it has been observed in some iPSC cell lines that not all the epigenetic changes that occurred during differentiation are reversed during dedifferentiation.

Flow Cytometry

Flow cytometry is a commonly used technique used for the analysis of stem cells. Fluorescently tagged monoclonal antibodies can be used to distinguish specific cell populations, and cell sorting can be used to physically separate different stem cell populations. For example, when a population of somatic cells is being reprogrammed, iPSCs can be separated from cells either not reprogrammed or not fully reprogrammed. Another common use of cell sorting is during maintenance and expansion of stem cell culture. The development of mosaic populations is common in stem cells stocks. Sorting can be used to remove unwanted cell types and retain homogeneity of the population.

There are a large number of well-characterized cell surface markers that can be used both to determine cellular phenotypes and to separate mixed populations. Additionally, stem cells express a number of cell surface proteins that are used as markers of pluripotency and lineage. For example, the glycolipid antigens SSEA-3 and SSEA-4 and the keratin sulfate antigens TRA-1-60 and TRA-1-81 are commonly used antigens for identifying and sorting ESC populations [52]. Fixed stem cells can also be sorted based on the expression of the transcription factors such as Oct3/4 and NANOG. As stem cells differentiate, the loss of these markers and expression of new markers can be used to track the lineages and level of differentiation of the population.

Immunocytochemistry and Immunohistochemistry

Immunocytochemistry is predominantly used to confirm the presence or absence of protein expression in stem cells. Cells are incubated with antibodies targeting specific proteins prior to fluorescent imaging. In addition to providing information about the numbers of cells expressing particular proteins, information can be obtained about relative expression levels and the subcellular localization of the target proteins. Immunohistochemistry is also used to visualize stem cells in tissues, including the engraftment of stem cells after transplantation.

In animal models, stem cells can be transduced to express a protein that can be easily tracked. The visualization of a protein that is not expressed in a given tissue or organism permits discrimination between transplanted stem cells and any resident stem cells. Green fluorescent protein (GFP) can be tracked either directly by green fluorescence.

RT-PCR, RT-qPCR, and Digital PCR

Reverse transcription quantitative PCR (RT-qPCR) and RT-digital PCR provide rapid, sensitive, and quantitative methods for monitoring the gene expression profile of any cell population. The levels of transcription factors such as Oct4, NANOG, SOX2, and other synergistic factors determine whether pluripotency is maintained. Multiplexed RT-qPCR assays and panels provide an efficient way to screen and quantify transcription factors, kinases, and other molecules involved in both maintaining pluripotency and in differentiation. The ability to do multiplex transcriptome analysis increases the sensitivity of determining the status of cells.

Digital PCR can increase the accuracy of measurements with lower amounts of input nucleic acid. Droplet digital PCR allows for absolute quantification of transcripts and detection of rare sequences. In addition to being useful for transcriptome analysis, digital PCR can accurately detect CNVs and SNPs.

Western Blotting

Western blotting is useful for determining the success of transfection experiments. When a gene is either introduced into the cell or knocked down using antisense RNA or RNAi molecules, detection and quantitation of the presence and level of protein expression can be analyzed by western blotting. Importantly, western blots can be used to study the effects of transfection on downstream protein expression.

Biomarker Analysis

Biological processes can be controlled by protein interactions, relative protein levels, and post-translational protein modification. Not only can the presence or absence of proteins or signaling peptides be assessed, but relative levels can be determined. This technique can be used to evaluate pluripotency, lineages, and differentiation. The ability to multiplex allows for fine analysis of stem cells and differentiation pathways. Biological reactions and signaling networks can be monitored in real time. The use of bead-based multiplex immunoassays allows for simultaneous monitoring of protein, cytokines, and chemokines and can also assess the phosphorylation states of many proteins involved in signal transduction.

Embryoid Body and Teratoma Formation

Embryoid body (EB) and teratoma formation are used as indicators of stemness. Both will differentiate and contain cells from all three germ layers, demonstrating pluripotency. EBs will develop spontaneously in ESC or iPSC cultures upon withdrawal of factors that maintain pluripotency, when cultures become very dense or when suspended in media that promotes formation of EBs. Generation of teratomas is a method for functional analysis of pluripotent stem cells in vivo. Teratomas are tumors containing differentiated and undifferentiated cells from all three germ layers.

3.2.3 Cellular Differentiation Mechanisms

Specialized cell in an organism expresses a subset of all the genes that constitute the genome. Each cell type is defined by its particular pattern of regulated gene expression. Cell differentiation is thus a transition of a cell from one cell type to another, and it involves a switch from one pattern of gene expression to another. Cellular differentiation during development can be understood as the result of a gene regulatory network. A regulatory gene and its cis-regulatory modules are nodes in a gene regulatory network [53]. The systems biology approach to developmental biology emphasizes the importance of investigating how developmental mechanisms interact to produce predictable patterns.

A few conserved types of molecular processes are often involved in the cellular mechanisms that control these switches. The major types of molecular processes that control cellular differentiation involve cell signaling. The signal molecules that convey information from cell to cell during the control of cellular differentiation are called growth factors. The specific signal transduction pathways share the following general steps. A ligand produced by one cell binds to a receptor in the extracellular region of another cell. The shape of the cytoplasmic domain of the receptor changes, and the receptor acquires enzymatic activity. The receptor then catalyzes reactions that phosphorylate other proteins, activating them. A cascade of phosphorylation reactions eventually activates a dormant transcription factor or cytoskeletal protein, thus contributing to the differentiation process in the target cell [54]. Signal induction refers to cascades of signaling events.

Other important mechanisms fall under the category of asymmetric cell divisions. Asymmetric cell divisions can occur because of asymmetrically expressed maternal cytoplasmic determinants [55]. The distinct daughter cells are created during cytokinesis because of an uneven distribution of regulatory molecules in the parent cell. A well-studied example of pattern formation by asymmetric divisions is body axis patterning in *Drosophila*. RNA molecules are an important type of intracellular differentiation control signal.

3.2.3.1 Epigenetics in Stem Cell Differentiation

Since each cell possesses the same genome, determination of cell type must occur at the level of gene expression. While the regulation of gene expression can occur through cis- and trans-regulatory elements including a gene's promoter and enhancers, the problem arises as to how this expression pattern is maintained over numerous generations of cell division. As it turns out, epigenetic processes play a crucial role in regulating the decision to adopt a stem, progenitor, or mature cell fate.

Epigenetic Control

One of the important gene regulatory networks for determination of cell fate is of epigenetic processes. A paper by Lister et al. [56] is reported on aberrant epigenomic programming in human-induced pluripotent stem cells. As induced pluripotent stem cells (iPSCs) are thought to mimic embryonic stem cells in their pluripotent properties, few epigenetic differences should exist between them. To test this prediction, the authors conducted whole-genome profiling of DNA methylation patterns in several human embryonic stem cells (ESCs), iPSC, and progenitor cell lines.

Adipose cells, lung fibroblasts, and foreskin fibroblasts were reprogrammed into induced pluripotent state with the OCT4, SOX2, KLF4, and MYC genes. Patterns of DNA methylation in ESCs, iPSCs, somatic cells were compared. Results observed significant resemblance in methylation levels between embryonic and induced pluripotent cells. Around 80% of CG dinucleotides in ESCs and iPSCs were methylated; the same was true of only 60% of CG dinucleotides in somatic cells. In addition, somatic cells possessed minimal levels of cytosine methylation in non-CG dinucleotides, while induced pluripotent cells possessed similar levels of methylation as embryonic stem cells. Thus, consistent with their respective transcriptional activities, DNA methylation patterns, at least on the genomic level, are similar between ESCs and iPSCs. Thus, epigenetic processes are heavily involved in cell fate determination, as seen from the similar levels of cytosine methylation between induced pluripotent and embryonic stem cells, consistent with their respective patterns of transcription.

Mechanisms of Epigenetic Regulation

Three transcription factors, OCT4, SOX2, and NANOG, are highly expressed in undifferentiated embryonic stem cells and are necessary for the maintenance of their pluripotency [57]. It is thought that these factors' alterations in chromatin structure, such as histone modification and DNA methylation, can restrict or permit the transcription of target genes.

In the process of gene silencing, Polycomb repressive complex 2 catalyzes the di- and tri-methylation of histone H3 lysine 27 (H3K27me2/me3) [57]. By binding to the H3K27me2/3-tagged nucleosome, PRC1 catalyzes the mono-ubiquitinylation of histone H2A at lysine 119 (H2AK119Ub1), blocking RNA polymerase II activity and resulting in transcriptional suppression. PcG knockout ES cells do not differentiate efficiently into the three germ layers, and deletion of the PRC1 and PRC2 genes leads to increased expression of lineage-affiliated genes and unscheduled differentiation. Presumably, PcG complexes are responsible for transcriptionally repressing differentiation and development-promoting genes.

PcG proteins are recruited to promoters of pluripotency transcription factors. PcG-deficient ES cells can begin differentiation but cannot maintain the differentiated phenotype. Simultaneously, differentiation and development-promoting genes are activated by trithorax group (TrxG) chromatin regulators and lose their repression. TrxG proteins are recruited at regions of high transcriptional activity, where they catalyze the trimethylation of histone H3 lysine 4 and promote gene activation through histone acetylation. PcG and TrxG complexes engage in direct competition and are thought to be functionally antagonistic, creating at differentiation and development-promoting loci what is termed a "bivalent domain" and rendering these genes sensitive to rapid induction or repression.

Gene expression regulation is more progressed through DNA methylation, in which the DNA methyltransferase-mediated methylation of cytosine residues in CpG dinucleotides maintains heritable repression by controlling DNA accessibility [58]. The majorities of CpG sites in embryonic stem cells are unmethylated and appear to be associated with H3K4me3-carrying nucleosomes. Upon differentiation, a small number of genes, including OCT4 and NANOG, are methylated and their promoters repressed to prevent their further expression. Consistently, DNA methylation-deficient embryonic stem cells rapidly enter apoptosis upon in vitro differentiation.

The DNA sequence of most cells of an organism is the same, but the binding patterns of transcription factors and the corresponding gene expression patterns are different. To a large extent, differences in transcription factor binding are determined by the chromatin accessibility of their binding sites through histone modification and/or pioneer factors. In particular, it is important to know whether a nucleosome is covering a given genomic binding site or not. Recent studies have elucidated the role of nucleosome positioning during stem cell development [59].

Cell Signaling in Epigenetic Control

The cell signaling can give influence to the epigenetic processes governing differentiation. The first major candidate is Wnt signaling pathway. The Wnt pathway is involved in all stages of differentiation, and the ligand Wnt3a can substitute for the overexpression of c-Myc in the generation of induced pluripotent stem cells [60]. On the other hand, disruption of ß-catenin, a component of the Wnt signaling pathway, leads to decreased proliferation of neural progenitors.

The second candidate of epigenetic regulators of cellular differentiation is growth factors. These morphogens are crucial for development and include bone morphogenetic proteins, transforming growth factors (TGFs) and fibroblast growth factors (FGFs). TGFs and FGFs have been shown to sustain expression of OCT4, SOX2, and NANOG by downstream signaling to Smad proteins. Depletion of growth factors promotes the differentiation of ESCs, while genes with bivalent chromatin can become either more restrictive or permissive in their transcription.

As other pathways, cytokine leukemia inhibitory factors are associated with the maintenance of mouse ESCs in an undifferentiated state. This is achieved through its activation of the Jak-STAT3 pathway, which has been shown to be necessary and sufficient toward maintaining mouse ESC pluripotency [61]. Retinoic acid can induce differentiation of human and mouse ESCs, and Notch signaling is involved in the proliferation and self-renewal of stem cells. Sonic hedgehog, in addition to its role as a morphogen, also promotes embryonic stem cell differentiation and the self-renewal of somatic stem cells [60].

Matrix Elasticity Effect on Differentiation

Stem cells are known to migrate from their niches to regenerate variety of tissues by adhering to new extracellular matrices (ECM) and differentiate. The ductility of these microenvironments is unique to different tissue types. The ECM surrounding brain, muscle, and bone tissues range from soft to stiff. The transduction of the stem cells into these cells types is not directed solely by chemokine cues and cell to cell signaling. The elasticity of the microenvironment can also affect the differentiation of mesenchymal stem cells. When MSCs are placed on substrates of the same stiffness as brain, muscle, and bone ECM, the MSCs take on properties of those respective cell types [62].

Matrix sensing requires the cell to pull against the matrix at focal adhesions, which triggers a cellular mechano-transducer to generate a signal to be informed what force is needed to deform the matrix. To determine the key players in matrix-elasticity-driven lineage specification in MSCs, different matrix microenvironments were mimicked. Researchers have obtained some success in inducing stem cell-like properties in HEK 239 cells by providing a soft matrix without the use of diffusing factors [63]. The stem cell properties appear to be linked to tension in the cells' actin network. One identified mechanism for matrix-induced differentiation is tension-induced proteins, which remodel chromatin in response to mechanical stretch [64].

3.2.4 Stem Cell Therapy and Research

Stem cells become a major part of regenerative medicine. There are an increasing number of stem cell therapies being researched and thousands of clinical trials.

3.2.4.1 Transplantation of Stem Cells

Transplantation of stem cells into the body entails assessment of either morphologic repair or restoration of function. There are a number of requirements for a successful stem cell therapy: 1) easy to obtain and can readily be sorted and characterized; 2) will survive in the body, remaining viable, and not trigger a severe immune response; 3) able to proliferate, preferably in the target tissue; 4) can differentiate into the target tissue; 5) can integrate into the targeted tissue; 6) engrafted cells function appropriately; and 7) will at least partially restore function and repair damaged area.

Blood stem cells are currently the most frequently used stem cells for therapy. For more than 50 years, doctors have been using bone marrow transplants to transfer blood stem cells to patients, and more advanced techniques for collecting blood stem cells are now being used to treat leukemia, lymphoma, and several inherited blood disorders. Umbilical cord blood, like bone marrow, is often collected as a source of blood stem cells and in certain cases is being used as an alternative to bone marrow transplantation. Additionally, some bone, skin, and corneal diseases or injuries can be treated by grafting tissues that are derived from or maintained by stem cells. These therapies have also been shown to be safe and effective.

Other stem cell treatments are still at very early experimental stages. For example, the mesenchymal stem cell, found throughout the body including in the bone marrow, can be directed to become bone, cartilage, fat, and possibly even muscle. In certain experimental models, these cells also have some ability to modify immune functions. These abilities have created considerable interest in developing ways of using mesenchymal stem cells to treat a range of musculoskeletal abnormalities, cardiac disease, and some immune abnormalities such as graft-versus-host disease following bone marrow transplant.

3.2.4.2 For Research

Although much of the initial focus has been on stem cell models for disease, stem cells provide tractable models for the study of all aspects of cell biology. Stem cell models are being used to study many cell functions including: mechanisms and signals for stem cell proliferation (symmetric and asymmetric), signals that trigger migration, signals and pathways involved in differentiation, pathways for differentiation to different phenotypes, sequences of epigenetic changes during differentiation into different lineages and cell types, cell functions of stem and somatic cells under normal nonstressed conditions, changes in cell functions under stress, changes in cell functions upon stimulation, apoptosis, cell death, functions and behaviors of stem cells, roles of stem cells in normal development, growth, and repair.

3.2.4.3 For Disease

The generation of stem cell models from somatic cells provides the ability to develop stem cell lines for many diseases. Comparison of normal and disease stem cells will identify disease phenotype(s), investigate the development of the disease phenotype, profile control and disease cells at transcriptome and protein levels to reveal differences, measure functional changes, determine which cell types are affected by the disease, and identify disease biomarkers and potential new diagnostics.

The major advantage of iPS cells is that they are a very good way to make pluripotent stem cell lines that are specific to a disease or even to an individual patient. Disease-specific stem cells are powerful tools for studying the cause of a particular disease and then for testing drugs or discovering other approaches to treat or cure that disease. The development of patient-specific stem cells is also very attractive for cell therapy, as these cell lines are from the patient themselves and may minimize some of the serious complications of rejection and immunosuppression that can occur.

3.2.4.4 For Drug Discovery

The ability to generate large populations of cells with defined phenotypes is having an impact on drug discovery and testing. Stem cell-derived model systems can be used for all aspects of drug discovery, including: screening compounds for changes in known biomarkers, screening compounds or pools of compounds for changes in disease phenotype, evaluating the mechanism(s) of alteration of disease by lead compounds, toxicity screening, screening different cell lines to ensure applicability to a whole population rather than small subpopulations, safety and effects in nontarget cell types, in vitro clinical trials, and reduced animal testing (and associated costs).

Researchers are able to grow up differentiated cell lines and then test new drugs on each cell type to examine possible interactions in vitro before performing in vivo studies. This is critical in the development of drugs for use in veterinary research because of the possibilities of species-specific interactions. The hope is that having these cell lines available for research use will reduce the need for research animals used because effects on human tissue in vitro will provide insight not normally known before the animal testing phase.

3.2.4.5 Clinical Translation

Clinical translation is the process used to turn scientific knowledge into real-world medical treatments. Researchers take what they have learned about how a tissue usually works and what goes wrong in a particular disease or injury and use this information to develop new ways to diagnose, stop, or fix what goes wrong. Before being marketed or adopted as standard of care, most treatments are tested through clinical trials. Sometimes, in attempting new surgical techniques or where the disease or condition is rare and does not have a large enough group of people to form a clinical trial, certain treatments might be tried on one or two people, a form of testing sometimes referred to as innovative medicine. In recent years, stem cell clinics have emerged that treat patients with their own bone marrow or adipose-derived adult stem cells as part of clinical trials or FDA-authorized same-day outpatient IRB programs.

3.2.4.6 Challenges

Identifying, isolating, and growing the right kind of stem cell, particularly in the case of rare adult stem cells, are painstaking and difficult processes. Pluripotent stem cells, such as embryonic stem cells, can be grown indefinitely in the lab and have the advantage of having the potential to become any cell in the body, but these processes are again very complex and must be tightly controlled. iPS cells, while promising, are also limited by these concerns. In both cases, considerable work remains to be done to ensure that these cells can be isolated and used safely and routinely. In addition, it is very important to have a close match between the donor tissue and the recipient; the more closely the tissue matches the recipient, the lower the risk of rejection. Being able to avoid the life-long use of immunosuppressants would also be preferable. The discovery of iPS cells has opened the door to develop patient-specific pluripotent stem cell lines that can later be developed into a needed cell type without the problems of rejection and immunosuppression that occur from transplants from unrelated donors. Also, a system for delivering the cells to the right part of the body must be developed. Once in the right location, the new cells must then be encouraged to integrate and function together with the body's other cells.

3.2.4.7 Conservation

Stem cells are being explored for use in conservation efforts. Spermatogonial stem cells have been harvested from a rat and placed into a mouse host, and fully mature sperm were produced with the ability to produce viable offspring. Currently research is underway to find suitable hosts for the introduction of donor spermatogonial stem cells. If this becomes a viable option for conservationists, sperm can be produced from high genetic quality individuals who die before reaching sexual maturity, preserving a line that would otherwise be lost [65].

Glossary

Adult stem cell See somatic stem cell

Astrocyte: A type of supporting (glial) cell found in the nervous system

Blastocoel: The fluid-filled cavity inside the blastocyst, an early, preimplantation stage of the developing embryo

Blastocyst: A very early embryo that has the shape of a ball and consists of approximately

150–200 cells. It contains the inner cell mass, from which embryonic stem cells are derived, and an outer layer of cells called the trophoblast that forms the placenta

Bone marrow stromal cells: A population of cells found in bone marrow that are different from blood cells

Bone marrow stromal stem cells (skeletal stem cells): A multipotent subset of bone marrow stromal cells able to form bone, cartilage, and stromal cells that support blood formation, fat, and fibrous tissue

Cell-based therapies: Treatment in which stem cells are induced to differentiate into the specific cell type required to repair damaged or destroyed cells or tissues

Cell culture: Growth of cells in vitro in an artificial medium for research

Cell division: Method by which a single cell divides to create two cells. There are two main types of cell division depending on what happens to the chromosomes: mitosis and meiosis

Cell line: Cells that can be maintained and grown in a dish outside of the body

Chromosome: A structure consisting of DNA and regulatory proteins found in the nucleus of the cell. The DNA in the nucleus is usually divided up among several chromosomes. The number of chromosomes in the nucleus varies depending on the species of the organism. Humans have 46 chromosomes

Clinical translation: The process of using scientific knowledge to design, develop, and apply new ways to diagnose, stop, or fix what goes wrong in a particular disease or injury

Clone (v): To generate identical copies of a region of a DNA molecule or to generate genetically identical copies of a cell or organism; (n) the identical molecule, cell, or organism that result from the cloning process

Culture medium: The liquid that covers cells in a culture dish and contains nutrients to nourish and support the cells. Culture medium may also include growth factors added to produce desired changes in the cells

Differentiation: The process whereby an unspecialized embryonic cell acquires the features of a specialized cell such as a heart, liver, or muscle cell. Differentiation is controlled by the interaction of a cell's genes with the physical and chemical conditions outside the cell, usually through signaling pathways involving proteins embedded in the cell surface

Directed differentiation: The manipulation of stem cell culture conditions to induce differentiation into a particular cell type

DNA: Deoxyribonucleic acid, a chemical found primarily in the nucleus of cells. DNA carries the instructions or blueprint for making all the structures and materials the body needs to function. DNA consists of both genes and non-gene DNA in between the genes

Ectoderm: The outermost germ layer of cells derived from the inner cell mass of the blastocyst; gives rise to the nervous system, sensory organs, skin, and related structures

Embryo: The early developing organism; this term denotes the period of development between the fertilized egg and the fetal stage

Embryonic bodies: Rounded collections of cells that arise when embryonic stem cells are cultured in suspension. Embryonic bodies contain cell types derived from all three germ layers

Embryonic germ cells: Pluripotent stem cells that are derived from early germ cells (those that would become sperm and eggs). Embryonic germ cells are thought to have properties similar to embryonic stem cells

Embryonic stem cell: Cells derived from very early in development, usually the inner cell mass of a developing blastocyst. These cells are self-renewing (can replicate themselves) and pluripotent (can form all cell types found in the body)

Embryonic stem cell line: Embryonic stem cells, which have been cultured under in vitro conditions that allow proliferation without differentiation for months to years

Endoderm: The innermost layer of the cells derived from the inner cell mass of the blastocyst; it gives rise to lungs, other respiratory structures, and digestive organs or generally "the gut"

Enucleated: Having had its nucleus removed

Epigenetic: The process by which regulatory proteins can turn genes on or off in a way that can be passed on during cell division

Feeder layer: Cells used in co-culture to maintain pluripotent stem cells. For human embryonic stem cell culture, typical feeder layers include mouse embryonic fibroblasts (MEFs) or human embryonic fibroblasts that have been treated to prevent them from dividing

Fertilization: The joining of the male gamete (sperm) and the female gamete (egg)

Fetus: In humans, the developing human from approximately 8 weeks after conception until the time of its birth

Gamete: An egg (in the female) or sperm (in the male) cell. See also somatic cell

Gastrulation: The process in which cells proliferate and migrate within the embryo to transform the inner cell mass of the blastocyst stage into an embryo containing all three primary germ layers

Gene: A functional unit of heredity that is a segment of DNA found on chromosomes in the nucleus of a cell. Genes direct the formation of an enzyme or other protein

Germ layers: After the blastocyst stage of embryonic development, the inner cell mass of the blastocyst goes through gastrulation, a period when the inner cell mass becomes organized into three distinct cell layers, called germ layers. The three layers are the ectoderm, the mesoderm, and the endoderm

Hematopoietic stem cell: A stem cell that gives rise to all red and white blood cells and platelets

Human embryonic stem cell (hESC): A type of pluripotent stem cell derived from early-stage human embryos, up to and including the blastocyst stage. hESCs are capable of dividing without differentiating for a prolonged period in culture and are known to develop into cells and tissues of the three primary germ layers

Induced pluripotent stem (iPS) cell: Induced pluripotent cells (iPS cells) are stem cells that were engineered ("induced") from nonpluripotent cells to become pluripotent. In other words, a cell with a specialized function (e.g., a skin cell) that has been "reprogrammed" to an unspecialized state similar to that of an embryonic stem cell

Innovative medicine: Treatments that are performed on a small number of people and are designed to test a novel technique or treat a rare disease. These are done outside of a typical clinical trial framework

In vitro: Latin for "in glass"; in a laboratory dish or test tube; an artificial environment

In vitro fertilization: A procedure in which an egg cell and sperm cells are brought together in a dish to fertilize the egg. The fertilized egg will start dividing and, after several divisions, forms the embryo that can be implanted into the womb of a woman and give rise to pregnancy

Inner cell mass (ICM): The cluster of cells inside the blastocyst. These cells give rise to the embryo and ultimately the fetus. The ICM may be used to generate embryonic stem cells

Long-term self-renewal: The ability of stem cells to replicate themselves by dividing into the same nonspecialized cell type over long periods (many months to years) depending on the specific type of stem cell

Meiosis: The type of cell division a diploid germ cell undergoes to produce gametes (sperm or eggs) that will carry half the normal chromosome number. This is to ensure that when fertilization occurs, the fertilized egg will carry the normal number of chromosomes rather than causing aneuploidy (an abnormal number of chromosomes)

Mesenchymal stem cells: Mesenchymal stem cells were originally discovered in the bone marrow, but have since been found throughout the body and can give rise to a large number of connective tissue types such as bone, cartilage, and fat

Mesoderm: Middle layer of a group of cells derived from the inner cell mass of the blastocyst; it gives rise to bone, muscle, connective tissue, kidneys, and related structures

Microenvironment: The molecules and compounds such as nutrients and growth factors in the fluid surrounding a cell in an organism or in the laboratory, which play an important role in determining the characteristics of the cell

Mitosis: The type of cell division that allows a population of cells to increase its numbers or to maintain its numbers. The number of chromosomes in each daughter cell remains the same in this type of cell division

Multipotent stem cells: Stem cells that can give rise to several different types of specialized cells, but in contrast to a pluripotent stem cell, are restricted to a certain organ or tissue types. For example, blood stem cells are multipotent cells that can produce all the different cell types that make up the blood but not the cells of other organs such as the liver or brain

Neural stem cell: A stem cell found in adult neural tissue that can give rise to neurons and glial (supporting) cells. Examples of glial cells include astrocytes and oligodendrocytes

Neurons: Nerve cells, the principal functional units of the nervous system. A neuron consists of a cell body and its processes: an axon and one or more dendrites. Neurons transmit information to other neurons or cells by releasing neurotransmitters at synapses

Oligodendrocyte: A supporting cell that provides insulation to nerve cells by forming a myelin sheath (a fatty layer) around axons

Parthenogenesis: The artificial activation of an egg in the absence of a sperm; the egg begins to divide as if it has been fertilized

Passage: In cell culture, the process in which cells are disassociated, washed, and seeded into new culture vessels after a round of cell growth and proliferation. The number of passages a line of cultured cells has gone through is an indication of its age and expected stability

Pluripotent stem cells: Stem cells that can become all the cell types that are found in an implanted embryo, fetus, or developed organism. Embryonic stem cells are pluripotent stem cells. Scientists demonstrate pluripotency by providing evidence of stable developmental potential, even after prolonged culture, to form derivatives of all three embryonic germ layers from the progeny of a single cell and to generate a teratoma after injection into an immunosuppressed mouse

Polar body: A polar body is a structure produced when an early egg cell, or oogonium, undergoes meiosis. In the first meiosis, the oogonium divides its chromosomes evenly between the two cells but divides its cytoplasm unequally. One cell retains most of the cytoplasm, while the other gets almost none, leaving it very small. This smaller cell is called the first polar body. The first polar body usually degenerates. The ovum, or larger cell, then divides again, producing a second polar body with half the amount of chromosomes but almost no cytoplasm. The second polar body splits off and remains adjacent to the large cell, or oocyte, until it (the second polar body) degenerates. Only one large functional oocyte, or egg, is produced at the end of meiosis

Preimplantation: With regard to an embryo, preimplantation means that the embryo has not yet implanted in the wall of the uterus. Human embryonic stem cells are derived from preimplantation-stage embryos fertilized outside a woman's body (in vitro)

Proliferation: Expansion of the number of cells by the continuous division of single cells into two identical daughter cells

Regenerative medicine: A field of medicine devoted to treatments in which stem cells are induced to differentiate into the specific cell type required to repair damaged or destroyed cell populations or tissues. (See also cell-based therapies)

Reproductive cloning: The process of using somatic cell nuclear transfer (SCNT) to produce a normal, full grown organism (e.g., animal) genetically identical to the organism (animal) that donated the somatic cell nucleus. In mammals, this would require implanting the resulting embryo in a uterus where it would undergo normal development to become a live independent being. The first mammal to be created by reproductive cloning was Dolly the sheep, born at the Roslin Institute in Scotland in 1996. See also somatic cell nuclear transfer (SCNT)

Self-renewal: The process by which a cell divides to generate another cell that has the same potential

Stem cells: Cells that have both the capacity to self-renew (make more stem cells by cell division) and to differentiate into mature, specialized cells

Signals: Internal and external factors that control changes in cell structure and function. They can be chemical or physical in nature

Somatic cell: Any body cells other than gametes (egg or sperm); sometimes referred to as "adult" cells

Somatic cell nuclear transfer (SCNT): A technique that combines an enucleated egg and the nucleus of a somatic cell to make an embryo. SCNT can be used for therapeutic or reproductive purposes, but the initial stage that combines an enucleated egg and a somatic cell nucleus is the same. See also therapeutic cloning and reproductive cloning

Somatic (adult) stem cell: A relatively rare undifferentiated cell found in many organs and differentiated tissues with a limited capacity for both self-renewal (in the laboratory) and differentiation. Such cells vary in their differentiation capacity, but it is usually limited to cell types in the organ of origin. This is an active area of investigation

Stem cells: Cells with the ability to divide for indefinite periods in culture and to give rise to specialized cells

Stromal cells: Connective tissue cells found in virtually every organ. In bone marrow, stromal cells support blood formation

Subculturing: Transferring cultured cells, with or without dilution, from one culture vessel to another

Surface markers: Proteins on the outside surface of a cell that are unique to certain cell types and that can be visualized using antibodies or other detection methods

Teratoma: A multi-layered benign tumor that grows from pluripotent cells injected into mice with a dysfunctional immune system. Scientists test whether they have established a human embryonic stem cell (hESC) line by injecting putative stem cells into such mice and verifying that the resulting teratomas contain cells derived from all three embryonic germ layers

Therapeutic cloning: The process of using somatic cell nuclear transfer (SCNT) to produce cells that exactly match a patient. By combining a patient's somatic cell nucleus and an enucleated egg, a scientist may harvest embryonic stem cells from the resulting embryo that can be used to generate tissues that match a patient's body. This means the tissues created are unlikely to be rejected by the patient's immune system. See also somatic cell nuclear transfer (SCNT)

Tissue-specific stem cells (also known as adult or somatic stem cells): Stem cells found in different tissues of the body that can give rise to some or all of the mature cell types found within the particular tissue or organ from which they came, i.e., blood stem cells can give rise to all the cells that make up the blood, but not the cells of organs such as the liver or brain

Totipotent stem cells: Stem cells that, under the right conditions, are wholly capable of generating a viable embryo (including the placenta) and, for humans, exist until about 4 days after fertilization, prior to the blastocyst stage from which embryonic stem cells are derived

Transdifferentiation: The process by which stem cells from one tissue differentiate into cells of another tissue

Trophoblast: The outer cell layer of the blastocyst. It is responsible for implantation and develops into the extraembryonic tissues, including the placenta, and controls the exchange of oxygen and metabolites between mother and embryo

Umbilical cord blood stem cells: Stem cells collected from the umbilical cord at birth that can produce all of the blood cells in the body. Cord blood is currently used to treat patients who have undergone chemotherapy to destroy their bone marrow due to cancer or other blood-related disorders

Undifferentiated: A cell that has not yet developed into a specialized cell type

References

1. Schöler R. The potential of stem cells, vol. 28. Farnham: Ashgate Publishing; 2007.
2. Thomson IE, Shapiro J, Waknitz SS, et al. Blastocysts embryonic stem cell lines derived from human. Science. 1998;282:1145–7.
3. Thomson JA. Homologous recombination in human embryonic stem cells. Nat Biotechnol. 2003;21:319–21.
4. Parish CL, Arenas E. Stem-cell-based strategies for the treatment of Parkinson's disease. Neurodegener Dis. 2007;4:339–47.
5. Abdul MB. Designer's microglia with novel delivery system in neurodegenerative diseases. Med Hypotheses. 2014;83(4):510–2.

6. Nichols Y, Chambers J, Smith IA. BMP induction of id proteins suppresses differentiation and sustains embryonic stem cell self-renewal in collaboration with STAT3. Cell. 2003;115:281–92.
7. Söderdahl T, Küppers-Munther B, Heins N, et al. Glutathione transferases in hepatocyte-like cells derived from human embryonic stem cells. Toxicol In Vitro. 2007;21:929–37.
8. Davila JC, Cezar GG, Thiede M, et al. Use and application of stem cells in toxicology. Toxicol Sci. 2004;79:214–23.
9. Kuliev A, Verlinsky Y. Place of preimplantation diagnosis in genetic practice. Am J Med Genet A. 2005:134A;105–10.
10. Tichy ED, Pillai R, Deng L, et al. Mouse embryonic stem cells, but not somatic cells, predominantly use homologous recombination to repair double-strand DNA breaks. Stem Cells. 2010;19:1699–711.
11. Knoepfler PS. Deconstructing stem cell tumorigenicity: a roadmap to safe regenerative medicine. Stem Cells. 2009;27:1050–6.
12. Bharadwaj S, Liu G, Shi Y, et al. Multipotential differentiation of human urine-derived stem cells: potential for therapeutic applications in urology. Stem Cells. 2013;31:1840–56.
13. Yehezkel S, Rebibo-Sabbah A, Segev Y, et al. Reprogramming of telomeric regions during the generation of human induced pluripotent stem cells and subsequent differentiation into fibroblast-like derivatives. Epigenetics. 2011;6:63–75.
14. Dabir DV, Hasson SA, Setoguchi K, et al. A small molecule inhibitor of redox-regulated protein translocation into mitochondria. Dev Cell. 2013;25:81–92.
15. Lindgren AG, Natsuhara K, Tian E, et al. Loss of pten causes tumor initiation following differentiation of murine pluripotent stem cells due to failed repression of Nanog. PLoS One. 2011;6:e16478.
16. Ohnishi K, Semi K, Yamamoto T, et al. Premature termination of reprogramming in vivo leads to cancer development through altered epigenetic regulation. Cell. 2014;156:663–77.
17. Yamanaka S, Blau H. Nuclear reprogramming to a pluripotent state by three approaches. Nature. 2010;465:704–12.
18. Tachibana M, Amato P, Sparman M, et al. Human embryonic stem cells derived by somatic cell nuclear transfer. Cell. 2013;153:1228–38.
19. Paull D, Emmanuele V, Weiss K, et al. Nuclear genome transfer in human oocytes eliminates mitochondrial DNA variants. Nature. 2012;493:632–7.
20. Shinagawa T, Takagi T, Tsukamoto D, et al. Histone variants enriched in oocytes enhance reprogramming to induced pluripotent stem cells. Cell Stem Cell. 2014;14:217–27.
21. Han DW, Tapia N, Hermann A, et al. Direct reprogramming of fibroblasts into neural stem cells by defined factor. Cell Stem Cell. 2012;10:465–72.
22. Wei S, Zou Q, Lai S, et al. Conversion of embryonic stem cells into extraembryonic lineages by CRISPR-mediated activators. Sci Rep. 2016;6:19648.
23. Zhang H, Wilson IA, Lerner RA. Selection of antibodies that regulate phenotype from intracellular combinatorial antibody libraries. PNAS. 2012;109:15728–33.
24. Liu X, Ory V, Chapman S, et al. ROCK inhibitor and feeder cells induce the conditional reprogramming of epithelial cells. Am J Pathol. 2012;180:599–607.
25. Kurian L, Sancho-Martinez I, Nivet E, et al. Conversion of human fibroblasts to angioblast-like progenitor cells. Nat Methods. 2012;10:77–83.
26. Suila H, Hirvonen T, Ritamo I, et al. Extracellular o-linked n-acetylglucosamine is enriched in stem cells derived from human umbilical cord blood. BioRes Open Access. 2014;3:39–44.
27. Downing TL, Soto J, Morez C, et al. Biophysical regulation of epigenetic state and cell reprogramming. Nat Mater. 2013;12:1154–62.
28. Sun Y, Koh AY, Meng V, et al. Hippo/YAP-mediated rigidity-dependent motor neuron differentiation of human pluripotent stem cells. Nat Mater. 2014;13:599–604.
29. Caiazzo M, Okawa Y, Ranga A, et al. Defined three-dimensional microenvironments boost induction of pluripotency. Nat Mater. 2016;15:344–52.
30. Takahashi K, Yamanaka S. Induction of pluripotent stem cells from mouse embryonic and adult fibroblast cultures by defined factors. Cell. 2006;124:663–76.
31. Stadtfeld M, Nagaya M, Utikal J, et al. Induced pluripotent stem cells generated without viral integration. Science. 2008;322:949–9.
32. Seki T, Yuasa S, Fukuda K. Generation of induced pluripotent stem cells from a small amount of human peripheral blood using combination of activated T cells and Sendai virus. Nat Protoc. 2012;7:718–28.
33. Fusaki N, Ban H, Nishiyama A, et al. Efficient induction of transgene-free human pluripotent stem cells using a vector based on Sendai virus, an RNA virus that does not integrate into the host genome. Proc Jpn Acad Ser B Phys Biol Sci. 2009;85:348–62.
34. Anokye-Danso F, Trivedi CM, Juhr D, et al. Highly efficient miRNA-mediated reprogramming of mouse and humansomatic cells to pluripotency. Sell Stem Cell. 2011;8:376–88.
35. Judson RL, Babiarz JE, Venere M, et al. Embryonic stem cell-specific microRNAs promote induced pluripotency. Nat Biotechnol. 2009;27:459–61.
36. Huangfu D, Maehr R, Guo W, et al. Induction of pluripotent stem cells by defined factors is greatly improved by small-molecule compounds. Nat Biotechnol. 2008;26:795–7.
37. Zhang M, Zhang J, Chen X, et al. Glycogen synthase kinase 3 promotes p53 mRNA translation viaphosphorylation of RNPC1. Genes Dev. 2012;27:2246–58.
38. Okita K, Nakagawa M, Hyenjong H, et al. Generation of mouse induced pluripotent stem cells without viral vectors. Science. 2008;322:949–53.
39. Brighton CT, Hunt RM. Early histological and ultrastructural changes in medullary fracture callus. J Bone Joint Surg. 1991;73:832–47.

40. Haniffa MA, Collin MP, Buckley CD, et al. Mesenchymal stem cells: the fibroblasts' new clothes? Haematologica. 2009;94:258–63.
41. Hematti P. Mesenchymal stromal cells and fibroblasts: a case of mistaken identity? Cytotherapy. 2012;14:516–21.
42. Dominici M, Le Blanc K, Mueller I, et al. Minimal criteria for defining multipotent mesenchymal stromal cells. The international society for cellular therapy position statement. Cytotherapy. 2006;8:315–7.
43. Jiang Y, Jahagirdar BN, Reinhardt RL, et al. Pluripotency of mesenchymal stem cells derived from adult marrow. Nature. 2002;418:41–9.
44. Ryan JM, Barry FP, Murphy JM, Mahon BP. Mesenchymal stem cells avoid allogeneic rejection. J Inflamm. 2005;2:8.
45. Ryan JM, Barry F, Murphy JM, Mahon BP. Interferon-γ does not break, but promotes the immunosuppressive capacity of adult human mesenchymal stem cells. Clin Exp Immunol. 2007;149:353–63.
46. Gronthos S, Graves SE, Ohta S, Simmons PJ. The STRO-1+ fraction of adult human bone marrow contains the osteogenic precursors. Blood. 1994;84:4164–73.
47. Iudicone P, Fioravanti D, Bonanno G, et al. Pathogen-free, plasma-poor platelet lysate and expansion of human mesenchymal stem cells. J Transl Med. 2014;12:28.
48. Fauza D. Amniotic fluid and placental stem cells. Best Pract Res Clin Obstet Gynaecol. 2004;18:877–91.
49. Inzunza J. Comparative genomic hybridization and karyotyping of human embryonic stem cells reveals the occurrence of an isodicentric X chromosome after long-term cultivation. Mol Hum Reprod. 2004;10:461–6.
50. Liehr T. Multicolor fish probe sets and their applications. Histol Histopathol. 2004;19:229–37.
51. Kallioniemi A. Comparative genomic hybridization for molecular cytogenetic analysis of solid tumors. Science. 1992;258:818–21.
52. Adewumi O. Characterization of human embryonic stem cell lines by the international stem cell initiative. Nat Biotechnol. 2007;25:803–16.
53. Ben-Tabou S, Davidson EH. Gene regulation: gene control network in development. Annu Rev Biophys Biomol Struct. 2007;36:191–212.
54. Knisely K, Gilbert SF. Developmental biology. 8th ed. Sunderland, Mass: Sinauer Associates; 2009. p. 147.
55. Rudel S. The evolution of developmental mechanisms. Dev Biol. 2003;264:15–37.
56. Lister R. Hotspots of aberrant epigenomic reprogramming in human induced pluripotent stem cells. Nature. 2011;471:68–73.
57. Christophersen NS, Helin K. Epigenetic control of embryonic stem cell fate. J Exp Med. 2010;207:2287–95.
58. Meissner A. Epigenetic modifications in pluripotent and differentiated cells. Nat Biotechnol. 2010;28:1079–88.
59. Teif VB, Vainshtein Y, Caudron-Herger M, et al. Genome-wide nucleosome positioning during embryonic stem cell development. Nat Struct Mol Biol. 2012;19:1185–92.
60. Mohammad HP, Baylin SB. Linking cell signaling and the epigenetic machinery. Nat Biotechnol. 2010;28:1033–8.
61. Niwa H, Burdon T, Chambers I, Smith A. Self-renewal of pluripotent embryonic stem cells is mediated via activation of STAT3. Genes Dev. 1998;12:2048–60.
62. Engler AJ, Sen S, Sweeney HL, Discher DE. Matrix elasticity directs stem cell lineage specification. Cell. 2006;126:677–89.
63. Guo J, Wang Y, Sachs F, Meng F. Actin stress in cell reprogramming. Proc Natl Acad Sci. 2014;111:E5252–61.
64. Guilak F, Cohen DM, Estes BT, et al. Control of stem cell fate by physical interactions with the extracellular matrix. Cell Stem Cell. 2009;5:17–26.
65. Dobrinski A, Travis J. Germ cell transplantation for the propagation of companion animals, non-domestic and endangered species. Reprod Fertil Dev. 2007;19:732–9.

Bioreactors for Regenerative Medicine in Urology

4

In Kap Ko, Anthony Atala, and James J. Yoo

4.1 Introduction

The urogenital system is generally composed of the urinary and reproductive tissues and organs. The primary function of the urinary system, which includes the kidneys, ureters, bladder, and urethra, is to produce and excrete urine. The male reproductive system consists of the penis and testes, while the female reproductive organs include the vagina and uterus. Congenital disorders, infections, tumors, and defects in the urogenital system may result in tissue and organ damage or the complete loss of function [1]. Currently, these conditions may be treated using autografts of non-urological tissue, such as the skin and mucosa. However, this method is limited by donor site morbidity and poor survival of the grafted tissue [2]. Complicated injuries require extensive reconstructive surgery, which often incorporates biomaterial-derived products [3–6]. For instance, disorders in the pelvic muscles have been clinically treated with natural and synthetic material-based slings and meshes; however, this treatment is limited by several issues, including (1) a tendency to perforate the other urogenital organs, (2) retraction of the graft due to shrinkage in vivo, and (3) severe fibrosis resulting from foreign body reactions and other immune responses that inhibit tissue function. Therefore, an alternative treatment is needed to address these limitations.

Recent advances in tissue engineering and regenerative medicine have offered promising approaches for the treatment of damaged tissues and organs in the urogenital system [7–9]. Several strategies incorporate cell-based therapies that use either a single cell-type (cell therapy) or engineered implantable tissue constructs (bioengineering of functional tissue). Recent developments in cell biology, stem cell biology, and cell manipulation technology have enabled the identification, characterization, and expansion of therapeutic cells for use in treatments. Potential cell sources include tissue-specific primary cells, adult stem cells, and pluripotent stem cells. The engineering of functional urogenital tissue constructs in vitro by culturing tissue-specific cells seeded on a template (e.g., scaffold) and then implanting the resulting construct in vivo may provide a solution to current unmet medical needs [7–9]. In order to engineer functional tissues in vitro, the appropriate selection of cells, scaffold material and structure, and biological and mechanical cues is of the utmost importance. Scaffolding systems play a significant role in instructing cellular behavior and function, and they need to be properly selected in order to provide an appropriate environment for cells. Naturally derived or synthetically prepared

I.K. Ko (✉) • A. Atala • J.J. Yoo
Wake Forest Institute for Regenerative Medicine, Wake Forest School of Medicine, Medical Center Boulevard, Winston-Salem, NC 27157, USA
e-mail: iko@wakehealth.edu

B.W. Kim (ed.), *Clinical Regenerative Medicine in Urology*, DOI 10.1007/978-981-10-2723-9_4

materials [10–12] have been fabricated into scaffolds for three-dimensional (3D) cell culture. Additionally, biological (e.g., cytokines, growth factors) and mechanical cues need to be applied to the newly formed tissues, thus promoting more functional and tissue-specific characteristics. While numerous studies have produced functional engineered tissue constructs that have been tested in preclinical studies, the successful application of these technologies for clinical translation requires several components, including a large-scale system for producing therapeutic cells, a well-designed three-dimensional (3D) culture system with sufficient oxygen and nutrient distribution, and appropriate biological and mechanical stimulation to the tissue constructs that have been fabricated in vitro. Automated and well-established bioreactor systems may provide a comprehensive solution.

Bioreactor systems have been used for industrial fermentation, food processing, and the production of biological and pharmaceutical drugs by mass culture [13]. Recently, bioreactor-based mass tissue culture has been successfully established as a method of producing sufficient numbers of therapeutic cells, while conventional cell culture on culture dishes is still incapable of producing the required cell quantities [14]. Bioreactor systems have also been used to produce 3D tissue constructs in vitro [15]. Several conditions of the bioreactor system, including pH, temperature, and nutrient and oxygen distribution in the culture medium, need to be controlled in an efficient manner in order to maintain cell viability and support cellular maturation [13]. For instance, efficient delivery of nutrients and oxygen to seeded cells during bioreactor culture is crucial for cell survival. Appropriate mechanical stimulation is another essential cue that allows preconditioning of the newly formed tissue under physiological conditions. Such bioreactor design enables the control of environmental conditions for cell culture, thus allowing cells within the scaffold to undergo proper maturation and differentiation under physiological environments that are similar to in vivo conditions [13]. In this book chapter, basic components and design of bioreactor systems will be discussed, and potential uses of bioreactor systems toward specific urogenital applications will be presented.

4.2 Fundamental Design and Types of Bioreactor Systems

4.2.1 Spinner Flask Bioreactor System

Spinner flask bioreactors have been used for the culture of single cells (e.g., suspension cells) and small-sized tissue constructs (Fig. 4.1a). To provide efficient and homogeneous mass transfer to cells during culture, culture medium is agitated by a stir bar, and the agitation speed can be easily controlled by altering the magnetic forces used by the device [16]. Several types of mammalian cells have been tested in spinner flask bioreactors for efficient cell expansion, including neural stem cells [17], hematopoietic stem cells [18], embryonic stem cells [19], and induced pluripotent stem cells (iPSCs) [20, 21] (Fig.4.1b–d). 3D engineered tissue constructs, such as cartilage tissue constructs [22], have also been successfully created using this type of bioreactor. While the spinner flask bioreactor has been successfully applied to in vitro cell culture, uncontrolled convection flow within the system may affect cell viability, so flow parameters need to be well adjusted and controlled within specific cell culture systems [16].

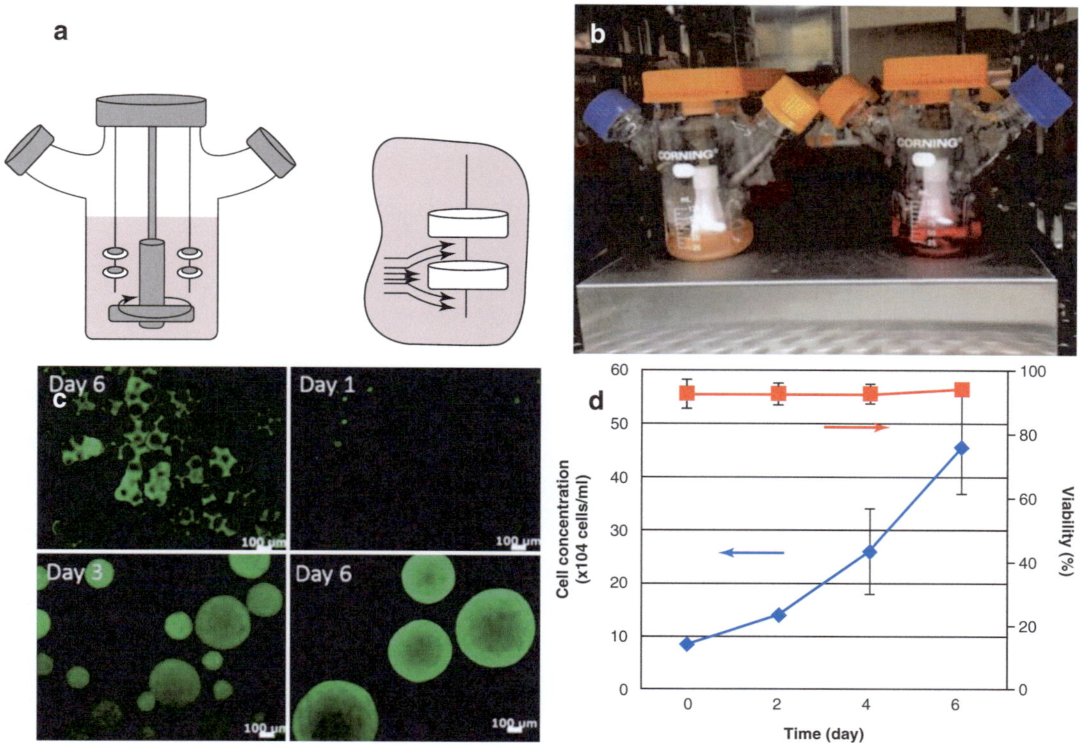

Fig. 4.1 Spinner-flask bioreactor. (**a**) Schematic design [13]. (**b**) Human iPSC cultures in a spinner-flask bioreactor and the cultured iPSCs on the Matrigel-coated polystyrene beads and cell aggregations (**c**) and cell growth and viability during the culture (**d**) [21]. Reprinting of each image was permitted from publishers

4.2.2 Rotating Wall Vessel (RWV) Bioreactor System

Unlike the spinner flask bioreactor, the rotating wall vessel (RWV) bioreactor is designed to provide low shear stress and a homogeneous dynamic cell culture environment [22] (Fig. 4.2a). The RWV bioreactor system was originally developed by scientists at the National Aeronautics and Space Agency (NASA) for cell culture under low- or no-gravity conditions for targeted use in space. Briefly, the RWV system consists of culture medium that is horizontally rotated while connected to an oxygen and gas supply. The gravitational force experienced by single cells is minimal (approximately 10^{-2} g), thus reducing damage to cells and promoting cell viability and tissue maturation during in vitro culture. The efficiency of RWV bioreactors for in vitro cell culture has been evaluated through histological, biochemical, and biomechanical analysis of several cell types that have been cultured in RWV bioreactors, including tumor cells [23, 24] and stem cells [25, 26] (Fig. 4.2b–d), and engineered tissue constructs, such as cartilage tissue constructs [27, 28].

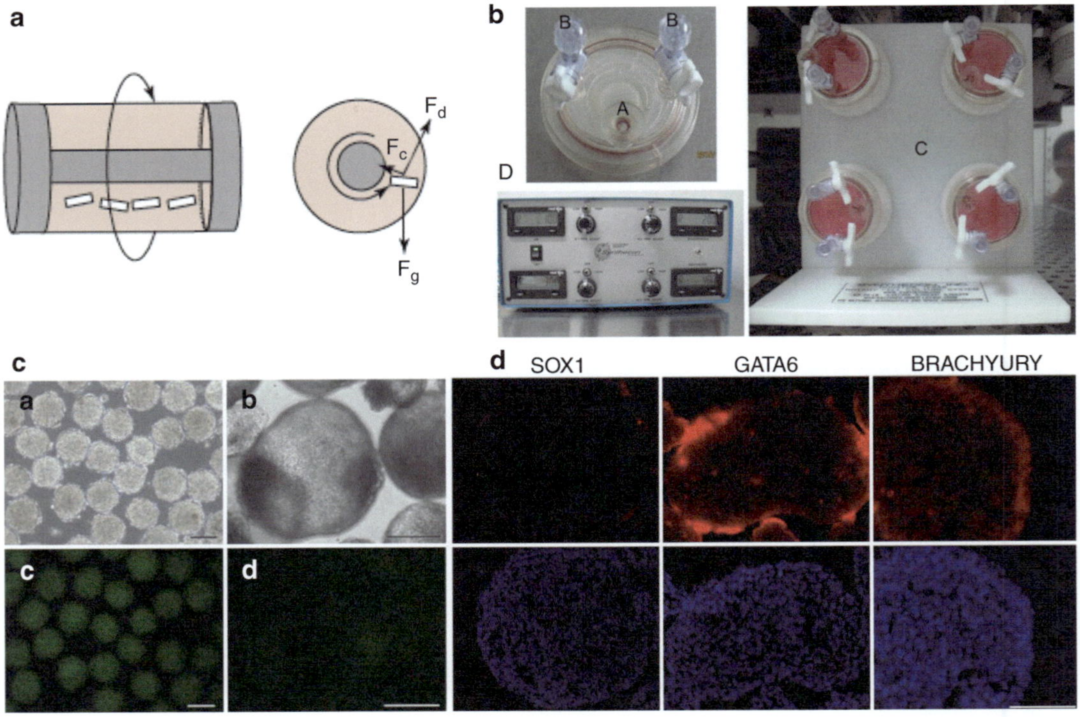

Fig. 4.2 Rotating wall vessel bioreactor. (**a**) Schematic design [13]. (**b**) A rotary bioreactor used for embryoid body (EB) formation (**c**) and characterization of EBs using several markers (**d**) [26]. Reprinting of each image was permitted from publishers

4.2.3 Hollow Fiber System

One representative application of hollow fiber bioreactors is used as a hemodialysis membrane (dialyzer) for treating patients with kidney failure [29]. A dialyzer contains a bundle of hollow fibers (≈10,000 fibers), each with an inner diameter of approximately 200 μm. As a patient's blood is perfused through the fibers, blood waste is removed through the fiber membrane, and the filtered blood is returned to the patient. Ever since the first dialysis membrane was developed by Lipps et al. using cellulose-derived hollow fibers [30], various types of materials have been developed to improve blood dialysis efficiency [31]. Dialysis efficiency primarily depends on the biocompatibility of the hollow fibers, which is influenced by the membrane components and flow properties within the dialyzer. In addition, several fiber parameters, such as diameter, molecular weight cutoff (MWCO), wall thickness, and material type (i.e., cellulose or synthetic materials), have been modified to improve dialysis efficiency [31].

4.2.4 Hollow Fiber-Incorporated Perfusion Bioreactor

Recent progress in cell culture techniques and stem cell research has offered great promise in the field of cell-based therapies, and large-scale production of therapeutic cells is required to address massive needs for clinical translation. To this end, hollow fiber bioreactors have been used to scale up mammalian cell culture [32]. When compared with stirred flask culture, cell culture within the hollow fiber system has several advantages. Hollow fibers provide a large surface area for cell inoculation, and cells can be seeded and cultured within the hollow fibers, thus creating cultures with high cell densities. Appropriate perfusion through the hollow fibers is needed to efficiently provide oxygen and nutrients to seeded cells. During bioreactor culture, shear stress must be maintained at an appropriate level that does not harm the cells in order to preserve cellular viability during long-term cell culture. Hybridoma cell lines have been used to test the feasibility of large-scale antibody production [32], and recent studies have reported in vitro mass production of therapeutic cells [33] and stem cells [34], including embryonic stem cells (ESCs) [35] and mesenchymal stem cells (MSCs) [36, 37]. The hollow fiber system, in particular, has been used as an appropriate environment for stem cell differentiation [34]. Several types of cells, including hematopoietic stem cells (HSCs) [38], ESCs [39], and iPSCs [40], have been efficiently differentiated into blood cells [38], dopamine-producing neurons [39], and hepatocytes [40], respectively.

Another application of the hollow fiber-incorporated perfusion system includes the development of bioartificial organs as a promising strategy for treating organ failures. One example is the use of bioartificial livers to temporarily support liver function. In end-stage liver failure, hemodialysis and liver-assist systems are limited in their ability to remove toxins from the blood and cannot replicate several other functions of the liver. However, liver-assist systems are developing in order to incorporate viable cells and thereby support metabolic function. One early study, performed by Sussman et al., used hollow fiber systems to temporarily support patients with liver failure [41]. The bioartificial liver developed in this study, known as the extracorporeal liver-assist device (ELAD®), was designed to seed immortalized liver cells into the hollow fibers of the system, maintain cell viability, and provide metabolic function to blood perfused throughout the system. After ELAD® showed promises in numerous preclinical trials [42], the safety and efficiency of ELAD® in supporting liver function were confirmed in clinical trials [41], and the device is currently available for liver failure patients [43] (Fig. 4.3b).

Likewise, kidney-assist devices using viable kidney cells have been developed to address the limitations of blood dialysis in the conventional hemodialysis system [44]. To support renal biological functions that cannot be replicated by conventional blood dialysis, Humes et al. [45] developed a bioartificial kidney device by seeding pig-derived proximal renal cells into a hollow fiber kidney-assist device and implanting the device into dogs with renal failure (Fig. 4.3a). The bioartificial kidney system was able to facilitate improved filtration capability, plasma parameters, and metabolism compared to a hemodialysis

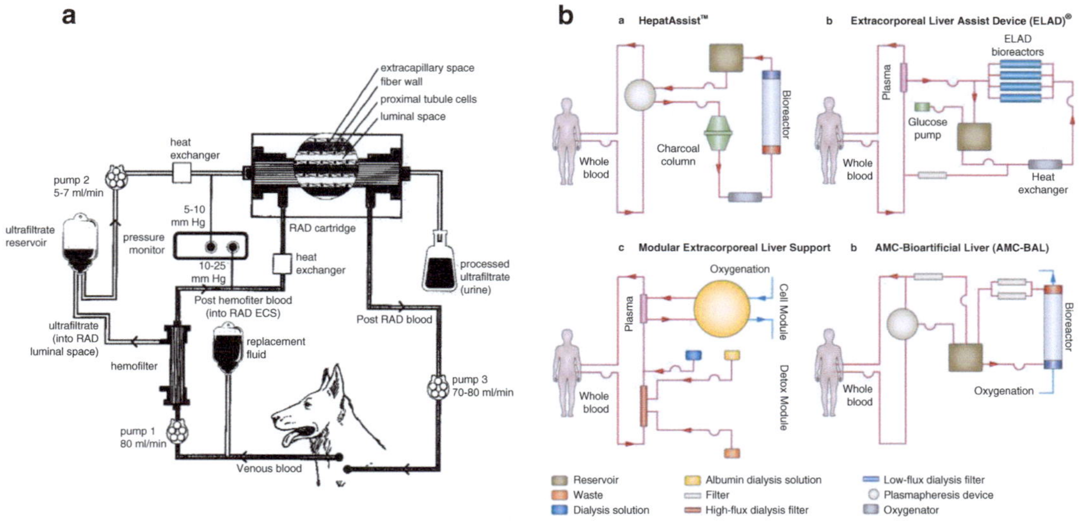

Fig. 4.3 Hollow fiber-incorporated perfusion bioreactor. (**a**) An extracorporeal perfusion unit for bioartificial kidney [45], (**b**) various types of bioartificial livers [43]. Reprinting of each image was permitted from publishers

system that did not incorporate cells. These results demonstrate that it is possible to develop an artificial kidney by incorporating viable renal cells into the conventional kidney dialysis system and that such a cell-based artificial kidney may contribute to the improvement of renal function in kidney failure patients.

4.3 Application of Bioreactors for the Engineering of Urogenital Constructs

As described previously, bioreactor systems have been applied to the reconstruction of 3D engineered tissues, and numerous types of urogenital constructs, including bladder, urethra, and kidney constructs, have been fabricated in vitro for implantation in vivo. While most early studies did not use bioreactor systems for 3D culture, the need to develop functional large-scale tissue constructs that mimic in vivo environments has necessitated the creation of well-designed bioreactor systems.

4.3.1 Bladder

End-stage bladder disease is frequently treated using grafts of native gastrointestinal segments; however, the incorporation of the graft tissue into the host urinary tract often causes several complications, including metabolic disturbances, increase of mucosa production, and malignant diseases. Alternatively, natural and synthetic biomaterials have been used to address these issues, but several disadvantages, such as foreign body reactions, perforation of adjacent urogenital tissues, severe fibrosis, and poor host tissue integration, still remain unresolved [46]. Cell-based approaches utilizing viable bladder cells have allowed researchers to recreate bladderlike structures in vitro, thus enabling functional bladder reconstruction by in vivo implantation. Numerous studies have reported the creation of bioengineered bladderlike constructs using different types of scaffolds, such as those made from biological [47–51] and synthetic materials [7, 52], and a majority of early studies were conducted in 3D culture systems without in vitro bioreactor systems. In an early study, we developed an engineered bladder tissue construct using dog-derived bladder-derived cells [51]. We used canine-derived allogeneic submucosa as a scaffold and then seeded urogenital and smooth muscle cells into the scaffold for in vitro culture. The implantation of this autologous cell-seeded bladder construct resulted in an approximately 99% increase in bladder urodynamic capacity, while the implantation of cell-free scaffolds increased bladder size by only 30%. Functional and histological analysis demonstrated that the implantation of the cell-seeded construct facilitated normal bladder compliance, and the seeded cells showed normal cellular organization and phenotypic properties along the harvested host tissue. These results demonstrate that a small biopsy ($\approx$1 cm^2) is sufficient to obtain enough cells for engineering bladder tissue and that the implantation of cell-seeded tissue constructs produces more bladder augmentation than the implantation of cell-free scaffolds [51].

The most critical challenge in the implantation of cell-seeded constructs is to accelerate host vascularization that is able to provide seeded cells with sufficient nutrients and oxygen for better cell survival within the implant. Schoeller et al. tested whether pre-vascularization of the scaffold matrix would enhance bladder reconstruction [49]. To achieve pre-vascularization, the authors utilized native host regenerative capabilities to fabricate a capsule pouch with host vascularization. Pre-vascularized scaffolds were created by implanting silicone blocks into the groins of rats for 1 week. Autologous urothelial cells were then seeded into these vascularized capsules, and the cell-seeded constructs were transposed into bladder wall defect sites. The group treated with the pre-vascularized cell-seeded scaffolds experienced a higher survival rate than the scaffold-only (80% mortality) and saline groups (100% mortality). Histological analysis of the cell-seeded constructs with pre-fabricated capsule structures showed multilayered urothelial cell lining along the transplantation site. These results suggest that it is possible to pre-vascularize cell-seeded constructs using the host's regenerative capacities, which may also contribute to accelerating urogenital tissue regeneration in vivo [49]. Another study utilizing a pre-vascularization strategy for the engineering of bladder constructs was conducted by Schultheiss et al. [47], who seeded autologous urothelial and smooth muscle cells onto scaffolds made of decellularized bowel segments. To provide the constructs with pre-vascularization capabilities, endothelial progenitor cells were also incorporated into the engineered bladder constructs, which underwent 3D in vitro culture before implantation. The implantation study revealed that bladder constructs seeded with endothelial cells efficiently prevented blood thrombosis when compared to a control without pre-vascularization capabilities, indicating that pre-vascularization of an engineered bladderlike construct can be used for the reconstruction of a defective bladder.

More recently, our group successfully developed an approach to create engineered bladder constructs containing autologous bladder cell sources for treating patients with end-stage bladder disease [7]. Autologous urothelial and smooth muscle cells were cultured, seeded, and matured within collagen-based scaffolds for several weeks. The engineered bladder constructs were then implanted into myelomeningocele patients using an omental wrap. In follow-up studies, we found that bladder functions, such as leak point pressures and compliance, were improved in patients who were implanted with our engineered bladder constructs. These results suggest that our bioengineered bladderlike construct can be used to treat patients who need cystoplasty.

Since the bladder tissue repeatedly undergoes dynamic circumstances (such as fill-void cycles) daily, reconstruction of functional bladder tissue in vitro is desirable in order to retain structural stability against mechanical stimulation [53]. To this end, mechanical preconditioning can be applied to the bladder construct to create a functional bladder that has mechanical properties similar to those of bladders in in vivo environments [15, 53] (Fig. 4.4a). Several approaches have been used in order to improve the mechanical properties of engineered bladder constructs. One early study attempted to test the feasibility of simulating normal bladder physiology using a bioreactor system [54]. Wallis et al. developed a bioreactor

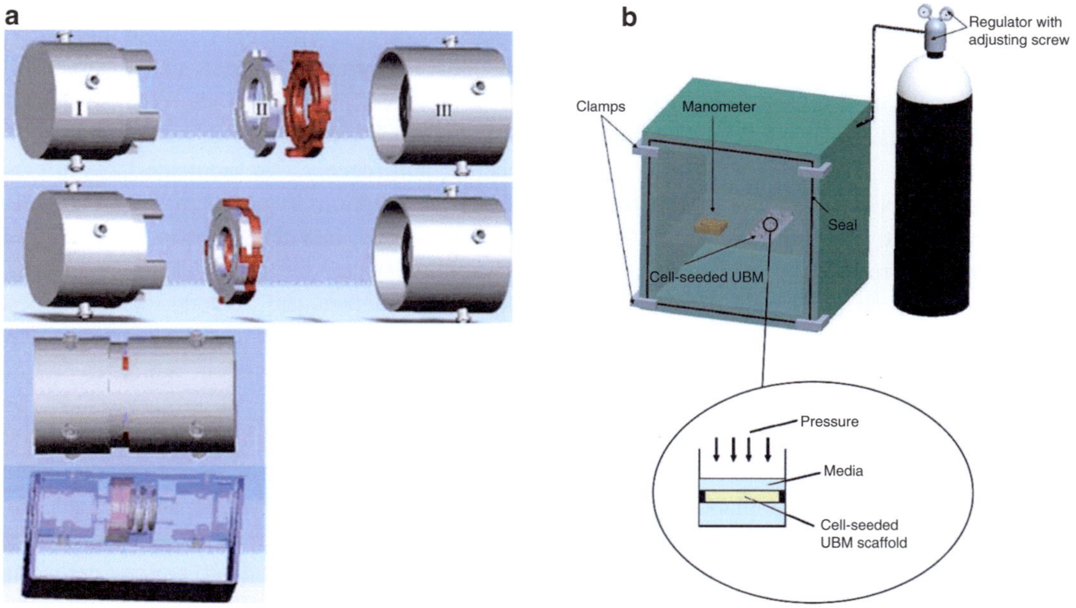

Fig. 4.4 In vitro bioreactor systems for engineering bladder tissue constructs. (**a**) [15] and (**b**) [55]. Reprinting of each image was permitted from publishers

system that simulated mechanical conditions occurring in normal bladder tissue, thereby allowing cell-seeded scaffolds to undergo cyclic mechanical stimulation. The scaffolds, which were seeded with bladder cells, were based on acellular porcine matrix or commercially available small intestinal submucosa (SIS). Histological analysis demonstrated that the seeded urothelial and smooth muscle cells were evident during bioreactor culture and that cellular structure aligned along the direction of applied pressure. The authors suggest that the bioreactor developed in this study was successful in promoting the formation of mechanically sound constructs and could be used to test the role of mechanical stimulation on bladderlike tissue [54].

In another study, Davis et al. [55] tested the effects of physiological bladder dynamics on cell viability using a bioreactor system (Fig. 4.4b). To engineer a bladderlike structure, they seeded human urothelial cells into a porcine bladder matrix and compared cell viability and proliferative capability in the resulting construct with that of constructs grown in conventional static culture. Cell viability analysis demonstrated that dynamic culture in the bioreactor significantly improved cell viability and proliferation after 4 days of culture. The authors suggest that a bioreactor system that provides mechanical stimulation is beneficial in recreating functional bladder tissue in vitro. Similar studies using different bioreactors have consistently confirmed these results [25, 56].

Several studies have also utilized the microenvironments of native tissues or organs as "in vivo bioreactors" that retain their original regenerative capabilities for cell proliferation. While a few studies have already used the in vivo environment for pre-vascularization of scaffolding systems [49], Campbell et al. [57] used the peritoneal cavity as an "in vivo bioreactor" to bioengineer visceral organs, such as the bladder and uterus. To produce tissue grafts, they implanted appropriate templates for different organs into rats or rabbits for 2–3 weeks. The produced tissue, which was found to contain a myofibroblast-rich cell population, was then harvested and reimplanted into the bladders (of what animals?). After 14 months, the implanted bladders developed morphologically normal bladder structure, suggesting that the in vivo environment may be used as a functioning bioreactor system to reconstruct bladder tissue for treating bladder augmentation cystoplasty.

In another study, Kajbafzadeh et al. used native bladder tissue as an in vivo bioreactor to determine appropriate scaffolding systems for bladder reconstruction [58]. They implanted several scaffolds (such as pericardium, biofilm, and polyglycolic acid (PGA) scaffolds or a combination of each type of scaffold) between bladder mucosa layers for several weeks and examined the immunological response and vascularization of the implants. Histological analysis revealed that PGA-coated pericardium was the most effective scaffold in terms of reducing inflammation and promoting efficient vascularization. This result suggests that a combination of biodegradable materials with acellular matrices will produce appropriate scaffolding systems that optimize cell attachment in vitro and bladder tissue reconstruction in vivo [58].

4.3.2 Urethra and Ureter

Urethral defect is a common disease that occurs secondary to urinary injuries, and treatment of these defects is limited by a shortage of implantable tissue [59]. Recent advances in tissue engineering and regenerative medicine have offered alternative solutions by producing functional urethral constructs in vitro. Although the cell-free scaffolding system has potential for use in treating urethral defects [60], several reports suggest that the use of cell-based urethral constructs may provide better results without any graft failure and stricture formation. In an early study, De Filippo et al. developed a 3D culture method to engineer urethral constructs. Autologous rabbit bladder cells were expanded and seeded into bladder submucosa, and the engineered urethral constructs were implanted into urethral defects in rabbits. To examine the effects of the seeded cells on urethral recovery, some animals were implanted with the scaffold alone. Histological and molecular analysis demonstrated that the seeded cells formed normal urethral structures 1 month after implantation, and the neoformations retained epithelial and smooth muscle cell phenotypes, as determined by Western blotting. Organ bath analysis confirmed contractility of the harvested urethral tissue, indicating efficient neural integration. However, implantation of the scaffold alone caused graft failure, characterized by strictures and poor tissue formation. These results suggest that a tubularized tissue construct with bladder cells can be fabricated and that such a construct may contribute to the reconstruction of defective urethral tissues. Different cell sources have also been used to engineer urethral tissue constructs. Since the epidermis seems to play an important role in the reconstruction of urethral tissue, Li et al. [61, 62] established a cell culture method for the expansion of oral keratinocytes. They isolated keratinocytes from the epidermis of rabbits, incorporated the keratinocytes into constructs seeded with various other cell types for engineering urethral tissues, and implanted the engineered constructs into urethral mucosa defect sites. Histological analysis showed that the constructs with keratinocytes facilitated better integration with host epithelium than those without keratinocytes. In fact, animals that were given the implants without keratinocytes developed inflammation, resulting in graft failure. These results suggest that the incorporation of epidermal layers into urethral constructs is helpful for promoting efficient mucosal integration with the host [62].

In addition to the formation of epidermal layers within engineered urethral constructs, mechanical stimulation has been applied to create mature, functional, urethral constructs. Fu et al. developed a robust bioreactor system that permits cyclic mechanical stimulation (Fig. 4.5a) [59]. To create their constructs, they incorporated several cell types, such as adipose-derived stem cells (ADSC) and oral epithelial cells, for muscular and mucosal layers, respectively. The epithelial cells were purified to obtain cells with higher proliferative capacity. Both types of cells were seeded onto fibrous PGA mesh using a layer seeding technique. The seeded construct was then placed into the bioreactor, where cyclic mechanical stimulation was applied to produce extended constructs. Histological analysis showed higher cell densities within these extended constructs compared to those not given mechanical stimulation (Fig. 4.5b). An in vivo study using dogs also revealed that implantation of the extended constructs resulted in improved urethral functional outcomes and normal structural morphology similar to that of autologous urethral tissue (Fig. 4.5c–d). These results suggest that mechanical stimulation in combination with the use of appropriate cell types is necessary for engineering functional urethral constructs and that mechanically stimulated implants better facilitate urethral reconstruction following implantation.

Various approaches for reconstructing ureteral tissues have been developed in order to treat ureteral defects and lesions due to surgical dissection and traumatic injury. While a few studies have utilized cell-free scaffolding systems, such as biologic scaffolds [63, 64], more recent trials have focused on cell-based strategies to engineer ureteral tissue constructs for replacement of ureteral defects [65, 66]. In several trials, functional ureteral tissues preconditioned with mechanical stimulation have been fabricated using bioreactor systems. Vardar et al. developed a flow bioreactor system that mimics normal flow in the human ureter [67]. Cyclic mechanical stimuli were applied to collagen-based scaffolds seeded with human urothelial and smooth muscle cells, and phenotypic changes were examined by histological and molecular analysis. Dynamic culture under mechanical stimulation was found to significantly improve tissue formation and facilitate normal muscle tissue-specific phenotypes. These results also emphasize the importance of mechanical cues for the reconstruction of functional ureteral tissue, where a flow bioreactor system is necessary to provide cyclic mechanical stimulation. In another study using a bioreactor system, an automated bioreactor system able to control physiological conditions and cell culture parameters, such as pH and temperature, was established and used to produce tubular 3D tissue constructs [68].

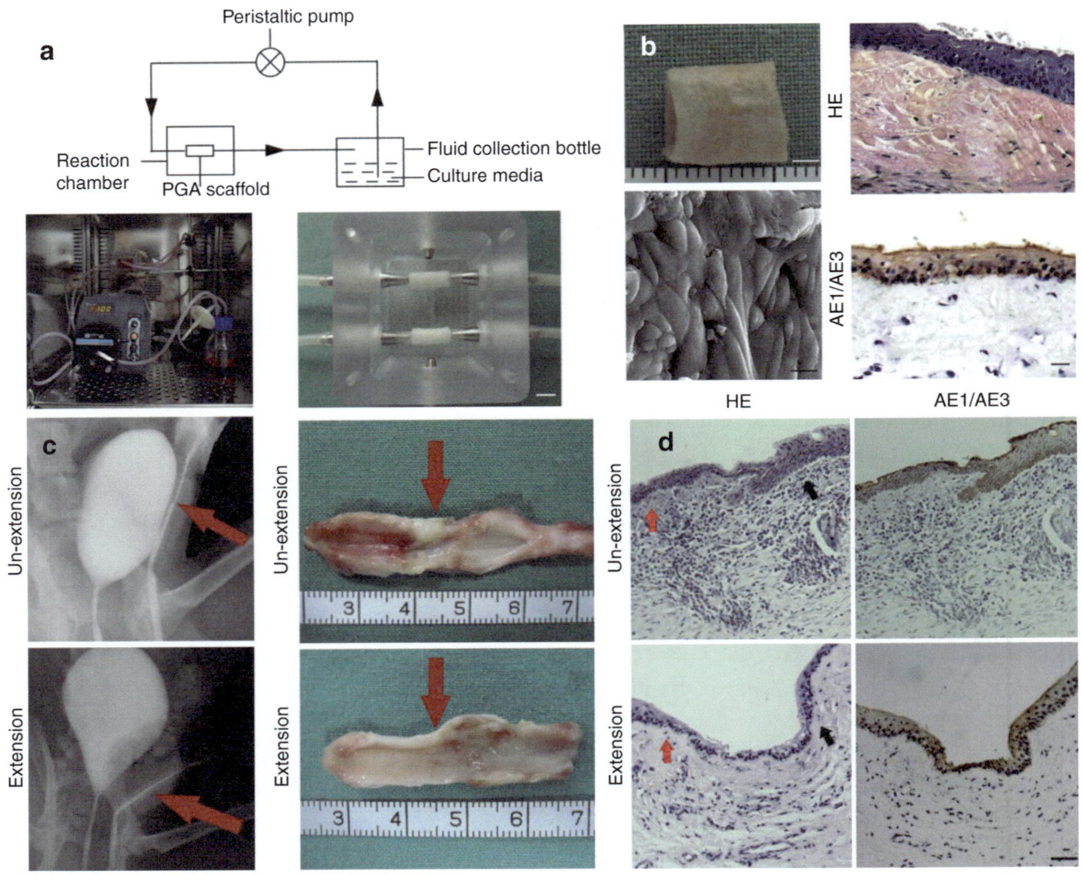

Fig. 4.5 (**a**) An in vitro bioreactor system for engineering urethral constructs [59]. (**b**) Gross and histological images of the engineered urethral tissue construct in vitro. (**c**) Urethrography (*left*) and the harvested tissue images (*right*). (**d**) Histological evaluation of the harvested tissue. Reprinting of each image was permitted from publisher

4.3.3 Kidney

Kidney transplantation is the only definitive treatment for end-stage renal disease, but the lack of transplantable kidneys has resulted in an increase in the number of patients waiting for treatment. Recent advances in the field of regenerative medicine and tissue engineering have enabled scientists to recreate kidney tissue constructs in vitro, and preclinical trials have been performed to test the feasibility of restoring renal function using these functional constructs. Important components of this treatment strategy include (1) the development of scaffolding systems, (2) the establishment of kidney cell culture systems and 3D culture conditions, and (3) the designing of in vitro bioreactor systems. Early efforts have focused on the engineering of kidney tissue constructs using renal cells and scaffolding systems without bioreactor systems. Collagen-based biomaterials have often been used as scaffolds due to their biocompatibility and mechanical properties when seeded with renal epithelial and mesangial cells [69] and neonatal renal cells [70]. Our group has also developed a collagen gel-based culture system to create 3D renal constructs using primary renal cells [71]. To improve the mechanical properties of the engineered constructs, we used materials such as hyaluronic acid [72], synthetic polycarbonate material [73], and PGA scaffolds [74]. When the engineered renal constructs were tested in preclinical studies, the implants maintained renal-specific structures [73, 74] and, interestingly, produced urine-like fluids [73]. More recently, the development of induced pluripotent stem cells (iPSCs) has enabled the recreation of kidney tissues in vivo [75–78]. By combining an appropriate scaffolding system with iPSCs, researchers can produce large-scale renal constructs in vitro that can be used to treat renal failure in clinical trials. Although the use of implantable segments is a promising treatment for kidney failure, engineered segments may not able to replicate functions performed by a whole-sized kidney. To this end, an interesting approach to producing organ-sized kidney constructs has been developed for whole-kidney transplantation.

The concept of engineering whole-kidney constructs originated from utilizing native kidney architecture as a cell-seeding template [79–81]. The engineering process is composed of two techniques. First, cellular components are removed from the native kidney tissue in order to eliminate potential immune responses when the kidney construct is implanted into a normal, non-immunocompromised host [82]. This process, called "decellularization," is usually completed by dynamically perfusing detergents through the entire kidney tissue. The optimal decellularization process completely removes cell components, maintains an intact extracellular matrix, and preserves vascular integrity. Preservation of the intact vascular network is particularly critical because it allows for efficient blood perfusion following implantation [79, 82]. The decellularized kidney scaffold is then recellularized in order to produce a functional renal scaffold. In the recellularization process, cells with kidney phenotypes are repopulated within the acellular collagen-based kidney scaffold. When undifferentiated stem cells are used to recellularize the kidney scaffolds, the seeded cells need to undergo the differentiation process for renal lineages. The use of a bioreactor is essential in the recellularization process for the creation of functional whole-kidney constructs in vitro. Unlike the previously mentioned trials that produced small scaffold segments with low cell-seeding densities, whole-kidney scaffolds are seeded with a high number of cells and therefore require efficient perfusion to maintain cellular viability throughout the entire scaffold. Several types of bioreactors have thus been developed to produce recellularized whole-kidney constructs [79–81, 83–85].

An early study confirmed the importance of bioreactor culture for repopulation of a whole-kidney construct. Ross et al. developed a bioreactor system to recellularize acellular rat whole-kidney constructs using ESCs [81]. The use of ESCs for this purpose is advantageous due to the high proliferative capacity and multi-differentiative ability of ESCs; however, several ethical and safety issues need to be addressed prior to using ESCs in clinical trials [86]. To examine cellular viability and tissue formation, the authors first cultured ESCs seeded within whole-kidney scaffolds under static conditions. However, most of the cells underwent apoptosis within a few days of culture, so the authors then established a perfusion culture system based on a bioreactor system. The perfusion culture system was equipped with a peristaltic pump, a tubing, a pressure monitoring system, and a cyclic beating system that maintained constant physiologic pressure (120/80 mmHg) with periodic beating (270–300 beats/min), constant CO_2, and constant temperature. Over 10 days of perfusion culture, the whole-kidney constructs, which were seeded with ESCs through the renal artery or ureter, maintained viable cells and facilitated proliferation of seeded ESCs within tubular, vascular, and glomerular structures. A majority of the ESCs lost their pluripotency during bioreactor culture, implying that they may have undergone renal differentiation. In a subsequent study, the authors also demonstrated that the seeded ESCs were able to differentiate into the endothelial cell lineage [83]. Interestingly, the attachment of mouse ESCs induced remodeling of the basement membrane in the rat-derived kidney matrix, suggesting a possible strategy for clinical xenotransplantation. Overall, these results suggest that acellular whole-kidney scaffolds can be repopulated with pluripotent cells using a bioreactor system with simulated physiological conditions in order to form renal tissue.

Other studies focus on monitoring physical and biochemical function during perfusion bioreactor culture. The examination of viable and functional kidney constructs for implantation is preferably done in a noninvasive manner. Uzarski et al. developed a bioreactor system that allowed them to examine renal function in terms of cell viability and renal-specific behaviors [84] (ref). Using the bioreactor system, the authors were able to measure decellularization efficiency in a noninvasive manner, monitor (arterial?) pressure changes due to the level of recellularization, and evaluate recellularization efficiency by measuring cell viability (resazurin) and renal function (alumin, kidney injury molecule-1) through sampling the culture medium. The results demonstrate that this bioreactor system may be successfully used to scale up and broaden the applications of whole-kidney technology.

In addition to in vitro recreation of whole-kidney constructs, the possibility of implanting whole-kidney constructs that were created in vitro needs to be tested prior to clinical translation. Song et al. fabricated a whole-kidney construct using rat-derived kidney tissue and tested it in a preclinical rat study [80]. They developed a bioreactor system (Fig. 4.6a) to use endothelial and kidney cells to recellularize kidney scaffolds made from decellularized native rat, pig, and human renal tissue. The endothelial and kidney cells were delivered via the renal artery and ureter, respectively. The seeded kidney constructs were then cultured in a bioreactor system, where constant flow was maintained through the renal artery at 1.5 ml/min and 5% CO_2 was applied with passive drainage through the renal vein and ureter. Histological and functional analysis demonstrated that the seeded kidney constructs induced partial re-endothelialization and renal tubule formation, and tissue formation was confirmed by endothelial and renal functional tests. When the kidney construct was orthotopically implanted in a rat, cannulation between the engineered construct and the host animal was successfully performed, blood perfusion was maintained for several hours, and production of urine-like fluid was confirmed. These results suggest that the bioengineering of a whole-kidney construct is possible and that the developed kidney construct may be translatable for use in clinical trials. For clinical translation, several issues need to be addressed, including (1) scaling-up, (2) functional and homogeneous re-endothelialization, and (3) kidney-specific recellularization of engineered kidney constructs using clinically available cell sources [87].

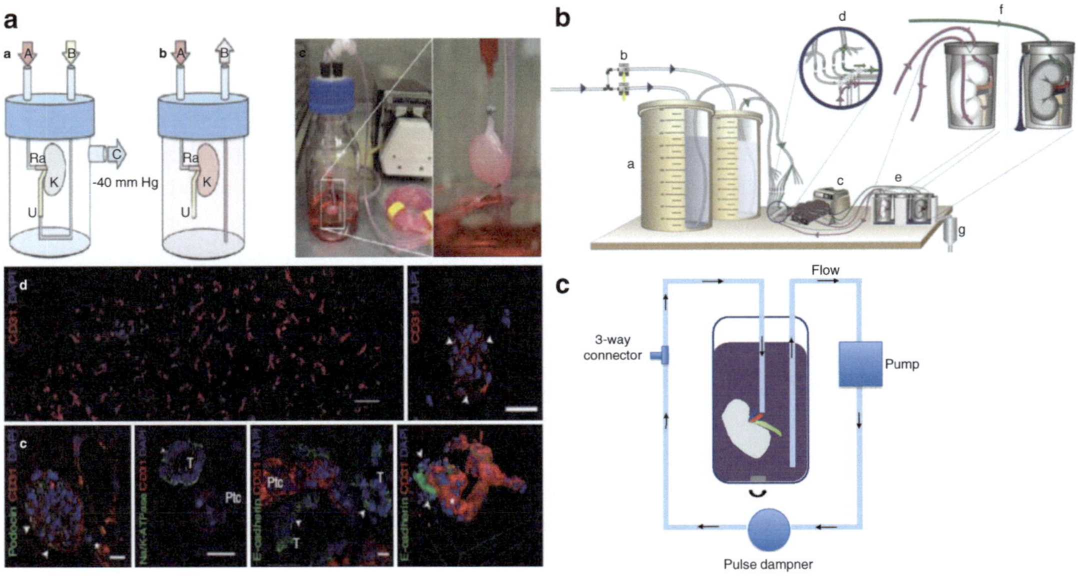

Fig. 4.6 In vitro bioreactor systems for engineering kidney constructs. (**a**) In vitro recellularization of a rat kidney scaffold with endothelial and kidney cells [80]. (**b**) High-throughput system for pig kidney decellularization [90] and (**c**) perfusion bioreactor system for recellularization of a whole-pig-kidney construct [85]. Reprinting of each image was permitted from publisher

To address these challenges, our group first worked on the decellularization of clinical-sized kidney scaffolds, particularly those from pig-derived kidney sources. Pig and human kidneys are similarly sized, and porcine-derived tissue materials, such as the heart valve [88] and small intestinal submucosa (SIS) [89], have been used in the clinic. Through decellularization of pig kidneys, we successfully developed acellular kidney scaffolds that maintained vascular architecture and an intact extracellular matrix (ECM) [90] (Fig. 4.6b). One critical challenge in developing this technology was maintaining long-term blood flow so that perfused blood would be able to provide nutrients and oxygen to the seeded kidney cells following scaffold implantation. To provide antithrombogenic capacity to the acellular kidney scaffolds, we tested a method for re-endothelialization of the porcine kidney constructs using a bioreactor system [79]. Our bioreactor system allowed static and dynamic culture for functional and efficient re-endothelialization of the acellular kidney constructs, as confirmed by in vitro and in vivo studies. Since we observed endothelial cell detachment, which caused further blood clots on the vascular surface, we introduced CD31 antibody conjugation technology into the vascular lumen in order to better maintain re-endothelialization of the constructs. We hypothesized that the conjugated CD31 antibody would firmly retain endothelial cells by interacting with the antigen on the endothelial cells. Our in vivo implantation study demonstrated that re-endothelialization in combination with CD31 antibody conjugation facilitated better blood perfusion than re-endothelialization alone, indicating the feasibility of using this anticoagulation strategy when implanting whole-kidney constructs [79].

In a parallel study, we also developed a recellularization method for creating functional renal constructs [85]. To test the feasibility of expanding cells from a clinically relevant source, we established a culture method that allowed us to expand primary renal cells isolated from human renal tissues (Fig. 4.6c) [71]. Based on histological analysis, a majority of the cell population demonstrated proximal tubular phenotypes. We used these proximal tubular cells to develop a recellularization method to efficiently form tubular structures within the kidney constructs [85]. Cells were seeded through an extravascular route by direct injection into the kidney parenchyma, and the seeded construct was maintained in the bioreactor system to promote the formation of renal structures. The newly formed tubules showed functional characteristics, such as electrolyte and protein readsorption, amino acid transport capability, and erythropoietin production. These results suggest that our repopulation method in combination with bioreactor culture will help promote the formation of tubular structures within the recellularized porcine kidney constructs. The ongoing work is investigating the long-term implantation of engineered whole-kidney constructs that are seeded with autologous endothelial and renal cells.

4.4 Future Perspectives

To date, the application of bioreactors in the fields of regenerative medicine and tissue engineering has focused on mass cell culture and the creation of functional urological tissue constructs in vitro. Dynamic cell culture in bioreactors has enabled better tissue grafting and functional outcomes in vivo. To apply bioreactor systems for clinical translation, several things need to be considered (Fig. 4.7) [91]. First, bioreactor systems need to produce engineered functional tissue products in a consistent manner. One possible method is the development of sensor-based bioreactors that may minimize variations in terms of product controls. Second, in terms of regulatory aspects, bioreactor systems for clinical purposes need to satisfy safety guidelines and offer traceability. The overall processes of bioreactor systems need to be in compliance with clear specifications. For commercialization of engineered tissue products, an automated bioreactor system needs to be built for cost-effective, standardized production and scaling-up of engineered tissue constructs [91].

Other applications of bioreactors for urological regenerative medicine include (1) a microarray bioreactor for drug screening (Fig. 4.8a) [92, 93] and (2) a microfluidic bioreactor for maintenance of reproductive tissue (e.g., sperm)

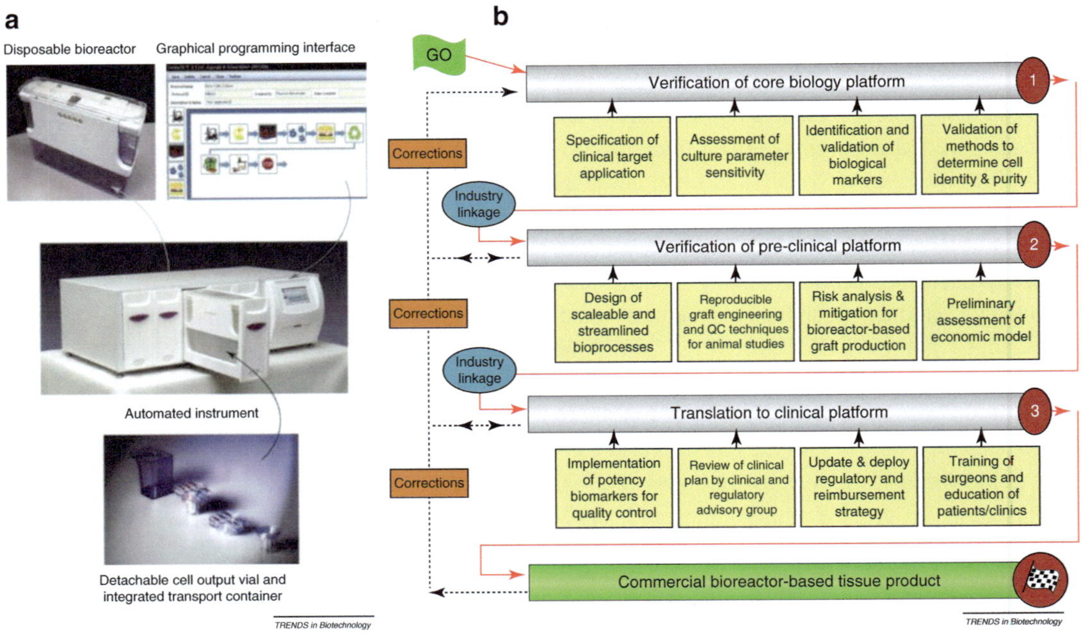

Fig. 4.7 Translation of tissue engineering strategies [91]. (**a**) Potential model of a commercial bioreactor-based system. (**b**) Proposed road map for translating bioreactor-based engineered products into the clinic. Reprinting of each image was permitted from publisher

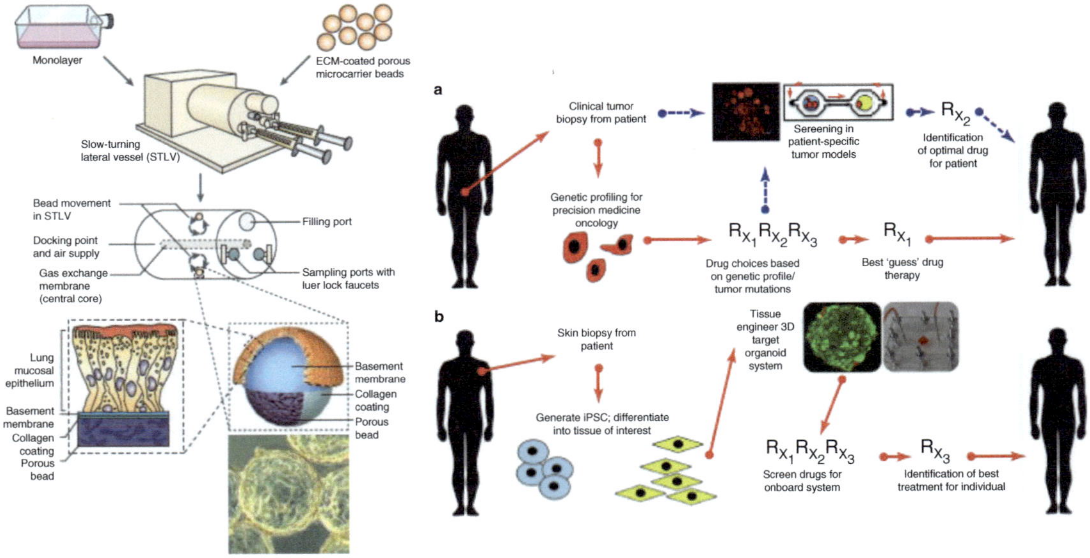

Fig. 4.8 Various urological applications using bioreactor systems. (**a**) Rotating wall vessel bioreactor system for studying host-pathogen interactions using organotypic 3D cell culture models [93]. (**b**) Microfluidic bioreactor system to use biofabricated tissues in personalized medicine [98]. Reprinting of each image was permitted from publisher

[94, 95]. For example, there has been particular interest in using bioreactors for drug screening and developing patient-specific treatments [96–98] (Fig. 4.8b). More active research thrusts focused on new applications of bioreactors for treating urological diseases, and complications will be expected in the future.

Acknowledgment I wish to thank Ms. Alicia Lee for editorial assistance with this manuscript.

References

1. Ko IK, Atala A, Yoo JJ. Tissue engineering of the urogenital system. In: Tissue engineering: principles and practices. Taylor & Francis Group, vol. 34: CRC Press; 2012. p. 31–9.
2. Ninkovic M, Dabernig W. Flap technology for reconstructions of urogenital organs. Curr Opin Urol. 2003;13:483–8.
3. Keys T, Badlani G. The scientific rationale for using biomaterials in stress urinary incontinence and pelvic organ prolapse. Curr Urol Rep. 2011;12:393–5.
4. Comiter CV. Surgery insight: management of failed sling surgery for female stress urinary incontinence. Nat Clin Pract Urol. 2006;3:666–74.
5. Devaseelan P, Fogarty P. The role of synthetic mesh in the treatment of pelvic organ prolapse. Obstet Gynaecol. 2009;11:1–9.
6. Baessler K, Maher CF. Mesh augmentation during pelvic-floor reconstructive surgery: risks and benefits. Curr Opin Obstet Gynecol. 2006;18:560–6.
7. Atala A, Bauer SB, Soker S, Yoo JJ, Retik AB. Tissue-engineered autologous bladders for patients needing cystoplasty. Lancet. 2006;367:1241–6.
8. Raya-Rivera A, Esquiliano DR, Yoo JJ, Lopez-Bayghen E, Soker S, Atala A. Tissue-engineered autologous urethras for patients who need reconstruction: an observational study. Lancet. 2011;377:1175–82.
9. Raya-Rivera AM, Esquiliano D, Fierro-Pastrana R, Lopez-Bayghen E, Valencia P, Ordorica-Flores R, Soker S, Yoo JJ, Atala A. Tissue-engineered autologous vaginal organs in patients: a pilot cohort study. Lancet. 2014;384:329–36.
10. Khodabukus A, Baar K. Regulating fibrinolysis to engineer skeletal muscle from the C2C12 cell line. Tissue engineering part C. Methods. 2009;15:501–11.
11. Huang YC, Dennis RG, Larkin L, Baar K. Rapid formation of functional muscle in vitro using fibrin gels. J Appl Physiol. 2005;98:706–13.
12. Levenberg S, Rouwkema J, Macdonald M, Garfein ES, Kohane DS, Darland DC, Marini R, van Blitterswijk CA, Mulligan RC, D'Amore PA, Langer R. Engineering vascularized skeletal muscle tissue. Nat Biotechnol. 2005;23:879–84.
13. Martin I, Wendt D, Heberer M. The role of bioreactors in tissue engineering. Trends Biotechnol. 2004;22:80–6.
14. Martin Y, Vermette P. Bioreactors for tissue mass culture: design, characterization, and recent advances. Biomaterials. 2005;26:7481–503.
15. Farhat WA, Yeger H. Does mechanical stimulation have any role in urinary bladder tissue engineering? World J Urol. 2008;26:301–5.
16. Ismadi MZ, Hourigan K, Fouras A. Experimental characterization of fluid mechanics in a spinner flask bioreactor. PRO. 2014;2:753–72.
17. Kallos MS, Sen A, Behie LA. Large-scale expansion of mammalian neural stem cells: a review. Med Biol Eng Comput. 2003;41:271–82.
18. Cabrita GJ, Ferreira BS, da Silva CL, Goncalves R, Almeida-Porada G, Cabral JM. Hematopoietic stem cells: from the bone to the bioreactor. Trends Biotechnol. 2003;21:233–40.
19. Zandstra PW, Bauwens C, Yin T, Liu Q, Schiller H, Zweigerdt R, Pasumarthi KB, Field LJ. Scalable production of embryonic stem cell-derived cardiomyocytes. Tissue Eng. 2003;9:767–78.
20. Ismadi MZ, Gupta P, Fouras A, Verma P, Jadhav S, Bellare J, Hourigan K. Flow characterization of a spinner flask for induced pluripotent stem cell culture application. PloS One. 2014;9:e106493.
21. Ashok P, Fan Y, Rostami MR, Tzanakakis ES. Aggregate and microcarrier cultures of human pluripotent stem cells in stirred-suspension systems. Methods Mol Biol. 2016;1502:35–52.
22. Unsworth BR, Lelkes PI. Growing tissues in microgravity. Nat Med. 1998;4:901–7.
23. Rhee HW, Zhau HE, Pathak S, Multani AS, Pennanen S, Visakorpi T, Chung LW. Permanent phenotypic and genotypic changes of prostate cancer cells cultured in a three-dimensional rotating-wall vessel. In Vitro Cell Dev Biol Anim. 2001;37:127–40.
24. Licato LL, Prieto VG, Grimm EA. A novel preclinical model of human malignant melanoma utilizing bioreactor rotating-wall vessels. In Vitro Cell Dev Biol Anim. 2001;37:121–6.
25. Wei X, Li DB, Xu F, Wang Y, Zhu YC, Li H, Wang KJ. A novel bioreactor to simulate urinary bladder mechanical properties and compliance for bladder functional tissue engineering. Chin Med J. 2011;124:568–73.
26. Lei X, Deng Z, Duan E. Uniform Embryoid body production and enhanced Mesendoderm differentiation with murine embryonic stem cells in a rotary suspension bioreactor. Methods Mol Biol. 2016;1502:63–75.
27. Vunjak-Novakovic G, Martin I, Obradovic B, Treppo S, Grodzinsky AJ, Langer R, Freed LE. Bioreactor cultivation conditions modulate the composition and mechanical properties of tissue-engineered cartilage. J Orthopaedic Res. 1999;17:130–8.
28. Pei M, Solchaga LA, Seidel J, Zeng L, Vunjak-Novakovic G, Caplan AI, Freed LE. Bioreactors

mediate the effectiveness of tissue engineering scaffolds. FASEB J. 2002;16:1691–4.
29. Sakai K. Dialysis membranes for blood purification. Front Med Biol Eng. 2000;10:117–29.
30. Lipps B, Stewart RD, Perkins H, Holmes GW, McLain EA, Rolfs MR, Oja P. The hollow fiber artificial kidney. Trans Am Soc Artif Intern Organs. 1967;13:200–7.
31. Clark WR. Hemodialyzer membranes and configurations: a historical perspective. Semin Dial. 2000;13:309–11.
32. Patankar D, Oolman T. Wall-growth hollow-fiber reactor for tissue culture: I. Preliminary experiments Biotechnology and bioengineering. 1990;36:97–103.
33. Storm MP, Sorrell I, Shipley R, Regan S, Luetchford KA, Sathish J, Webb S, Ellis MJ. Hollow fiber bioreactors for in vivo-like mammalian tissue culture. J Vis Exp. 2016;111:53431.
34. Wung N, Acott SM, Tosh D, Ellis MJ. Hollow fibre membrane bioreactors for tissue engineering applications. Biotechnol Lett. 2014;36:2357–66.
35. Roberts I, Baila S, Rice RB, Janssens ME, Nguyen K, Moens N, Ruban L, Hernandez D, Coffey P, Mason C. Scale-up of human embryonic stem cell culture using a hollow fibre bioreactor. Biotechnol Lett. 2012;34:2307–15.
36. Nold P, Brendel C, Neubauer A, Bein G, Hackstein H. Good manufacturing practice-compliant animal-free expansion of human bone marrow derived mesenchymal stroma cells in a closed hollow-fiber-based bioreactor. Biochem Biophys Res Commun. 2013;430:325–30.
37. Usuludin SB, Cao X, Lim M. Co-culture of stromal and erythroleukemia cells in a perfused hollow fiber bioreactor system as an in vitro bone marrow model for myeloid leukemia. Biotechnol Bioeng. 2012;109:1248–58.
38. Housler GJ, Miki T, Schmelzer E, Pekor C, Zhang X, Kang L, Voskinarian-Berse V, Abbot S, Zeilinger K, Gerlach JC. Compartmental hollow fiber capillary membrane-based bioreactor technology for in vitro studies on red blood cell lineage direction of hematopoietic stem cells. Tissue Eng Part C Methods. 2012;18:133–42.
39. Yamazoe H, Iwata H. Efficient generation of dopaminergic neurons from mouse embryonic stem cells enclosed in hollow fibers. Biomaterials. 2006;27:4871–80.
40. Amimoto N, Mizumoto H, Nakazawa K, Ijima H, Funatsu K, Kajiwara T. Hepatic differentiation of mouse embryonic stem cells and induced pluripotent stem cells during organoid formation in hollow fibers. Tissue Eng Part A. 2011;17:2071–8.
41. Sussman NL, Kelly JH. Extracorporeal liver support: cell-based therapy for the failing liver. Am J Kidney Dis. 1997;30:S66–71.
42. Kelly JH, Koussayer T, He D, Chong MG, Shang TA, Whisennand HH, Sussman NL. Assessment of an extracorporeal liver assist device in anhepatic dogs. Artif Organs. 1992;16:418–22.
43. Struecker B, Raschzok N, Sauer IM. Liver support strategies: cutting-edge technologies. Nat Rev Gastroenterol Hepatol. 2014;11:166–76.
44. Jansen J, Fedecostante M, Wilmer MJ, van den Heuvel LP, Hoenderop JG, Masereeuw R. Biotechnological challenges of bioartificial kidney engineering. Biotechnol Adv. 2014;32:1317–27.
45. Humes HD, Buffington DA, MacKay SM, Funke AJ, Weitzel WF. Replacement of renal function in uremic animals with a tissue-engineered kidney. Nat Biotechnol. 1999;17:451–5.
46. Gleeson MJ, Griffith DP. The use of alloplastic biomaterials in bladder substitution. J Urol. 1992;148:1377–82.
47. Schultheiss D, Gabouev AI, Cebotari S, Tudorache I, Walles T, Schlote N, Wefer J, Kaufmann PM, Haverich A, Jonas U, Stief CG, Mertsching H. Biological vascularized matrix for bladder tissue engineering: matrix preparation, reseeding technique and short-term implantation in a porcine model. J Urol. 2005;173:276–80.
48. Danielsson C, Ruault S, Basset-Dardare A, Frey P. Modified collagen fleece, a scaffold for transplantation of human bladder smooth muscle cells. Biomaterials. 2006;27:1054–60.
49. Schoeller T, Lille S, Stenzl A, Ninkovic M, Piza H, Otto A, Russell RC, Wechselberger G. Bladder reconstruction using a prevascularized capsular tissue seeded with urothelial cells. J Urol. 2001;165:980–5.
50. Kanematsu A, Yamamoto S, Noguchi T, Ozeki M, Tabata Y, Ogawa O. Bladder regeneration by bladder acellular matrix combined with sustained release of exogenous growth factor. J Urol. 2003;170:1633–8.
51. Yoo JJ, Meng J, Oberpenning F, Atala A. Bladder augmentation using allogenic bladder submucosa seeded with cells. Urology. 1998;51:221–5.
52. Pattison MA, Wurster S, Webster TJ, Haberstroh KM. Three-dimensional, nano-structured PLGA scaffolds for bladder tissue replacement applications. Biomaterials. 2005;26:2491–500.
53. Korossis S, Bolland F, Ingham E, Fisher J, Kearney J, Southgate J. Review: tissue engineering of the urinary bladder: considering structure-function relationships and the role of mechanotransduction. Tissue Eng. 2006;12:635–44.
54. Wallis MC, Yeger H, Cartwright L, Shou Z, Radisic M, Haig J, Suoub M, Antoon R, Farhat WA. Feasibility study of a novel urinary bladder bioreactor. Tissue Eng Part A. 2008;14:339–48.
55. Davis NF, Mooney R, Piterina AV, Callanan A, McGuire BB, Flood HD, McGloughlin TM. Construction and evaluation of urinary bladder bioreactor for urologic tissue-engineering purposes. Urology. 2011;78:954–60.
56. Davis NF, Callanan A. Development of a bladder bioreactor for tissue engineering in urology. Methods Mol Biol. 2016;1502:213–21.
57. Campbell GR, Turnbull G, Xiang L, Haines M, Armstrong S, Rolfe BE, Campbell JH. The peritoneal cavity as a bioreactor for tissue engineering visceral

organs: bladder, uterus and vas deferens. J Tissue Eng Regen Med. 2008;2:50–60.
58. Kajbafzadeh AM, Esfahani SA, Sadeghi Z, Elmi A, Monajemzadeh M. Application of different scaffolds for bladder wall regeneration: the bladder as a natural bioreactor. Tissue Eng Part A. 2012;18:882–7.
59. Fu Q, Deng CL, Zhao RY, Wang Y, Cao Y. The effect of mechanical extension stimulation combined with epithelial cell sorting on outcomes of implanted tissue-engineered muscular urethras. Biomaterials. 2014;35:105–12.
60. le Roux PJ. Endoscopic urethroplasty with unseeded small intestinal submucosa collagen matrix grafts: a pilot study. J Urol. 2005;173:140–3.
61. Li C, Xu Y, Song L, Fu Q, Cui L, Yin S. Preliminary experimental study of tissue-engineered urethral reconstruction using oral keratinocytes seeded on BAMG. Urol Int. 2008;81:290–5.
62. Li C, Xu YM, Song LJ, Fu Q, Cui L, Yin S. Urethral reconstruction using oral keratinocyte seeded bladder acellular matrix grafts. J Urol. 2008;180:1538–42.
63. Dahms SE, Piechota HJ, Nunes L, Dahiya R, Lue TF, Tanagho EA. Free ureteral replacement in rats: regeneration of ureteral wall components in the acellular matrix graft. Urology. 1997;50:818–25.
64. Osman Y, Shokeir A, Gabr M, El-Tabey N, Mohsen T, El-Baz M. Canine ureteral replacement with long acellular matrix tube: is it clinically applicable? J Urol. 2004;172:1151–4.
65. Shen J, Fu X, Ou L, Zhang M, Guan Y, Wang K, Che Y, Kong D, Steinhof G, Li W, Yu Y, Ma N. Construction of ureteral grafts by seeding urothelial cells and bone marrow mesenchymal stem cells into polycaprolactone-lecithin electrospun fibers. Int J Artif Organs. 2010;33:161–70.
66. Shi JG, Fu WJ, Wang XX, Xu YD, Li G, Hong BF, Wang Y, Du ZY, Zhang X. Tissue engineering of ureteral grafts by seeding urothelial differentiated hADSCs onto biodegradable ureteral scaffolds. J Biomed Mater Res A. 2012;100:2612–22.
67. Vardar E, Engelhardt EM, Larsson HM, Mouloungui E, Pinnagoda K, Hubbell JA, Frey P. Tubular compressed collagen scaffolds for ureteral tissue engineering in a flow bioreactor system. Tissue Eng Part A. 2015;21:2334–45.
68. Seifarth V, Gossmann M, Janke HP, Grosse JO, Becker C, Heschel I, Artmann GM, Temiz Artmann A. Development of a bioreactor to culture tissue engineered ureters based on the application of tubular OPTIMAIX 3D scaffolds. Urol Int. 2015;95:106–13.
69. Wang PC, Takezawa T. Reconstruction of renal glomerular tissue using collagen vitrigel scaffold. J Biosci Bioeng. 2005;99:529–40.
70. Lu SH, Lin Q, Liu YN, Gao Q, Hao T, Wang Y, Zhou J, Wang H, Du Z, Wu J, Wang CY. Self-assembly of renal cells into engineered renal tissues in collagen/Matrigel scaffold in vitro. J Tissue Eng Regen Med. 2012;6:786–92.
71. Guimaraes-Souza NK, Yamaleyeva LM, AbouShwareb T, Atala A, Yoo JJ. In vitro reconstitution of human kidney structures for renal cell therapy. Nephrol Dial Transplant. 2012;27:3082–90.
72. Rosines E, Johkura K, Zhang X, Schmidt HJ, Decambre M, Bush KT, Nigam SK. Constructing kidney-like tissues from cells based on programs for organ development: toward a method of in vitro tissue engineering of the kidney. Tissue Eng Part A. 2010;16:2441–55.
73. Yoo JJ, Ashkar S, Atala A. Creation of functional kidney structures with excretion of kidney-like fluid in vivo. Pediatrics. 1996;98(suppl):605.
74. Kim SS, Park HJ, Han J, Choi CY, Kim BS. Renal tissue reconstitution by the implantation of renal segments on biodegradable polymer scaffolds. Biotechnol Lett. 2003;25:1505–8.
75. Taguchi A, Kaku Y, Ohmori T, Sharmin S, Ogawa M, Sasaki H, Nishinakamura R. Redefining the in vivo origin of metanephric nephron progenitors enables generation of complex kidney structures from pluripotent stem cells. Cell Stem Cell. 2014;14:53–67.
76. Sharmin S, Taguchi A, Kaku Y, Yoshimura Y, Ohmori T, Sakuma T, Mukoyama M, Yamamoto T, Kurihara H, Nishinakamura R. Human induced pluripotent stem cell-derived podocytes mature into vascularized glomeruli upon experimental transplantation. J Am Soc Nephrol. 2016;27:1778–91.
77. Lam AQ, Freedman BS, Morizane R, Lerou PH, Valerius MT, Bonventre JV. Rapid and efficient differentiation of human pluripotent stem cells into intermediate mesoderm that forms tubules expressing kidney proximal tubular markers. J Am Soc Nephrol. 2014;25:1211–25.
78. Morizane R, Monkawa T, Itoh H. Differentiation of murine embryonic stem and induced pluripotent stem cells to renal lineage in vitro. Biochem Biophys Res Commun. 2009;390:1334–9.
79. Ko IK, Abolbashari M, Huling J, Kim C, Mirmalek Sani SH, Moradi M, Orlando G, Jackson JD, AbouShwareb T, Soker S, Yoo JJ, Atala A. Enhanced re-endothelialization of acellular kidney scaffolds for whole organ engineering via antibody conjugation of vasculatures. Technology. 2014;2:243–53.
80. Song JJ, Guyette JP, Gilpin SE, Gonzalez G, Vacanti JP, Ott HC. Regeneration and experimental orthotopic transplantation of a bioengineered kidney. Nat Med. 2013;19:646–51.
81. Ross EA, Williams MJ, Hamazaki T, Terada N, Clapp WL, Adin C, Ellison GW, Jorgensen M, Batich CD. Embryonic stem cells proliferate and differentiate when seeded into kidney scaffolds. J Am Soc Nephrol. 2009;20:2338–47.
82. Arenas-Herrera JE, Ko IK, Atala A, Yoo JJ. Decellularization for whole organ bioengineering. Biomed Mater. 2013;8:014106.
83. Ross EA, Abrahamson DR, John PS, Clapp WL, Williams MJ, Terada N, Hamazaki T, Ellison GW, Batich CD. Mouse stem cells seeded into decellularized rat kidney scaffolds endothelialize and remodel basement membranes. Organogenesis. 2012;8:49–55.

84. Uzarski JS, Bijonowski BM, Wang B, Ward HH, Wandinger-Ness A, Miller WM, Wertheim JA. Dual-purpose bioreactors to monitor noninvasive physical and biochemical markers of kidney and liver scaffold recellularization. Tissue Eng C Methods. 2015;21:1032–43.
85. Abolbashari M, Agcaoili SM, Lee MK, Ko IK, Aboushwareb T, Jackson JD, Yoo JJ, Atala A. Repopulation of porcine kidney scaffold using porcine primary renal cells. Acta Biomater. 2016;29:52–61.
86. Martin GR. Isolation of a pluripotent cell line from early mouse embryos cultured in medium conditioned by teratocarcinoma stem cells. Proc Natl Acad Sci U S A. 1981;78:7634–8.
87. Kim IH, Ko IK, Atala A, Yoo JJ. Whole kidney engineering for clinical translation. Curr Opin Organ Transplant. 2015;20:165–70.
88. Lawton JS, Moazami N, Pasque MK, Moon MR, Damiano RJ Jr. Early stenosis of Medtronic mosaic porcine valves in the aortic position. J Thorac Cardiovasc Surg. 2009;137:1556–7.
89. Iannotti JP, Codsi MJ, Kwon YW, Derwin K, Ciccone J, Brems JJ. Porcine small intestine submucosa augmentation of surgical repair of chronic two-tendon rotator cuff tears. A randomized, controlled trial. J Bone Joint Surg Am. 2006;88:1238–44.
90. Sullivan DC, Mirmalek-Sani SH, Deegan DB, Baptista PM, Aboushwareb T, Atala A, Yoo JJ. Decellularization methods of porcine kidneys for whole organ engineering using a high-throughput system. Biomaterials. 2012;33:7756–64.
91. Martin I, Smith T, Wendt D. Bioreactor-based roadmap for the translation of tissue engineering strategies into clinical products. Trends Biotechnol. 2009;27:495–502.
92. Radtke AL, Herbst-Kralovetz MM. Culturing and applications of rotating wall vessel bioreactor derived 3D epithelial cell models. J Vis Exp. 2012;62:3868.
93. Barrila J, Radtke AL, Crabbe A, Sarker SF, Herbst-Kralovetz MM, Ott CM, Nickerson CA. Organotypic 3D cell culture models: using the rotating wall vessel to study host-pathogen interactions. Nat Rev Microbiol. 2010;8:791–801.
94. Komeya M, Kimura H, Nakamura H, Yokonishi T, Sato T, Kojima K, Hayashi K, Katagiri K, Yamanaka H, Sanjo H, Yao M, Kamimura S, Inoue K, Ogonuki N, Ogura A, Fujii T, Ogawa T. Long-term ex vivo maintenance of testis tissues producing fertile sperm in a microfluidic device. Sci Rep. 2016;6:21472.
95. McCormack MC, McCallum S, Behr B. A novel microfluidic device for male subfertility screening. J Urol. 2006;175:2223–7. discussion 2227
96. Schneider M, Schuler J, Hofflin R, Korzeniewski N, Grullich C, Roth W, Teber D, Hadaschik B, Pahernik S, Hohenfellner M, Duensing S. Phenotypic drug screening and target validation for improved personalized therapy reveal the complexity of phenotype-genotype correlations in clear cell renal cell carcinoma. Urol Oncol. 2014;32:877–84.
97. Skardal A, Devarasetty M, Forsythe S, Atala A, Soker S. A reductionist metastasis-on-a-chip platform for in vitro tumor progression modeling and drug screening. Biotechnol Bioeng. 2016;113:2020–32.
98. Skardal A, Shupe T, Atala A. Organoid-on-a-chip and body-on-a-chip systems for drug screening and disease modeling. Drug Discov Today. 2016;21:1399–411.

3D Bioprinting for Tissue Engineering

5

Sujin Noh, Noehyun Myung, Myeongji Park, Seulgi Kim, Sung-Uk Zhang, and Hyun-Wook Kang

Abbreviations

BMSC	Bone marrow stem cells
CTZ	Cetirizine HCl
DAT	Decellularized adipose tissue
DPD	Diphenylhydramine HCl
ECM	Endothelial cell medium
FM	Fibroblast medium
GelMA	Gelatin methacrylate
GS	Gentamicin sulfate
HA	Hyaluronic acid
hASCs	Human adipose tissue-derived mesenchymal stem cells
hCMPCs	Human cardiac-derived cardiomyocyte progenitor cells
hESCs	Human embryonic stem cells
HFF-1	Neonatal human foreskin fibroblasts
hiPSCs	Human induced pluripotent stem cells
HUVECs	Human umbilical vein endothelial cells
HUVSMC	Human umbilical vein smooth muscle cells
IBU	Ibuprofen
ihMSCs	Immortalized human mesenchymal stem cells
MSCs	Mesenchymal stromal cells
MTX	Methotrexate
MWNTs	Multi walled carbon nanotubes
n-HA	*nano*-Hydroxyapatite
NHDFs	Normal human dermal fibroblasts
NHLFs	Normal human lung fibroblasts
PCL	Polycaprolactone
PDLLA	Poly(D,L-lactic acid or D,L-lactide)
PDMS	Polydimethylsiloxane
PLGA	Poly(lactic-co-glycolic acid)
PLLA	Poly-L-lactide
PU	Polyurethane
PVA	Polyvinyl alcohol
SCs	Schwann cells
SF-PGA	Silk fibroin-poly glutamic acid
SF-PLL	Silk fibroin-poly-L-lysine
SMCM	Smooth muscle cell medium
SMCs	Smooth muscle cells

S. Noh • N. Myung • M. Park • S. Kim • H.-W. Kang (✉)
School of Life Science, Ulsan National Institute of Science and Technology, 50 UNIST-gil, Ulju-gun, 44919 Ulsan, South Korea
e-mail: hkang@unist.ac.kr

S.-U. Zhang
Department of Mechanical Engineering, Dong-Eui University, 176 Eomgwangro, Busanjin-gu, 47340 Busan, South Korea

5.1 Introduction

Tissue engineering has become a promising technology for the functional recovery of tissues or organs damaged by accidents or diseases. According to the 2014 annual report of the

B.W. Kim (ed.), *Clinical Regenerative Medicine in Urology*, DOI 10.1007/978-981-10-2723-9_5

Scientific Registry of Transplant Recipients (SRTR), an imbalance remains between the supply and demand for organ transplantation [1]. The shortage of donor organs has resulted in the development of alternatives, such as artificial mechanical organs, xenotransplantation, and tissue engineering. Using artificial mechanical organs may produce many unwanted side effects and markedly reduces patient quality of life. Xenotransplantation uses organs from transgenic animals with reduced capacity. This technology has the intrinsic problem of an acute immune response after transplantation. To address this, immunosuppressive therapy is used with xenotransplantation but the potential for spreading animal viruses and long-term psychological problems are still concerns. Tissue engineering, a candidate technology to avoid these problems, provides a solution for repairing injured organs by artificially regenerating the damaged organs (Fig. 5.1). Moreover, engineered tissues and organs can be used for drug testing and screening and as personalized medicine. However, this approach still has many obstacles to overcome to achieve these goals, such as vascularization, scale-up, and composite tissue regeneration [2, 3].

Recently, bioprinting has emerged as a technology within tissue engineering; it is a two-dimensional (2D) or three-dimensional (3D) fabrication method that uses living cells, biomaterials, and bioactive molecules. Using bioprinting technology, tissue- or organ-like architectures, composed of multiple cell types, can be produced. Moreover, such printing processes can be automated using computer-aided design templates. Many researchers working in regenerative medicine have tried to develop biomimetic architectures with this technology. In this chapter, current bioprinting technologies and their applications are introduced.

5.2 Bioprinting Techniques

Bioprinting techniques can be categorized into three types: jetting-, extrusion-, and laser-based bioprinting (LBB). In this section, the three techniques are described in terms of their working principles, components, advantages, disadvantages, and applicable bio-inks.

Table 5.1 summarizes achievable specifications for each bioprinting technology, in terms of

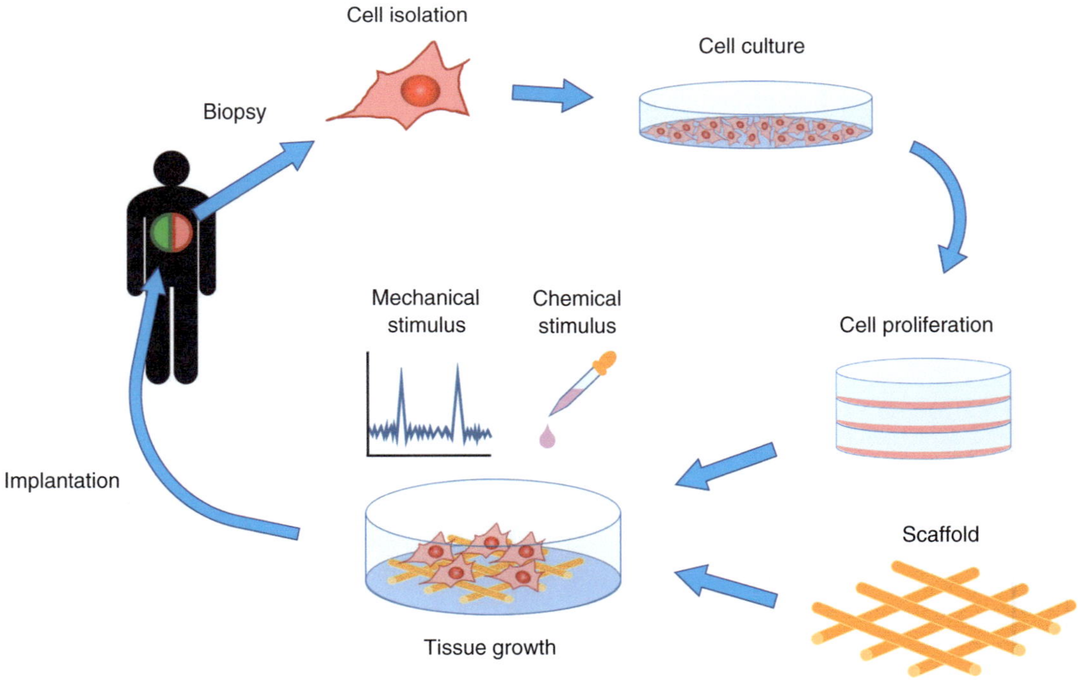

Fig. 5.1 The classical tissue engineering approach

Table 5.1 Summary of bioprinting techniques

Printing technique	Jetting-based bioprinting	Extrusion-based bioprinting	Laser-based printing		References
			Stereo-lithography	Laser-assisted bioprinting	
Resolution or droplet size	<85 μm	5 μm ~	<1.2 μm	8.7–140 μm	[22, 30, 34, 42, 43]
Printing speed	10 μm/s–50 mm/s	2–5 mm/s	50 μm/s–5,000 mm/s (laser scanning rates)	200–2,000 mm/s (laser scanning rates)	[13, 24, 35, 36, 43, 45, 46]
Cell viability	>89.8%	70.0–93.0%	>85.0%	63.8–96.0%	[8, 19, 30, 41, 44, 47]
Cell density	10^6 cells/mL	<10^7 cells/mL	10^6 cells/mL	10^7–10^8 cells/mL	[13, 36, 48, 49]
Material	Hydrogel, media, cells, proteins and drug	Hydrogel, cell aggregates (w/ or w/o media), bioceramics, proteins and drug	Photocurable hydrogel, cells, and bioceramics	Hydrogel, cell aggregates (w or w/o media), peptides, and bioceramics	[5–7, 13–15, 26, 27, 35, 38, 41, 50, 51]
Material viscosity	1–18 mPa/s	30,000–50,000 mPa/s	2.05–5,000 mPa/s	40–120 mPa/s	[8, 12, 27, 36, 52, 53]
Gelation method	Ionic, thermal, enzymatic, and photo crosslinking	Ionic, thermal, enzymatic, photo, chemical and pH crosslinking	Photo crosslinking	Ionic and pH crosslinking	[10, 13, 41, 42, 47, 54–60]
Advantages	Low printer cost, control droplet size, availability to many biomaterials and control cell concentration by changing drop density	Capacity to extrude cells up to high densities and wide range of viscosity	Control pore distribution, pore size and shape of a scaffold, high resolution and less effect of temperature during the printing process	Nozzle-free, no clogging and wide range of viscosity	[4, 25]
Disadvantages	Frequent nozzle clogging, mechanical and thermal stress to cells, difficulty in cell encapsulation, limitations of material viscosity, requirement for crosslinkers which are cytotoxic and low cell density than other techniques	Extrusion pressure and crosslinkers decrease cell viability	Limited printing materials	Long preparation time, poor cell positioning/targeting, contamination by metallic residues and high cost	[4, 61]

resolution, printing speed, cell viability and density, materials, applicable viscosity, gelation method, and advantages and disadvantages.

5.2.1 Jetting-Based Bioprinting

Jetting-based bioprinting generates and delivers small droplets to a desired site to produce cell patterning; it is also called "drop-on-demand" printing (Fig. 5.2). This method is easy to develop because it needs only minor modifications from a "traditional" inkjet printer. The first modification is to replace the ink with a bio-ink; the second is that an elevator is placed under the printing head to achieve a layer-by-layer process for the construction of a 3D structure. Thus, a jetting-based bio-printer consists essentially of the printing head and elevator. The printing head produces droplets using a thermal piezoelectric actuator, or a solenoid valve. The first method involves a heater attached to the printer head. The heater generates thermal energy using electric current pulses. The energy induces vapor bubbles within the ink. Then, these bubbles push the ink out of the nozzle, so that droplets can be generated. The second method uses a piezoelectric actuator on the printer head. The actuator is triggered by applying a voltage pulse, so that the actuator deforms and squeezes ink out of the nozzle. The third method is solenoid valve-based bioprinting. The valve is controlled by a microcontroller, so that the size of droplets can be changed by the valve-open duration.

Jetting-based bioprinting has several advantages. First, this method is relatively inexpensive in comparison with extrusion- based bioprinting and laser-based bioprinting. Second, the printed concentration of cells can be adjusted readily by controlling for the density of droplets. Finally, this methodology can use a variety of biomaterials as bio-ink [4]. For example, Horváth et al. [5] developed an in vitro air–blood tissue barrier using hy926 endothelial cells and A549 epithelial cells with Matrigel. Scoutaris et al. [7] reported printing results with cetirizine HCl (CTZ), diphenhydramine HCl (DPD), and ibuprofen (IBU).

However, jetting-based bioprinting has an intrinsic limitation in terms of the biomaterials; these should have a low viscosity (1–18 mPa/s) because the technique requires the generation of small droplets [8]. The viscosity also affects the speed of the process and the cell viability. To cre-

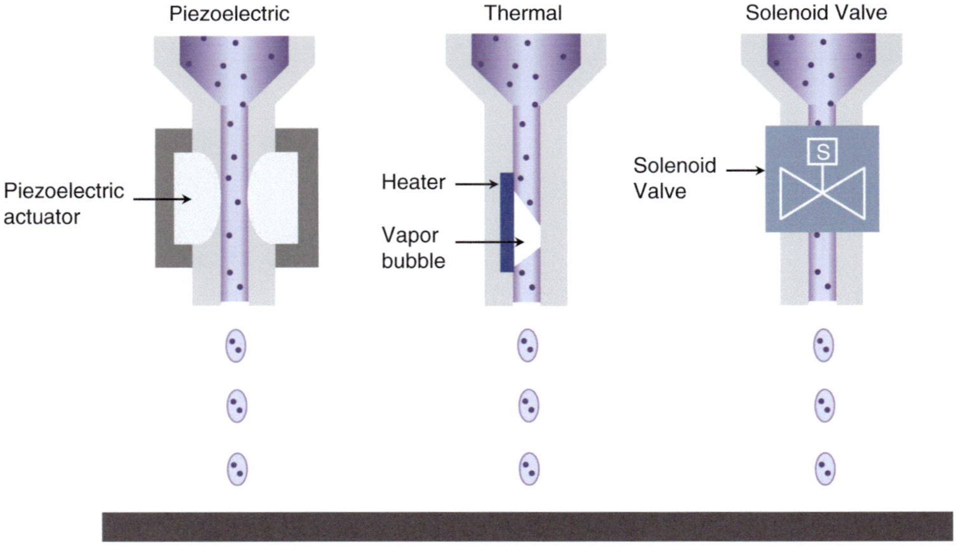

Fig. 5.2 Jetting-based bioprinting (using piezoelectric pressure (*left*), thermally induced air-pressure (*middle*), and solenoid valve (*right*))

ate a rigid scaffold using low-viscosity materials, a time-consuming process for crosslinking is typically required after printing. When droplets are created from the printer head, cell viability can be influenced by thermal or mechanical stress. Moreover, the available cell density is limited, because nozzle clogging is an issue with this technology [4].

Recently, several new methods have been introduced to overcome these difficulties. Kador et al. [9] suggested a design for a retinal model produced by thermal inkjet printing. They overcame the limitations of the electrospinning approach, in terms of cell positioning, by combined use of jetting-based bioprinting. Xu et al. [10] prepared functional cartilage tissue with electrospinning. The electrospinning technique was used to address the disadvantages of jetting-based bioprinting in terms of mechanical properties. Moreover, Binder [11] developed an in situ skin bioprinter for enhancing the healing of skin wounds. They directly printed skin cells, including fibroblasts and keratinocytes, on a skin wound.

5.2.2 Extrusion-Based Bioprinting

Extrusion-based bioprinting is a technique to create 3D cellular constructs by dispensing continuous filaments with a cell-laden hydrogel. The printer consists of a cartridge, deposition stage, and printer head, composed of a syringe and nozzle. According to the dispensing system, extrusion-based bioprinting is categorized into two types: pneumatic and mechanical extrusion systems (Fig. 5.3). Pneumatic extrusion-based bioprinting prints bio-ink out of the nozzle by applying air pressure. The amount of bio-ink is adjusted by controlling the air pressure. This method can use relatively high-viscosity materials, in comparison with jetting-based methods. Mechanical extrusion-based bioprinting uses a piston or screw for printing. Piston-based bioprinting controls the volume of ink extruded directly. Screw-based dispensing can produce much higher forces, so the system can handle materials of much higher viscosity.

As mentioned above, extrusion-based bioprinting can handle high-viscosity materials, of about 30,000–50,000 mPa/s [12]. Thus, high concentrations of cells can be used in such a system. A cell concentration of ~1×10^7 cells/mL has been reported [13]. Moreover, a wide range of biomaterials has been used in extrusion-based systems. Luo et al. [14] created a scaffold with alginate, Pluronic F127, and a bioceramic ($Ca_7Si_2P_2O_{16}$) powder for bone tissue engineering. Latza et al. [15] studied a device for wearable electronics using sucker ring teeth (SRT; from squid) protein, water, and glycerol. Yang

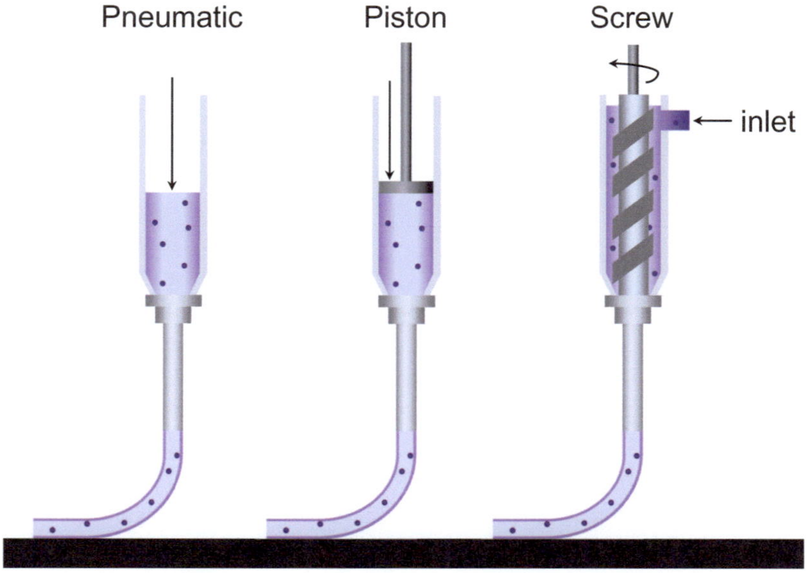

Fig. 5.3 Extrusion-based bioprinting (using pneumatic (*left*) and mechanical (piston (*middle*), and screw (*right*)) pressure)

et al. [13] produced osteochondral biphasic grafts using alginate, cartilage-derived ECM, PLGA, hydroxyapatite (HA), and cells.

Recently, various functional bio-inks have been developed for extrusion-based bioprinting. Rees et al. [17] suggested a nanocellulose (carboxymethylation and periodate oxidation)-based bio-ink that could prevent bacterial growth. The material was used to print a wound dressing. Highley et al. [16] developed a supramolecular self-healing hydrogel, which was cured by applying a physical stimulus. The hydrogel was restored when the physical stimulus was removed. The material could be used to construct complex structures with high resolution. Cho's group used a decellularization process to develop tissue-specific bio-inks with adipose, cartilage, and heart tissue. These bio-inks provided enhanced biocompatibility for tissue growth in a 3D structure [18].

Extrusion-based scaffold-free bioprinting has also been described, in which cell aggregates having spherical or rod shapes are printed directly with no additional biomaterials [20, 21]. Using the bio-printed cell aggregates, organ-specific biological features could be mimicked. Moreover, tissue-like cell densities can be achieved. For example, Itoh et al. [20] introduced a tubular structure with multicellular spheroids. However, the structure did not easily stand by itself without supporting materials, so a fine architecture could not be formed. To overcome this, Kucukgul et al. [21] introduced a new process. They defined a designed structure and empty spaces. In the empty spaces, a supporting structure was printed with agarose gel. Then, computer-designed structures were plotted with the rod-shaped cell aggregates.

5.2.3 Laser-Based Bioprinting (LBB)

The LBB technique involves constructing a structure by solidifying or depositing bio-ink with photon energy. This technique can be divided into stereolithography (SLA) (Fig. 5.4) and laser-assisted bioprinting (LAB) (Fig. 5.5).

5.2.3.1 Stereolithography (SLA)

SLA produces 3D structures with a scanning laser beam on a photo-curable material. Thus, the system consists primarily of a laser and optics, a photopolymer container, and an elevator. The laser beam irradiates the photopolymer. Galvanometric mirrors and lenses are used to control the position of the beam and its focusing. Transferred photon energy initiates the photo-curing process and a 2D pattern is generated by scanning the beam. Then, a 3D structure can be produced, by stacking the patterns layer-by-layer.

SLA technology has good spatial resolution, of <1.2 μm [22]. Recently, higher resolutions have been obtained using two-photon polymerization (TPP). TPP uses a femtosecond laser to obtain a high-intensity near-infrared beam that induces localized excitation by two-photon absorption. This then induces nano-scale photopolymerization. Haske et al. [23] achieved a resolution of 65 nm. Thus, SLA can control the shape of a scaffold, its pore distribution, and overall size very precisely. Other advantages of SLA technology include the speed and temperature of the fabrication process. The speed is high because the laser scanning rate can be up to 5000 mm/s [24]. The temperature during the fabrication is lower than with other techniques, so that cells are less affected [25]. SLA can use several types of biomaterials. Lee et al. [26] introduced a customized poly(propylene fumarate)/diethyl fumarate photopolymer and built 3D scaffolds for bone tissue engineering. Jansen et al. [27] created porous structures with a well-defined gyroid architecture for tissue engineering using pentaerythritol tetra-acrylate resin, ceramic suspensions, and PDLLA3-FAME/NVP resins. However, the number of applicable biomaterials for SLA is still small in comparison with the other techniques, because SLA requires a photocurable polymer. Recently, Shanjani et al. [28] reported a hybrid printing system combining SLA and extrusion-based bioprinting to allow use of "softer" materials as the bio-ink.

Recently, researchers have studied building 3D cellular scaffolds using SLA [29–31]. Soman et al. [29] created an in vitro model with PEG material

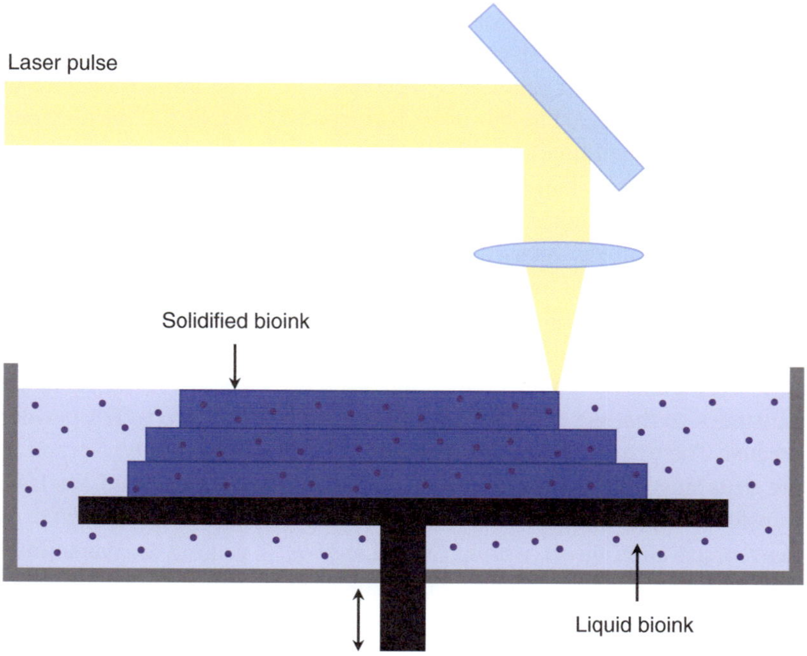

Fig. 5.4 Laser-based bioprinting (stereolithography (SLA))

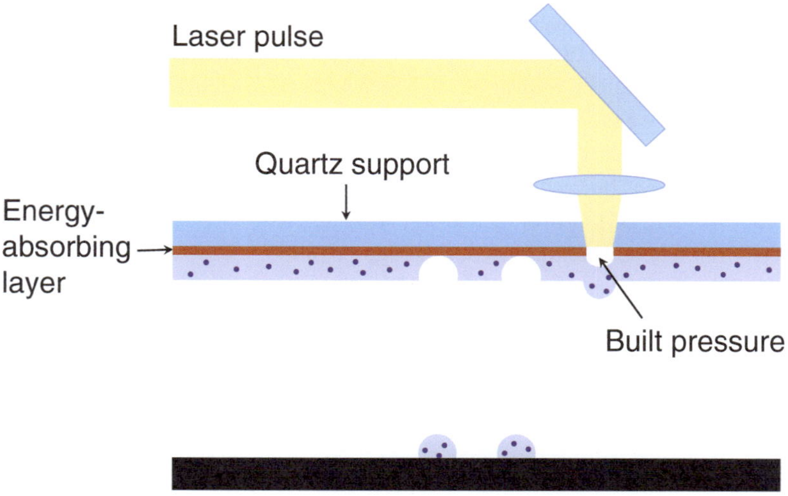

Fig. 5.5 Laser-based bioprinting (laser-assisted bioprinting (LAB))

and cancer cells to mimic a metastatic process. SLA has also been used to construct microfluidic chips, because the method is cost-efficient and the process can be automated. Shallan et al. [31] developed a complex monolithic device using a transparent resin. Au et al. [32] evaluated a commercially available SLA system to produce computer-designed microfluidic chips. It was validated by culturing and imaging with Chinese hamster ovary (CHO) cell on the chip.

5.2.3.2 Laser-Assisted Bioprinting (LAB)

LAB involves primarily a laser, bio-ink-coated ribbon, and the collector substrate. Lasers can be divided into pulsed and continuous devices. The ribbon consists of a glass or quartz plate, a metal layer, typically of gold, silver, or titanium and an energy absorbing layer (interface layer) made of a sacrificial hydrogel. The bio-ink is then coated on the energy absorbing layer. The collector substrate is coated with biopolymer and cell culture medium to enhance adhesion of the cells.

The laser pulse is focused and irradiated on the bio-ink-coated ribbon using the optics. Then, the energy absorbing layer is expanded thermally and generates vapor bubbles to form and jet droplets of bio-ink. The droplets fall on the receiving substrates so that the designed 2D cell pattern is created. The two most common methods of LAB are absorbing film-assisted laser-induced forward transfer (AFA-LIFT) and matrix-assisted pulsed laser evaporation direct writing (MAPLE-DW). AFA-LIFT is modified from LIFT for bioprinting. AFA-LIFT uses a sacrificial metal, such as gold or titanium, for the energy absorbing layer. It uses a high-power laser system and the metal layer can induce metallization of the ejected ink. This can affect the cytotoxicity of the process. To overcome this, MAPLE-DW uses a lower-power laser and a sacrificial hydrogel, such as Matrigel, for the energy absorbing layer [33].

LAB uses a laser pulse, so that relatively high resolution can be achieved, from 8.7 μm to 140 μm [31, 34]. Laser scanning rates up to 2000 mm/s have been reported; this high printing speed can be advantageous [35]. LAB uses a ribbon, so a nozzle is not necessary. Thus, nozzle clogging is not a concern, so materials of various viscosities can be used for printing. Additionally, the cell density can be increased, up to 10^8 cells/mL [36]. However, the ribbon is used over a relatively long time and high-precision cell patterning cannot be guaranteed. LAB is expensive because of the laser and it has limitations in terms of scaling up the structure [37].

Several bio-inks have been introduced for use with LAB technology. Pagès et al. [35] created 3D constructs for complex tissue substitutes using collagen and cells in medium. Dinca et al. [38] introduced a controllable self-assembly technique for peptides for use in biosensors and tissue engineering. Cheng et al. [39] developed a unique process to coat HA powders on titanium substrates.

Recently, several advanced processes have been reported for LAB. Catros et al. [40] optimized experimental conditions using inorganic nano-hydroxyapatite (*n*-HA) and human osteoprogenitor cells. They showed that cell viability, proliferation and phenotype were maintained without changes in the physicochemical properties of *n*-HA. Xiong et al. [41] printed straight and Y-shaped tubes using a ribbon composed of alginate and cells in medium. Their preliminary results suggest the technique is valuable for 3D tubular constructs and organ printing.

5.3 Applications

Recently, 3D bioprinting has been used widely in applications such as tissue engineering, body-on-a-chip, and drug delivery systems. Table 5.2 shows a summary of bioprinting applications sorted by printing methodology, cell source/protein, and biomaterials.

Table 5.2 Bioprinting applications for blood vessel

Printing method	Cell source/ protein	Material	References
Extrusion-based	HUVECs, SMCs, and NHDFs	ECM, SMCM, FM, and gelatin	[20]
Extrusion-based	HCASMCs	Alginate, MWCNT	[64]
Extrusion-based	L929 mouse fibroblasts	Alginate	[65]
Extrusion-based	HUVECs, NHLFs	Collagen, gelatin, fibrin	[66]
Laser-based	Breast cancer cell (MDA-MB-231 and MCF-7) and fibroblasts (BJ5ta)	Gelatin	[59]
Laser-based	HUVSMCs	Alginate	[41]

5.3.1 Tissue Engineering

As mentioned above, the "traditional" scaffold-based approach has difficulties in terms of vascularization and developing a complex tissue or organ. In tissue engineering, 3D bioprinting is considered a breakthrough technology. Bioprinting processes can use multiple cell types and precisely control scaffold geometry in developing complex tissues. This subsection introduces artificial vessels, bones, cartilage, and organs produced with 3D bioprinting technology.

5.3.1.1 Blood Vessels

Blood vessels are the part of the cardiovascular system that transfers blood in circulation, transporting nutrients, oxygen, carbon dioxide, hormones, and blood cells to and from tissues. To prepare scalable organs, the artificial formation of blood vessels is one of the most important issues in tissue engineering. The vessel is tube-shaped, having a lumen, endothelium, and basement membrane. Printing a blood vessel is a challenge because of its complexity.

Many studies have been reported on the construction of artificial vessels. Itoh et al. [20] produced vascular prostheses with human umbilical vein endothelial cells (HUVECs), human aortic smooth muscle cells (HASMCs), and human normal dermal fibroblasts (HNDFBs) using a scaffold-free bioprinting technique (Fig. 5.6). The tubular tissues were implanted in the aortas of male F344-rnu/rnu athymic nude rats. They observed endothelial cells on the inner surface of the tubular tissues and normal flow rates in the aortas using percutaneous ultrasonography. Also, thrombosis did not occur in any rat.

Miller et al. [62] used a printable, cytocompatible, and sacrificial material to produce micro-channels incorporating a hydrogel structure. After printing a 3D channel network, cell-laden hydrogel was infiltrated into the sacrificial mold and cross-linked (Fig. 5.7). Then, the mold was dissolved to produce hollow channels. An artificial vascular structure was produced by seeding HUVECs in the channel and culturing them. Then, the construct was implanted in the circulatory system of a rat. The results showed that the rat femoral artery could sustain physiological blood pressure, from 80 to 100 mmHg and a normal level of flow was observed [63].

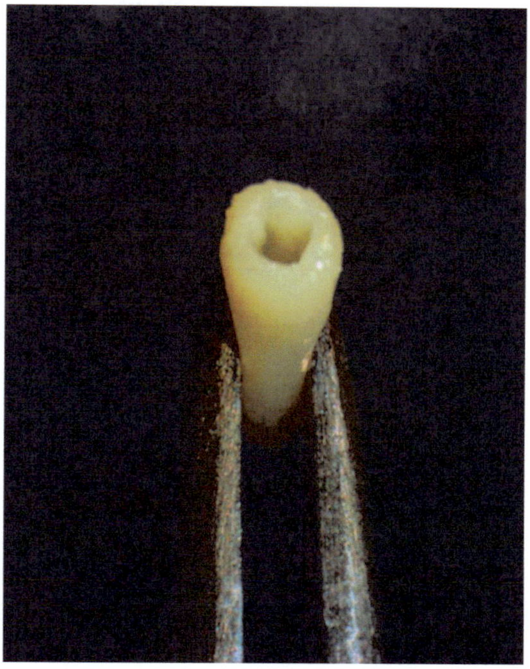

Fig. 5.6 Scaffold-free vascular graft with MSCs. Adapted from Itoh et al. PLoS One 2015;10(9):1–15 [20]

Dolati et al. [64] used a coaxial bioprinting process and printed vascular conduits directly. They prepared alginate reinforced with carbon nanotubes and then encapsulated human coronary artery smooth muscle cells (HCASMCs) in the material for printing. Their method had flexibility in terms of scale, from micro- to sub-millimeter. Similarly, Gao et al. [65] made a built-in micro channel structure by using layer-by-layer printing. They used a coaxial nozzle and printed hollow calcium alginate filaments, encapsulating L929 mouse fibroblasts.

Thus, arteries have been printed successfully. However, a capillary network at the single-cell scale is much harder to print. To address this, several researchers have studied perfusable capillary networks to induce angiogenesis. Lee et al. [66] used HUVECs and biological matrices and formed larger fluidic vascular channels. The capillary network was constructed by angiogenesis, induced from the printed channels.

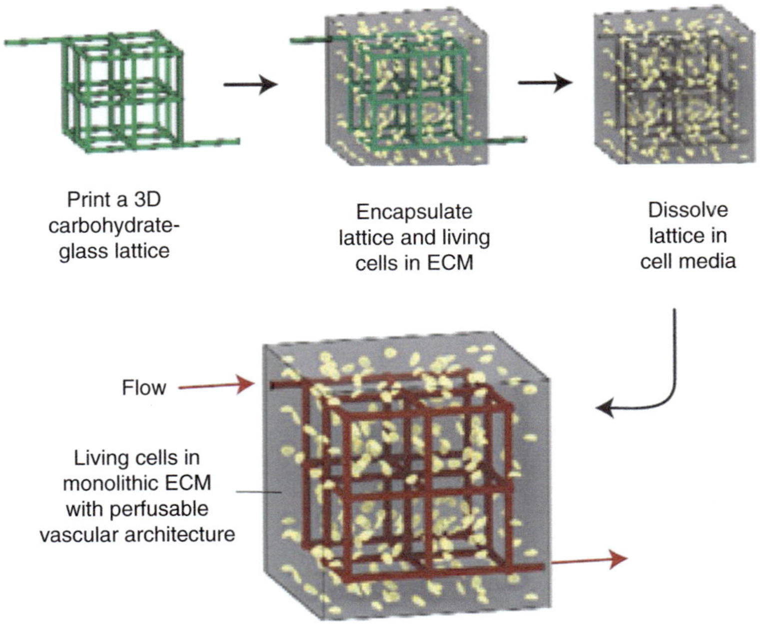

Fig. 5.7 A schematic overview of the casting of perfusable vascular architectures by using bioprinted sugar lattice. Reprinted from NATURE MATERIALS, Miller JS, Stevens KR, Yang MT, Baker BM, Nguyen D-HT and Cohen DM, Rapid casting of patterned vascular networks for perfusable engineered three-dimensional tissues, 2012;11(7):768–74, Copyright © 2012, Rights Managed by Nature Publishing Group [62]

Table 5.3 Bioprinting applications for bone

Printing method	Cell source/protein	Material	References
Extrusion-based	Osteoblast precursor cells (MC3T3-E1)	PLGA and PLGA/*n*-HAp/β-TCP	[67]
Extrusion-based	Osteoblast precursor cells (MC3T3-E1)	PCL, cryogel, and gelatin	[68]
Extrusion-based	N/A	Bioceramics, Pluronic F-127, and alginate	[14]
Extrusion-based	ihMSCs	PLGA-PEG-PLGA triblock copolymer hydrogel	[84]
Extrusion-based	Human AFSCs	Gelatin, fibrinogen, HA, glycerol, and PCL	[70]
Laser-based	MG-63 cells	*n*-HA and Matrigel	[40]

5.3.1.2 Bone and Cartilage

Many studies have been reported on the regeneration of bone and cartilage. Tissues can be printed using patient data, acquired with imaging techniques. This has been shown to be an effective method for reconstructing structures damaged by osteoporosis and bone fractures, for example.

Roh et al. [67] produced scaffolds using oxygen plasma treatment and a PLGA/*n*-HAp/β-TCP composite. Based on the scaffold, they enhanced the initial adhesion, proliferation, and differentiation of preosteoblasts. Van Rie et al. [68] printed scaffolds using cryogel-PCL composite and seeded MC3T-E1 osteoblast precursor cells. They observed colonized cells attached to the porous cryogel network.

Luo et al. [14] described a hollow struts-packed (HSP) bioceramic scaffold using a coaxial 3D printing strategy. The HSP bioceramic scaffold enhanced the proliferation and attachment of cells. When this structure was implanted in the damaged femur of a rabbit, recovery of a large bone defect was observed because of the implant's osteo-conductivity (Table 5.3).

Izadifar et al. [69] performed a layer-by-layer process with two bio-inks and a PCL. The first bio-ink was composed of alginate and primary chondrocytes extracted from embryonic chick cartilage. The

Table 5.4 Bioprinting applications for cartilage

Printing method	Cell source/ protein	Material	Reference
Extrusion-based	Rabbit ear chondrocytes	Gelatin, fibrinogen, HA, glycerol, PCL	[70]
Extrusion-based	Primary chondrocytes (from embryonic chick cartilage), ATDC5 mouse cell lines	Polycaprolactone (PCL) and alginate	[69]
Extrusion-based	Human nasoseptal chondrocyte	Nanocellulose-alginate	[82]
Extrusion-based	Equine chondrocytes and MSCs	Polycaprolactone (PCL), gelatin methacrylate (GelMA), and GelMA-gellan hydrogel	[83]

second contained ATDC5 mouse cell with alginate. The structure showed outstanding performance in terms of proliferation, cell viability, and secretion of cartilage ECM. Moreover, the real thickness of human articular cartilage could be achieved.

Kang et al. [70] developed an integrated tissue-organ printer (ITOP) and applied it to the regeneration of bone, cartilage, and muscle tissue. They printed a cell-laden hydrogel and a biodegradable polymer (PCL), anchored in sacrificial Pluronic F127, into a single structure. Their method increased the mechanical stability of the scaffold and overcame limitations in terms of structural integrity, size, and shape. To obtain a more precise structure, tissues were constructed using medical image data (Table 5.4). They implanted the structure in nude rats and observed the functionality of the structure (Fig. 5.8).

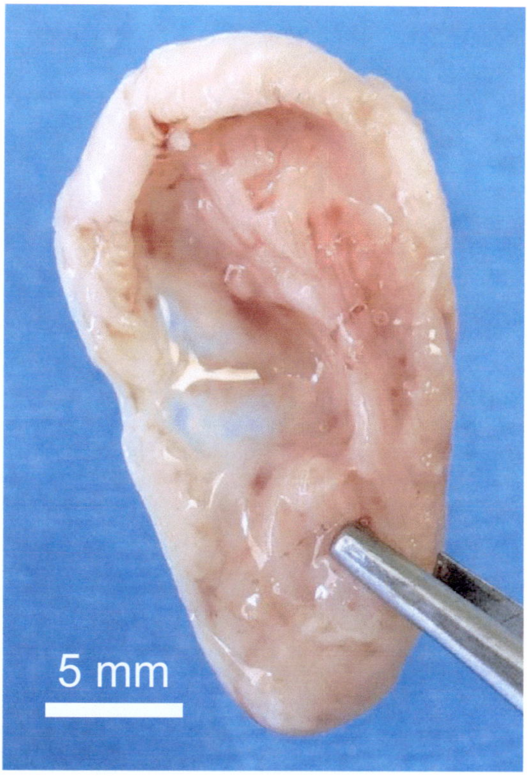

Fig. 5.8 Gross appearance of a bioprinted ear construct at 1 month after implantation. Reproduced from Kang et al. Nat Biotechnol. 2016;34(3):312–9 [70]

5.3.1.3 Other Organs

Faulkner-Jones et al. [71] printed human-induced pluripotent stem cells (hiPSCs) to create an artificial liver. Gaetani et al. [72] reported a bioprinted cardiac construct. They used human cardiomyocyte progenitor cells (hCMPCs) mixed with alginate for printing. After printing the scaffold, they observed high cell viability, of 89%, for 7 days. They increased the expression of cardiac transcription factors and a sarcomeric protein through 3D culture. This was an important achievement with respect to the goal of creating a fully differentiated cardiac construct.

Marchioli et al. [73] made a porous construct using INS1E β-cells and human and mouse islets mixed with alginate/gelatin mixture. The scaffold showed a high surface area to volume ratio so it obtained high oxygen and nutrient diffusion. They reported that the functionality of the scaffold was fully restored by vascularization in the pores.

Pati et al. [74] produced biomimetic 3D adipose tissues with a bioprinting technique to reconstruct soft tissues. They constructed flexible dome-shaped structures encapsulating human adipose tissue-derived mesenchymal stem cells (hASCs) with a decellularized adipose tissue (DAT) matrix-based bio-ink (Fig. 5.9). They also showed a viable clinical result for soft tissue regeneration in a patient-specific manner.

Damaged sweat glands cannot be recovered because of their low regenerative potential and lack of a definite niche. Huang et al. [75] made a 3D biomimetic scaffold with gelatin and alginate. The 3D bioprinted scaffold had micro-environmental characteristics, so that a restrictive niche could be regenerated and epidermal progenitor could be induced to regenerate sweat gland cells (Table 5.5).

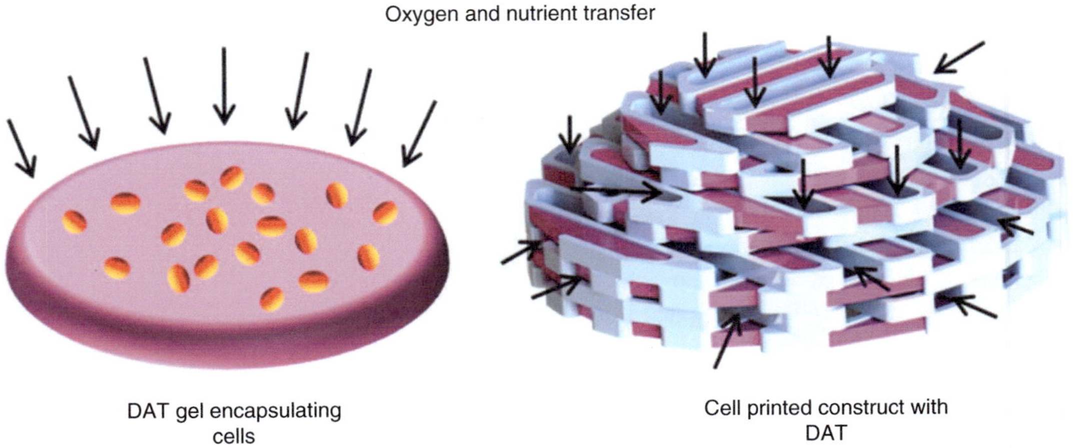

Fig. 5.9 Cell-printed constructs were prepared by layer-by-layer deposition of synthetic polymer and decellularized adipose tissue (DAT) bio-ink, maintaining a porous structure. An open porous construct enhances nutrient and oxygen transfer to the core of the construct. Reprinted from Biomaterials, Pati F, Ha DH, Jang J, Han HH, Rhie JW and Cho DW, A 3D bioprinting system to produce human-scale tissue constructs with structural integrity, 2015;62:164–75, Copyright © 2015 Elsevier Ltd. All rights reserved [74]

Table 5.5 Bioprinting applications for other organs

Organs	Printing method	Cell source/protein	Material	References
Skin and sweat gland	Extrusion-based	NHFF-1	Polyelectrolyte gelatin-chitosan	[85]
	Extrusion-based	Mouse derived epidermal progenitor cells	Gelatin and alginate	[75]
	Laser-based	NIH3T3 fibroblasts and HaCaT keratinocytes	Collagen	[37]
Muscle	Extrusion-based	neonHFF-1	Thermoplastic PU and PCL	[86]
Pancreas	Extrusion-based	INISIE β-cell line (from rat insulinoma), human and mouse islets	Alginate/gelatin mixture	[73]
Heart	Extrusion-based	hCMPCs	Alginate and Matrigel	[72]
Neural tissue	Extrusion-based	Murine neural stem cells	PU, PLLA, PCL, and PDLLA	[51]
	Extrusion-based	Mouse derived BMSC and SCs	Collagen and agarose	[87]
Liver	Jetting-based	hiPSCs and hESCs	Alginate	[71]
Breast	Extrusion-based	hASCs	Decellularized adipose tissue (DAT) hydrogel	[74]

5.3.2 Bio-Chip

A bio-chip is a miniaturized system for testing biological reactions; 3D bioprinting technology has been used in the development of bio-chip systems. This subsection introduces 3D bioprinted bio-chip systems in terms of an organoid for a body-on-a-chip and biosensors.

5.3.2.1 Body-on-a-Chip

The so-called body-on-a-chip refers to a multi-channel microfluidic biochip that includes human cells. The body-on-a-chip will be used to simulate human body responses. In the near future, this may become an alternative drug screening model, replacing some animal studies. Recently, bioprinting technologies have been applied to produce 3D organoids for studies that will ultimately lead to a body-on-a-chip.

Chang et al. [76] developed 3D liver micro-organ devices for an in vitro drug metabolism model using HepG2 liver cells and alginate. Liver tissue could be precisely patterned using bioprinting and was then integrated with a microfluidic system. They showed that the approach could control cellular-level differentiation and tissue-level function. Tourlomousis and Chang [77] also introduced a micro-organ device as a drug screening model.

Snyder et al. [78] applied a precision extrusion deposition (PED)/replica molding process to produce a cell-laden device. They simplified the fabrication process and enhanced leak-resistance (up to 2.0 mL/min) and saw pervasive diffusion in the device. They also showed that multi-nozzle printing was adequate for a device including multiple cell types.

5.3.2.2 Biosensor

A biosensor is an analytical device that detects an analyte of interest using biological components, such as protein or bacteria. Bioprinting technology has been applied in the field of biosensors. Fuchiwaki et al. [6] developed a microchip for enzyme-linked immunosorbent assays (ELISAs) using anti-C-reactive protein antibody. They developed a quartz crystal microbalance (QCM) sensor system to reduce background noise in a process with jetting-based bioprinting. Drachuk et al. [79] introduced a prototype of a cell-based thin-film biosensor that was printed using a dual array of silk fibroin-poly(L-lysine) (SF-PLL) and silk fibroin-poly(glutamic acid) (SF-PGA). Each array included *E. coli* coated with different fluorescent markers (Fig. 5.10). They sought to detect

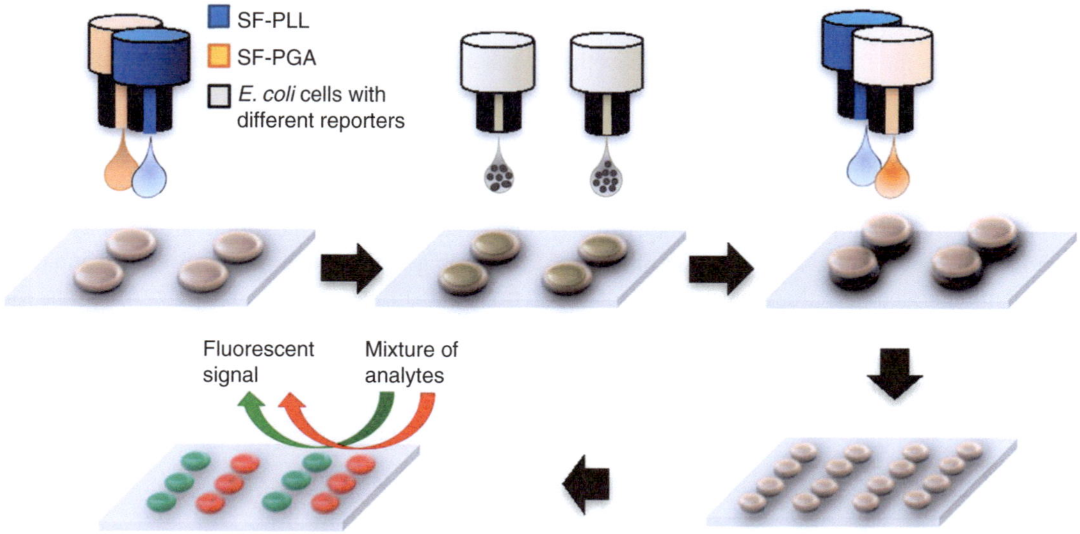

Fig. 5.10 Schematics of the inkjet printing process for dual-color cell–cell arrays. Reprinted from ACS Biomaterials Science & Engineering, Drachuk I, Suntivich R, Calabrese R, Harbaugh S, Kelley-Loughnane N and Kaplan DL, Printed Dual Cell Arrays for Multiplexed Sensing, 2015;150,410,104,710,005, Copyright © 2015 American Chemical Society [79]

Table 5.6 Bioprinting applications for bio-chips

Bio-chip	Printing method	Cell source/ protein	Material	References
Body-on-a-chip	Extrusion-based	HepG2 cells	Alginate	[77]
	Extrusion-based	HepG2 cells	Alginate, PCL	[78]
	Extrusion-based	HepG2 cells	Alginate	[76]
	Jetting-based	Human breast cancer cells (MCF-7)	Gelatin	[88]
Biosensor	Jetting-based	*E. coli* cells	SF-PLL and SF-PGA	[79]
	Jetting-based	Anti-C-reactive protein antibody	Anti-C-reactive protein antibody aqueous solution	[6]
	Laser-based	Mouse, rabbit IgG and human cDNA	Polylysine	[89]

multiple target analytes using a dual-type sensor and studied chemical and pharmaceutical screening and environmental monitoring with the system (Table 5.6).

5.3.3 Drug Delivery System

Many researchers have applied 3D printing techniques to drug delivery systems. Scoutaris et al. [7] created a mucosal thin film for taste masking using jetting-based bioprinting. As it contained the bitter substances CTZ, DPD, and IBU, the film could be used as a test to distinguish bitter tastes. Many other drug–polymer solutions could be used so that the film could be adapted for other applications.

Weisman et al. [50] produced antibiotic and chemotherapeutic drugs with pellets containing gentamicin sulfate (GS) and methotrexate (MTX). They verified the functionality of drug products using bacteria and osteosarcoma cells and demonstrated the thermal stability of the drug preparation made using 3D printing. The shape of the drug product could be of various forms, such as filaments, discs, and beads.

Khaled et al. [80] printed polypills including multiple active drugs (Fig. 5.11). The polypills include multiple active ingredients for specific diseases. For example, they used aspirin, hydrochlorothiazide, atenolol, pravastatin sodium, and ramipril for cardiovascular disease and high blood pressure. They used a 3D printing technique to control the amounts of ingredients, and their distributions and geometries, so that they could make patient-specific polypills.

Goyanes et al. [81] used hot-melt extrusion-based 3D printing to produce computer-designed drug forms, such as cubes, pyramids, cylinders, spheres, and a torus. Traditional manufacturing techniques, such as powder compaction, cannot be used to control the shape of the drug product in this way. The printed drug products were composed of polyvinyl alcohol (PVA) and paracetamol. They measured drug release rates according to the shape of the drug form and confirmed that the drug release rate increased as the surface-to-volume ratio increased (Table 5.7).

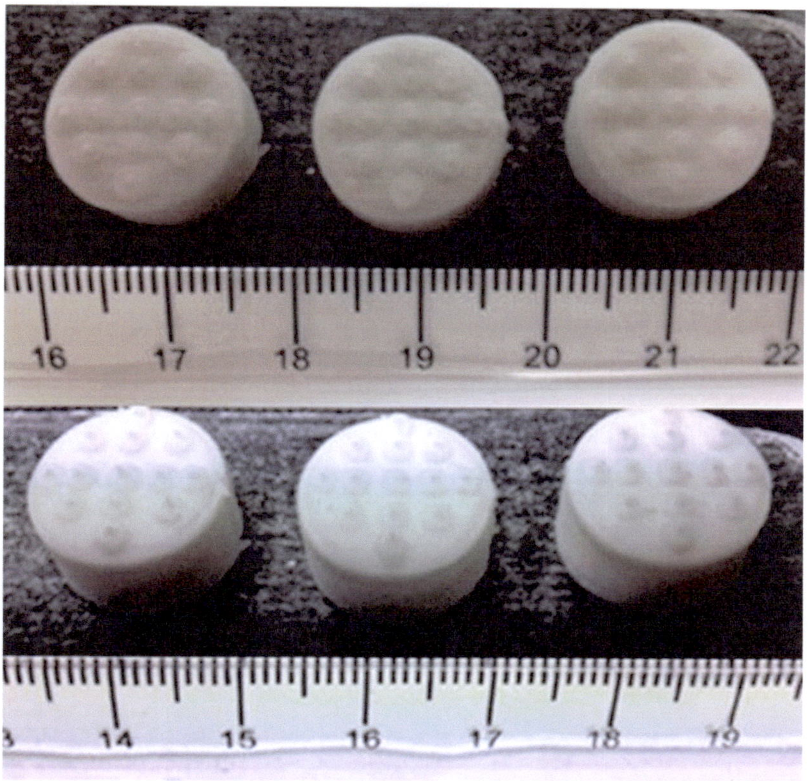

Fig. 5.11 Multi-active drug tablets (polypills). Reprinted from Journal of Controlled Release, Khaled SA, Burley JC, Alexander MR, Yang J and Roberts CJ, 3D printing of five-in-one dose combination polypill with defined immediate and sustained release profiles, 2015;217:308–14, Copyright © 2015 Elsevier B.V. All rights reserved [80]

Table 5.7 Bioprinting applications for drug delivery system

Method	Cell source/ protein	Material	References
Extrusion-based	N/A	PLA, gentamicin sulfate, and methotrexate	[50]
Extrusion-based	N/A	Aspirin, hydrochlorothiazide, atenolol, pravastatin, and ramipril	[80]
Extrusion-based	N/A	Paracetamol and PVA	[81]
Jetting-based	N/A	Three-powder blend (cornstarch, dextran, and gelatin)	[90]
Jetting-based	N/A	Cetirizine HCl, diphenhydramine HCl, and ibuprofen	[7]

Conclusions

Bioprinting technology can produce computer-designed 3D structures not only with cell-laden bio-ink, but also with various biomaterials and biomolecules. This processability confers great benefits in regenerative medicine. In this chapter, we reviewed bioprinting technology and some of its applications have been introduced. According to the printing mechanism, three types of bioprinting technology—jetting-, extrusion-, and laser-based bioprinting—were described in terms of their components, applicable bio-inks, and pros and cons. Although these bioprinting technologies are recent, many researchers have reported progress in various applications, including artificial tissue/organ regeneration, body-on-a-chip, biosensors, and drug delivery systems. Continued advances in bioprinting may soon move this technology into clinical sites and help in advancing medical technology.

References

1. Hart A, Smith JM, Skeans MA, Gustafson SK, Stewart DE, Cherikh WS, Wainright JL, Boyle G, Snyder JJ, Kasiske BL, Israni AK. OPTN/SRTR Annual Data Report 2014: Kidney. Am J Transplant. 2016;16(Suppl 2):11–46.
2. Nerem RM, Seliktar D. Vascular tissue engineering. Annu Rev Biomed Eng. 2001;3:225–43.
3. Derby B. Printing and prototyping of tissues and scaffolds. Science. 2012;338(6109):921–6.
4. Murphy SV, Atala A. 3D bioprinting of tissues and organs. Nat Biotechnol. 2014;32(8):773–85.
5. Horváth L, Umehara Y, Jud C, Blank F, Petri-Fink A, Rothen-Rutishauser B. Engineering an in vitro air-blood barrier by 3D bioprinting. Sci Rep. 2015;5:7974.
6. Fuchiwaki Y, Tanaka M, Ooie T. Inexpensive and reliable monitoring of the microdeposition of biomolecules. Anal Lett. 2016;2719:921–8.
7. Scoutaris N, Snowden M, Douroumis D. Taste masked thin films printed by jet dispensing. Int J Pharm. 2015;494(2):619–22.
8. Demirci U, Montesano G. Single cell epitaxy by acoustic picolitre droplets. Lab Chip. 2007;7(9):1139–45.
9. Kador KE, Grogan SP, Dorthé EW, Venugopalan P, Malek MF, Goldberg JL. Control of retinal ganglion cell positioning and neurite growth: combining 3D printing with radial. Tissue Eng Part A. 2016;22:286–94.
10. Xu T, Zhao W, Zhu JM, Albanna MZ, Yoo JJ, Atala A. Complex heterogeneous tissue constructs containing multiple cell types prepared by inkjet printing technology. Biomaterials. 2013;34(1):130–9.
11. Binder KW. In situ bioprinting of the skin. ProQuest Dissertations Publishing. 2011;3458142.
12. Tan EYS, Yeong WY. Concentric bioprinting of alginate-based tubular constructs using multi-nozzle extrusion-based technique. Int J Bioprinting. 2015;1(1):49–56.
13. Yang SS, Choi WH, Song BR, Jin H, Lee SJ, Lee SH. Fabrication of an osteochondral graft with using a solid freeform fabrication system. Tissue Eng Regen Med. 2015;12(4):239–48.
14. Luo Y, Zhai D, Huan Z, Zhu H, Xia L, Chang J. Three-dimensional printing of hollow-struts-packed bioceramic scaffolds for bone regeneration. ACS Appl Mater Interfaces. 2015;7(43):24377–83.
15. Latza V, Guerette PA, Ding D, Amini S, Kumar A, Schmidt I. Multi-scale thermal stability of a hard thermoplastic protein-based material. Nat Commun. 2015;6:8313.
16. Highley CB, Rodell CB, Burdick JA. Direct 3D printing of shear-thinning hydrogels into self-healing hydrogels. Adv Mater. 2015;27(34):5075–9.
17. Rees A, Powell LC, Chinga-Carrasco G, Gethin DT, Syverud K, Hill KE. 3D bioprinting of carboxymethylated-periodate oxidized nanocellulose constructs for wound dressing applications. Biomed Res Int. 2015;2015:1–7.
18. Pati F, Jang J, Ha D-H, Won Kim S, Rhie J-W, Shim J-H. Printing three-dimensional tissue analogues with decellularized extracellular matrix bioink. Nat Commun. 2014;5:3935.
19. Tabriz AG, Hermida MA, Leslie NR, Shu W. Three-dimensional bioprinting of complex cell laden alginate hydrogel structures. Biofabrication. 2015;7(4):045012.
20. Itoh M, Nakayama K, Noguchi R, Kamohara K, Furukawa K, Uchihashi K. Scaffold-free tubular tissues created by a bio-3D printer undergo remodeling and endothelialization when implanted in rat aortae. PLoS One. 2015;10(9):1–15.
21. Kucukgul C, Ozler SB, Inci I, Karakas E, Irmak S, Gozuacik D. 3D bioprinting of biomimetic aortic vascular constructs with self-supporting cells. Biotechnol Bioeng. 2015;112(4):811–21.
22. Zhang X, Jiang XN, Sun C. Micro-stereolithography of polymeric and ceramic microstructures. Sensors Actuators A Phys. 1999;77(2):149–56.
23. Haske W, Chen VW, Hales JM, Dong W, Barlow S, Marder SR. 65 nm feature sizes using visible wavelength 3-D multiphoton lithography. Opt Express. 2007;15(6):3426–36.
24. Alharbi N, Osman R, Wismeijer D. Effects of build direction on the mechanical properties of 3D-printed complete coverage interim dental restorations. J Prosthet Dent. 2016;11:760–7.
25. Lan PX, Lee JW, Seol YJ, Cho DW. Development of 3D PPF/DEF scaffolds using micro-stereolithography and surface modification. J Mater Sci Mater Med. 2009;20(1):271–9.

26. Lee JW, Kang KS, Lee SH, Kim JY, Lee BK, Cho DW. Bone regeneration using a microstereolithography-produced customized poly(propylene fumarate)/diethyl fumarate photopolymer 3D scaffold incorporating BMP-2 loaded PLGA microspheres. Biomaterials. 2011;32(3):744–52.
27. Jansen J, Melchels FPW, Grijpma DW, Feijen J. Fumaric acid monoethyl ester-functionalized poly (D,L-lactide)/N-vinyl-2-pyrrolidone resins for the preparation of tissue engineering Scaffolds by stereolithography. Biomacromolecules. 2009;10(2):214–20.
28. Shanjani Y, Pan CC, Elomaa L, Yang Y. A novel bioprinting method and system for forming hybrid tissue engineering constructs. Biofabrication. 2015;7(4):045008.
29. Soman P, Kelber JA, Lee JW, Wright TN, Vecchio KS, Klemke RL. Cancer cell migration within 3D layer-by-layer microfabricated photocrosslinked PEG scaffolds with tunable stiffness. Biomaterials. 2012;33(29):7064–70.
30. Hribar KC, Meggs K, Liu J, Zhu W, Qu X, Chen S. Three-dimensional direct cell patterning in collagen hydrogels with near-infrared femtosecond laser. Sci Rep. 2015;5(October):17203.
31. Shallan AI, Smejkal P, Corban M, Guijt RM, Breadmore MC. Cost-effective three-dimensional printing of visibly transparent microchips within minutes. Anal Chem. 2014;86(6):3124–30.
32. Au AK, Lee W, Folch A. Mail-order microfluidics: evaluation of stereolithography for the production of microfluidic devices. Lab Chip. 2014;14(7):1294–301.
33. Larson C, Shepher R. 3D bioprinting technologies for cellular engineering. In: Singh A, Gaharwar AK, editors. Microscale technologies for cell engineering. Cham: Springer; 2015.
34. Koch L, Kuhn S, Sorg H, Gruene M, Schlie S, Gaebel R. Laser printing of skin cells and human stem cells. Tissue Eng Part C Methods. 2010;16(5):847–54.
35. Pagès E, Rémy M, Kériquel V, Correa MM, Guillotin B, Guillemot F. Creation of highly defined mesenchymal stem cell patterns in three dimensions by laser-assisted bioprinting. J Nanotechnol Eng Med. 2015;6(2):021005.
36. Guillotin B, Souquet A, Catros S, Duocastella M, Pippenger B, Bellance S. Laser assisted bioprinting of engineered tissue with high cell density and microscale organization. Biomaterials. 2010;31(28):7250–6.
37. Michael S, Sorg H, Peck CT, Koch L, Deiwick A, Chichkov B. Tissue engineered skin substitutes created by laser-assisted bioprinting form skin-like structures in the dorsal skin fold chamber in mice. PLoS One. 2013;8(3):e57741.
38. Dinca V, Ranella A, Farsari M, Kafetzopoulos D, Dinescu M, Popescu A. Quantification of the activity of biomolecules in microarrays obtained by direct laser transfer. Biomed Microdevices. 2008;10(5):719–25.
39. Cheng GJ, Pirzada D, Cai M, Mohanty P, Bandyopadhyay A. Bioceramic coating of hydroxyapatite on titanium substrate with Nd-YAG laser. Mater Sci Eng C. 2005;25(4):541–7.
40. Catros S, Fricain J-C, Guillotin B, Pippenger B, Bareille R, Remy M. Laser-assisted bioprinting for creating on-demand patterns of human osteoprogenitor cells and nano-hydroxyapatite. Biofabrication. 2011;3:025001.
41. Xiong R, Zhang Z, Chai W, Huang Y, Chrisey DB. Freeform drop-on-demand laser printing of 3D alginate and cellular constructs. Biofabrication. 2015;7(4):45011.
42. Cui X, Breitenkamp K, Finn MG, Lotz M, D'Lima D. Direct human cartilage repair using 3D bioprinting technology. Tissue Eng. 2012;18(11–12):1304–12.
43. Smith CM, Stone AL, Parkhill RL, Stewart RL, Simpkins MW, Kachurin AM. Three-dimensional bioassembly tool for generating viable tissue-engineered constructs. Tissue Eng. 2004;10(9):1566–76.
44. Kolesky DB, Truby RL, Gladman AS, Busbee TA, Homan KA, Lewis JA. 3D bioprinting of vascularized, heterogeneous cell-laden tissue constructs. Adv Mater. 2014;26(19):3124–30.
45. Mapili G, Lu Y, Chen S, Roy K. Laser-layered microfabrication of spatially patterned functionalized tissue-engineering scaffolds. J Biomed Mater Res - Part B Appl Biomater. 2005;75(2):414–24.
46. Duan B, Hockaday LA, Kang KH, Butcher JT. 3D Bioprinting of heterogeneous aortic valve conduits with alginate/gelatin hydrogels. J Biomed Mater Res A. 2013;101(A(5)):1255–64.
47. Wang Z, Abdulla R, Parker B, Samanipour R. A simple and high-resolution stereolithography-based 3D bioprinting system using visible light crosslinkable bioinks. Biofabrication. 2015;7(4):1–29.
48. Xu T, Jin J, Gregory C, Hickman JJ, Boland T. Inkjet printing of viable mammalian cells. Biomaterials. 2005;26(1):93–9.
49. Liu Tsang V. Chen a a, Cho LM, Jadin KD, Sah RL and DeLong S, fabrication of 3D hepatic tissues by additive photopatterning of cellular hydrogels. FASEB J. 2007;21(3):790–801.
50. Weisman JA, Nicholson JC, Tappa K, Jammalamadaka U, Wilson CG, Mills D. Antibiotic and chemotherapeutic enhanced three-dimensional printer filaments and constructs for biomedical applications. Int J Nanomedicine. 2015;10:357–70.
51. Hsieh FY, Lin HH, Hsu SH. 3D bioprinting of neural stem cell-laden thermoresponsive biodegradable polyurethane hydrogel and potential in central nervous system repair. Biomaterials. 2015;71:48–57.
52. Bajaj P, Marchwiany D, Duarte C, Bashir R. Patterned three-dimensional encapsulation of embryonic stem cells using Dielectrophoresis and Stereolithography. Adv Healthc Mater. 2013;2(3):450–8.
53. Catros S, Guillotin B, Bačáková M, Fricain JC, Guillemot F. Effect of laser energy, substrate film thickness and bioink viscosity on viability of endothelial cells printed by laser-assisted bioprinting. Appl Surf Sci. 2011;257(12):5142–7.
54. Jakab K, Norotte C, Damon B, Marga F, Neagu A, Besch-Williford CL. Tissue engineering by self-

assembly of cells printed into topologically defined structures. Tissue Eng. 2007;14(3):110306233438005.
55. Irvine SA, Agrawal A, Lee BH, Chua HY, Low KY, Lau BC. Printing cell-laden gelatin constructs by free-form fabrication and enzymatic protein cross-linking. Biomed Microdevices. 2015;17(1):1–8.
56. Duarte Campos DF, Blaeser A, Weber M, Jäkel J, Neuss S, Jahnen-Dechent W. Three-dimensional printing of stem cell-laden hydrogels submerged in a hydrophobic high-density fluid. Biofabrication. 2013;5(1):015003.
57. Xu T, Binder KW, Albanna MZ, Dice D, Zhao W, Yoo JJ. Hybrid printing of mechanically and biologically improved constructs for cartilage tissue engineering applications. Biofabrication. 2013;5(1):015001.
58. Skardal A, Zhang J, Prestwich GD. Bioprinting vessel-like constructs using hyaluronan hydrogels crosslinked with tetrahedral polyethylene glycol tetracrylates. Biomaterials. 2010;31(24):6173–81.
59. Phamduy TB, Sweat RS, Azimi MS, Burow ME, Murfee WL, Chrisey DB. Printing cancer cells into intact microvascular networks: a model for investigating cancer cell dynamics during angiogenesis. Integr Biol (Camb). 2015;7:1068–78.
60. Pescosolido L, Schuurman W, Malda J, Matricardi P, Alhaique F, Coviello T. Hyaluronic acid and dextran-based semi-IPN hydrogels as biomaterials for bioprinting. Biomacromolecules. 2011;12(5):1831–8.
61. Ronca A, Ambrosio L, Grijpma DW. Preparation of designed poly(d,l-lactide)/nanosized hydroxyapatite composite structures by stereolithography. Acta Biomater. 2013;9(4):5989–96.
62. Miller JS, Stevens KR, Yang MT, Baker BM, Nguyen D-HT, Cohen DM. Rapid casting of patterned vascular networks for perfusable engineered three-dimensional tissues. Nat Mater. 2012;11(7):768–74.
63. Sooppan R, Paulsen SJ, Han J, Ta AH, Dinh P, Gaffey AC. In vivo anastomosis and perfusion of a three-dimensionally-printed construct containing microchannel networks. Tissue Eng Part C Methods. 2016;22(1):1–7.
64. Dolati F, Yu Y, Zhang Y, De Jesus AM, Sander EA, Ozbolat IT. In vitro evaluation of carbon-nanotube-reinforced bioprintable vascular conduits. Nanotechnology. 2014;25(14):145101.
65. Gao Q, He Y, Fu J, Liu A, Ma L. Coaxial nozzle-assisted 3D bioprinting with built-in microchannels for nutrients delivery. Biomaterials. 2015;61:203–15.
66. Lee VK, Lanzi AM, Ngo H, Yoo SS, Vincent PA, Dai G. Generation of multi-scale vascular network system within 3D hydrogel using 3D bio-printing technology. Cell Mol Bioeng. 2014;7(3):460–72.
67. Roh H-S, Jung S-C, Kook M-S, Kim B-H. In vitro study of 3D PLGA/n-HAp/β-TCP composite scaffolds with etched oxygen plasma surface modification in bone tissue engineering. Appl Surf Sci. 2016;388:321–30.
68. Van Rie J, Declercq H, Van Hoorick J, Dierick M, Van Hoorebeke L, Cornelissen R. Cryogel-PCL combination scaffolds for bone tissue repair. J Mater Sci Mater Med. 2015;26(3):124.
69. Izadifar Z, Chang T, Kulyk WM, Chen D, Eames BF. Analyzing biological performance of 3D-printed, cell-impregnated hybrid constructs for cartilage tissue engineering. Tissue Eng Part C Methods. 2016;22(3):173–88.
70. Kang H-W, Lee SJ, Ko IK, Kengla C, Yoo JJ, Atala A. A 3D bioprinting system to produce human-scale tissue constructs with structural integrity. Nat Biotechnol. 2016;34(3):312–9.
71. Faulkner-Jones A, Fyfe C, Cornelissen D-J, Gardner J, King J, Courtney A. Bioprinting of human pluripotent stem cells and their directed differentiation into hepatocyte-like cells for the generation of mini-livers in 3D. Biofabrication. 2015;7(4):044102.
72. Gaetani R, Doevendans PA, Metz CHG, Alblas J, Messina E, Giacomello A. Cardiac tissue engineering using tissue printing technology and human cardiac progenitor cells. Biomaterials. 2012;33(6):1782–90.
73. Marchioli G, van Gurp L, van Krieken PP, Stamatialis D, Engelse M, van Blitterswijk CA. Fabrication of three-dimensional bioplotted hydrogel scaffolds for islets of Langerhans transplantation. Biofabrication. 2015;7(2):025009.
74. Pati F, Ha DH, Jang J, Han HH, Rhie JW, Cho DW. Biomimetic 3D tissue printing for soft tissue regeneration. Biomaterials. 2015;62:164–75.
75. Huang S, Yao B, Xie J, Fu X. 3D bioprinted extracellular matrix mimics facilitate directed differentiation of epithelial progenitors for sweat gland regeneration. Acta Biomater. 2016;32:170–7.
76. Chang R, Emami K, Wu H, Sun W. Biofabrication of a three-dimensional liver micro-organ as an in vitro drug metabolism model. Biofabrication. 2010;2(4):045004.
77. Tourlomousis F, Chang RC. Numerical investigation of dynamic microorgan devices as drug screening platforms. Part I: macroscale modeling approach & validation. Biotechnol Bioeng. 2016;113(3):612–22.
78. Snyder J, Rin Son A, Hamid Q, Sun W. Fabrication of microfluidic manifold by precision extrusion deposition and replica molding for cell-laden device. J Manuf Sci Eng. 2015;138(4):041007.
79. Drachuk I, Suntivich R, Calabrese R, Harbaugh S, Kelley-Loughnane N, Kaplan DL. Printed dual cell arrays for multiplexed sensing. ACS Biomater Sci Eng. 2015;1:287–94.
80. Khaled SA, Burley JC, Alexander MR, Yang J, Roberts CJ. 3D printing of five-in-one dose combination polypill with defined immediate and sustained release profiles. J Control Release. 2015;217:308–14.
81. Goyanes A, Robles Martinez P, Buanz A, Basit AW, Gaisford S. Effect of geometry on drug release from 3D printed tablets. Int J Pharm. 2015;494(2):657–63.
82. Markstedt K, Mantas A, Tournier I, Martínez Ávila H, Hägg D, Gatenholm P. 3D bioprinting human chondrocytes with nanocellulose-alginate bioink for cartilage tissue engineering applications. Biomacromolecules. 2015;16(5):1489–96.
83. Visser J, Peters B, Burger TJ, Boomstra J, Dhert WJA, Melchels FPW. Biofabrication of multi-material

anatomically shaped tissue constructs. Biofabrication. 2013;5(3):035007.
84. Sawkins MJ, Mistry P, Brown BN, Shakesheff KM, Bonassar LJ, Yang J. Cell and protein compatible 3D bioprinting of mechanically strong constructs for bone repair. Biofabrication. 2015;7(3):035004.
85. Ng WL, Yeong WY, Naing MW. Polyelectrolyte gelatin-chitosan hydrogel optimized for 3D bioprinting in skin tissue engineering. Int J Bioprinting. 2016;2 doi:10.18063/IJB.2016.01.009.
86. Merceron TK, Burt M, Seol Y-J, Kang H-W, Lee SJ, Yoo JJ. A 3D bioprinted complex structure for engineering the muscle–tendon unit. Biofabrication. 2015;7(3):035003.
87. Owens CM, Marga F, Forgacs G, Heesch CM. Biofabrication and testing of a fully cellular nerve graft. Biofabrication. 2013;5(4):045007.
88. Ling K. Bioprinting-based high-throughput fabrication of three-dimensional MCF-7 human breast cancer cellular spheroids. Engineering. 2015;1(2):269–74.
89. Duocastella M, Fernandez-Pradas JM, Morenza JL, Zafra D, Serra P. Novel laser printing technique for miniaturized biosensors preparation. Sensors Actuators B Chem. 2010;145(1):596–600.
90. Lam CXF, Mo XM, Teoh SH, Hutmacher DW. Scaffold development using 3D printing with a starch-based polymer. Mater Sci Eng C. 2002;20(1–2):49–56.

Decellularization

6

Taekmin Kwon and Kyung Hyun Moon

6.1 Introduction

The main components used for tissue engineering are living cells, a scaffolding system based on biomaterials, bioactive factors, and appropriate microenvironments that facilitate cellular behavior. A well-orchestrated combination of these components is of critical significance in creating engineered tissues or organs for the development of functional substitutes [1]. The principal function of a scaffolding system for tissue engineering is to provide a "template" to direct cellular behavior, which includes cell migration, proliferation, and differentiation, and to maintain cell type-specific phenotypes [2]. Generally, three classes of biomaterials are used for tissue engineering purposes: acellular tissue matrices, natural polymers, and synthetic polymers [3]. Body organs are complex structures composed of extracellular matrix (ECM) and cell ingredients. In the field of regenerative medicine, organs are decellularized, i.e., cellular components are removed to produce an acellular ECM, or decellularized scaffolds. The use of decellularized tissue matrices as scaffolds is very attractive, because naturally derived polymers and synthetic polymers are unable to replicate the precise spatial organization of complex structures. These biologic scaffolds made of ECM, typical in reconstructive surgery, are used in a variety of applications, including regenerative medicine strategies for increasing tissue and organ replacement [4]. In in vitro studies, relying on the bioreactor, researchers have examined the effect of these scaffolds on cell proliferation and organ structure. In in vivo implantations of decellularized scaffolds, the effect of the scaffold on promoting angiogenesis and regional regeneration was explored (Fig. 6.1). The clinical use of decellularized scaffold blood vessels has been documented for applications such as heart valves and renal bladder. Nevertheless, the current application is limited to anatomically simple organs, but will eventually provide the foundation for complex organ regeneration in the future. The use of decellularized scaffold in regenerative medicine has recently made a breakthrough. Despite variations in the organ to be fashioned and used, these scaffolds have a proven ability to promote regeneration. In this review, the most commonly used decellularization methods are discussed, and their effect upon the resulting ECM structure and composition is presented. We hope this discussion will provide useful information regarding the decellularization methods.

T. Kwon • K.H. Moon (✉)
Department of Urology, Ulsan University Hospital, University of Ulsan College of Medicine, Ulsan, South Korea
e-mail: urofirst@gmail.com

B.W. Kim (ed.), *Clinical Regenerative Medicine in Urology*, DOI 10.1007/978-981-10-2723-9_6

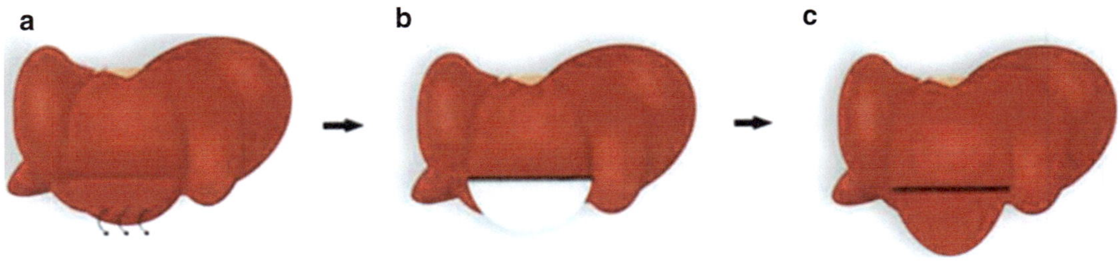

Fig. 6.1 Schematic diagram of the liver regeneration hypothesis using decellularized scaffolds. (**a**) Partial resection of one hepatic lobule is performed. (**b**) The defective part is replaced with decellularized liver scaffold. (**c**) Cells in the native liver cross the suture border and regenerate on the liver scaffold. (From Yu et al. [2])

6.2 Rationale for Decellularization

Xenogeneic and allogeneic cellular antigens are recognized as foreign by the host and therefore induce an inflammatory response or an immune-mediated rejection of the tissue. However, the components of ECM are generally conserved among species and are tolerated well, even by xenogeneic recipients [5, 6]. ECM from a variety of tissues, including heart valves [7, 8], blood vessels [9, 10], skin [11], nerves [12], skeletal muscle [13], tendons [14], ligaments [15], small intestinal submucosa [16], urinary bladder [17], and liver [18], has been reported for use in tissue engineering and regenerative medicine applications. The purpose of decellularization protocols, though minimizing adverse effects on the mechanical integrity of the remaining ECM, is to efficiently remove all cells and nuclear material (Fig. 6.2).

With the purpose of removing the cells, any of the processing steps changes the native three-dimensional structure of ECM. The most common method used for the decellularization of an organ involves a combination of physical and chemical processing. Physical treatment can include agitation or sonication, mechanical massage and pressure, or a freeze–thaw process. Using these methods, the cell contents are removed, the cell membrane is easily destroyed by subsequent rinsing and removal of the cell content from the ECM. However, although these physical processes achieve complete decellularization, they are insufficient, and must be combined with chemical treatment. Enzymatic treatments, such as trypsin, ionic detergents and chemical treatments, can disrupt charge coupling among the cell membrane, cell, and extracellular component. Tissue is composed of both cell materials, and ECM has a variable degree of miniaturization according to the tissue source. ECM is sufficient to allow proper exposure of all the cells to the chaotropic agents. To provide a path for removal from the organ, the cellular material must be destroyed during the decellularization process. The purpose of the main part of the decellularization process is to retain the original mechanical and biological properties and minimize disruption.

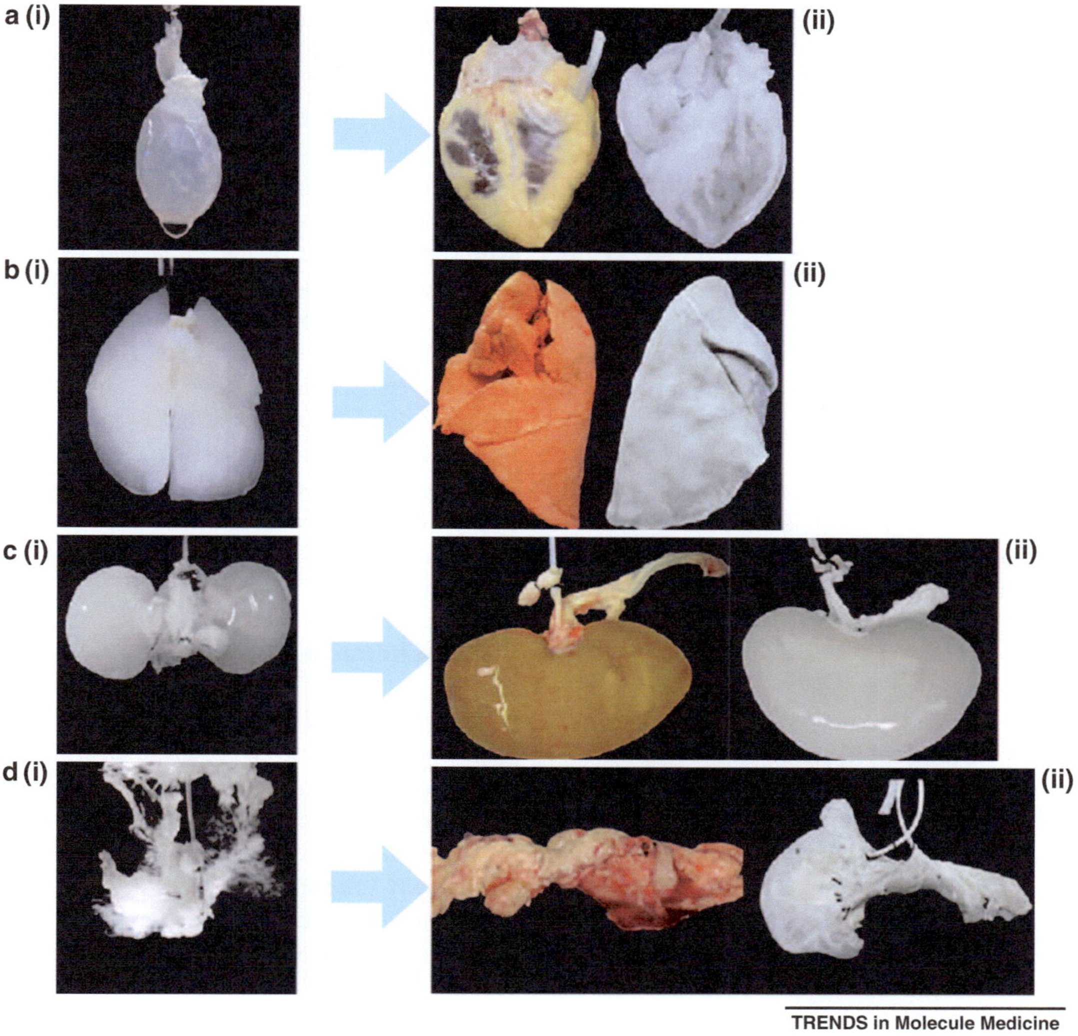

Fig. 6.2 Whole-organ scaffolds decellularized using perfusion. The native extracellular matrixes (ECMs) of cadaveric organs can be isolated by perfusion of the native vascular system with detergent solutions. The resulting scaffolds are acellular, but maintain the structure of the native organ. (**a**) (*i*) Rat heart scaffold generated from a cadaveric heart using perfusion decellularization. The ascending aorta was cannulated for perfusion. (*ii*) Cadaveric human heart before and after perfusion decellularization. (**b**) (*i*) Rat lung scaffold generated from cadaveric lung by perfusion decellularization. Perfusion was performed via the pulmonary artery. (*ii*) Cadaveric sheep lung before and after decellularization. (**c**) (*i*) Rat kidney scaffold generated from cadaveric kidney using perfusion decellularization. The abdominal aorta was cannulated for perfusion. (*ii*) Porcine kidney before and after perfusion decellularization. (**d**) (*i*) Rat pancreas scaffold generated from cadaveric pancreas by perfusion decellularization. The abdominal aorta was cannulated for perfusion. (*ii*) Human pancreas before and after perfusion decellularization. (From Song and Ott [19])

6.3 Decellularization Protocols

The most robust and effective physical decellularization protocol uses a combination of chemicals, and includes an enzymatic approach. Decellularization protocols, typically enzyme treatment, surface-active agents, cytoplasmic, solubilization of cellular components of the nucleus, and the final removal of cellular debris from tissue, cellular components from the ECM, involve physical treatment or ionic solution. Following separation, the dissolution of the cell membrane begins. To enhance effectiveness, these steps can be combined with mechanical stirring. Following decellularization, all the residual chemicals must be removed to avoid the response of chemically adverse host tissue. The efficiency of the storage of decellularization and ECM can be assessed using one of several methods.

6.3.1 Physical Methods

6.3.1.1 Temperature

The freeze–thaw process lyses the cells of the tissues and organs effectively; unless removed by subsequent processing, the resulting membranous and intracellular content remains. By using a single freeze–thaw cycle, a harmful immune response, such as leukocyte infiltration into the ECM scaffold, can be reduced [20]. Multiple freeze–thaw cycles may be used during decellularization [21] and do not significantly increase the loss of ECM proteins from the tissue [22]. Freeze–thaw processing does produce minor disruptions of the ECM ultrastructure [23] and should therefore be used only when such effects are acceptable in the final ECM product. The effect of freeze–thaw treatment on the mechanical, load-bearing properties is minimal for a mechanically robust organ [24].

6.3.1.2 Force and Pressure

Cells on the surface of a tissue or organ (e.g., urinary bladder, small intestine, skin, amnion) can be effectively removed using mechanical abrasion in combination with enzymes, hypertonic saline, or chelating agents, all of which facilitate the dissociation of cells from their subjacent basement membrane [23]. However, the integrity of the underlying hyperfine structure and the basement membrane can been damaged by direct applications of mechanical force. The formation of barium crystals may disrupt the ultrastructural hydrostatic pressure of the ECM and requires relatively little time compared with enzymes used to remove cells from detergents and blood vessels. It may also be effective in corneal tissue [25]. The temperature rises at the time of the pressure of decellularization, but to prevent the formation of crystals, it is possible to destroy the ECM because of the associated increase in entropy, which can be mitigated by a colloid such as dextran [26].

6.3.2 Chemical Methods

6.3.2.1 Acids and Bases

Acids and bases cause and catalyze the hydrolysis of biological molecules. Peracetic acid, a common disinfectant that by removing residual nucleic acids has a minimal impact on the composition and structure of ECM, also serving as a decellularization agent [27, 28].

Acetic acid damages and removes collagens, thus decreasing ECM strength, but sulfated glycosaminoglycans (sGAG) are not affected. Bases (e.g., calcium hydroxide, sodium sulfide, and sodium hydroxide) are used to remove hair from dermis samples during the early stages of decellularization [29].

6.3.2.2 Hypotonic and Hypertonic Solutions

Hypertonic saline dissociates DNA from proteins [30]. Hypotonic solutions can readily cause cell lysis because of simple osmotic effects, with minimal changes in matrix molecules and architecture [31]. For a maximum osmotic effect for dipping alternately hyper- and hypotonic solution through a few cycles in the organ is common. Hypertonic and hypotonic solution also helps to rinse the cell debris from within the organ after dissolution.

6.3.2.3 Detergents

Ionic, non-ionic, and zwitterionic detergents solubilize cell membranes, isolating DNA from the protein and removing cellular material effectively from the tissue [30]. However, these agents also disrupt and dissociate proteins in the ECM, as demonstrated by their use in the protein extraction procedures of tissue proteomics [22]. The removal of ECM proteins and DNA by detergents increases with exposure time [11] and varies with organ subunit, tissue type, and donor age [22]. Combining multiple detergents increases ECM protein loss [32], but also allows for more complete detergent removal from ECM after decellularization [33]. Triton X-100 can effectively remove cell residues from thicker tissues, such as valve conduits, where enzymatic and osmotic methods are insufficient, with concomitant ECM protein loss accompanied by a decreased adverse immune response in vivo [34]. Sodium dodecyl sulfate (SDS) appears to be more effective than Triton X-100 at removing nuclei from dense tissues and organs such as the kidney and temporomandibular joint, while preserving tissue mechanics [35].

6.3.2.4 Alcohols

Alcohol such as glycerol acid, decellularizes by dehydrating and lysing the cells [29]. Phospholipids in valve leaflets and conduits contribute to prosthesis calcification and failure and can be extracted using alcohols [36]. In fact, isopropanol, ethanol, and methanol are more effective than lipase at the removal of lipids from adipose tissue in a relatively short period of time [37]. Methanol has been used in combination with chloroform during tissue delipidation [29]. Caution should be used when treating tissues with alcohols such as ethanol and methanol because of their use as tissue fixatives in histology, their ability to precipitate proteins [38], and the damage they cause to the ECM ultrastructure [39].

6.3.3 Biologic Agents

6.3.3.1 Enzymes

Enzymes that have been reported in the decellularized protocol include nuclease, trypsin, collagenase, lipase, dispase, thermolysin, and α-galactosidase. Enzymes provide a high specificity for the removal of cell debris or undesirable ECM components. However, it is difficult to effect complete cell removal in a single enzyme treatment; enzyme residues impair recellularization, and can induce a harmful immune response. Nucleases (e.g., DNase and RNase) cleave nucleic acid sequences and thus help in the removal of nucleotides after cell lysis [26].

Endonucleases such as benzonase [40] may be more effective than exonucleases because they cleave nucleotides mid-sequence and thereby more effectively fragment DNA to prepare for its removal. Similarly, nonrestriction endonucleases fragment the DNA more effectively than sequence-dependent counterparts. Trypsin is a serine protease that is used as an enzymatic decellularization agent. However, ECM proteins such as collagens have limited resistance to trypsin cleavage [41], and tissue exposure to trypsin should therefore be used with caution. Trypsin is slower at cell removal than detergents. Elastin and collagen are removed from the cell, but although trypsin is more destructive, it does show better preservation of the glycosaminoglycan (GAG) content [42, 43]. Collagenase decellularization may be employed, but only if maintenance of the ultrastructure and of collagen is not critical to the intended clinical applications of the resulting ECM.

6.3.3.2 Nonenzymatic Agents

Ethylenediaminetetraacetic acid (EDTA) and ethylene glycol tetraacetic acid (EGTA) are chelating agents that aid cell dissociation from ECM proteins by sequestering metal ions [44]. Chelating agents are likely to contribute to subtle disruptions in protein–protein interactions using the same mechanism [45]. Chelating agents alone are not sufficient for superficial cell removal, even if agitated [23], and are thus usually used with enzymes such as trypsin [46] or with detergents [20]. The effectiveness of the chelate plus a simple hyper- or hypotonic solution for decellularization is unknown [20]. Toxins such as latrunculin enable the researcher to take advantage of generating a natural cytotoxic agent for decellularization . Gillies et al.

[47] demonstrated removal of DNA and intracellular proteins from dense tissue, the tibialis anterior, using only latrunculin B, hyper- and hypotonic solutions, and DNase treatments. This method provided superior removal of DNA and retention of GAG to the enzymes and detergents for decellularization. The properties of passive mechanical testing of the ECM scaffold were similar.

6.4 Changes in the Mechanical Properties of Decellularized ECM During Constructive Remodeling

Facilitating the replacement of the ECM graft with functional host tissue is considered a constructive remodeling response. ECM scaffold changes during such a remodeling response are dependent on factors such as the local tissue microenvironment, the rate of scaffold degradation, forces present within the mechanical environment, and the rate and extent to which the infiltrating cells deposit new ECM [48, 49]. Decellularized scaffolds retain a variety of collagens, growth factors, glycosaminoglycans, and other matrix proteins, and the arrangement of the proteins is minimally disturbed through the process [50–53]. Upon implantation, mononuclear cells migrate into the decellularized scaffold and degrade it, while concomitantly depositing new site-appropriate tissue [54, 55]. The structural changes that occur during in vivo remodeling of ECM scaffolds are associated with marked changes in scaffold strength. In the early phase of remodeling, degradation occurs quite rapidly, before the newly deposited ECM can become fully organized. However, once the infiltrating cells have established residence and begin producing new site-specific ECM, the remodeling process progresses over time toward a tissue construct that has similar mechanical properties very similar to those of native tissue [54, 56].

6.5 Sterilization of Decellularized Scaffolds

Clinical application of a decellularized scaffold requires donor scaffold sterilization. The sterilization process primarily eliminates endotoxins and intact viral and bacterial DNA that may induce unwanted inflammation upon scaffold implantation [57]. Decellularized scaffolds may be sterilized using simple treatments such as incubation in acids [58] or solvents [59], but such methods may not provide sufficient penetration or may damage key ECM constituents [60]. Sterilization methods such as ethylene oxide exposure, gamma irradiation, and electron beam irradiation achieve effective sterilization, but are known to alter ECM microstructure and mechanical properties [61, 62].

6.5.1 Ethylene Oxide Exposure

Ethylene oxide sterilization does not affect cell attachment to ECM or the stimulation of growth factor secretion by fibroblasts in vitro [28]. Ethylene oxide may substantially change ECM mechanical properties [63] or leave them unaltered [24, 62], and ethylene oxide treatment can cause undesirable host immune responses that impair proper function of the biologic scaffold after implantation [64]. Therefore, the use of ethylene oxide for the sterilization of decellularized scaffolds should be considered carefully.

6.5.2 Gamma Irradiation Exposure

Degradation of ECM during gamma irradiation is at least partially attributed to spontaneous denaturation of key structural proteins such as collagen that occurs even at relatively low doses at body temperature [65]. As the irradiation dose increases, denaturation of collagen continues to increase, resulting in dose-dependent alterations in mechanical properties [66]. Even at a dose as low as 2 kGy, gamma irradiation increases tissue stiffness and tensile strength, but at doss higher than 15 kGy, mechanical properties decrease in a dose-dependent manner [67]. Gamma irradiation causes residual lipids to become cytotoxic and accelerates enzymatic degradation [67, 68].

6.5.3 Supercritical Carbon Dioxide

Supercritical carbon dioxide has become a mainstay of the sterilization of food and pharmaceuticals. It has recently been investigated as an alternative method of sterilizing decellularized scaffolds. Supercritical carbon dioxide is versatile for sterilization owing to its nonreactive nature, its ability to penetrate cells and tissues, its reduced energy usage, and its improved quality of retention of heat-sensitive substrates [69]. Supercritical carbon dioxide sterilization with peracetic acid sterilant causes only minor changes in the susceptibility of porcine acellular matrix to collagenase digestion, in tensile or tear strength, indicating limited alteration of the tissue structure following supercritical carbon dioxide sterilization [70]. However, supercritical carbon dioxide sterilization is relatively new and requires further investigation.

6.6 Effects and Evaluation of Decellularized Scaffolds

The major concerns of all decellularization protocols remain ECM disruption, immunogenicity, and thrombogenicity. The effectivity of decellularization and the alterations to the ECM vary according to the source, composition, and density of the tissue, and to other factors. Decellularization agents that are very efficient at removing cellular components can also cause damage to the collagen structure and remove growth factors and other important ECM components. However, because of insufficient decellularization, residual cellular components within ECM may contribute to in vitro cytotoxicity and adverse host responses in vivo upon reintroduction of cells [71–73]. There are conflicting results on the effect of decellularization techniques. Grauss et al. [74] found that chemically induced decellularization using Triton X-100 or trypsin resulted in changes in the ECM constitution, which could lead to problems in valve functionality and cell growth and migration, whereas Schenke-Layland et al. [75] used a decellularization method, in which pulmonary valves were dissected from porcine hearts, placed in a solution of trypsin-EDTA, incubated at 37°C for 24 h, followed by a 24-h wash in phosphate-buffered solution (PBS) , and showed acceptable removal of cell components from a porcine pulmonary valve. Cartmell and Dunn [14] compared the methods of decellularization for rat tail tendon using three extraction chemicals (*t*-octylphenoxy-polyethoxyethanol [Triton X-100], tri(*n*-butyl) phosphate [TnBP], and sodium dodecyl sulfate [SDS]). They showed that treatment of tendons with 1% Triton X-100 for 24 h disrupted the collagen fiber structure, but did not remove cells. Treatment with 1% SDS for 24 h or 1% TnBP for 48 h resulted in an acellular tendon matrix with retention of near-normal structure and mechanical properties. Woods and Gratzer [15] also compared the effect of decellularization using one of the three protocols incorporating the surfactant lauryl sulfate (SDS), Triton X-100, and/or an organic solvent, tri(n-butyl)phosphate (TnBP). They showed that all three treatments were effective at removing cells and preserving the mechanical properties of the graft, with subtle, but notable, differences. Most detergents, when used as agents for decellularization, cause some removal of GAG from the scaffold, which has varying degrees of a negative effect on the tensile viscoelasticity of the scaffold [76]. Another factor that can influence the mechanical properties of the scaffold is the duration of exposure to decellularization agents [77]. There is practically no consensus regarding the effects of any of the decellularization methods. Therefore, it is very important to optimize the decellularization method for obtaining acceptable cell removal according to the source, composition and density of the tissue, and other factors.

Evaluation of the residual materials and cellular components within the decellularized ECM is a pivotal step in translating them into a clinically practicable, transplantable substance. There are a number of methods available to determine the effectivity of the elimination of cellular material from tissues. Although the decellularization process cannot remove 100% of cell material, it is crucial in quantifying cellular components such as double-stranded DNA (dsDNA), mitochondria, or membrane-associated molecules such as phospholipids in a decellularized scaffold. Crapo et al. [57] suggested the following minimal criteria sufficient to satisfy the intent of decellularization: (1) no visible nuclei upon histological evaluation, (2) the remaining dsDNA content should not exceed 200 base pairs in length, and (3) the amount of dsDNA should not exceed 50 ng/mg of dry weight of the material. The first criterion is easily evaluated using standard histological staining with hematoxylin and eosin or immunofluorescent methods such as DAPI or Hoechst. The second and third criteria are easily evaluated using commercially available dsDNA intercalators, such as PicoGreen®, propidium iodide, or bisbenzimide, and by gel electrophoresis respectively. Jackson and Simon [78] demonstrated the fate of DNA from allografts after transplantation using a DNA probe technique that clearly distinguished donor cells from host cells and could also be utilized to evaluate whether any DNA is present in the decellularized scaffold in the Spanish goat model. Electron microscopic methods or polymerase chain reaction (PCR) are usable, but not typically used to evaluate the presence of residual nuclear material or cytoplasmic debris because of the technical complexity and expense of these procedures for routine work [79]. Schmitt et al. [80] studied the cell viability, cell migration parameters, and biomechanical properties of human adipose-derived stem cells following reseeding on human tendon scaffolds in vivo using bioluminescent imaging and immunohistochemistry. They showed that reseeded cells remained viable on the implanted constructs at up to 4 weeks using bioluminescent imaging, and histological evaluation showed host cell invasion and proliferation of the repair sites. Biomechanical testing revealed no significant difference in ultimate load to failure and a 2-mm gap force. This movement of host cells and human adipose-derived stem cells into the scaffold suggests the biointegration of the allograft. Lin et al. [81] evaluated the biocompatibility of the decellularized scaffolds using enzyme–detergent methods for the cell removal of mouse skeletal muscle tissue by histological staining (sections stained with H&E, Van Gieson's, and DAPI) and DNA quantification. They showed that mouse skeletal muscles combined with an enzyme–detergent mixture (trypsin and Triton X-100) can yield an intact matrix devoid of cells, depleting more than 93% of the nuclear component and exhibiting comparable biomechanical properties to those of native tissue, and that infiltration of inflammatory cells increased into the scaffold initially and then decreased gradually until day 30. Although the above methods provide important information regarding the effectiveness of decellularization methods, the biologic consequences of small amounts of nuclear material or cytoplasmic debris within the decellularized scaffold are unclear. Crapo et al. [57] suggested that a standard for tissue decellularization might provide numerous benefits, including (1) allowing investigators and ECM product manufacturers to evaluate the effectiveness of a protocol when reporting new decellularization techniques or describing products composed of ECM derived from decellularized tissue; (2) enabling congruous comparison of different ECM products; (3) eliminating variations in cell and host responses to ECM products caused by variations in residual DNA, thereby facilitating interpretation and comparison of in vitro and in vivo results; and (4) promoting the rapid and effective development of additional clinical applications for ECM products within the field of regenerative medicine and tissue engineering.

6.7 Removal of Residual Chemicals

The decellularization methods involve a wide variety of chemicals used because of their capability to damage cells. The strong chemicals can thoroughly remove the cell components, but the internal fiber scaffold structure may also be seriously damaged, whereas mild chemicals can preserve the internal fiber scaffold structure. However, the removal effect of cell components is far from satisfying, which is the messy problem of chemical extraction. The presence of some residual decellularization agents within ECM biomaterials is toxic to host cells and may be responsible for discouraging cellular ingrowth [43, 82]. There is a need for the development of assays to quantify the presence of residual chemicals in the decellularized scaffold material and the removal methods of residual chemicals.

6.8 Eight Decellularized Scaffolds as a Platform for Bioengineered Organs

6.8.1 Liver

Liver tissue engineering provides an insight into liver regeneration; it has seen remarkable progress in recent years [83, 84]. In 2010, liver scaffolding of transferred and intact acellular liver has been developed by perfusion of various chemical detergents into the portal vein of the rat. These scaffolds maintain the function of the microvasculature and the three-dimensional structure of ECM components (Fig. 6.3) [84]. Scaffolding of decellularized liver has demonstrated the ability to efficiently support in vitro decellularization with subsequent perfusion of primary liver cells [85, 86]. In in vivo microinjection of scaffolds from decellularized liver with microscopic vascular anastomosis, the scaffold showed seeding cells. However, thrombosis formation was noticed shortly post-transplantation [87]. To address thrombus formation, heparin was perfused in the multilayer on the inside

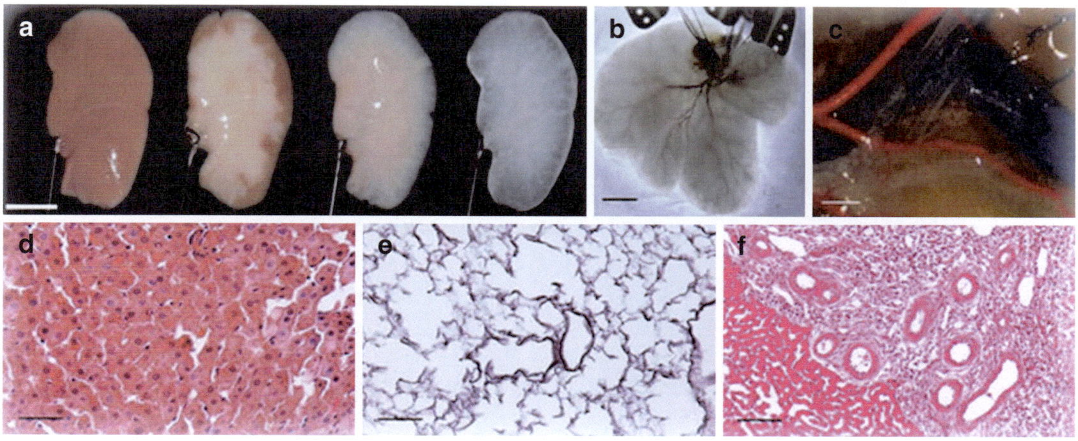

Fig. 6.3 Fabrication, vascular cast, light microstructure, and implantation of the decellularized liver scaffolds. (**a**) Decellularization of a single lobe of rat liver progressing under continuous detergent perfusion. Scale bar 10 mm. (**b**) Decellularized whole-liver scaffold with the hepatic artery intact. Scale bar 20 mm. (**c**) Vessel corrosion casting of the microstructure of the hepatic portal vein (*blue*), the hepatic artery (*red*), and the hepatic duct (transparent). Scale bar 2 mm. (**d**) H&E staining of liver matrix shows the existence of blue-stained nuclei in intact liver, but not in (**e**) decellularized liver scaffold. (**f**) H&E staining results show the border between the liver parenchyma and the implanted decellularized scaffold. Scale bar 100 μm. (From Yu et al. [2])

surface of the scaffold [88, 89]. Despite the short-term effectiveness of this intervention, long-term effectiveness requires further experiments.

6.8.2 Heart

The first decellularized cardiac scaffolds were produced from rats in 2008 [14]. These scaffolds were perfused with heart muscle cells and vascular endothelial cells in vitro to mimic the cell composition of the heart. This construct successfully ran the pump function after transplantation [90]. Induced pluripotent stem cells (iPSCs) of human origin were seeded onto decellularized mouse heart in vitro. The seeded iPSCs migrated, proliferated, and differentiated into functional cardiomyocytes after implanting, enabling the constructed cardiac tissues to demonstrate contractility [91].

Recently, these studies focused on the repair of myocardial infarction after myocardial ischemia. Bone marrow mesenchymal stem cells (MSCs) were promoted to cardiac repair after myocardial ischemia or infarction [92]. Transplantation of stem cells improves the tissue condition and overall cardiac function of infarcted tissue [93].

6.8.3 Lung

The study of lung tissue regeneration has passed through two stages. The basic concept of the regeneration of the lung stem cell is possible to play the lung tissue, to join the synthetic material and lung stem cells to build functional pulmonary units (alveoli). Based on such a proposal, lung stem cells were seeded in synthetic material in vivo and in vitro. The construct could not be formed, owing to the poor integration and tissue compatibility and the possibility of complete failure of the respiratory function caused by surgical infection [94, 95]. Recently, lung tissue engineering has been facilitated by decellularized scaffold in vivo and in vitro. Although the cellular components are removed during decellularization, the relevant cytokines and ECM are retained [96].

6.8.4 Kidney

Porcine kidneys have been successfully decellularized, proposing the possibility of using these transplantable scaffolds to construct tissue-engineered kidney that is clinically applicable (Fig. 6.4) [97]. Whole porcine kid-

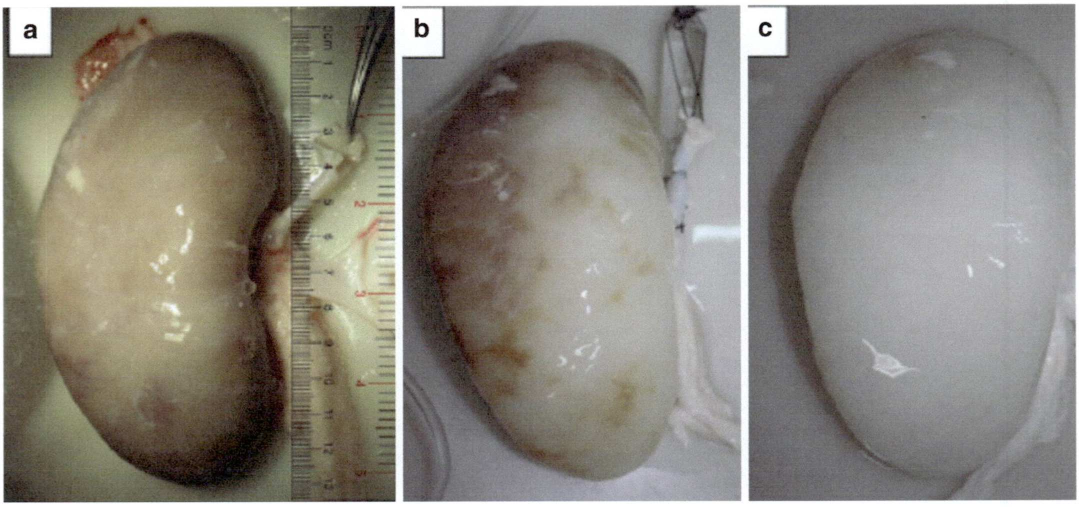

Fig. 6.4 Gross appearance of a porcine kidney during the decellularization process (**a**) Beginning of decellularization. (**b**) After 16 h of perfusion, the kidney became more translucent. (**c**) After 32 h, the kidney is fully decellularized and is white. (From Guan et al. [97])

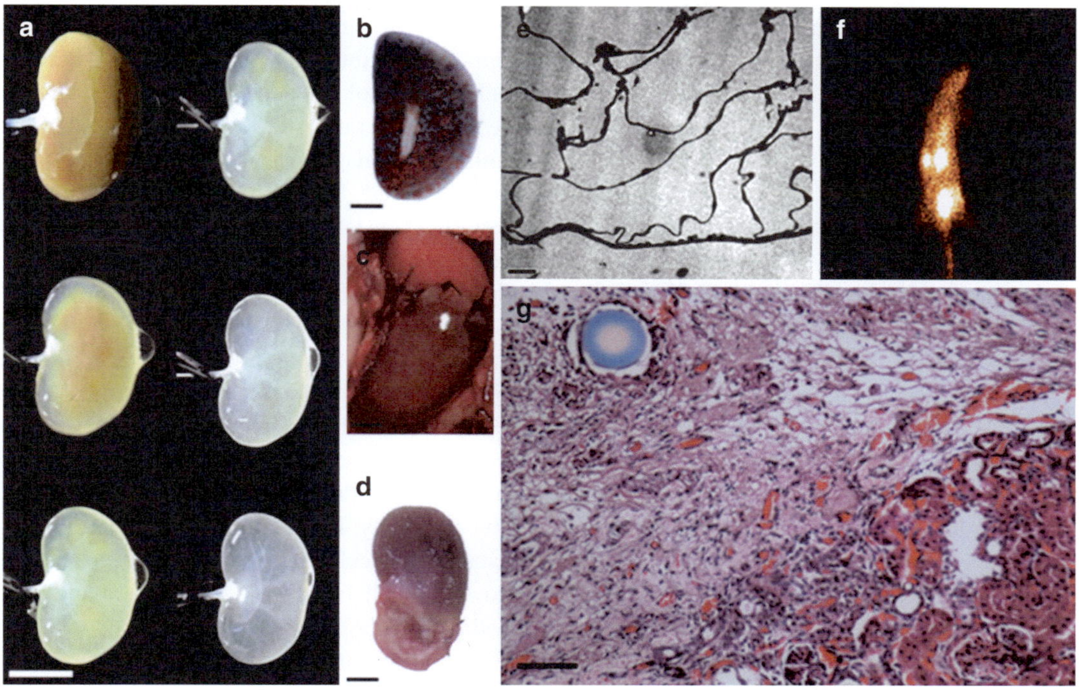

Fig. 6.5 Fabrication and implantation of decellularized kidney scaffolds. (**a**) With continuous detergent perfusion, the rat decellularized kidney scaffold shows different gross appearances. Scale bar 10 mm. (**b**) Casting model of decellularized kidney scaffolds shows intact microvessels. (**c**) Decellularized scaffolds were sutured to a rat that underwent partial nephrectomy. (**d**) Macroscopic images show longitudinal cross-sections of whole experimental kidneys observed under a stereoscopic microscope. Scale bar 20 mm. (**e**) Electron microscopy observation shows intact ECM in decellularized kidney scaffold. Scale bar 2 μm. (**f**) Radionuclide scanning analysis of experimental kidneys. (**g**) H&E staining shows the border between the renal parenchyma and implanted decellularized scaffold. Scale bar 100 μm. (From Yu et al. [2])

neys were decellularized and then orthotopically transplanted in vivo, and then anticoagulant was administered as prophylaxis. Inflammatory cells in the pericapsular region and thrombosis occurred owing to the lack of endothelial cells [98]. In recent findings, successfully engineered renal tissue absorbed by the proximal tubule was shown to have the ability of metabolic and endocrine functions [99]. Yu et al. [4] successfully reported that renal decellularized scaffolds induce the regeneration of injured kidney (Fig. 6.5). There is a possibility that various cytokines in the scaffold play an important role in the recovery of postoperative renal function after partial nephrectomy.

6.8.5 Pancreas

Early studies on pancreatic regeneration focused on synthetic materials such as cross-linked collagen matrix liquid scaffold [100, 101]. First, decellularized pancreatic scaffolds were produced from a porcine model in 2013. These scaffolds were then seeded with human amniotic fluid-derived stem cells (hAFSCs) and porcine pancreatic islets. The scaffolds promoted cell proliferation and maintained cellular function [102]. Pancreatic acinar cells and β cells were used to construct decellularized pancreatic scaffolds in vitro and resulted in an increased insulin level post-subcutaneous transplantation [103]. In recent studies, after implantation into mice in in vivo, constructs were shown that synthesized pancreas using an artificial three-dimensional material with β cells that regulate the blood glucose level [104].

6.8.6 Spinal Cord and Brain

First, spinal scaffolds made the cellular structure, myelin, and nervous process of rats disappeared, whereas most ECM structural proteins were preserved [105]. the weakness of the immunogenicity of the subcutaneous spinal cord scaffold positioning, when implanted, and infiltration of CD4+ and CD8+ cells were not obvious. Rat spinal scaffolds were combined with mesenchymal stem cells from human umbilical cord blood and then implanted into rat spinal cords. The results showed that nerve cells migrated into the scaffold, accompanied by the formation of new myelinated axons, resulting in motor function recovery [106]. Decellularized cerebral scaffolds, derived from porcine brains, failed to maintain the original structure, but the ECM, containing GAGs, was successfully preserved [107].

The study suggests that ECM could be used for cell culture because of nerve biocompatibility. Nerve cells derived from human-iPSCs have grown and mature on the matrix. Brain matrix can also be processed into an injectable hydrogel fiber structure. Porcine brain, spinal cord, and optic nerve were decellularized using a combination of freeze–thaw, trypsin digestion, and chemical detergents. The generated cross-linked scaffolds preserved various growth factors, were cultured with pc12 cells, and demonstrated an ability to promote cell proliferation, migration, and differentiation. ECM from the central nervous system is likely to be more effective than bladder ECM at promoting the proliferation and differentiation of neural cells [108].

6.8.7 Visceral Organs

Decellularized scaffolds derived from the bladder and mucosal layer of the small intestine were widely adapted to be used for the treatment of defects in the internal organs, before the clinical application of internal organ scaffolding. Bladder acellular scaffold was used to repair a defect in the bladder in rats in 1996 [109]. Because of the similarity in anatomical simplicity, bladder acellular scaffold has also been used in the reconstruction of other visceral organs, such as the tympanic membrane [110], esophagus [111], trachea [112], larynx [113], glottis [114], thoracic wall [115], ventricular wall [116], small intestine [117], and arteries [118]. The process based on the scaffolding of the internal organs, in addition to mobility, requires adequate blood supply to support the repair of the structure and components of the organ [119]. In the presence of a blood supply, i.e., scaffold can promote transplanted stem cells to enhance recovery and the growth of cell function [120, 121]. In addition, modified scaffold can inhibit the inflammatory response for better integration into the recipient site [122].

6.8.8 Skin

The development of bioengineered products of different skin layers—including the tissue-engineered epidermis, dermis, and composite skin—has provided innovative tools for clinical applications. Cultured epithelial autograft (CEA), an approach to obtaining epidermal grafts, has been used in the repair of major burns. Tissue-engineered epithelial cells prepared by culturing human self-epidermal keratinocytes in vitro have been transported to two patients to repair burns [123].

However, absence of dermis layer and wound contracture and may lead poor cells adhesion and subsequent survival. In addition, swelling and scar contracture have been reported in the later stages. Artificial skin, developed through extensive experimentation, comprising a layer of silastic (epidermis) and a porous bovine collagen–chondroitin 6-sulfate (dermis), was used physiologically to repair extensive burn injuries constituting 50–95% of the body surface area [124]. Compared with engineered epidermis, scaffolds of skin promote the migration of fibroblasts and angiogenesis and have the ability to provide optimal mechanical and physicochemical properties necessary for healing.

6.9 Concluding Remarks

The galloping field of decellularization holds great promise for the creation of biologic scaffolds for tissue engineering and regenerative medicine. It is unlikely that any combination of methods will eliminate all cell components from a tissue or organ. However, it seems obvious that methods that can eliminate most or all the cellular components result in biologic scaffolds that are safe for transplantation. The beneficial effects of biologic scaffolds in the field of tissue engineering and regenerative medicine can be achieved if optimal methods of decellularization are applied. Furthermore, these biologic scaffolds are clinically relevant and important in the continued development of decellularization methods.

References

1. Langer R, Vacanti JP. Tissue engineering. Science. 1993;260:920–6.
2. Yu Y, Alkhawaji A, Ding Y, Mei J. Decellularized scaffolds in regenerative medicine. Oncotarget. 2016;7:58671–83.
3. Pariente JL, Kim BS, Atala A. In vitro biocompatibility assessment of naturally derived and synthetic biomaterials using normal human urothelial cells. J Biomed Mater Res. 2001;55:33–9.
4. Yu YL, Shao YK, Ding YQ, et al. Decellularized kidney scaffold-mediated renal regeneration. Biomaterials. 2014;35:6822–8.
5. Exposito JY, D'Alessio M, Solursh M, Ramirez F. Sea urchin collagen evolutionarily homologous to vertebrate pro-alpha 2(I) collagen. J Biol Chem. 1992;267:15559–62.
6. Constantinou CD, Jimenez SA. Structure of cDNAs encoding the triple-helical domain of murine alpha 2 (VI) collagen chain and comparison to human and chick homologues. Use of polymerase chain reaction and partially degenerate oligonucleotide for generation of novel cDNA clones. Matrix. 1991;11:1–9.
7. Booth C, Korossis SA, Wilcox HE, et al. Tissue engineering of cardiac valve prostheses I: development and histological characterization of an acellular porcine scaffold. J Heart Valve Dis. 2002;11:457–62.
8. Kasimir MT, Rieder E, Seebacher G, et al. Comparison of different decellularization procedures of porcine heart valves. Int J Artif Organs. 2003;26:421–7.
9. Conklin BS, Richter ER, Kreutziger KL, Zhong DS, Chen C. Development and evaluation of a novel decellularized vascular xenograft. Med Eng Phys. 2002;24:173–83.
10. Dahl SL, Koh J, Prabhakar V, Niklason LE. Decellularized native and engineered arterial scaffolds for transplantation. Cell Transplant. 2003;12:659–66.
11. Chen RN, Ho HO, Tsai YT, Sheu MT. Process development of an acellular dermal matrix (ADM) for biomedical applications. Biomaterials. 2004;25:2679–86.
12. Hudson TW, Liu SY, Schmidt CE. Engineering an improved acellular nerve graft via optimized chemical processing. Tissue Eng. 2004;10:1346–58.
13. Borschel GH, Dennis RG, Kuzon WM Jr. Contractile skeletal muscle tissue-engineered on an acellular scaffold. Plast Reconstr Surg. 2004;113:595–602. discussion 3-4
14. Cartmell JS, Dunn MG. Effect of chemical treatments on tendon cellularity and mechanical properties. J Biomed Mater Res. 2000;49:134–40.
15. Woods T, Gratzer PF. Effectiveness of three extraction techniques in the development of a decellularized bone-anterior cruciate ligament-bone graft. Biomaterials. 2005;26:7339–49.
16. Wang ZQ, Watanabe Y, Toki A. Experimental assessment of small intestinal submucosa as a small bowel graft in a rat model. J Pediatr Surg. 2003;38:1596–601.
17. Kropp BP, Eppley BL, Prevel CD, et al. Experimental assessment of small intestinal submucosa as a bladder wall substitute. Urology. 1995;46:396–400.
18. Lin P, Chan WC, Badylak SF, Bhatia SN. Assessing porcine liver-derived biomatrix for hepatic tissue engineering. Tissue Eng. 2004;10:1046–53.
19. Song JJ, Ott HC. Organ engineering based on decellularized matrix scaffolds. Trends Mol Med. 2011;17(8):424–32. doi:10.1016/j.molmed.2011.03.005.
20. Lehr EJ, Rayat GR, Chiu B, et al. Decellularization reduces immunogenicity of sheep pulmonary artery vascular patches. J Thorac Cardiovasc Surg. 2011;141:1056–62.
21. Cortiella J, Niles J, Cantu A, et al. Influence of acellular natural lung matrix on murine embryonic stem cell differentiation and tissue formation. Tissue Eng A. 2010;16:2565–80.
22. Patel N, Solanki E, Picciani R, et al. Strategies to recover proteins from ocular tissues for proteomics. Proteomics. 2008;8:1055–70.
23. Hopkinson A, Shanmuganathan VA, Gray T, et al. Optimization of amniotic membrane (AM) denuding for tissue engineering. Tissue Eng Part C Methods. 2008;14:371–81.
24. Jackson DW, Grood ES, Wilcox P, et al. The effects of processing techniques on the mechanical

properties of bone-anterior cruciate ligament-bone allografts. An experimental study in goats. Am J Sports Med. 1988;16:101–5.
25. Sasaki S, Funamoto S, Hashimoto Y, et al. In vivo evaluation of a novel scaffold for artificial corneas prepared by using ultrahigh hydrostatic pressure to decellularize porcine corneas. Mol Vis. 2009;15:2022–8.
26. Funamoto S, Nam K, Kimura T, et al. The use of high-hydrostatic pressure treatment to decellularize blood vessels. Biomaterials. 2010;31:3590–5.
27. Gilbert TW, Wognum S, Joyce EM, et al. Collagen fiber alignment and biaxial mechanical behavior of porcine urinary bladder derived extracellular matrix. Biomaterials. 2008;29:4775–82.
28. Hodde J, Janis A, Ernst D, et al. Effects of sterilization on an extracellular matrix scaffold. I. Composition and matrix architecture. J Mater Sci Mater Med. 2007;18:537–43.
29. Prasertsung I, Kanokpanont S, Bunaprasert T, Thanakit V, Damrongsakkul S. Development of acellular dermis from porcine skin using periodic pressurized technique. J Biomed Mater Res B Appl Biomater. 2008;85:210–9.
30. Cox B, Emili A. Tissue subcellular fractionation and protein extraction for use in mass-spectrometry-based proteomics. Nat Protoc. 2006;1:1872–8.
31. Xu CC, Chan RW, Tirunagari N. A biodegradable, acellular xenogeneic scaffold for regeneration of the vocal fold lamina propria. Tissue Eng. 2007;13:551–66.
32. Alhamdani MS, Schroder C, Werner J, et al. Single-step procedure for the isolation of proteins at near-native conditions from mammalian tissue for proteomic analysis on antibody microarrays. J Proteome Res. 2010;9:963–71.
33. Cebotari S, Tudorache I, Jaekel T, et al. Detergent decellularization of heart valves for tissue engineering: toxicological effects of residual detergents on human endothelial cells. Artif Organs. 2010;34:206–10.
34. Meyer SR, Chiu B, Churchill TA, et al. Comparison of aortic valve allograft decellularization techniques in the rat. J Biomed Mater Res A. 2006;79:254–62.
35. Lumpkins SB, Pierre N, McFetridge PS. A mechanical evaluation of three decellularization methods in the design of a xenogeneic scaffold for tissue engineering the temporomandibular joint disc. Acta Biomater. 2008;4:808–16.
36. Levy RJ, Vyavahare N, Ogle M, et al. Inhibition of cusp and aortic wall calcification in ethanol- and aluminum-treated bioprosthetic heart valves in sheep: background, mechanisms, and synergism. J Heart Valve Dis. 2003;12:209–16. discussion 16
37. Brown BN, Freund JM, Han L, et al. Comparison of three methods for the derivation of a biologic scaffold composed of adipose tissue extracellular matrix. Tissue Eng Part C Methods. 2011;17:411–21.
38. Jamur MC, Oliver C. Cell fixatives for immunostaining. Methods Mol Biol. 2010;588:55–61.
39. Cole MB Jr. Alteration of cartilage matrix morphology with histological processing. J Microsc. 1984;133:129–40.
40. Petersen TH, Calle EA, Zhao L, et al. Tissue-engineered lungs for in vivo implantation. Science. 2010;329:538–41.
41. Waldrop FS, Puchtler H, Meloan SN, Younker TD. Histochemical investigations of different types of collagen. Acta Histochem Suppl. 1980;21:23–31.
42. Yang M, Chen CZ, Wang XN, Zhu YB, Gu YJ. Favorable effects of the detergent and enzyme extraction method for preparing decellularized bovine pericardium scaffold for tissue engineered heart valves. J Biomed Mater Res B Appl Biomater. 2009;91:354–61.
43. Rieder E, Kasimir MT, Silberhumer G, et al. Decellularization protocols of porcine heart valves differ importantly in efficiency of cell removal and susceptibility of the matrix to recellularization with human vascular cells. J Thorac Cardiovasc Surg. 2004;127:399–405.
44. Gailit J, Ruoslahti E. Regulation of the fibronectin receptor affinity by divalent cations. J Biol Chem. 1988;263:12927–32.
45. Maurer P, Hohenester E. Structural and functional aspects of calcium binding in extracellular matrix proteins. Matrix Biol. 1997;15:569–80. discussion 81
46. Wainwright JM, Czajka CA, Patel UB, et al. Preparation of cardiac extracellular matrix from an intact porcine heart. Tissue Eng Part C Methods. 2010;16:525–32.
47. Gillies AR, Smith LR, Lieber RL, Varghese S. Method for decellularizing skeletal muscle without detergents or proteolytic enzymes. Tissue Eng Part C Methods. 2011;17:383–9.
48. Badylak S, Kokini K, Tullius B, Whitson B. Strength over time of a resorbable bioscaffold for body wall repair in a dog model. J Surg Res. 2001;99:282–7.
49. Badylak SF, Freytes DO, Gilbert TW. Extracellular matrix as a biological scaffold material: structure and function. Acta Biomater. 2009;5:1–13.
50. Hodde JP, Badylak SF, Brightman AO, Voytik-Harbin SL. Glycosaminoglycan content of small intestinal submucosa: a bioscaffold for tissue replacement. Tissue Eng. 1996;2:209–17.
51. Hodde JP, Record RD, Liang HA, Badylak SF. Vascular endothelial growth factor in porcine-derived extracellular matrix. Endothelium. 2001;8:11–24.
52. Hodde J, Record R, Tullius R, Badylak S. Fibronectin peptides mediate HMEC adhesion to porcine-derived extracellular matrix. Biomaterials. 2002;23:1841–8.
53. McDevitt CA, Wildey GM, Cutrone RM. Transforming growth factor-beta1 in a sterilized tissue derived from the pig small intestine submucosa. J Biomed Mater Res A. 2003;67:637–40.
54. Badylak S, Kokini K, Tullius B, Simmons-Byrd A, Morff R. Morphologic study of small intestinal

submucosa as a body wall repair device. J Surg Res. 2002;103:190–202.
55. Gilbert TW, Stewart-Akers AM, Simmons-Byrd A, Badylak SF. Degradation and remodeling of small intestinal submucosa in canine Achilles tendon repair. J Bone Joint Surg Am. 2007;89:621–30.
56. Badylak SF, Vorp DA, Spievack AR, et al. Esophageal reconstruction with ECM and muscle tissue in a dog model. J Surg Res. 2005;128:87–97.
57. Crapo PM, Gilbert TW, Badylak SF. An overview of tissue and whole organ decellularization processes. Biomaterials. 2011;32:3233–43.
58. Hodde J, Hiles M. Virus safety of a porcine-derived medical device: evaluation of a viral inactivation method. Biotechnol Bioeng. 2002;79:211–6.
59. Gorschewsky O, Klakow A, Riechert K, Pitzl M, Becker R. Clinical comparison of the Tutoplast allograft and autologous patellar tendon (bone-patellar tendon-bone) for the reconstruction of the anterior cruciate ligament: 2- and 6-year results. Am J Sports Med. 2005;33:1202–9.
60. Gorschewsky O, Puetz A, Riechert K, Klakow A, Becker R. Quantitative analysis of biochemical characteristics of bone-patellar tendon-bone allografts. Biomed Mater Eng. 2005;15:403–11.
61. Keane TJ, Swinehart IT, Badylak SF. Methods of tissue decellularization used for preparation of biologic scaffolds and in vivo relevance. Methods. 2015;84:25–34.
62. Freytes DO, Stoner RM, Badylak SF. Uniaxial and biaxial properties of terminally sterilized porcine urinary bladder matrix scaffolds. J Biomed Mater Res B Appl Biomater. 2008;84:408–14.
63. Rosario DJ, Reilly GC, Ali Salah E, et al. Decellularization and sterilization of porcine urinary bladder matrix for tissue engineering in the lower urinary tract. Regen Med. 2008;3:145–56.
64. Jackson DW, Windler GE, Simon TM. Intraarticular reaction associated with the use of freeze-dried, ethylene oxide-sterilized bone-patella tendon-bone allografts in the reconstruction of the anterior cruciate ligament. Am J Sports Med. 1990;18:1–10. discussion-1
65. Sun WQ, Leung P. Calorimetric study of extracellular tissue matrix degradation and instability after gamma irradiation. Acta Biomater. 2008;4:817–26.
66. Bailey AJ, Tromans WJ. Effects of ionizing radiation on the ultrastructure of collagen fibrils. Radiat Res. 1964;23:145–55.
67. Gouk SS, Lim TM, Teoh SH, Sun WQ. Alterations of human acellular tissue matrix by gamma irradiation: histology, biomechanical property, stability, in vitro cell repopulation, and remodeling. J Biomed Mater Res B Appl Biomater. 2008;84:205–17.
68. Moreau MF, Gallois Y, Basle MF, Chappard D. Gamma irradiation of human bone allografts alters medullary lipids and releases toxic compounds for osteoblast-like cells. Biomaterials. 2000;21:369–76.
69. Sikin AM, Rizvi SS. Recent patents on the sterilization of food and biomaterials by supercritical fluids. Recent Pat Food Nutr Agric. 2011;3:212–25.
70. Qiu QQ, Leamy P, Brittingham J, et al. Inactivation of bacterial spores and viruses in biological material using supercritical carbon dioxide with sterilant. J Biomed Mater Res B Appl Biomater. 2009;91:572–8.
71. Nagata S, Hanayama R, Kawane K. Autoimmunity and the clearance of dead cells. Cell. 2010;140:619–30.
72. Manfredi AA, Capobianco A, Bianchi ME, Rovere-Querini P. Regulation of dendritic- and T-cell fate by injury-associated endogenous signals. Crit Rev Immunol. 2009;29:69–86.
73. Brown BN, Valentin JE, Stewart-Akers AM, McCabe GP, Badylak SF. Macrophage phenotype and remodeling outcomes in response to biologic scaffolds with and without a cellular component. Biomaterials. 2009;30:1482–91.
74. Grauss RW, Hazekamp MG, Oppenhuizen F, et al. Histological evaluation of decellularised porcine aortic valves: matrix changes due to different decellularisation methods. Eur J Cardio-Thoracic Surg. 2005;27:566–71.
75. Schenke-Layland K, Vasilevski O, Opitz F, et al. Impact of decellularization of xenogeneic tissue on extracellular matrix integrity for tissue engineering of heart valves. J Struct Biol. 2003;143:201–8.
76. Lovekamp JJ, Simionescu DT, Mercuri JJ, et al. Stability and function of glycosaminoglycans in porcine bioprosthetic heart valves. Biomaterials. 2006;27:1507–18.
77. Conconi MT, De Coppi P, Di Liddo R, et al. Tracheal matrices, obtained by a detergent-enzymatic method, support in vitro the adhesion of chondrocytes and tracheal epithelial cells. Transplant Int. 2005;18:727–34.
78. Jackson DW, Simon TM. Donor cell survival and repopulation after intraarticular transplantation of tendon and ligament allografts. Microsc Res Tech. 2002;58:25–33.
79. Gilbert TW, Sellaro TL, Badylak SF. Decellularization of tissues and organs. Biomaterials. 2006;27:3675–83.
80. Schmitt T, Fox PM, Woon CY, et al. Human flexor tendon tissue engineering: in vivo effects of stem cell reseeding. Plastic Reconstructive Surg. 2013;132:567e–76e.
81. Lin CH, Yang JR, Chiang NJ, Ma H, Tsay RY. Evaluation of decellularized extracellular matrix of skeletal muscle for tissue engineering. Int J Artif Organs. 2014;37:546–55.
82. Wang Q, Zhang C, Zhang L, et al. The preparation and comparison of decellularized nerve scaffold of tissue engineering. J Biomed Mater Res A. 2014;102:4301–8.
83. Uygun BE, Soto-Gutierrez A, Yagi H, et al. Organ reengineering through development of a transplantable recellularized liver graft using decellularized liver matrix. Nat Med. 2010;16:814–20.
84. Shupe T, Williams M, Brown A, Willenberg B, Petersen BE. Method for the decellularization of intact rat liver. Organogenesis. 2010;6:134–6.

85. Barakat O, Abbasi S, Rodriguez G, et al. Use of decellularized porcine liver for engineering humanized liver organ. J Surg Res. 2012;173:e11–25.
86. Soto-Gutierrez A, Zhang L, Medberry C, et al. A whole-organ regenerative medicine approach for liver replacement. Tissue Eng Part C Methods. 2011;17:677–86.
87. Zhang H, Zhang Y, Ma F, Bie P, Bai L. Orthotopic transplantation of decellularized liver scaffold in mice. Int J Clin Exp Med. 2015;8:598–606.
88. Zhou P, Lessa N, Estrada DC, et al. Decellularized liver matrix as a carrier for the transplantation of human fetal and primary hepatocytes in mice. Liver Transplant. 2011;17:418–27.
89. Bao J, Shi Y, Sun H, et al. Construction of a portal implantable functional tissue-engineered liver using perfusion-decellularized matrix and hepatocytes in rats. Cell Transplant. 2011;20:753–66.
90. Ott HC, Matthiesen TS, Goh SK, et al. Perfusion-decellularized matrix: using nature's platform to engineer a bioartificial heart. Nat Med. 2008;14:213–21.
91. Lu TY, Lin B, Kim J, et al. Repopulation of decellularized mouse heart with human induced pluripotent stem cell-derived cardiovascular progenitor cells. Nat Commun. 2013;4:2307.
92. Venugopal JR, Prabhakaran MP, Mukherjee S, et al. Biomaterial strategies for alleviation of myocardial infarction. J Roy Soc Interface Roy Soc. 2012;9:1–19.
93. Wu YF, Zhang J, Gu YQ, et al. Reendothelialization of tubular scaffolds by sedimentary and rotative forces: a first step toward tissue-engineered venous graft. Cardiovasc Revascularization Med. 2008;9:238–47.
94. Cegielski M, Calkosinski I, Dziegiel P, Zabel M. The search for stem cells of the epithelium in pulmonary alveoli. Folia Morphol (Warsz). 2004;63:221–3.
95. Kim J, Yaszemski MJ, Lu L. Three-dimensional porous biodegradable polymeric scaffolds fabricated with biodegradable hydrogel porogens. Tissue Eng Part C Methods. 2009;15:583–94.
96. White ES. Lung extracellular matrix and fibroblast function. Ann Am Thorac Soc. 2015;12(Suppl 1)):S30–3.
97. Guan Y, Liu S, Liu Y, et al. Porcine kidneys as a source of ECM scaffold for kidney regeneration. Mater Sci Eng C Mater Biol Appl. 2015;56:451–6.
98. Orlando G, Farney AC, Iskandar SS, et al. Production and implantation of renal extracellular matrix scaffolds from porcine kidneys as a platform for renal bioengineering investigations. Ann Surg. 2012;256:363–70.
99. Song JH, Humes HD. The bioartificial kidney in the treatment of acute kidney injury. Curr Drug Targets. 2009;10:1227–34.
100. Coronel MM, Stabler CL. Engineering a local microenvironment for pancreatic islet replacement. Curr Opin Biotechnol. 2013;24:900–8.
101. Hosseini-Tabatabaei A, Jalili RB, Hartwell R, et al. Embedding islet in a liquid scaffold increases islet viability and function. Can J Diabetes. 2013;37:27–35.
102. Mirmalek-Sani SH, Orlando G, McQuilling JP, et al. Porcine pancreas extracellular matrix as a platform for endocrine pancreas bioengineering. Biomaterials. 2013;34:5488–95.
103. Goh SK, Bertera S, Olsen P, et al. Perfusion-decellularized pancreas as a natural 3D scaffold for pancreatic tissue and whole organ engineering. Biomaterials. 2013;34:6760–72.
104. Yuan X, Huang Y, Guo Y, et al. Controlling the blood glucose of type 1 diabetes mice by co-culturing MIN-6 beta cells on 3D scaffold. Pediatr Transplant. 2015;19:371–9.
105. Guo SZ, Ren XJ, Wu B, Jiang T. Preparation of the acellular scaffold of the spinal cord and the study of biocompatibility. Spinal Cord. 2010;48:576–81.
106. Liu J, Chen J, Liu B, et al. Acellular spinal cord scaffold seeded with mesenchymal stem cells promotes long-distance axon regeneration and functional recovery in spinal cord injured rats. J Neurol Sci. 2013;325:127–36.
107. DeQuach JA, Yuan SH, Goldstein LS, Christman KL. Decellularized porcine brain matrix for cell culture and tissue engineering scaffolds. Tissue Eng A. 2011;17:2583–92.
108. Crapo PM, Medberry CJ, Reing JE, et al. Biologic scaffolds composed of central nervous system extracellular matrix. Biomaterials. 2012;33:3539–47.
109. Sutherland RS, Baskin LS, Hayward SW, Cunha GR. Regeneration of bladder urothelium, smooth muscle, blood vessels and nerves into an acellular tissue matrix. J Urol. 1996;156:571–7.
110. Parekh A, Mantle B, Banks J, et al. Repair of the tympanic membrane with urinary bladder matrix. Laryngoscope. 2009;119:1206–13.
111. Nieponice A, McGrath K, Qureshi I, et al. An extracellular matrix scaffold for esophageal stricture prevention after circumferential EMR. Gastrointest Endosc. 2009;69:289–96.
112. Gilbert TW, Gilbert S, Madden M, Reynolds SD, Badylak SF. Morphologic assessment of extracellular matrix scaffolds for patch tracheoplasty in a canine model. Ann Thoracic Surg. 2008;86:967–74. discussion 74
113. Huber JE, Spievack A, Simmons-Byrd A, Ringel RL, Badylak S. Extracellular matrix as a scaffold for laryngeal reconstruction. Ann Otol Rhinol Laryngol. 2003;112:428–33.
114. Kitamura M, Hirano S, Kanemaru SI, et al. Glottic regeneration with a tissue-engineering technique, using acellular extracellular matrix scaffold in a canine model. J Tissue Eng Regen Med. 2014;10:825–32.
115. Gilbert TW, Nieponice A, Spievack AR, et al. Repair of the thoracic wall with an extracellular matrix scaffold in a canine model. J Surg Res. 2008;147:61–7.
116. Robinson KA, Li J, Mathison M, et al. Extracellular matrix scaffold for cardiac repair. Circulation. 2005;112:I135–43.
117. Badylak S, Meurling S, Chen M, Spievack A, Simmons-Byrd A. Resorbable bioscaffold for esophageal repair in a dog model. J Pediatr Surg. 2000;35:1097–103.

118. Kajitani M, Wadia Y, Hinds MT, et al. Successful repair of esophageal injury using an elastin based biomaterial patch. ASAIO J. 2001;47:342–5.
119. Totonelli G, Maghsoudlou P, Fishman JM, et al. Esophageal tissue engineering: a new approach for esophageal replacement. World J Gastroenterol. 2012;18:6900–7.
120. Pokrywczynska M, Jundzill A, Bodnar M, et al. Do mesenchymal stem cells modulate the milieu of reconstructed bladder wall? Arch Immunol Ther Exp. 2013;61:483–93.
121. Hori Y, Nakamura T, Kimura D, et al. Functional analysis of the tissue-engineered stomach wall. Artif Organs. 2002;26:868–72.
122. Zhou HY, Zhang J, Yan RL, et al. Improving the antibacterial property of porcine small intestinal submucosa by nano-silver supplementation: a promising biological material to address the need for contaminated defect repair. Ann Surg. 2011;253:1033–41.
123. Wainwright D, Madden M, Luterman A, et al. Clinical evaluation of an acellular allograft dermal matrix in full-thickness burns. J Burn Care Rehabil. 1996;17:124–36.
124. Burke JF, Yannas IV, Quinby WC Jr, Bondoc CC, Jung WK. Successful use of a physiologically acceptable artificial skin in the treatment of extensive burn injury. Ann Surg. 1981;194:413–28.

Part III

Cell and Tissue Applications

Kidney

7

Bum Soo Kim and Hyun Tae Kim

7.1 Introduction

The kidney is a vital organ which plays various important roles in humans' body, including excretion of waste products, production of erythropoietin, and maintenance of systemic blood pressure through regulation of fluid volume and electrolytes. The nephrons, which are the functional units of the kidney, maintain these renal functions. Specific conditions, such as hypertension, diabetes, glomerulonephritis, and nephrotoxic drugs, can cause the damage of nephrons, and if damages persist to progress, renal function gradually deteriorates, and the kidney finally enters the state of renal failure.

Renal failure is a worldwide health issue, constituting about 8–16% of the adult population suffering from chronic kidney disease, which is defined as a reduced glomerular filtration rate and increased urinary albumin excretion [1]. Another type of renal failure, acute kidney injury, which is defined as a sudden increase in serum creatinine concentration and decreased urine output can often become permanent and progress to chronic kidney disease [2, 3]. There are two main therapeutic options for the current treatment of end-stage renal disease. One is lifelong dialysis, which has demonstrated therapeutic effects on improved renal functions and increased survival. However, dialysis compromises patients' quality of life and cannot substitute other renal functions, such as erythropoietin production and activation of vitamin D, while it replaces function of filtration by removing toxic substances from the blood [4]. Kidney transplantation, which remains the only potential curative treatment, is another option, but critical shortage of organs is a limitation. In addition, despite advances in renal transplant immunology, approximately 20% of recipients experience an episode of acute rejection within 5 years of kidney transplantation, and about 40% of recipients lose graft function or die within 10 years [5, 6].

These limitations of current therapies for renal failure have encouraged researchers to explore a better treatment for end-stage renal disease that could improve, restore, or replace either partial or total renal function. Tissue engineering and regenerative medicine represent two of the promising and innovative alternatives. The basis of tissue engineering is that cells can be expanded and differentiated in vitro, placed on a scaffold made of feasible biomaterial, and finally implanted into the host. Moreover, xenoembryo transplantation approach and blastocyst complementation methods were suggested and showed promising results for potential kidney regeneration [7].

In 1999, Chan et al. reported the first attempt to develop a whole functional renal unit using

B.S. Kim (✉) • H.T. Kim
Department of Urology, School of Medicine, Kyungpook National University, Daegu, South Korea
e-mail: urokbs@knu.ac.kr; ht-kim1212@hanmail.net

B.W. Kim (ed.), *Clinical Regenerative Medicine in Urology*, DOI 10.1007/978-981-10-2723-9_7

pronephros from animal caps in *Xenopus* [8]. Transplantation of this pronephros improved the edema in bilaterally nephrectomized tadpoles, and they could survive for up to 1 month. However, the pronephros, which was used in this study, was too primitive for clinical application in humans. Since this trial, many investigations have been performed to regenerate whole kidneys de novo. Advances in stem cell research over the past decades accelerated significant progress in the field of kidney regeneration. During the last decade, studies have shown that it is possible to seed cultured cells into a three-dimensional scaffold system in vitro and implant the cell-seeded scaffold in vivo to restore renal functions, and other studies also have shown the possibility of growing new kidneys in living hosts or integrated new nephrons into immature kidneys [4, 9].

However, kidney regeneration is more complex than regeneration of other solid organs. Despite recent advancement of tissue engineering and stem cell therapy, we could not achieve the appropriate functional kidney regeneration due to the complicated architecture and function of kidneys as well as the lack of precise understanding of molecular mechanisms for cell differentiation and development of kidneys. Although the regeneration of an entire kidney may be difficult, functional recovery of as little as 10% of renal filtration function can allow for the withdrawal of dialysis in patients with end-stage renal disease, thus improving patients' quality of life [10]. Therefore, research in the field of regeneration of the kidney is continuing to make progress, with the aim of not only making a whole functional kidney but also improving partial renal function for renal failure.

There are four major steps for the research of kidney regeneration: (1) identification of renotropic factors and signaling pathway, which involve cell differentiation and growth; (2) identification of renal stem/progenitor cells in the adult or embryonic kidney; (3) cell therapies using mesenchymal stem cells, such as bone marrow or adipose tissue-derived cells, and amniotic fluid stem cells; and (4) reconstruction of artificial kidney or renal components using stem cells and biomaterial scaffold [11].

Since emerging technologies in the field of kidney regeneration still have limitations, various approaches were attempted, and strategies can be broadly classified into several categories. Current major strategies for kidney regeneration are as follows: (1) stem/progenitor cell-based therapy, (2) developmental biology, (3) bioartificial kidney, and (4) whole-organ reconstruction (Fig. 7.1). In this chapter, all of the contents above related to kidney regeneration will be discussed in detail.

7.2 Mechanisms of Kidney Regeneration

7.2.1 Cells Involved in Kidney Repair

Various types of cells are considered to be able to either contribute directly to renal repair after damage or substantially improve renal injury without direct contributing to the renal epithelium. The proper regulation of these cells and successful delivery are important for the development of kidney regeneration. There are four possible main origins for cells contributing to kidney repair: (1) interstitial cell transdifferentiation to epithelium [12], (2) recruitment of stem cells from the bone marrow [13–17], (3) tubular cell dedifferentiation and proliferation in response to injury [12, 18, 19], and (4) repopulation of the renal tubules by a resident kidney stem/progenitor cells [20–22]. The first two options involve nonepithelial cell transdifferentiation into renal epithelium, whereas the other two options involve epithelial cells within the renal epithelium itself. The theory that renal repair involves cells within the renal epithelium of the nephron was supported by lineage studies showing no evidence of dilution of the cap mesenchyme-derived (Six2+) tubular epithelium with a nonepithelial cell source [23]. However, this study did not answer whether any cell within

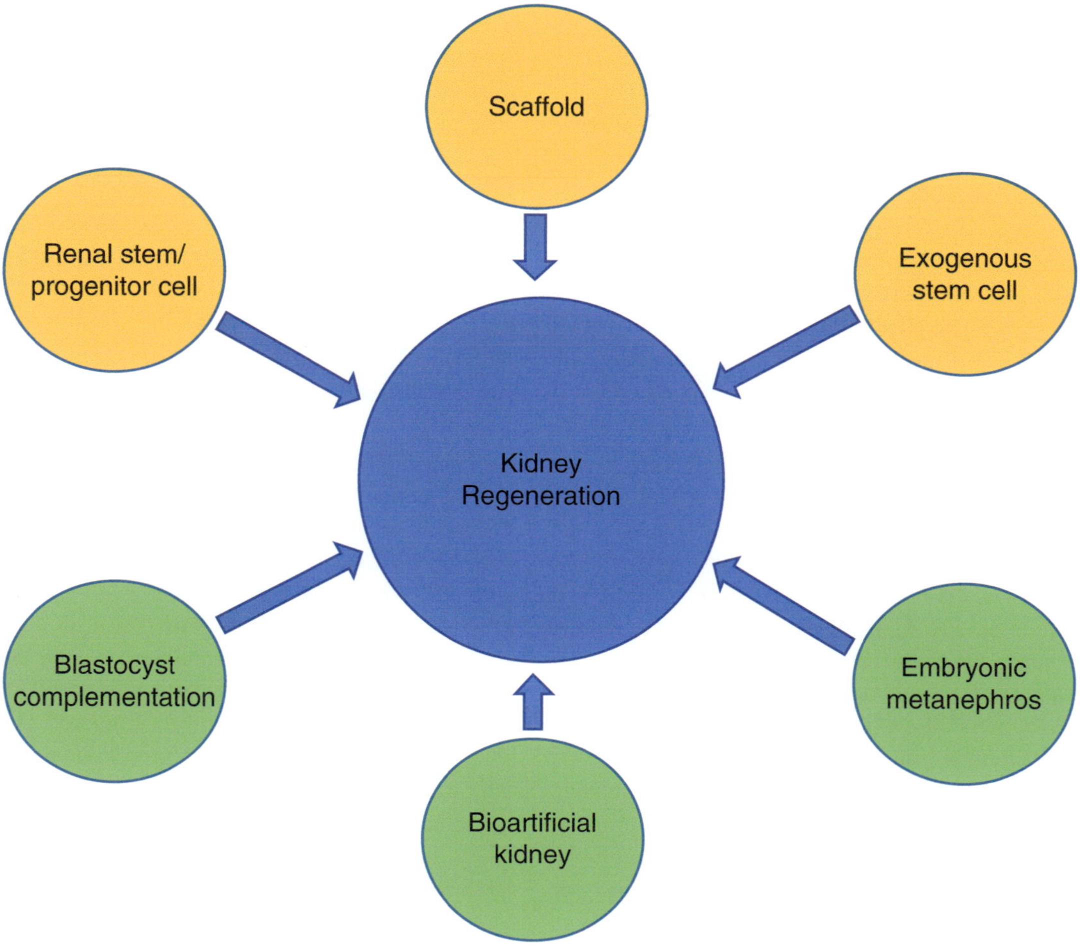

Fig. 7.1 Schematic representation of the current major strategies for kidney regeneration

the epithelium can contribute to repair or whether repair relied on a resident stem cell population within the tubules.

7.2.1.1 Renal Progenitor Cells

Although many studies have investigated the existence, location, and contribution of renal progenitor cells to epithelial repair [21, 22, 24, 25], existence and identification of renal stem or progenitor cells in the adult kidney tissues have been somehow controversial [3]. Grobstein first demonstrated the presence of renal progenitor cells in the metanephric blastema [26, 27]. Discrete populations of these renal progenitors have been shown to persist into adulthood, and numerous studies have identified renal progenitor cells in the proximal tubule [28], tubular epithelial cells [30], Bowman's capsule [24], glomeruli [31], renal papilla [32], and cortical stroma. These renal progenitor cells possess high cloning efficiency, self-renewal potential, and differentiate into specific renal cell lineages, suggesting a role in renal regeneration [24]. Several transcriptional regulators, such as Six2 and Foxd+1, are essential for the persistence of these progenitor cell populations beyond development, while Pax2 is required for later differentiation [33, 34].

Renal progenitor cells could be harvested and isolated from adult human kidneys as well as human urine, expanded in vitro, maintain their renal phenotypes through several passages, and reimplanted [20, 35, 36]. Bussolati et al. showed that intravenously injected renal progenitor cells migrated to the site of tubular injury and contributed to renal repair [20]. These results may support that differentiated renal epithelial cells isolated from kidney tissue have proliferation capacity through dedifferentiation of the surviving epithelia during cell culture.

In contrast to the stem cells, renal progenitor cells can only differentiate to specific cellular lineage and display limited self-renewal capacity [37]. Renal progenitor cells are identified by cell marker CD133 and CD24. CD133 is a marker of several types of adult tissue stem cells, while CD24 is a surface molecule which is expressed in human metanephric mesenchyme [38, 39]. Renal progenitor cells can differentiate into glomerular as well as tubular epithelial cells. Tubular progenitor cells represent about 2–6% of all tubular epithelial cells in healthy adult kidneys and express CD133, CD24, vimentin, cytokeratins 7 and 19, Pax2, and nestin that are not expressed by differentiated tubular epithelial cells [20]. Glomerular progenitor cells localize within the Bowman's capsule and can differentiate toward podocyte phenotype. These glomerular progenitor cells can express CD106, whereas tubular progenitor cells cannot express this surface marker.

Renal progenitor cells have higher resistance to injury compared to other differentiated cells of the kidney. When injected in severe combined immunodeficient mice affected by rhabdomyolysis-induced acute kidney injury, these cell populations displayed the capacity to integrate into the tubules, generate new tubular epithelial cells, and improve renal function [22]. It can be an attractive strategy if renal progenitor cells can be induced via lineage-instructive reprogramming or differentiated from extrarenal stem cells, such as mesenchymal, embryonic, and induced pluripotent stem cells. These regenerated exogenous renal progenitor cells may enhance renal repair combined with a small population of endogenous renal progenitor cells. In comparison to stem cells, renal progenitor cells have several advantages, such as no need of intermediate cell culture conditions and direct transit from one phenotype to another. However, a reliable method of inducing extrarenal stem cells to differentiate into renal progenitor cells has not been appropriately established [40, 41].

7.2.1.2 Proliferation of Mature Renal Epithelial Cells

Lineage tracing studies in mice suggested that the proliferating populations after kidney injury were mature renal epithelial cells rather than a specific progenitor cell population [42], whereas analysis of cell division using unbiased DNA labeling also suggested randomness in the cells reentering cell division within the tubule [43]. Anatomical reanalysis of human tubular epithelium has showed a distinct subpopulation of cells within the tubules referred to as scattered tubular cells [44]. The first study of scattered tubular cells performed by Berger et al provided the evidence that this cell population is not a defined intratubular progenitors of fixed morphology but may arise from any tubular cell as a result of spontaneously dedifferentiation triggered by local injury [45]. Indeed, it has been known that cells within the proximal tubule can express mesenchymal cell markers (vimentin) after kidney injury, suggesting a dedifferentiation process can occur in response to injury [19].

7.2.1.3 Mesenchymal Stem Cells

For decades, many studies demonstrated that cells from the bone marrow could migrate into the sites of injury around the body, including the kidney [17]. Although bone marrow-derived cells were initially considered to have a capacity to generate new tissue even in remote injured sites, several studies have showed it may not occur in some cases [46, 47]. Mesenchymal stem cells, a population of cells within the bone marrow, are clogenic, able to undergo proliferation, and to be differentiated into various cell types, including bone, cartilage, and fat [48]. Although they were originally investigated as a support population within the bone marrow, they have also been investigated for the therapeutic potential with respect to mesenchymal tissue replacement. Recent several studies suggested that bone marrow-derived mesenchymal stem cells are able to improve deteriorated renal function in response to kidney injury and to reduce allograft rejection [49–52]. However, the mechanisms of how these mesenchymal stem cells can improve renal function were not well established. This is not regarded as a result of integration into the damaged epithelium or transdifferentiation into an epithelial cell type. These beneficial effects are rather considered to be induced by paracrine effects and immunomodulatory capacity of mesenchymal stem cells [48, 53–59].

Although mesenchymal stem cells were originally identified in the bone marrow, mesenchymal stem cell-like cells have been also identified to reside in many adult tissues, including the kidney. It may represent a perivascular cell population involved in normal tissue repair [53, 54, 57]. Recently, a population of endogenous kidney mesenchymal stem cells was characterized from the adult murine kidney [60]. It showed the characteristic gene and protein expression profile, clonogenicity, and potential to differentiate to other cell types, such as bone, cartilage, and fat. This cell population showed enrichment for kidney-specific genes, compared to bone marrow-derived mesenchymal stem cells.

7.2.2 Renotropic Factors

Soluble factors involved in kidney development have been identified by gene targeting techniques, in vitro tubulogenesis models, and organ culture systems. The regeneration process after renal injury resembles the kidney organogenesis. Most of these factors have been proved to regulate renal recovery as potential renotropic factors by using animal kidney injury models. These renotropic factors are as follows: hepatocyte growth factor (HGF) [61], epidermal growth factor (EGF) [62], heparin-binding EGF-like growth factor (HB-EGF) [63, 64], bone morphogenetic protein-7 (BMP-7) [65, 66], insulin-like growth factor-I (IGF-I) [67, 68], platelet-derived growth factor (PDGF) [69], and uterine sensitization-associated gene-1 (USAG-1) [70]. In addition, several key regulatory molecules required for renal organogenesis, such as Pax-2, leukemia inhibitory factor, and Wnt4, are reactivated in regenerating tubular cells after ischemic renal injury [71–73].

Recently, the role of these renotropic factors in kidney injury has been demonstrated. Chen et al. showed the importance of EGF receptor activation in the recovery phase after acute kidney injury through mice with a specific EGF receptor deletion in renal proximal tubules [74]. Deletion of the BMP receptor activin-like kinase 3 in the tubular epithelium enhances epithelial damage and fibrosis [75]. Conditional knockout mice lacking the HGF receptor, *c-met*, specifically in renal tubules demonstrated the antiapoptotic of anti-inflammatory effect of *c-met* signaling in renal protection after acute kidney injury [76].

A negative regulatory factor of kidney repair has also been identified. Several studies using transgenic mice expressing truncated activin type II receptor, an in vitro tubulogenesis model, the Wolffian duct culture, and isolated rat embryonic kidney culture showed that activin A is an endogenous inhibitor of renal organogenesis [77–83]. Activin A is also a potent inhibitor of renal regeneration after kidney injury [84].

Most of these renotropic factors can induce tubular cell proliferation after kidney injury when exogenously administered. However, it is still not well known whether these factors are involved in cell maturation, modulation of renal blood flow, neutrophil infiltration, and restoration of polarity. It will be interesting to examine if these renotropic factors promote renal regeneration via the activation of intrinsic renal stem cells.

7.2.3 Macrophage

Generally, macrophages have a function to clear foreign pathogens in the body that can disrupt tissue homeostasis [85] and also play an important role to mediate kidney repair. Although macrophages are well known to promote the fibrotic and inflammation reaction during renal disease, they have been also demonstrated to have a protective role in acute and chronic kidney diseases [86]. Moreover, Wyburn et al. reported that macrophages can ameliorate allograft rejection [87].

Macrophages represent a heterogeneous population with distinct functions depending on resident tissue and phase of tissue injury or disease. In some cases, macrophages display classical activation through the release of nitric oxide, reactive oxygen species, and proinflammatory cytokines, termed an M1 response or phenotype [88, 89]. In another stage of disease, macrophages can play antifibrotic and anti-inflammatory roles, termed M2 phenotype [90, 91]. These M2 macrophages have a reparative function in ischemia-reperfusion injury and obstructive nephropathy, induced by unilateral ureteric obstruction [92–95]. Many studies reported that changes in the local microenvironment are important to trigger an M1 phenotype macrophages, such as macrophage colony-stimulating factor, which can be released by tubular epithelial cells to stimulate this process [96, 97]. Based on these findings, macrophages were directed toward an M2 phenotype via genetic manipulation or cytokine alteration, such as overexpression of interleukin 4 (IL-4), IL-10, inducible nitric oxide synthase, or heme oxygenase, resulting in improved renal function [98–106]. In terms of regenerative medicine, macrophages are considered to have therapeutic potential and be one of the options for cellular or cytokine/growth factor-based therapies in kidney diseases.

7.2.4 Signaling Pathways in Mediating the Regenerative Process

Many cells, such as renal progenitor cells, endogenous renal tubular epithelial cells, bone marrow-derived stem cells, and macrophages, are involved in kidney regeneration. However, the exact mechanism of cell to cell interaction and influencing factors for cell reaction involving regenerative process after kidney injury is still not exactly known. Although several studies suggested possible pathways involved in kidney regeneration, further research to identify the factors in specific signaling pathways will be needed.

7.2.4.1 PI3K/AKT/mTOR Pathway

It has been demonstrated that growth factors, such as EGF, IGF-1, and HGF, can accelerate recovery of renal function after acute kidney injury [107]. These growth factors activate a lipid kinase (phosphatidylinositol-3 kinase, PI3K) that phosphorylates phosphatidylinositol 4,5-bisphosphate to yield phosphatidylinositol 3,4,5-trisphosphate. The phosphatidylinositol 3,4,5-trisphosphate phosphorylates and activates Akt. After being activated, Akt stimulates mammalian target of rapamycin (mTOR) by regulating the activity of intermediary kinases. The activation of mTOR leads to phosphorylation of downstream substrates and then induces cell regeneration. It is demonstrated that inhibition of mTOR by rapamycin delays recovery of renal function [108]. Akt can activate antiapoptotic genes and might inactivate proapoptotic factors, such as Bcl-2-associated death promoter, forkhead family transcription factors, and procaspase-9 [109]. Deletion of the EGF receptor in renal proximal tubular epithelial cells impairs phosphotidylinositol-3 kinase/Akt signaling and delays recovery from acute kidney injury [74].

7.2.4.2 Wnt/GSK3/β-Catenin Pathway

The Wnts are a family of secreted and glycosylated protein ligands. Wnt signals can inhibit glycogen synthase kinase 3 (GSK3) by phosphorylation. When GSK3 is inhibited, β-catenin is stabilized and translocates into the nucleus to act as a transcriptional coactivator of the T-cell factor/lymphoid enhancer-binding factor family or transcription factors and drive the expression of its target genes [110]. This pathway is involved in the regulation of cell fate, cell mobility, proliferation, survival, protein synthesis, and glycogen metabolism [111]. Acute kidney injury induces activation of Wnt pathway, whereas genetic inactivation of Wnt signaling can deteriorate renal function recovery and kidney regeneration [112]. Among the increased Wnt ligands after kidney injury, macrophage-derived Wnt7b can promote tubular epithelial cell regeneration and kidney repair. In the downstream of Wnt signaling, GSK3 is inhibited by Wnt ligands, IGF, EGF, and fibroblast growth factors 16, 19, and 23 [113]. In acute kidney injury, GSK3 can promote the systemic inflammatory response and participate in a number of apoptotic signaling pathways by phophorylating transcription factors that regulate apoptosis [114]. TDZD-8, a GSK3β inhibitor, can inactivate ischemia-induced activation of GSK3, caspase 3, and Bax, ameliorate tubular epithelial cell apoptosis, and protect renal function [115]. Acute kidney injury can induce the expression of β-catenin, and renal tubule-specific knockout of endogenous β-catenin aggravates kidney injury in mice [116]. It is also reported that inhibition of GSK3β can ameliorate nonsteroidal anti-inflammatory drug-induced acute kidney injury [117]. Consequently, it is considered that activation of the Wnt/GSK3/β-catenin pathway is beneficial for kidney disease and GSK3 inhibitor can be a target of therapeutic agents in the future.

7.2.4.3 JAK/STAT Pathway

When EGF binds to the EGF receptor, Janus-activated kinase (JAK) is activated and phosphorylates the intracellular domain of the receptor and allows recruitment and phosphorylation of STAT. Salahudeen et al. demonstrated that erythropoiesis-stimulating proteins suppress renal tubular cell apoptosis in vitro and enhance renal recovery in the cisplatin-induced acute kidney injury through the activation of JAK/STAT pathway [118]. In contrast, Ucero et al. reported that inhibition of the JAK/STAT pathway can decrease tubular epithelial cell apoptosis and kidney inflammation in murine acute kidney injury model [119]. (Therefore, further researches will be needed to clarify the effect of JAK/STAT pathway on renal repair.)

7.2.4.4 MAPK/ERK Pathway

Mitogen-activated protein kinases (MAPKs) are a family of kinases that have been commonly investigated on the kidney disease. There are four different MAPK pathways: extracellular signal-regulated kinase-1 and extracellular signal-regulated kinase-2 (ERK1/2), extracellular signal-regulated kinase-5 (ERK5), c-Jun N-terminal kinase (JNK), and p38MAPK [120, 121]. ERK is mainly activated by mitogenic stimuli, such as growth factor, and ERK1/2 pathway has been widely studied in kidney regeneration. It was demonstrated that activation of ERK pathway can enhance renal epithelial cell survival during oxidative injury in vitro [122]. Activation of the signal transducer and activator of transcription-3 (STAT3) during oxidative stress can attenuate EGF receptor-mediated ERK activation and renal tubular cell survival [123]. It was also reported that inhibition of ERK pathway reduces kidney regeneration in rats with myoglobinuric acute kidney injury in vivo [124].

7.3 Improving Renal Function After Kidney Injury

7.3.1 Cell-Based Approach

Early studies using stem cells for kidney regeneration usually have shown that the predominant recovery mechanism of the acute kidney injury was mediated through the transdifferentiation of the infused mesenchymal stem cells into specific renal cell types, especially podocytes, renal tubular cells, glomerular cells, and mesangial cells [13, 125–127]. More recent studies support the concept that cytokines and/or growth factors secreted from the injected mesenchymal stem cells stimulate endogenous cells to proliferate and regenerate the infused renal tissues [128–131]. Several studies suggested that the effect of mesenchymal stem cells in acute kidney injury models is associated with paracrine mechanisms, such as immune-related response and mitogenic, antiapoptotic, and vasoprotective effect [132, 133]. It was evidenced by the results of several studies that bone marrow-derived mesenchymal stem cell-cultured medium contains microvesicles and growth factors that reduce inflammation and enhance renal repair through interactions with renal progenitors [15, 125, 134, 135].

Bone marrow-derived mesenchymal stem cells are multipotent, migrate across tissues, are easily harvested, are produced throughout life, and contribute to the repair of organs [136]. The administration of bone marrow stem cells has been shown to promote neovascularization, reduce inflammation, inhibit apoptosis, and stimulate differentiation and proliferation in multiple systems [137]. Poulsom et al. first reported the effect of engrafted bone marrow stem cells into the damaged kidney, improving repair of renal tissues after acute kidney injury [17]. However, despite these benefits of bone marrow stem cells, clinical application is still in question because negative results were also reported. Bone marrow stem cells have been shown to promote interstitial fibrosis in mouse models, and direct renal injection of bone marrow stem cells induced the development of angiomyeloproliferative lesions of unknown neoplastic potential [138, 139]. Therefore, further safety studies are necessary before clinical applications to ensure safe and successful outcomes for kidney regeneration using bone marrow stem cells.

Another promising cell source of mesenchymal stem cells is adipose-derived stem cells. Several studies reported the effectiveness of adipose-derived stem cells to improve renal functions in animal models, including mouse, rat, and pig [140–143]. Administration of adipose-derived stem cells into an acute kidney injury model reduced renal fibrosis at 6 weeks [144], and intrarenal injection of adipose-derived stem cells showed improved angiogenesis and preserved the renal structure integrity, which restored renal function at 14 days [145].

Amniotic fluid-derived stem cells are easily harvested and cultured with lower risk of tumorigenicity and similar multipotency to bone marrow stem cells. Rota et al. demonstrated the therapeutic potential of human amniotic fluid stem cells [146]. In their study, engrafted amniotic fluid stem cells localized primarily to the peritubular region, limiting tubular damage, improving renal function, and prolonging survival of the animals. They also reported that amniotic fluid stem cells that were preconditioned with glial cell line-derived neurotrophic factor were observed to promote better functional recovery by contributing to renal regeneration in acute kidney injury models when compared to amniotic fluid stem cells without treatment. The authors suggested that a combination of transdifferentiation of amniotic fluid stem cells and paracrine signaling involved the regenerative effect through this study. Hauser et al. reported that intravenous infusion of amniotic fluid stem cells induced more rapid recovery of renal function in an animal model of acute kidney injury, when compared to that of bone marrow stem cells. In this study, bone marrow stem cells showed higher potential for proliferation, whereas amniotic fluid stem cells showed better capacity for antiapoptosis and persistence within the peritubular capillaries. In addition, they isolated different cytokines and growth factors from bone marrow and amniotic fluid stem cells, suggesting different modes of action between both cell sources.

Induced pluripotent stem cells (iPSCs) have gained increasing interest in the field of regenerative medicine, as they possess pluripotent capability and potentially provide an inexhaustible source of patient- and tissue-specific stem cells. Recently, iPSCs have been successfully generated from human renal sources, such as kidney mesangial cells, proximal tubular cells, podocytes, and epithelial cells derived from urine [147–149]. IPSCs have several advantages compared with embryonic stem cell, such as absence of ethical issues related to cell sourcing, fewer immune rejection, and less potential for abnormal tissue formation. Moreover, they also have more capabilities to facilitate targeted, organ-specific differentiation due to retaining the epigenetic pattern of the parent cell [150]. Recent studies reported the improvement of renal functions in an acute kidney injury rodent model after administration of iPSCs [151, 152]. However, iPSCs have been shown to express abnormal genes and induce T-cell-dependent immune reaction in syngeneic mice [153]. This unexpected immune response is a big hurdle of iPSCs for clinical application.

Aforementioned renal progenitor cells are also promising cell sources for the treatment of acute kidney injury. It has been demonstrated that CD24+/CD133+/CD106- renal progenitor cells are resistant to apoptosis and promote recovery of tubular injury in the rhabdomyolysis-induced acute kidney injury rat model [22]. Neural cell adhesion molecule (NCAM)-positive cells that were identified from human kidney tissue showed clonogenic and renal progenitor properties [154]. After engraftment into the chick embryo, NCAM+ human nephron progenitor cells formed a renal-like structure. Administration of human nephron progenitor cells into the kidneys of acute kidney injury rat model induced inhibition of disease progression and improvement of renal function. Taken together, these results support the concept that renal progenitor cells have the capacity to migrate and proliferate to improve structural and functional damage after acute kidney injury.

7.3.2 Developmental Biology

Although advancement of stem cell technology has showed promising results in the field of kidney regeneration, stem cells themselves have limitations to maintain the structure of the kidney. Thus, many researchers tried to develop the cellular approach by incorporating principles of developmental biology. This investigative field sought to create, implant, and maintain a renal structure that mimics the physical and physiological characteristics of the native kidney.

The concept of early investigations in this field endeavored to infuse new nephrons to developing kidneys by implantation of embryonic metanephric tissue into the renal cortex of neonatal mice [155]. Metanephros is the embryological precursor of the adult kidney, characterized by mesenchyme and ureteric buds (Fig. 7.2) [156]. Transplantation of primitive metanephric tissue over fully developed kidney may have some advantages, such as reduced immunogenicity due to the absence of native vasculature and antigen-presenting cells [156, 157]. Thus, metanephros can be maintained in vivo without host immunosuppression. Although transplanted metanephric tissue was successfully differentiated and developed into functional nephrons in the host kidney of neonate, it was not initially reproduced in host kidneys of adult due to lack of differentiation and acute graft rejection [158]. However, another study reported successful implantation and differentiation of metanephroi into adult hosts, with demonstrated glomerular filtration and plasma clearance after modifying the protocol [156]. Encouraged by these results, further studies to

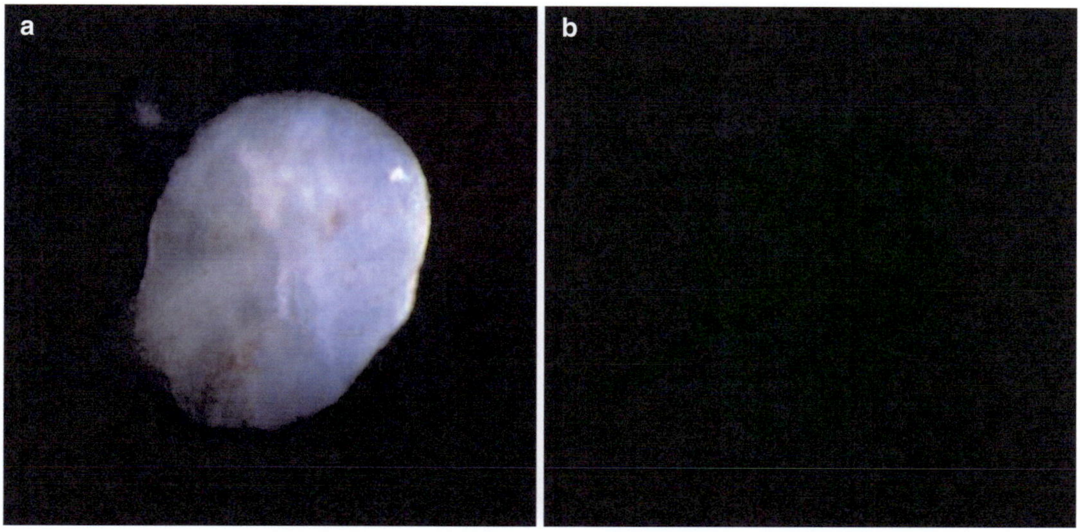

Fig. 7.2 Isolated metanephros from embryo of E15 transgenic mouse in which a green fluorescent protein is expressed in the ureteric bud under the control of the Hoxb7 promoter. (**a**) Only the ureter can be distinguished from metanephric mesenchyme under the vision of light microscope. (**b**) Ureter and branches of ureteric bud can be seen with green color under the vision of immunofluoro-microscope

investigate the possibility of xenotransplantation have been performed. More information related to xenotransplantation will be discussed in another section.

7.4 De Novo Organ Regeneration

The basic concept for regeneration of solid organ is to collect stem or progenitor cells from appropriate cell sources and expand these cells in culture with or without differentiation. Subsequently, cells are seeded into a scaffold and placed into a bioreactor to promote cell acclimation, attachment, and maturation. Then, finally this cell-seeded scaffold can be implanted in the host [159, 160].

7.4.1 Artificial Scaffold

Extracellular matrix is the naturally occurring scaffold material secreted and manufactured by the resident cells of each organ, and it plays a critical role in development and repair of organ [161, 162]. The extracellular matrix is a complex three-dimensional framework of structural and functional proteins in a state of dynamic equilibrium within the surrounding tissues and provides the means by which cells communicate with each other and the external environment [163]. The extracellular matrix contains growth factors and bioinductive cytokines, which facilitate the reparative process, cell attachment, and tissue integration [164, 165]. Consequently, the roles of extracellular matrix for organogenesis and regeneration can be categorized by providing a three-dimensional scaffold for the spatial organization of cells, secreting and storing growth factors and cytokines, and regulating signal transduction [166].

7.4.1.1 Bioengineered Scaffold

Development of biomaterial engineering has produced bioengineered scaffolds that facilitate improved differentiation of transplanted cells. The first histocompatible functional kidney using bioengineered scaffold was generated by Lanza et al. in 2002 [167]. They used a nuclear transplantation technique in which dermal fibroblasts isolated from an adult cow were transferred into enucleated bovine oocytes and then transferred into progestin-synchronized recipients. Metanephroi taken from embryos were dissociated to single-cell suspension using collagenase, and the cells were expanded in vitro. Then, the cells were seeded onto a specialized polymer tube as the artificial scaffold. The cell-seeded scaffold was implanted into the same cow from which the cells had been cloned and produced urine-like liquid. Histologic analysis showed that it had well-differentiated kidney-like structure, including organized glomerulus-like, tubular-like, and vascular components. This investigation established the possibility of kidney regeneration using bioengineered scaffolds.

Synthetic biodegradable and biocompatible polyesters, such as polyglycolic acid (PGA), polylactic acid-polyglycolic acid (PLGA), and polycaprolactone (PCL), also can be used as a scaffold for kidney regeneration. Although these synthetic polymers are a US Food and Drug Administration (FDA)-approved biocompatible material, they have critical drawbacks, such as producing acidic products during degradation and inducing an inflammatory response in the host, which restricts the clinical applications [168–172]. To overcome these adverse effects of synthetic biomaterial, several investigations are in progress.

7.4.1.2 Decellularized Cadaveric Scaffold

Artificial scaffold also can be generated from organs of animal or human's cadaver through decellularization [173–176]. The decellularization process can remove DNA, cellular material, and cell surface antigens from the extracellular matrix scaffold while preserving the structural and functional characteristics of vascular channels. The process of decellularization includes the repeated irrigation of cadaveric organs with acids or detergents through the innate vasculature, although organs with higher fat component, such as pancreas, often require the addition of lipid solvents [177]. To generate cadaveric scaffold, complete decellularization must be achieved because residual cellular material may contain antigenic epitopes that trigger inflammatory reactions and compromise subsequent recellularization [178, 179]. Ethylene oxide or paracetic acid can effectively sterilized the extracellular matrix without denaturing the extracellular matrix proteins or growth factors after decellularization [180, 181].

Using this technique, a functional artificial rat heart using a cadaveric heart as a scaffold was successfully generated [182]. For this, a whole-heart scaffold with intact three-dimensional geometry and vasculature was created via coronary perfusion with detergents into the cadaveric heart, followed by repopulation with neonatal cardiac cells or rat aortic endothelial cells and cultured under physiological conditions to stimulate organ maturation. In this study, the injected neonatal cardiac cells formed a contractile myocardium, which performed the stroke function. Decellularized cadaveric scaffolds have also been used to generate other organs, such as the liver, respiratory tract, and urinary bladder [183–185].

Recent advancement of decellularization-recellularization technique enabled to generate more complex organs, such as the kidney. Ross et al. first reported successful regeneration of an entire kidney using the decellularized rat kidney as a scaffold [186]. After fabricating intact scaffold, murine embryonic stem cells were injected through the innate vasculature. In this study, embryonic stem cells were used for seeding population because of their high doubling capacity, pluripotency, and potential to differentiate and integrate into primordial kidney cultures. The kidney scaffold successfully supported the growth and migration of the seeded embryonic stem cells within glomerular, vascular, and tubular structures while inducing differentiation down renal cell lines. Immunohistochemical analysis indicated that the seeded embryonic stem cells had lost their embryonic appearance and had shown gross morphological changes consistent with mature kidney cells, as well as expression of immunohistochemical markers of renal differentiation, such as Pax-2, Ksp-cadherin, and pan-cytokeratin. Through this study, the authors suggested several points to advance this research field that pretreatment of the embryonic stem cell with prodifferentiation agents, such as retinoic acid, activin-A, and BMP-7, may promote implantation and proliferation by providing a more kidney-specific lineage. Moreover, while injected cells through innate vascular channels can be evenly distributed in the renal cortex, the cells may not localize in the collecting system. Thus, retrograde seeding via ureter was attempted to resolve this problem, but this method caused uneven cell dispersion. Furthermore, murine scaffolds are not feasible for adult human kidney regeneration in terms of size and structure. Therefore, organs of larger animals, such as pig, or primates had to be considered to be investigated using this technique to meet human renal requirements.

However, application of this decellularization and recellularization technique to larger and complex organs is a great challenge, because larger organs with greater parenchymal mass require stronger detergents, higher perfusion pressures, and prolonged detergent exposure time that may damage the native vasculature and extracellular matrix proteins [187]. The first trial to decellularize primate kidney performed by Nakayama et al. confirmed effective decellularization with preservation of native extracellular matrix architecture and protein complement [188]. After recellularization, immunohistochemical analysis showed effective local migration and attachment of renal cells to the border of scaffold, demonstrating the feasibility of primate kidneys for decellularization and recellularization technique.

More recently, intact, whole-organ extracellular matrix scaffolds were generated from porcine kidneys [187, 189]. These studies reported successful decellularization using detergent perfusion through the innate vasculature and confirmed total cell clearance with preserving the architecture of scaffold, including glomeruli, tubules, and vasculatures. Despite these advances of generating decellularized scaffold technique, regeneration of a whole functional kidney using this method is still not achieved. Although several attempts have been performed, effective seeding population of cells was not established. Moreover, it is unclear whether larger kidney scaffold can support the retrograde ureteric perfusion to recellularize the collecting system. It is also not certain whether extracellular matrix of the decellularized kidney scaffold can support the proliferation and differentiation of stem cells and whether recellularization by perfusion through the innate vasculature does not cause thrombogenic complications after reimplantation. Furthermore, even after reliable protocols of recellularization are established, various functions of regenerated kidney, such as homeostasis, resorption, endocrine, metabolism, and immunomodulation, should be confirmed.

7.4.2 Isolation of Cells

The kidney has approximately 20 different specific cell types. For efficient kidney regeneration, all cells must be characterized for repopulation. Although many attempts have been performed to establish a reliable cell source and optimal cell growth conditions to provide adequate enrichment for achieving stable renal cell expansion systems, it is still challenging. It is essential to isolate feasible stem or progenitor cells for cell expansion and differentiation in vitro, and renal stem/progenitor cells are one of the good candidates. There are five different methods which have been used to isolate renal stem/progenitor cells from the adult kidney.

7.4.2.1 Label-Retaining Cells

One of the characteristics that distinguish stem cells from differentiated cells is their infrequent cell division. To conserve the proliferation capacity for a lifetime and to prevent genetic injuries during mitosis, the cycle of stem cells is very slow. Renal stem/progenitor cells were isolated using this property. When these cells are labeled with markers such as 5-bromo-2-deoxyuridine (BrdU) and histone 2B-GFP (H2B-GFP), they can retain the label for a long time. Then, these slow-cycling label-retaining cells (LRCs) can be detected following a chase period. Oliver et al. administered BrdU to rat and mouse neonates and found BrdU-retaining cells localized in the papilla, after chasing them for more than 2 months [32]. After the induction of ischemia, LRCs disappeared from papilla, whereas most of them expressed Ki67, which is a proliferation marker. Through this study, the authors demonstrated that the renal papilla is a niche for adult kidney stem cells and that LRCs proliferate and migrate following ischemic kidney injury. They also showed similar results in another study using H2B-GFP transgenic mice [190]. H2B-GFP is more stable and easily detected than BrdU, and it can label the cells in a cell cycle regardless of the stage, whereas BrdU can only label the cells in the S phase [191]. In this study, H2B-GFP-positive cells proliferated and migrated to renal tubules after kidney injury.

Maeshima et al. identified LRCs using BrdU predominantly in proximal tubules of the adult rat kidney [30]. After ischemic injury, most of the LRCs were found in tubules and were positive for PCNA, indicating that LRCs proliferate after injury. The LRCs were successfully isolated and cultured, and they initially expressed vimentin, a mesenchymal stem cell marker, and eventually became positive for E-cadherin, an epithelial cell marker, after multiple cell divisions. In another study, they demonstrated that LRCs became positive for proximal tubule and ureteric bud markers when transplanted into the metanephric kidney of embryonic rat, indicating that LRCs were integrated into epithelial components of the nephrons in the process of kidney development [192].

Although labeling with BrdU is an effective technique to detect stem cells, considerable problems were also suggested. The BrdU label may be released from dying cells and taken up by adjacent dividing cells [193]. Moreover, the possibility that BrdU-retaining cells are not stem/progenitor cells but fully differentiated cells was proposed [194]. Vogetseder et al. reported that LRCs expressed differentiated tubule markers, such as Na-K-ATPase, NaPi-IIa, PMP70, and megalin. Similarly, H2B-GFP labeling has also drawbacks because the labeling is diluted after massive cell proliferation. To obtain a better understanding of LRCs, a more accurate purification and characterization technique are required.

7.4.2.2 Side Population Cells

Since stem cells extrude dyes, such as Rhodamine 123 and Hoechst 33342 via ATP-binding cassette transporters, they are located in a unique position in fluorescent-activated cell sorting (FACS) analysis and are called side population (SP) cells [195]. SP cells are characterized as Hoechst low cells and are isolated from several tissues, including the heart, liver, and skeletal muscle. Several groups isolated and characterized SP cells from the kidney. Iwatani et al. isolated SP cells from adult rat kidney, but these SP cells were not involved in kidney repair following gentamicin-induced nephropathy [196]. Hishikawa et al. isolated SP cells from adult mouse kidney [197]. These cells expressed Musculin/MyoR, a transcription factor found in skeletal muscle precursors. Musculin/MyoR-positive cells reside in the renal interstitium and are decreased following cisplatin-induced acute kidney injury. Systemic injection of kidney SP cells into mice improved renal function in cisplatin-induced acute tubular injury model. In addition, SP cells expressed a high level of leukemia inhibitory factor (LIF) and renoprotective factors, such as HGF, vascular endothelial growth factor (VEGF), and BMP-7 in a cisplatin-induced acute kidney injury model. Challen et al. also isolated SP cells from adult mouse kidney, and these cells expressed genes involved in Notch signaling. These SP cells are located predominantly in the proximal tubules and integrated into the mesenchyme- and ureteric bud-derived structures when injected into the embryonic kidney, suggesting that they have multilineage differentiation potential. Administration of these cells improved renal function in adriamycin-induced kidney injury model, but these cells were barely incorporated into the renal tissues after infusion. Taken together, SP cells are considered to release renoprotective factors in paracrine fashion, but the exact function or composition of SP cells is still unclear.

7.4.2.3 Cell Surface Markers

Another approach for isolating putative stem cells is utilizing stem cell markers, such as CD133, CD24, and stem cell antigen-1 (Sca-1). Although CD133 is not a specific marker for kidney stem cells, it is a universal marker for stem cells in other tissues, such as hematopoietic stem cells, cancer stem cells, and vascular endothelial progenitor cells [198, 199]. Bussolati et al. isolated CD133+ cells from normal adult human kidney [20]. These cells expressed Pax-2 (embryonic kidney marker), but not CD34, CD45 (hematopoietic lineage marker), CD 90, or c-Kit (stem cell markers), indicating that they were originated from renal tissues. They were capable of self-expansion and limited self-renewal and could be induced to differentiate into epithelial and endothelial cells in vitro and in vivo. When CD133+ cells were injected intravenously in a glycerol-induced acute kidney injury model of severe combined immunodeficiency mice, they migrated to the injured kidney and were incorporated predominantly into the proximal and distal tubules. Sagrinati et al. isolated CD24+ and CD133+ cells from parietal epithelial cells in the adult human kidney after culturing glomeruli [24]. These cells also expressed the stem cell-specific transcription factors, such as Oct-4 and BmI-1. These cells had a high self-renewal potential, and when cultured in appropriate conditions, these cells could be differentiated into tubular epithelial cells, osteocytes, adipocytes, and neuronal cells in vitro. These cells were incorporated into regenerating tubules, and renal function was improved when they were administered intravenously into severe combined immunodeficiency mice with glycerol-induced acute kidney injury. Dekel et al. isolated Sca-1-positive cells from adult mouse kidney [200]. The Sca-1+ cells were located mainly in the papilla and could differentiate into myogenic, osteogenic, adipogenic, and neural lineages in vitro. When injected into the renal parenchyma of ischemia-induced acute kidney injury model, some of these cells were integrated into renal tubules.

7.4.2.4 Cell Culture

Selective culture conditions are useful to isolate stem cells [201]. A unique cell population can be isolated during the culture of dispersed cells derived from the adult kidney. Gupta et al. isolated progenitor-like cells from adult rat kidneys by using culture conditions similar to those used for bone marrow-derived cells [202]. These cells were characterized by long-term self-renewal without senescence and the expression of vimentin, CD90, Pax-2, and Oct-4 without any markers of major histocompatibility complex (MHC) class or other differentiated cell markers. These cells can also be induced to multiple lineages in vitro, including hepatocytes, neurons, and endothelial cells. The cells could differentiate into renal tubular cells when injected under the capsule of the kidney or intra-arterially in uninjured kidney or ischemia-reperfusion injury of the kidney. It is proposed that these progenitor cells isolated by using selective culture condition participate in the regenerative response of the kidney to acute injury. However, possibility of contamination of differentiated kidney cells and blood cells is a limitation of this method.

7.4.2.5 Parietal Epithelial Cells

Mature podocytes are highly differentiated non-dividing cells, and it is not clear whether damage of podocyte in adulthood can be repaired. Glomerular epithelial stem cells may reside in the kidney and may be capable of regenerating podocytes, and it was suggested that glomerular epithelial stem cells exist in the parietal epithelial cells [203]. It was reported that some of the parietal epithelial cells localized in the Bowman's capsule of adult human kidney have the characteristics of stem cells [24]. In this study, parietal epithelial cells were divided into three subpopulations according to the locations and characters. First, cells localized at the urinary pole expressing CD24 and CD133, but not podocalyxin (PDX), a differentiated podocyte marker (CD24+, CD133+, PDX- cells), could regenerate both tubular cells and podocytes. Second, cells localized between the urinary pole and the vascular pole expressing both progenitor and podocytes markers (CD24+, CD133+, PDX+ cells) could only regenerate to podocytes. Finally, cells localized at the vascular pole did not express progenitor markers but displayed phenotypic features of differentiated podocytes (CD24-, CD133-, PDX+ cells). Moreover, the in vivo properties of the cells were also evaluated by Ronconi et al [204]. Administration of CD24+, CD133+, and PDX- cells, but not CD24+, CD133+, PDX+ or CD24-, CD133-, and PDX+ cells, into mice with adriamycin-induced kidney injury reduced proteinuria and improved glomerular damage, suggesting that CD24+, CD133+, and PDX- cells are potential therapeutic targets for glomerular disorders characterized by podocyte injury. Appel et al. reported that the CD133+ parietal epithelial cells have the capacity to proliferate, migrate along the glomerular tuft, and differentiate to mature podocytes after podocyte injuries by using immunostaining, label retention, and transgenic mice approaches [205]. The cells at the base of the vascular pole adjacent to the podocytes were stained with both the parietal epithelial cell marker (claudin-1) and podocyte-specific markers (nestin, dipeptidyl peptidase IV, aminopeptidase A). The idea that parietal epithelial cells migrate to become podocytes was also supported by BrdU staining of rat parietal epithelial cells and genetic tagging of parietal epithelial cells utilizing parietal epithelial cell-specific promoter (hPODXL1). The researchers demonstrated that the cells at the urinary pole of the Bowman's capsule migrate down to the proximal tubules and are involved in the turnover of tubular epithelium. Finally, they concluded that the cells localized at the border of different compartments have features of progenitor cells.

7.4.3 Potential Cell Sources for Renal Regeneration

Recent advances in stem cell biology and cell culture techniques have facilitated the development of cell therapy for clinical translation [206, 207]. Compared with other approaches for kidney regeneration, this method using cellular approach can be more practically applied to kidney tissue regeneration due to the relatively simple cell manipulation process, easy access to the target site in a less invasive manner, and effective integration of administered cells with the host tissues. Consequently, many studies have been performed using this cellular approach to treat renal failure [3, 41, 206]. In this section, cellular approach for kidney tissue regeneration, including various cell sources used as well recent advances made in preclinical and clinical studies, will be discussed.

7.4.3.1 Kidney Tissue-Derived Cells

Although multipotent stem cells have been known to play important roles for various tissues and organs of humans, no definite evidence to date establishes the existence of a pluripotent, self-renewing cell population in the adult kidney. During kidney development, condensed mesenchyme around the tips of the ureteric bud possesses self-renewing cells capable of generating all other elements of the nephrons, interstitium, and vasculature via a mesenchyme-epithelial transition [208]. The cells of condensed mesenchyme are regarded as the renal stem cell population. These stem cells of condensed mesenchyme cease asymmetric division and self-renewal and then exhibit spontaneous commitment to mesenchyme-epithelial transition. However, these cells can be exhausted before the perinatal stage [209]. This phenomenon indicates that complete regeneration involving a complete replacement of the lost nephron does not occur in the mammalian kidney. Nevertheless, researchers have discovered the stem cell-like pluripotent cells in adult kidney, which are called renal progenitor cells, and there are two major theories to explain the origin of proliferating tubular epithelia after acute kidney injury. The first one is that stem or progenitor cells in renal tubule can generate new epithelia, and the other one is that fully differentiated epithelia may dedifferentiate, reenter cell cycle, and generate new epithelial cells through self-duplication.

Primary Kidney Cells

Kidney tissue consists of more than 20 specialized cell types that are structurally organized into functionally and anatomically distinct compartments. Primary renal cells can be harvested from both normal and diseased kidney tissues and expanded in vitro while maintaining the phenotype and function which they are derived. Among these primary renal cells, proximal tubular cells play important roles in kidney functions [210, 211]. The main roles of proximal tubular cells are reabsorption of proteins and electrolytes, production of erythropoietin, and hydrolase activity. Proximal tubular cells occupy the highest percentage of renal cell population in normal kidney; thus, isolation and expansion of functional primary proximal tubular cells from renal tissues can be considered as an attractive renal cell source for cell-based kidney regeneration [212]. Primary proximal tubular cell cultures have several advantages and are more representative of normal physiology of proximal tubular cells than immortalized cell lines, but primary renal cells, including proximal tubular cells, can lose expression of specific genes during in vitro culture and are limited to only two to five passages [213]. The optimal combination of high purity, well differentiation, and consistent proliferation is observed at passage 2–3 [214].

Culture method of primary kidney cells from human kidney tissues for cell expansion has been established [35]. In histological analyses, the majority of the cultured cells retained a proximal tubular phenotype, while distal tubular cells and podocytes were present in a lower percentage of the entire cell population. In addition, when the expanded cells were cultured under a three-dimensional environment, tubule-like structures with functional properties were formed from the cultured cells. These results indicated that the established cell harvesting and culture method may potentially be developed as an effective cell-based approach for kidney regeneration.

Several protocols for other kidney cell types, except proximal tubular cells, also have been established. Presnell et al. established a culture method for primary cell cultures isolated from all major compartments of the kidney, especially the tubular cell-enriched subpopulation and the erythropoietin-producing subpopulation that were reproducibly developed from both normal and diseased kidneys [215]. Recent advances of immunomagnetic cell isolation technique enabled to isolate additional subtypes of kidney-derived epithelial cells. Renal epithelial cells were successfully separated from the ascending limb and the distal tubule of the human kidney using glycoprotein-coated magnetic beads [216]. This result demonstrated that ascending limb of Henle's loop and distal tubule can be a promising cell source for cell-based kidney regeneration and establishment of in vitro cell culture system for various cell types of the kidney can be realized.

Renal Stem/Progenitor Cells

Stem/progenitor cells isolated from many adult organs demonstrate clonogenic, self-renewing ability and can give rise to terminally differentiated cells of the original tissue. Renal stem/progenitor population disappears in the adult kidney, possibly due to the loss of its niche [209, 217]. However, renal stem/progenitor cells still exist in the adult kidney and are located in specific locations, such as renal papilla, tubular epithelial cells, Bowman's capsule, and the S3 segment of the proximal tubules [24, 28–30, 32].

A recent study showed that renal stem/progenitor cells derived from the S3 segment of adult rat kidney nephrons were able to reconstitute a three-dimensional kidney-like structure in vitro [218]. In this study, kidney-like structures were formed when a cluster of renal stem/progenitor cells was suspended in an extracellular matrix gel and cultured in the presence of several growth factors, such as glial cell-derived neurotrophic factor (GDNF), basic fibroblast growth factor (bFGF), HGF, EGF, and BMP-7. The clusters from dissociated S3 segment-derived cells were induced by the hanging drop method in three-dimensional culture, while two-dimensional culture conditions were not able to reconstruct kidney-like structures. The reconstructed kidney-like structures included most of renal substructures, such as glomeruli, proximal and distal tubules, loop of Henle, and collecting ducts, but not the vasculature. This study suggested that a cluster of tissue stem/progenitor cells may have the ability to reconstitute the minimum unit of its original organ by differentiating into specialized cells in the correct niche. Renal stem/progenitor cells derived from the S3 segment of adult rat kidney expressed stem cell markers, such as Sca-1, c-kit, Musashi-1, and nestin, as well as renal lineage markers, such as WT-1 and PAX-2. Although it was assumed that these cells were similar to metanephric mesenchymal cells, based on marker protein expression, the cluster could differentiate into collecting duct-like cells or mesangial-like cells, which are not considered to be derived from metanephric mesenchyme.

Resident mesenchymal stem cells have been described in many organs as a multipotent population adjacent to endothelial cells in the microvasculature and characterized by expression of PDGF, neuron glial antigen-2, CD146, and coexpression of other mesenchymal markers [53]. Isolation of resident mesenchymal stem cells from the murine kidney was also reported [54]. Subsequently, mesenchymal stem cells were isolated from human glomeruli and showed self-renewal capability, clonogenicity, and multipotency [219]. Furthermore, Wang et al. reported that renal mesenchymal stem cells isolated from adult murine kidney could differentiate into renin-producing cells [220]. The comparative study of gene expression in mesenchymal stem cells between kidney and bone marrow origin identified selected patterns of genes in renal mesenchymal stem cells possibly related to a memory of tissue origin [60]. This result suggests that renal mesenchymal stem cells might display organ-specific regenerative capacities that could overcome those of mesenchymal stem cells from other organs, such as the bone marrow or adipose tissue, and that could be exploited for therapeutic applications.

Although adult renal stem/progenitor cells still remain poorly understood, they have been reported to be nontumorigenic when injected into a mouse model, whereas pluripotent stem cells, such as embryonic stem cells or iPSCs, have a risk of potential tumorigenicity as a clinical source [221, 222]. In this regard, the results of aforementioned studies support the possibility that adult renal stem/progenitor cells may represent a safer clinical source than pluripotent stem cells. Therefore, renal stem/progenitor cells may be a promising cell source for kidney regeneration, if it is possible to establish renal stem/progenitor cells with multipotency for kidney reconstruction.

7.4.3.2 Mesenchymal Stem Cells

Mesenchymal stem cells are stromal cells that can be found and isolated from various tissues. They also have the capacity to differentiate into a mesenchymal tissue types, including bone, cartilage, muscle, tendon, fat, and other connective tissues. Unlike pluripotent stem cells, such as embryonic stem cells and iPSCs, mesenchymal stem cells can be practically applied for cell-based therapies with fewer safety concerns and ethical issues. These mesenchymal stem cells have been studied in recent decades as a therapeutic agent for various tissues and organs due to its relatively easy approach to harvest, multilineage differentiation potential, ability to migrate to the site of injury, and no ethical concerns.

Bone Marrow-Derived Stem Cells

Since 2000, bone marrow-derived stem cells have been used in experimental kidney disease models. Several studies have shown that treatment with bone marrow-derived stem cells can ameliorate some of injured renal tissues, such as tubular epithelial cells, podocytes, mesangial cells, and endothelial cells [15–17, 125, 127, 223–226]. Treatment using bone marrow-derived stem cells can contribute to the attenuation of renal fibrosis during progression of chronic kidney disease [138]. Although early studies suggested that bone marrow-derived stem cells could be differentiated into renal epithelial cells and mesangial cells [227–229], recent researches have supported that migration of bone marrow-derived stem cells into the kidney is very rare and their ability to direct differentiation is limited [47]. Thus, the beneficial effects of bone marrow-derived stem cells are considered to be mainly mediated by paracrine action of injected cells that have antiapoptotic, immunomodulatory, and proangiogenic effects [50, 51, 230]. Taking ischemic renal injury, for example, bone marrow-derived stem cells can initially ameliorate the injury either by direct inhibition of cell apoptosis or prevention of inflammatory cell influx. During the repair phase, bone marrow-derived stem cells secrete several factors that promote tubular epithelial cell dedifferentiation and proliferation [231]. The transdifferentiation of bone marrow-derived stem cells observed in early studies may reflect cell fusion. If bone marrow-derived stem cells fuse with resident cells in the kidney, they acquire a phenotype of resident cells. Therefore, it could appear as if bone marrow-derived stem cells were differentiated into resident cells. Indeed, it was reported that bone marrow-derived stem cells could fuse with other cell types [232, 233].

Togel et al. performed the first phase I clinical study to evaluate the efficacy and safety of allogeneic bone marrow-derived stem cell administration for the prevention of acute kidney injury after open heart surgery [234]. Sixteen patients who underwent cardiac surgery and who were at a high risk of postoperative acute renal failure due to underlying chronic kidney disease, advanced age, congestive heart failure, and diabetes mellitus were enrolled in this study. Allogeneic bone marrow-derived stem cells were injected into the suprarenal aorta after surgery. Preliminary data showed that renal function was well preserved after surgery for up to 16 months and none of the patients who received stem cell injection required hemodialysis, while 20% of case controls developed acute renal failure. The length of hospital stay and readmission rates of stem cell group reduced by 40% compared to control group, and no adverse events related to this therapy occurred.

Recently, two subsequent clinical trials were performed to evaluate the effect of bone marrow-derived stem cells in patients who underwent kidney transplantation [235, 236]. These clinical trials have shown that bone marrow-derived stem cells were safe and offered an early recovery of graft function. However, early posttransplant stem cell infusion was found to be possibly associated with acute graft dysfunction due to engraftment syndrome [235]. This adverse effect, which was accompanied by neutrophil recruitment and complement C3 deposition, was possibly attributable to the inflammatory condition occurring after kidney transplantation, leading to altered functions of mesenchymal stem cells [237].

More recently, a phase I clinical study using autologous bone marrow-derived mesenchymal stem cells for the treatment of allograft rejection after kidney transplantation from living donors was performed [238]. In this trial, bone marrow-derived stem cells were administered intravenously at intervals of 7 days when a protocol of renal biopsy at 1 or 6 months showed signs of rejection and/or an increase in interstitial fibrosis or tubular atrophy. Although only six patients were enrolled in this study, two renal biopsies after treatment with bone marrow-derived stem cells showed resolution of cell infiltrates, suggesting a regenerative and anti-inflammatory effect of the cell therapy on the renal tissue. The engraftment syndrome, one of the negative effects of mesenchymal stem cell administration, was not observed in this study possibly due to a late timing of stem cell administration, beyond 6 months, after transplantation.

Despite these encouraging progresses of these clinical trials, questions related to the establishment of the best way to apply mesenchymal stem cells to kidney disease still remain unsolved. In addition, therapeutic mechanism is also still poorly understood. Mesenchymal stem cells do not engraft in the kidney when injected intra-arterially or intravenously; rather, they can form emboli in the lung [239]. Thus, these studies are not cell transplantation but cell infusion that exerts effects through paracrine mechanisms. This implies that if we can identify the bioactive substances secreted by bone marrow-derived stem cells, infusion of these substances without cells would be enough. Indeed, conditioned media from mesenchymal stem cells protected kidneys from acute injury as well as injection of the cells did in preclinical study [49]. Incomplete understanding of the mechanism of mesenchymal stem cells for cell therapy is not a reason not to pursue these clinical trials, but more basic research to investigate the biology of mesenchymal stem cells for kidney protection and renal regeneration as well as further clinical trials devoted to testing efficacy of this therapy is required.

Adipose Tissue-Derived Stem Cells

Adipose tissue-derived stem cells are another type of mesenchymal stem cells residing in subcutaneous adipose tissues. They are an attractive source of cells with regenerative properties that are similar to those of bone marrow-derived stem cells. Moreover, adipose tissue-derived stem cells are promising stem cells for clinical use, because the subcutaneous adipose tissues are abundant in humans and can be easily harvested using liposuction. In addition, these cells have no ethical issues regarding the cell source and less concern about safety of allo- and xenografting. Furthermore, adipose tissue-derived stem cells have more potential of anti-inflammatory and immunomodulatory functions than bone marrow-derived stem cells. Adipose tissue-derived stem cells have the capacity to differentiate into other cell types, such as adipocytes, osteoblasts, neurons, and myocytes [145]. Several studies have demonstrated that cell therapy using adipose tissue-derived stem cells minimized kidney damage or improved renal dysfunction after renal damages, such as ischemic injury, progressive renal fibrosis in a mouse model, cisplatin- or folic acid-induced acute kidney injury, atherosclerotic renal artery stenosis in swine models, and anti-glomerular basement membrane disease in a rat model [140–143, 145, 240–242]. Adipose tissue-derived stem cells also seem to improve renal function mainly via paracrine effects, such as suppressing oxidative stress and inflammatory response.

Umbilical Cord Blood-Derived Stem Cells

The umbilical cord blood has been known to contain mesenchymal stem cells, and several studies evaluated the efficacy of administration of umbilical cord blood-derived stem cells in the restoration of renal function in animal acute kidney injury models [243, 244]. Injection of human umbilical cord blood-derived mesenchymal stem cells to immunodeficient mice with cisplatin-induced acute kidney injury was performed [243]. This study showed that umbilical cord blood-derived stem cells ameliorated tubular injury, resulting in the recovery of renal function. Coculturing of umbilical cord blood-derived stem cells with cisplatin-treated proximal tubular cells showed that the expression of HGF was particularly induced and that of interleukin 1-β and tumor necrosis factor-α was significantly decreased in the coculture system. These results indicated that the process of renal function improvement by umbilical cord blood-derived stem cells was also correlated to the modulation of paracrine factors in the kidney. Both umbilical cord blood- and bone marrow-derived stem cells expressed similar set of genes in the comparative study of gene expression profile between these two kinds of stem cells, but bone marrow-derived stem cells predominantly expressed a set of genes related to antimicrobial activity and osteogenesis, whereas umbilical cord blood-derived stem cells predominantly expressed genes related to matrix remodeling and angiogenesis [245]. This result suggested that umbilical cord blood- and bone marrow-derived stem cells may have distinct activities in vivo.

Amniotic Fluid Stem Cells

As multipotent stem cells at the fetal stage, amniotic fluid-derived stem cells have been considered an attractive cell source for regenerative medicine. Amniotic fluid-derived stem cells exhibit characteristics of both embryonic and adult stem cells, are easily harvested, and retain high self-renewal potential and multiple differentiation capacity without development of teratoma formation compared to embryonic stem cells or iPSCs [246]. Perin et al. isolated amniotic fluid-derived stem cells from human amniotic fluid obtained at 12–18 weeks gestation and showed that these cells expanded in vitro could survive, proliferated, and integrated into the embryonic kidney and underwent organ development, demonstrating a potential cell source for kidney regeneration [247]. Therapeutic effects of human amniotic fluid-derived stem cells have been demonstrated in kidney injury animal models, such as glycerol- or cisplatin-induced acute kidney injury, a mouse model of unilateral ureteral obstruction, and Alport syndrome [146, 248–251]. Initial studies into the properties of amniotic fluid-derived stem cells in kidney injury demonstrated that the injection of such cells into animal models of kidney disease delayed progression of fibrosis [250]. More recent studies have investigated whether amniotic fluid-derived stem cells can generate renal cell types in vitro, with evidence for a capacity to differentiate into podocyte-like cells [252]. In addition, it was proposed that these cells could be propagated in a nephron progenitor state for many passages in order to provide an expandable source of cells to give rise to this specific renal cell type. Comparison of the characteristics between amniotic fluid- and bone marrow-derived stem cells was performed [248]. In this study, amniotic fluid-derived stem cells had a more potent anti-apoptotic activity against renal tubular cells but lesser stimulatory activity for the proliferation of renal tubular cells compared with bone marrow-derived stem cells. It was also demonstrated that amniotic fluid- and bone marrow-derived stem cells expressed different sets of paracrine factors, indicating that both cell populations may have distinct activities in vivo. In another study, mammalian target of rapamycin was reported as an important factor in renal differentiation of amniotic fluid-derived stem cells [253].

Endothelial Progenitor Cells

Endothelial progenitor cells participate in the repair of tissues, including the kidney, under specific physiological and pathological conditions. Renal ischemia mobilized endothelial progenitor cells, and transplantation of endothelial progenitor cell-enriched cells from the medullopapillary parenchyma provided partial renoprotection after ischemic injury of the kidney [254]. Acute elevation of uric acid acts as an endogenous mediator of mobilization and renoprotection of endothelial progenitor cells [255]. Endothelial progenitor cells can be isolated from human peripheral blood using CD34 as a marker for positive selection [256]. The CD34+ mononuclear blood cells obtained the characteristics of vascular endothelial cells when cultured on fibronectin-coated dishes. Endothelial progenitor cells were incorporated in ischemic tissues in vivo and expressed vascular endothelial cell markers, such as CD31, when introduced into the circulation using a hindlimb ischemia model. The efficacy of administration of endothelial progenitor cells for the improvement of renal function was also reported in animal models of acute kidney injury and renal artery stenosis [254, 257, 258]. In a chronic renal artery stenosis model, a single intrarenal injection of autologous endothelial progenitor cells preserved microvascular architecture and decreased microvascular remodeling by preserving hemodynamics [257]. However, it was reported that the function of endothelial progenitor cells was deteriorated in chronic kidney disease patients, suggesting that the transplantation of autologous endothelial progenitor cells may not be feasible for the treatment of chronic kidney disease [259].

7.4.3.3 Pluripotent Stem Cell

Pluripotent stem cells have the potential to differentiate into any cell type in the body and self-assemble into heterogeneous tissues or organs, and they have indeed been shown to generate mature cells in vitro [260–262]. Pluripotent stem cells include both embryonic stem cells, which are derived from embryos and grown in primary culture, and iPSCs, which are produced from terminally differentiated cells using transfection factors, such as c-myc, Oct4, Klf4, and Sox-2. Pluripotent stem cells have been successfully differentiated into various types of cells and tissues, including hematologic, neural, cardiac, hepatic, pancreatic, and intestinal lineages [263–271]. Recent progress toward generating renal cell types from embryonic stem cells or iPSCs has generated human nephron progenitor cells, including intermediate mesoderm and metanephric mesenchyme cells [272–277]. However, the progress of kidney regeneration using pluripotent stem cells is notably slow compared to other fields, and there have been few efficient protocols in the kidney, whereas full protocols for the derivation of neurons, hepatocytes, and cardiac myocytes were already established [278–280]. Although different studies have used different growth factor protocols in human pluripotent stem cells, the Wnt agonist CHIR99021 was commonly used to promote mesoderm differentiation [273].

Recently, an efficient method has been developed to induce the differentiation of intermediate mesodermal cells from pluripotent stem cells using a combination of activin A and CHIR99021, followed by combined treatment with BMP-7 and CHIR99021 [272, 274]. Odd-skipped related 1 (OSR1)+ intermediate mesodermal cells were differentiated from human iPSCs using this protocol with an efficiency greater than 90% [274]. Several studies also have endeavored to generate metanephric nephron progenitor cells by direct differentiation from pluripotent stem cells. Takasato et al. induced a primitive streak from human embryonic stem cells using activin A (or CHIR99021) and BMP-4 [275]. The differentiated human embryonic stem cells formed renal vesicles combined with dissociated embryonic mouse kidney cells, involving the integration of human cells into mouse renal structures. In contrast, Taguchi et al. demonstrated more reasonable protocols to induce renal structures such as nephrons and proximal tubules [277]. They proposed the concept of "posteriorization" to induce nephron progenitor cells from pluripotent stem cells in a mouse and human model, indicating that nephron progenitor cells of the metanephric mesenchyme could be derived from posteriorly located intermediate mesoderm.

The ureteric bud lineage cells were also attempted to be differentiated from pluripotent stem cells by stepwise treatment of iPSCs or human embryonic stem cells with a combination of fibroblast growth factor 2 and BMP-4 for 2 days, followed by combined treatment with activin A and BMP-2 for another 2 days [276]. After the 4 days of treatment, intermediate mesoderm-like cells expressing PAX2, OSR1, WT1, and LHX1 were produced, and the ureteric bud markers, such as HOXB7, RET, and GFRA1, were upregulated in these cells after additional 2 days of differentiation, indicating that ureteric bud progenitor-like cells could be differentiated from pluripotent cells.

Embryonic Stem Cell

Embryonic stem cells are pluripotent stem cells and were initially derived from the inner cell mass of the blastocyst of mouse embryos [281].

These cells have the ability to differentiate into various cell types of the mesodermal, endodermal, and ectodermal lineages; thus, they have been considered to be used as an effective tool for kidney regeneration. Although research on human embryonic stem cells is limited due to ethical concerns, extensive research was conducted using murine embryonic stem cells to establish the methods to efficiently induce differentiation into renal cells [282]. Thomson et al. first established human embryonic stem cell line in 1998 [283], and subsequently human embryonic stem cell lines have been found to be capable of differentiating in vitro into extraembryonic and somatic cell lineages [284]. When human embryonic stem cells were cultured with a mixture of growth factors, including bFGF, transforming growth factor (TGF) β1, activin-A, BMP-4, HGF, EGF, β-nerve growth factor, and retinoic acid, they differentiated into cells expressing WT-1 and renin [285]. Moreover, it was shown that mouse embryonic stem cells stably transfected with Wnt4 differentiated into tubular-like structures that express aquaporin-2 when cultured with combination of HGF and activin-A [286]. The mixture of LY294002, CCG1423, and Janus-associated tyrosine kinase inhibitor 1 was also shown to enhance the differentiation of mouse embryonic stem cells into intermediate mesoderm and renal progenitor cells [287]. In another study investigating an ex vivo culture system, embryonic stem cells were microinjected into the developing metanephros, and this was cultured to analyze the differentiation capacity of embryonic stem cells into renal cells [288]. Tubule-like renal epithelial structures were identified with an efficiency of about 50%, and individual embryonic stem cells were occasionally observed in structures resembling glomerular tufts. Furthermore, when embryonic stem cells treated with retinoic acid, activin A, and BMP-7 were injected into a developing metanephros, they contributed to the tubular epithelia with almost 100% efficiency [289]. Despite these promising results, there is a concern about this technique that embryonic stem cells can develop teratoma at 2 to 4 weeks after transplantation into mouse [290]. In addition, the risk of uncontrolled growth as well as legal and ethical problems associated with the use of embryonic tissue is limitation of the embryonic stem cells. Although embryonic stem cells are a valuable cell source for investigating the mechanism of kidney regeneration, there are still several hurdles for clinical applications of these cells.

Induced Pluripotent Stem Cell

Another cell source that possesses pluripotency is the iPSC, which was first generated by Takahashi et al. through reprogramming human fibroblasts in 2006 [261]. These novel discoveries have completely changed view of the development and cellular specialization. After then, iPSCs have been successfully established from several mammalian species, including rat, rabbit, pig, monkey, and humans [262, 291–296]. Nevertheless, not all adult cells can be similarly reprogrammed, indicating that critical factors for reprogramming are cell dependent. For example, reprogramming of mature B cells from adult spleen to iPSCs required an additional factor, C/EBPα, as well as Oct4, Sox2, Klf4, and c-Myc [20]. Recently, generation of iPSCs has been reported from human mesangial cells, tubular cells, and epithelial cells derived from urine [147, 149, 297]. The iPSCs derived from renal tissues may have advantages for renal cell therapy or kidney regeneration, because a persistent genome-wide epigenetic memory of the somatic cell of origin can be retained in iPSCs [298]. Indeed, when iPSCs derived from human mesangial cells that were differentiated in podocyte were engrafted in a metanephric kidney, they were integrated in the developing glomeruli [147, 148]. In addition, there are several advantages of iPSCs for kidney regeneration, such as no ethical issues and no immune rejection especially compared to embryonic stem cells. However, increased risk of uncontrolled growth is one of the great concerns, because these cells are reprogrammed using Klf4 and c-Myc, which are oncogenic factors [261, 299]. Hopefully, iPSCs can be established from human renal proximal tubular cells without transfection of Klf4 and c-Myc, suggesting that the oncogenic risk associated with iPSC generation can be decreased by expressing only Oct4 and Sox2 [299]. Therefore, it is possible to prepare patient-specific pluripotent cells without manipulating germ cells because iPSCs are pluripotent and can be generated from adult somatic cells.

Recent study reported the therapeutic effect of iPSCs on renal ischemia [151]. In this study, iPSCs were generated without transfection of c-Myc and administered via intrarenal arterial route into kidneys of ischemia/reperfusion-induced acute kidney injury rat models. The administration of iPSCs reduced the expression of oxidative substances, proinflammatory cytokines, and apoptotic factors and eventually improved survivals of rats.

Nevertheless, there are still several hurdles of iPSCs-based therapy, such as no established differentiation protocols for moving from pluripotent state to functional renal cells, undefined optimal final culture conditions for target cell, and multiple steps each requiring different factors to induce a stepwise differentiation [41]. In addition, it was reported that some cells differentiated from iPSCs expressed abnormal gene and induced T-cell-dependent immune response even in syngeneic recipients [300]. Therefore, it is necessary to evaluate the immunogenicity of specific cells derived from iPSCs before they can be used in a clinical application.

7.4.4 In Vitro Kidney Regeneration Without Scaffolds

There have been many efforts to generate mature cells using pluripotent stem cells in vitro. Several studies reported successful generation of mature cells, which have characteristics of liver, pancreas, and intestine, from embryonic stem cells or iPSCs using stepwise protocols mimicking the mechanism of embryonic development [263, 265, 267, 270, 271]. Furthermore, autonomous formation of three-dimensional adenohypophysis, polarized cortical tissues, optic cap, and retina structures using three-dimensional culture systems of embryonic stem cells suggested the possibility of organogenesis for therapeutic regeneration using this approach [301–304].

It was demonstrated that a single multipotent progenitor cell from embryonic mouse kidney can express Sall-1 and differentiate into several types of renal cells, such as podocytes and tubular epithelial cells, and eventually reconstruct a three-dimensional kidney structure [305]. Another study also reported that single-cell suspensions from embryonic kidney reaggregated and formed organotypic renal structures [306]. Because intermediate mesoderm cells differentiated into renal progenitor cells and then into maturing renal cells, if pluripotent stem cells can be differentiated into intermediate mesoderm, all kinds of renal cells may be generated from embryonic stem cells or iPSCs. The protocols to differentiate human iPSCs into intermediate mesoderm cells were established using bacterial artificial chromosome-based vectors and single-nucleotide polymorphism array-based detection [274]. These cells expressed intermediated mesoderm marker genes and could mature into various types of cells including those found in organs, such as the kidney, adrenal gland, and gonad. These results suggested that three-dimensional kidney structure could be constructed if pluripotent stem cells can be made to generate renal progenitor cells.

7.4.5 Blastocyst Complementation

Blastocyst is an early developmental stage of the embryo that occurs 5 days after fertilization. Injection of pluripotent stem cells into blastocysts synchronizes the development of two line cells, and the combined blastocyst can generate a chimeric body. The chimera can also be produced by injecting pluripotent stem cells into xenoblastocyst, which lacks potential to form any particular cell lineage. The chimera is then transplanted into a foster uterus, and the deficient cells are exclusively derived from injected pluripotent stem cells. It was first reported that injection of normal embryonic stem cells into the blastocysts of recombination-activating gene 2-deficient mice, which have no mature B or T lymphocytes, generated somatic chimeras with embryonic stem cell-derived mature B and T cells [307]. Recent several studies demonstrated that several types of tissues and organs, such as the thymic epithelium, yolk sac, germ cells, hepatocytes, heart, pancreas, and kidney, could be reconstructed using this blastocyst complementation system [308–315]. Kobayashi et al. successfully generated a functional pancreas using this approach [310]. In this study, iPSCs of rat were injected into Pdx-/- (pancreatogenesis-disabled) blastocysts of mouse that produced newborn rat/mouse chimeras with a pancreas which originated almost entirely from the injected iPSCs of rat. The mouse and rat iPSC-derived pancreas produced insulin, and the transplantation of this iPSC-derived pancreas islets improved hyperglycemia in a diabetic rodent model [217]. This study suggested that iPSC-derived cellular progeny could aggregate and differentiate into the deficient organ in a vacant developmental niche. Moreover, these results also demonstrated that interspecific blastocyst complementation could be used to generate organs derived from donor pluripotent stem cells in vivo using a xenogeneic environment. This study group generated a Pdx-/- pig and succeeded in generating bigger pancreas using this technique, suggesting that human-scale organs could theoretically be generated [311]. Likewise, chimeric mice carrying a liver derived from injected iPSCs of mouse were also generated from blastocysts with fumarylacetoacetate hydrolase deficiency [308]. Hepatocytes derived from the injected iPSCs also had a proliferative characteristic of normal hepatocytes.

Kidney reconstruction using this blastocyst complementation system has been attempted. Usui et al. injected normal mouse iPSCs into blastocysts from kidney-deficient mice lacking the SAL-like 1 protein, which is essential for kidney development; eventually, the neonatal mice had kidneys derived from injected iPSCs [315]. However, the vascular and nervous systems were not constructed from iPSC-originated cells; thus, the kidney was not completely complemented. Immunohistochemical analysis of the regenerated kidney showed that the vascular system, including renal segmental, lobar, interlobar, arcuate, and interlobular arterioles, was a chimeric structure originating from both donor iPSCs and host cells. Whether urine was produced by this regenerated kidney was not certain due to the lack of precise urine analysis. In addition, generation of rat kidneys in mice failed after injection of rat iPSCs into kidney-deficient mouse blastocysts. This result indicates that the key molecules in mice involved in interactions between the metanephric mesenchyme and ureteric bud do not cross-react with those in rats. Therefore, the generation of xenogeneic organs using interspecific blastocysts requires a host animal strain lacking all renal lineages.

Although this strategy using blastocyst complementation has showed promising results for kidney regeneration, there are several limitations for clinical applications. It is difficult to generate interspecific chimeras in livestock. In addition, tissue rejection also should be overcome. Moreover, ethical issue is one of the most serious problems. There has been always a concern about the possibility of generating interspecific chimeras containing brain derived from injected pluripotent stem cells.

7.4.6 Embryonic Organ Transplantation

The embryonic metanephros (Fig. 7.3), which constitutes the primordial mammalian kidney, has been used for the investigation of kidney regeneration and showed promising results as a potential source for the generation of a functional whole kidney [156, 316–324]. If a metanephros is transplanted into the renal cortex of a host mouse, it can continue to grow [155]. The transplanted metanephros contains vascularized glomeruli and mature proximal tubules endowed with a capacity for glomerular filtration. Collecting duct-like structures also appeared and extended from the transplant toward the papilla of the host. Although there is no direct evidence that these collecting duct-like structures connect with the collecting system of the host or that the transplanted metanephros functions in a manner similar to that of native kidney, these results offer a rationale for the existence of renal stem cells in the metanephros

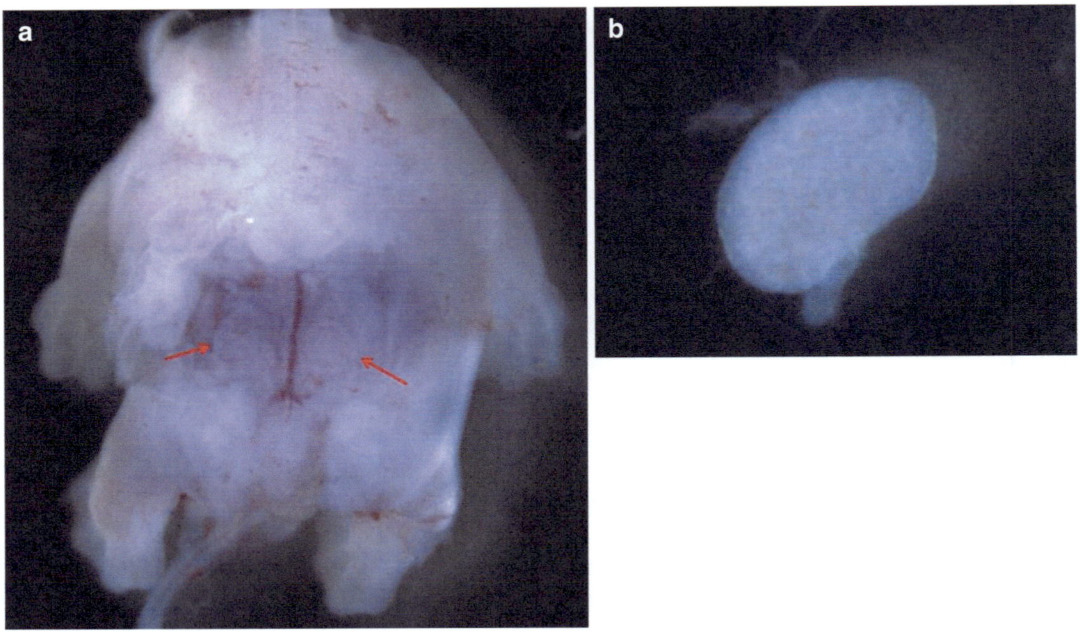

Fig. 7.3 Metanephroi of E13 mouse embryo. (**a**) Metanephroi, which are located just above aortic bifurcation, can be observed in E13 mouse embryo after removing lower half of the spinal cord and dorsal muscular layer of the embryo (*red arrow*, both metanephroi). (**b**) Intact metanephros can be isolated using microforceps in the vision of microscopy

from embryos, which can be a potential source of transplantable regenerated kidney. However, there are still several unsolved problems related to this technique, including questions as to the suitability of the renal capsule of patients with dialysis as a transplant site, given the significant disruptions to this area, such as the vasculature, and the fact that space limitations beneath the renal capsule may inhibit the growth of transplanted metanephros. To overcome these concerns, the metanephroi were transplanted into a host omentum in another study [156]. The transplanted metanephroi into the omentum were not confined by a tight organ capsule or disturbed by dialysis. The metanephroi transplanted into a host omentum also could differentiate into mature vascularized glomeruli and proximal tubules and produce urine. Moreover, after an intact ureteroureterostomy with the ureter of the native kidney that was removed, anephric host animals started to void and showed prolonged survival. These metanephroi developed in host animals exhibit metabolic renal function and produce erythropoietin and renin leading to detectable levels of these proteins in serum [319, 323, 324]. Furthermore, metanephroi from porcine embryos were transplanted into either the omentum of mice in which costimulation was blocked or area under the kidney capsules of immunodeficient mice and developed to generate fully functional nephrons [316, 320]. The average levels of creatinine and urea nitrogen were higher in the cystic fluid that arose from the transplanted tissue than in the sera of the transplanted host mice. These successful xenotransplantation of metanephroi between different species was based on previous studies showing minimal immunogenicity in tissues, including the metanephros harvested at earlier gestational ages [325]. In cases of allotransplantation, such as rat metanephros to rat omentum, transplants assumed a kidney-like shape in situ that was approximately one-third the diameter of the native kidney, and the transplants contained well-differentiated kidney structure in histological analysis. In addition, the transplantation can be performed without immunosuppression. However, in xenotransplantation, although transplanted metanephros grew and differentiated into renal tissue in the host omentum, immunosuppressants were required. Without these agents, the transplants disappeared soon after transplantation.

Transplantation of metanephros was also shown to maintain blood pressure in anephric rats with induced acute hypotension and reduce vascular calcification in rats with chronic renal failure, indicating that transplanted metanephroi have multiple renal functions as well as urine production [323, 324]. Although the omentum has been preferred because a tight renal capsule does not confine it and the host vessels can be preferentially integrated into the transplanted metanephros, the low hydrostatic pressure in the omentum may influence the acquisition of some renal function in the developed metanephros. Indeed, it was reported that the para-aortic area is better for the establishment of renin-producing tissue [319].

Recently, it was reported that xenotransplanted metanephros could supply endogenous mesenchymal stem cells with a niche for differentiation into erythropoietin-producing renal tissues [318]. Polymerase chain reaction (PCR) using species-specific primers and sequence analysis showed that xenotransplanted metanephroi expressed host animal-originated erythropoietin. This result implied that the erythropoietin-producing cells were originated from the host animals and developed to produce erythropoietin in the transplanted metanephros. It was also shown that the erythropoietin-producing cells were not originated from integrating vessels but from circulating host mesenchymal stem cells mobilized from the bone marrow.

Conventional transplantation of metanephros requires strong and persistent immunosuppression to avoid humoral rejection associated with the xenogeneic barrier, which can cause several adverse effects, such as carcinogenicity and severe rejection. Thus, for safety, when the xenotransplant is not needed anymore, it should be discarded by introducing a cell fate-regulating system in which a suicide gene is expressed on demand. To avoid the xenogeneic barrier, metanephroi isolated from transgenic ER-E2F1 mice that are expressing E2F1 under the control of tamoxifen can be used. E2F1 is a transcription factor which regulates cell proliferation and ectopic expression of which induces apoptosis. The administration of tamoxifen in host rats wherein metanephroi of ER-E2Fa mice were transplanted activated the suicide gene, and the xeno-compartments of transplanted metanephroi were ablated by apoptosis, leaving the autologous erythropoietin-producing tissues. In addition, it was also confirmed that the metanephros established by this method no longer needed immunosuppressants while maintaining erythropoietin production [318]. These results indicate that xenometanephroi per se could acquire some renal functions in the host omentum as well as supply a niche for host stem cells to regenerate renal tissues and then be rebuilt to constitute only host cell components. This technique can be helpful to reduce the adverse effects of long-term administration of immunosuppressants and to moderate the ethical issues surrounding xenotransplantation.

7.4.7 Renal Tissue Regeneration Using Metanephric Progenitor Cells

One of the most promising methods for kidney engineering is the generation of fetal or mature renal tissues using intrinsic proficiency of cells to organize themselves into fully self-derived sophisticated structures [301, 303]. Several early studies using this strategy demonstrated that renal epithelial cell lines cultured in extracellular matrix gels could be induced to create branching tubules either by the whole embryonic kidney or by the metanephric mesenchyme [326]. Advancement of this methodology showed that cells isolated from ureteric bud (Fig. 7.4) of mice at embryonic day 11.5 (E11.5) and propagated

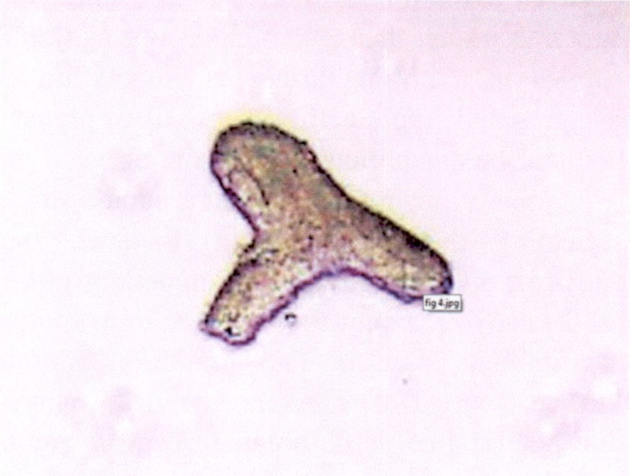

Fig. 7.4 Mechanically isolated ureteric bud from the metanephros of E11.5 mouse embryo

in vitro could undergo branching tubulogenesis when stimulated by a conditioned medium from a metanephric mesenchyme-derived cell line [327]. Similarly, a tubular epithelial cell line or mesenchymal stem cells generated glomerular and tubular structures in subcutaneous spaces when stimulated by a media from vascular endothelial and tubular cells [328]. These results suggested that simple cultures of cells derived from undifferentiated metanephric mesenchyme and unbranched ureteric bud could be used to create branching tubular structures similar to those that form in normal metanephros.

Generation of three-dimensional renal epithelia which contain glomerular-like structures in vitro was attempted using suspensions of murine metanephric mesenchymal E11.5 progenitor cells selected for a high expression of Sall-1, which is essential during kidney development [305]. This cell preparation required coculture with an exogenous spinal cord cell layer, as has been performed in traditional developmental investigations [329]. Although the functional capacity of these structures has not been investigated in vivo, this approach suggested that dispersed cell suspensions can be used as starting material to obtain rudimental three-dimensional kidney tissues. Further development of this method enabled the generation of fetal kidney tissue from suspensions of E11.5 kidney cells without using any exogenous tissue [306]. While this method successfully reproduced the structures and differentiation states of nephrons and stroma, collecting ducts developed as a multitude of very small collecting duct trees rather than a continuous one as would happen in a normal kidney. Combination of reaggregated single-cell suspensions of ureteric bud and metanephric mesenchyme allowed the formation of nephrons arranged around a single collecting duct tree [330]. However, this endeavor did not achieve development of glomeruli to any meaningful extent due to avascular environment in vitro. This limitation inhibits further progress of research related to maturation potential of tissue obtained from a simple culture or suspension of precursor cells. Therefore, the capacity of these cells to generate nephros with functional properties still remained unclear.

Using simple suspensions of singe kidney cells, "organoids" that could carry out renal functions when implanted into a living animal, such as glomerular filtration, tubular reabsorption of macromolecules, and production of erythropoietin, were finally constructed [322]. These organoids were generated from single-cell suspensions derived from E11.5 mouse kidneys and then implanted under the renal capsule of an athymic rat. The main procedure to obtain complete glomerulogenesis consisted of soaking the organoids in a solution containing VEGF, and then injection of VEGF into the recipient animals after the organoids had been implanted in the kidney. This method restituted the efficacy of the podocyte VEGF axis that is necessary for glomerular capillary endothelial induction and resulted in the formation of vascularized glomeruli with fully differentiated glomerular capillary walls, including endothelial fenestrae, slit diaphragms, and foot processes. These organoids could develop glomeruli with normal-appearing slits in vivo, wherein the proportion of diaphragms were comparable to that found in the adult glomerulus. When the ultrafiltering function of the intragraft nephrons was tested by injecting fluorescent dextrans of increasingly high molecular weight into the host blood system, the proximal tubular cells concentrated dextrans only of lower molecular weights from the lumen, indicating efficient ultrafiltration.

Approaches starting from the metanephros and dissociating down to the single-cell level offer a valuable source for performing pharmacologic or genetic modifications in individual cells prior to the in vitro aggregation step [322, 331]. The single-cell stage allows the use of genetic engineering approaches to humanize cells, by introducing immunomodulatory genes or by switching off others, to reduce the possibility of rejection and facilitate the xenotransplantation of animal renal tissues [331]. These self-forming organoids can be exploited to generate human renal tissue by constructing chimeric kidneys that combine animal progenitor cells and human stem cells, followed by the selective elimination of animal cells after transplantation. Previous studies have shown that the organoids could incorporate cells of another source, such as human amniotic fluid stem cells or iPSC-derived OSR1-positive cells, to build chimeric tissue in vitro [253, 274]. Although several technical problems remain to be overcome, these systems are useful to examine the mechanisms of renal progenitor differentiation and suggest the possibility of developing a whole kidney from a single stem cell.

7.4.8 Kidney Regeneration Using Xenoembryos

The idea of generating human organs in animals was previously suggested by several studies [332, 333]. Regeneration of a whole functional kidney that produced urine and renal hormones using a developing heterozoic embryo as an organ factory has been endeavored [334–336]. This strategy is based on the concept of "borrowing" the developing program of a growing xenoembryo by applying stem cells at the niche of organogenesis. During development of the metanephros, the metanephric mesenchyme initially forms from the caudal portion of the nephrogenic cord and secretes GDNF, which induces the nearby Wolffian duct to produce a ureteric bud [337]. Then, the metanephric mesenchyme consequently forms the glomerulus, proximal and distal tubule, loop of Henle, and the interstitium, as a result of reciprocal epithelial-mesenchymal induction between the ureteric bud and metanephric mesenchyme [338]. For this epithelial-mesenchymal induction to occur, GDNF must interact with its receptor, c-ret, which is expressed in the Wolffian duct. Thus, it can be hypothesized that GDNF-expressing stem cells can differentiate into kidney structures if they are placed at the budding site and stimulated by several factors spatially and temporally identical to those found in the developmental milieu.

To investigate this hypothesis, Yokoo et al. microinjected GDNF-expressing human mesenchymal stem cells into the developing metanephros in vitro, and the recipient embryo was grown in a whole-embryo culture system [335]. For the successful generation of the organ using this technique, human mesenchymal stem cells must be placed in a specific embryonic niche to allow their exposure to the nephrogenic signals before the metanephros begins to develop. This can be achieved by implanting human mesenchymal stem cells into the nephrogenic site of a developing embryo. Therefore, the researchers established a culture system in combination with a whole-embryo culture system, followed by metanephric organ culture. This relay culture enabled the development of the metanephros from structures present before budding until the occurrence of complete organogenesis ex utero. Using this system, embryos were isolated before budding and were grown in a culture bottle until the formation of rudimentary kidneys so that it could be further developed by organ culture in vitro (Fig. 7.5). It was also observed that rudimentary kidneys also continued to grow with fine tubulogenesis and ureteric bud branching in vitro, demonstrating that the metanephros can continue developing ex utero even if the embryo is dissected prior to sprouting of the ureteric bud (Fig. 7.6). In this study, before injection into the metanephros, the human mesenchymal stem cells were genetically engineered to express GDNF temporally using adenovirus and were labeled with the *LacZ* gene and dioctadecyl-3,3,3′,3′-tetramethylindocarbocyanine. However, viral-free manipulation can also be performed using a thermoreversible GDNF polymer during this process [339]. Consequently, donor human mesenchymal stem cells were integrated into the rudimentary metanephros and differentiated morphologically into tubular and glomerular epithelial cells and interstitial cells. Subsequently, they transplanted the developed metanephros into the omentum to allow vascular integration from the recipient to form a functional nephron. Finally, a human mesenchymal stem cell-derived neokidney was generated, which contained a human nephron associated with host vasculature [334]. The neokidney produced urine with higher concentrations of creatinine and urea nitrogen than those of sera of the recipient. This result suggested that the

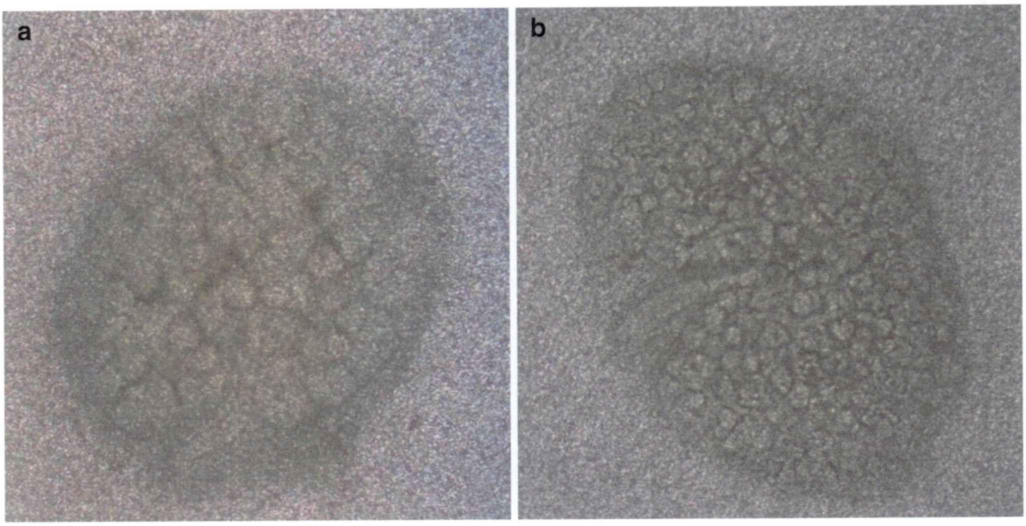

Fig. 7.5 Development of metanephros in vitro. Metanephros was isolated from E12 mouse and transferred into a culture dish. Then, it continued to grow. Compared to the morphology of metanephros at day 1 of culture (**a**), more branching of ureteric bud can be observed in the metanephros cultured in vitro for 5 days (**b**)

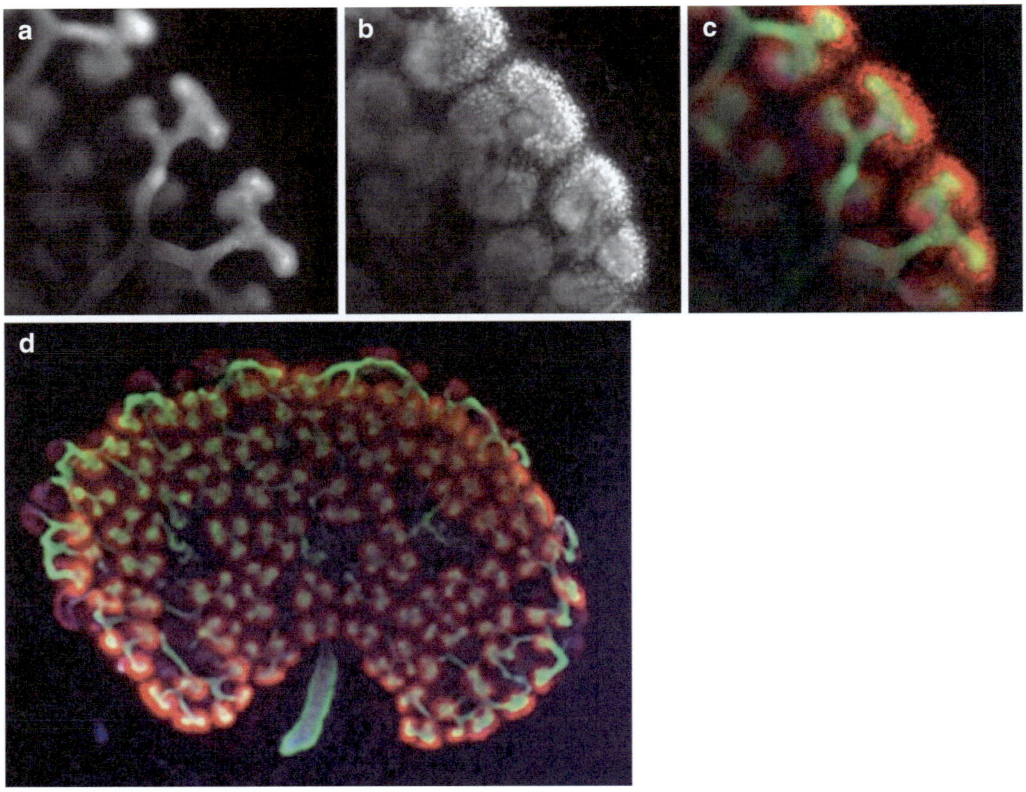

Fig. 7.6 Immunohistochemical staining of in vitro culture metanephros (*red*, Six2; *green*, cytokeratin 8; *blue*, E-cadherin). (**a**) Fine tubulogenesis and multiple branching of ureteric bud can be observed. (**b**) Six2 (nephron progenitor marker) expressing metanephric mesenchyme. (**c**) Merged image of ureteric bud and metanephric mesenchyme. (**d**) Kidney-shaped whole metanephros consisting of metanephric mesenchyme and multiple branches of ureteric bud

neokidney that developed in the host omentum possesses the ability to produce urine by filtering the host's blood. In addition, the human mesenchymal stem cell-derived neokidney secreted human erythropoietin, which was stimulated by the induction of anemia in the host animal [336]. This finding demonstrated that this system could preserve the normal hormonal regulation of erythropoietin levels. However, ureter or collecting ducts derived from ureteric bud could not be reconstructed using this system. Thus, another investigation to determine whether mesenchymal stem cells can differentiate into the ureteric bud progenitor cells was performed using chick embryo [340]. Human mesenchymal stem cells expressing Pax2 were injected into the chicken ureteric bud progenitor region, and then the cells migrated caudally with the elongating Wolffian duct and were integrated into the Wolffian duct epithelia and expressed LIM1. This result indicated that human mesenchymal stem cells can differentiate into Wolffian duct cells under the influence of local xenosignals. Taken together, a whole kidney may be reconstructed by transplanting human mesenchymal stem cells at an appropriate time and location to regenerate derivatives of the metanephric mesenchyme and ureteric bud.

7.5 Bioartificial Kidney

Although the final goal of the research for kidney regeneration is generation of a whole functional kidney, it still has many hurdles to be realized. Instead, ex vivo renal-supporting device, such as bioartificial kidney using various biomaterials, has been investigated. Current renal replacement therapy with either hemodialysis or peritoneal dialysis provides only filtration function and does not replace the homeostatic, metabolic, endocrine, and immunomodulatory functions of the kidney. Thus, when designing a bioartificial kidney for renal function replacement, these critical elements of renal function must be replaced. The functioning excretory unit of the kidney is composed of the filtering unit (glomerulus) and the regulatory or transport unit (tubule). Therefore, a bioartificial kidney requires these two main units, such as the glomerulus and the tubule, to replace both excretory and metabolic functions of the kidney [341, 342].

The development of bioartificial kidney represents the intersection of regenerative medicine and renal replacement therapy. Bioartificial kidney consists of a hemofilter used in conventional dialysis with a bioreactor unit containing human primary renal proximal tubule cells, termed a renal tubule assist device [343]. The addition of the bioreactor is intended to provide homeostatic, metabolic, endocrine, and immunomodulatory functions of the kidney that cannot be accomplished by dialysis. Moreover, these units were designed to be wearable or implantable to offer continuous renal support system with minimal deterioration of life quality. Since the bioartificial kidney was first proposed by Aebischer et al. in 1987, the technology has developed from concept to clinical trial [344]. Recently, Humes et al. reported the safe and efficacious use of bioartificial kidneys in patients with acute renal failure [345]. The bioartificial kidney demonstrated metabolic, endocrine, and immunomodulatory activities, including glutathione degradation, hydroxylation of calcifediol, and reduction of proinflammatory cytokines. More details of this progressive development of bioartificial kidney are presented in the following sections.

7.5.1 Renal Tubule Assist Device

Replacement of the various tubular functions of the kidney cannot be achieved with inanimate membrane devices, as has been accomplished with the renal ultrafiltration process, but requires the use of the naturally evolved biologic membranes of the renal tubular epithelium. In this regard, the tissue engineering of a bioartificial renal tubule as a cell therapy device to replace these missing functions can be conceived as a combination of living cells supported on synthetic scaffolds [346, 347]. A bioartificial tubule can be generated using progenitor cells of renal tubule culture on semipermeable hollow fiber membranes on which extracellular matrix has been layered to enhance the attachment and growth of epithelial cells [348, 349]. These hollow fiber synthetic membranes provide immunoprotection, which has been observed in the long-term implantation of the bioartificial pancreas in a xenogeneic host, as well as the architectural scaffold for these cells [350].

Subsequently, it was investigated whether the renal tubule assist device could maintain differentiated renal functional performance, similar to that observed in vitro, and viability in an extracorporeal hemoperfusion circuit in an acute uremic animal model [351]. The renal tubule assist device was accordingly placed in series with a standard hollow fiber hemofiltration cartridge. The two cartridges had ultrafiltrate and post-filtered blood connections to duplicate the structural anatomy of the nephron and to mimic the functional relationship between the glomerulus and tubule. In this study, fluid and electrolyte balances in the animal were adequately controlled with the bioartificial kidney, and especially plasma potassium and blood urea nitrogen levels were more easily controlled during the treatment using renal tubule assist device as compared with controls. In addition, active transport of potassium, bicarbonate, and glucose and successful metabolic functions, such as ammoniagenesis, were also observed.

An animal study to investigate whether treatment with the renal tubule assist device could alter the course of acute renal failure with sepsis was performed [352]. Mongrel dogs with nephrectomy-induced acute renal failure were treated with continuous venovenous hemofiltrations and either a renal tubule assist device containing cells or an identically prepared sham cartridge without cells. After 4 hours of therapy, endotoxin was infused intravenously to stimulate Gram-negative septic shock. Mean peak levels of an anti-inflammatory cytokine, IL-10, and mean arterial blood pressures were significantly higher in the animals treated by renal tubule assist device containing cells. This study group also performed similar experiments in a larger animal model [353]. *E. coli* was administered intraperitoneally into pigs with normal renal function to induce septic shock. Then, animals were treated using continuous venovenous hemofiltrations with either a renal tubule assist device containing cells or a sham cartridge without cells immediately following bacterial administration. Acute kidney injury model was successfully generated in all animals within 2–4 h after bacterial administration. Renal tubule assist device-treated group showed better cardiovascular performance, including cardiac output and renal blood flow, for longer periods than control group. Consequently, experimental group also survived longer than control groups. Treatment with renal tubule assist device was associated with significantly lower plasma circulating levels of IL-6, a proinflammatory cytokine, and interferon-γ. These findings indicate that septic shock results in early acute kidney injury and that treatment with renal tubule assist device in a bioartificial kidney extracorporeal circuit improves cardiovascular performance associated with changes in cytokine profiles and confers a significant survival advantage.

With these encouraging preclinical outcomes, phases I and II clinical studies using renal tubule assist device containing human cells were performed in patients with acute kidney injury receiving continuous renal replacement therapy. Human kidney cells were isolated from kidneys donated for cadaveric renal transplantation but found unsuitable due to anatomic of fibrotic defects. The results of phase I study in the first ten patients treated with renal tubule assist device containing cells demonstrated that this treatment modality can be delivered safely under study protocol guidelines for up to 24 h when used in conjunction with continuous venovenous hemofiltrations [345]. Cardiovascular stability of the patients was maintained with increased urine outputs. These results also demonstrated the viability, durability, and functionality of renal tubule assist device with cells in clinical setting. Of these ten patients who had high risk of mortality, six survived past 28 days with renal function recovery. Plasma cytokine levels implied that treatment with renal tubule assist device produces dynamic and individualized responses in patients, depending on their unique pathophysiologic conditions. Several cytokines, such as IL-6, IL-10, and granulocyte colony-stimulating factor, were significantly decreased after treatment with renal tubule assist device compared to baselines in patients who had excessive proinflammatory levels. These encouraging results led to perform randomized, controlled, open-label phase II trial, which was completed at ten clinical sites to assess the safety and early efficacy of this cell-based renal tubule assist device therapy.

In this phase II clinical trial, 58 patients with acute kidney injury requiring renal replacement therapy in the intensive care unit were enrolled and randomized to receive continuous renal replacement therapy with renal tubule assist device (n = 40) or renal replacement therapy alone (n = 18). Treatment with renal tubule assist device containing cells promoted a statistically significant survival advantage over 180 days of follow-up having a 67% survival rate compared to a 39% survival rate of the control group and demonstrated an acceptable safety profile. A follow-up phase IIb study to evaluate a commercial manufacturing process was not completed due to difficulties with the manufacturing process and clinical study design, which led to a suspension of the clinical development of this approach [354]. However, this suspended study has led to an innovative biomimetic membrane cell processing device to treat systemic inflammatory response syndrome and has advanced to a pivotal clinical trial in patients with multiorgan failure [355, 356].

Despite these successes of preclinical and clinical studies of the renal tubule assist device containing cells in renal replacement therapy for patients with acute kidney injury, there are several limitations for successful commercialization of cell-based therapeutic devices, such as the need for a reliable and consistent source of cells to manufacture thousands of these cell devices and the requirement to develop a cost-effective manufacturing, storage, and distribution process for these devices.

7.5.2 Bioartificial Renal Epithelial Cell System

The bioartificial renal epithelial cell system aims to address the important issues of cell cryopreservation, storage, and distribution for the widespread adoption of renal cell therapy [357]. The bioartificial renal epithelial cell system is a model of a cell therapeutic device that maintains a high density of renal cells for therapeutic application within an extracorporeal circuit. Compared to other cell therapies involving the direct application or injection of stem cells into the body, the extracorporeal approach of the bioartificial renal epithelial cell system allows for cell therapeutic application while maintaining immunoisolation using a series of filters. Compared to renal tubule assist device, bioartificial renal epithelial cell system enables mass fabrication, cryopreservation for storage, and distribution to meet on-demand clinical need.

In summary, the bioartificial renal epithelial cell system is a perfusion bioreactor that utilizes primary renal epithelial cells derived from the kidney, expanded from progenitor cells during in vitro culture. Cells are seeded on porous niobium-coated carbon disks, which are placed in a media flow path within the bioartificial renal epithelial cell system. Renal progenitor cells are directed toward a renal tubule cell fate and maintained in perfusion culture prior to therapeutic application. In vitro cell viability and metabolic activity are confirmed in the bioreactor renal epithelial cell system by measuring lactate production and oxygen consumption. Metrics of metabolism are consistent throughout the duration of perfusion culture, with an estimated total cell number of approximately 10^8 cells [357]. Renal epithelial cells in the bioartificial renal epithelial cell system maintain renal differentiated phenotypic characteristics over time in perfusion culture. Cryopreservation and thawing of renal epithelial cell in bioartificial renal epithelial cell system were accomplished using a commercially available cryopreservation media. Post-cryopreservation cell retention was optimized in combination with a controlled rate freezer, resulting in average cell retention of more than 80% and viability of more than 90% [357]. This system offers an all-in-one system for cell culture, cryostorage, and distribution of cell therapy for on-demand use for acute indications, such as acute kidney injury and sepsis.

7.5.3 Wearable or Implantable Bioartificial Kidney

The ideal bioartificial kidney suitable for long-term use in patients with end-stage renal disease would be capable of performing continuously to reduce risks from fluctuations in volume status, electrolytes, and solute concentrations and to maintain regulation of acid-base and uremic toxin. Such treatment requires the design and manufacture of a compact wearable or implantable dialysis unit and the development of compact renal tubular cell devices with long durability. Therefore, next-generation renal tubule assist device should be like that of an implantable device similar to the pacemaker.

Recently, a wearable bioartificial kidney has been suggested and evaluated in preclinical study utilizing either hemodialysis or peritoneal dialysis as the therapeutic circuit delivery route. The wearable bioartificial kidney is comprised of the use of sorbent-based technologies to replace the excretory function of the kidney and the compact bioartificial renal epithelial cell system to replace the metabolic function of the kidney. The first trial to develop wearable dialysis system generated a lightweight, wearable, continuous ambulatory ultrafiltration device consisting of a hollow fiber hemofilter, a battery-operated pulsatile pump, and two micropumps to control heparin infusion and ultrafiltration [358]. This device regenerates dialysate with activated carbon, immobilized urease, zirconium hydroxide, and zirconium phosphate. However, this approach requires continuous blood access to allow adequate ultrafiltration.

Another approach to utilize continuous regeneration of peritoneal dialytic fluid relying on continuous-flow peritoneal dialysis systems was also investigated [359, 360]. With the development of a sorbent-based wearable, continuous-flow recycling peritoneal dialysis, the integration of renal epithelial cell therapy in the peritoneal dialysis circuit to treat patients with end-stage renal disease is also being evaluated in large animal models with uremia. Although a cell therapy device requires a continuous support of nutrients and oxygen, the use of blood circuits for this nutrient stream has been avoided to prevent risks of clotting and infection. Similar to acute applications of cell-based therapies, the compact, cryopreservable design of bioartificial renal epithelial cell system and the enhanced propagation methods used to isolate and expand the renal cells in the bioartificial renal epithelial cell system are enabling advancements toward the development of a wearable bioartificial kidney for chronic applications which integrates a wearable sorbent dialysis system and bioartificial renal epithelial cell system with recycling peritoneal fluid. Further investigation and refinement according to this wearable bioartificial kidney are still going on.

For better patients' quality of life, there have been attempts to create a miniaturized version of the renal tubule assist device that would be suitable for implantation and offer the benefits of a transplanted kidney. An implantable bioartificial kidney has the potential to overcome both shortage of kidney donors and the burden of therapy of intensive maintenance dialysis. Microelectromechanical system technology has been utilized to develop and implantable renal tubule assist device. The silicon nanopore membranes have been used for both hemofilter and an immunoisolation scaffold for a cellular bioreactor [342, 361–364]. The ultrathin silicon nanopore membranes have a uniform slit pore design giving higher hydraulic permeability and selectivity compared to the roughly circular-shaped pores in standard hollow fiber membranes [365]. Since the silicon nanopore membranes can have pore sizes as small as 5 nm with less than 1% variability, these membranes can function like the glomerulus of the native kidney by selectively filtering solutes based on molecular weight cutoffs. The selectivity of silicon nanopore membranes has been shown in vitro using β2-microglobulin and several globular proteins. Silicon readily forms a thin oxide coating upon exposure to atmospheric oxygen. This silica film is negatively charged at physiologic pH, and then it absorbs plasma proteins and activates the coagulation cascade. It is necessary to modify the silicon surface with a highly hydrated polymer to reduce adverse blood-material interactions. The successful outcomes of grafting several organic polymers on silicon surfaces demonstrated the feasibility of using silicon nanopore membranes coated with antifouling films for hemofiltration, allowing for progress toward an implantable bioartificial kidney [366].

Despite these advancements in biocompatibility of silicon membranes, vascular perfusion of the bioartificial kidney will be needed to maintain filtration and cell viability with the use of long-term anticoagulation. In addition, there are still limitations to miniaturize and implant the hemodialysis circuit, such as the size and pump requirements of modern dialyzers and the water volume required for dialytic therapy. Hollow fiber polymer membranes have been immensely successful in treating renal failure with extracorporeal therapies but require super-physiologic driving pressures for blood circulation through the cartridge. The long cylindrical hollow fibers present a high resistance to blood flow, which rises further in the distal portion of the fiber as ultrafiltration increases hematocrit and viscosity, necessitating energy requiring roller pumps to circulate blood through the device. The pores in hollow fiber dialyzers are irregular, cylindrical, and multiform in size, and this polydispersity compromises the trade-off between permeability and selectivity. The majority of the membrane's pores should be maintained smaller than the desired cutoff target of the membrane to prevent albumin leakage through the largest pores in the membrane. Another limitation is the volume of dialytic water, which is most readily demonstrated by the estimated volume of water required during a 4-h dialysis with a dialysate flow of 600 ml/min, consuming 144 l of dialysate, being as much as additional 200 l of tap water [361, 365].

7.6 Future Perspectives

In this chapter, we have reviewed recent advances in renal regenerative therapy including renal stem/progenitor cells and exogenous stem cells to treat damaged renal tissue, de novo whole functional organ regeneration using cell-scaffold technique, blastocyst complementation and embryonic organ, and bioartificial kidney. Although there are many interesting approaches and new frontiers are being explored in the field of kidney regeneration, many unsolved problems still remain. In addition, determination of the optimal cell sources is also important for appropriate regeneration of kidney tissue. The use of renal stem/progenitor cells in kidney regeneration is still somehow limited due to their low prevalence and growth restriction. Mesenchymal stem cells are easily accessible and usually do not require technical manipulations, but mesenchymal stem cells from patients with chronic kidney disease may not be suitable for regenerative therapy, because uremia induces functional incompetence of mesenchymal stem cells. Although embryonic stem cells are pluripotent, there are ethical issues associated with manipulation of germ cells in producing embryonic stem cells. Whereas, iPSCs are pluripotent and free from ethical issues, but the use of retroviral transduction and our limited understanding of its effects still inhibit clinical applications of iPSCs.

However, new insights into the processes controlling endogenous renal repair and the capacity to isolate progenitor cells from the adult renal epithelium are providing possible options for the treatment of acute kidney injury. Major advances in the directed differentiation of stem cells to a large range of kidney cell types have also showed significant possibility of kidney regeneration. Moreover, cell-scaffold technology gave us more chance to reconstruct a three-dimensional whole functional kidney similar to the native human kidney. Furthermore, recent efforts to regenerate de novo a whole functional kidney using xenoembryo have suggested another potential strategy.

In the future, we should characterize gene expression profiles of regeneration associated cells and elucidate corresponding signaling molecules more clearly. Based on the knowledge of renal development, a more reliable technique should be established to manipulate stem/progenitor cells. Enhanced understanding of the mechanism of currently available stem cells with the capacity of renoprotection can also help us to find the novel pathway for kidney regeneration, and more specific cell or gene therapy should be discovered for different kidney diseases. Finally, continued collaborative efforts are required to achieve the generation of a bioengineered kidney that is capable of restoring renal function in patients with end-stage renal disease.

References

1. Jha V, Garcia-Garcia G. Global kidney disease – authors' reply. Lancet. 2013;382:1244.
2. McCampbell KK, Wingert RA. Renal stem cells: fact or science fiction? Biochem J. 2012;444:153–68.
3. Li Y, Wingert RA. Regenerative medicine for the kidney: stem cell prospects & challenges. Clin Transl Med. 2013;2:11.
4. Chung HC, Ko IK, Atala A, Yoo JJ. Cell-based therapy for kidney disease. Korean journal of urology. 2015;56:412–21.
5. Wolfe RA, Ashby VB, Milford EL, Ojo AO, Ettenger RE, Agodoa LY, et al. Comparison of mortality in all patients on dialysis, patients on dialysis awaiting transplantation, and recipients of a first cadaveric transplant. N Engl J Med. 1999;341:1725–30.
6. Katari R, Peloso A, Zambon JP, Soker S, Stratta RJ, Atala A, et al. Renal bioengineering with scaffolds generated from human kidneys. Nephron Exp Nephrol. 2014;126:119.
7. Fukui A, Yokoo T. Kidney regeneration using developing xenoembryo. Curr Opin Organ Transplant. 2015;20:160–4.
8. Chan TC, Ariizumi T, Asashima M. A model system for organ engineering: transplantation of in vitro induced embryonic kidney. Naturwissenschaften. 1999;86:224–7.
9. Xinaris C, Yokoo T. Reforming the kidney starting from a single-cell suspension. Nephron Exp Nephrol. 2014;126:107–12.
10. Locatelli F, Buoncristiani U, Canaud B, Kohler H, Petitclerc T, Zucchelli P. Dialysis dose and frequency. Nephrol Dial Transplant. 2005;20:285–96.
11. Maeshima A, Nakasatomi M, Nojima Y. Regenerative medicine for the kidney: renotropic factors, renal stem/progenitor cells, and stem cell therapy. Biomed Res Int. 2014;2014:1–10.
12. Strutz F, Okada H, Lo CW, Danoff T, Carone RL, Tomaszewski JE, et al. Identification and characterization of a fibroblast marker: FSP1. J Cell Biol. 1995;130:393–405.
13. Cornacchia F, Fornoni A, Plati AR, Thomas A, Wang Y, Inverardi L, et al. Glomerulosclerosis is transmitted by bone marrow-derived mesangial cell progenitors. J Clin Invest. 2001;108:1649–56.
14. Imasawa T, Utsunomiya Y, Kawamura T, Zhong Y, Nagasawa R, Okabe M, et al. The potential of bone marrow-derived cells to differentiate to glomerular mesangial cells. J Am Soc Nephrol. 2001;12:1401–9.
15. Ito T, Suzuki A, Imai E, Okabe M, Hori M. Bone marrow is a reservoir of repopulating mesangial cells during glomerular remodeling. J Am Soc Nephrol. 2001;12:2625–35.
16. Ito T, Suzuki A, Okabe M, Imai E, Hori M. Application of bone marrow-derived stem cells in experimental nephrology. Exp Nephrol. 2001;9:444–50.
17. Poulsom R, Forbes SJ, Hodivala-Dilke K, Ryan E, Wyles S, Navaratnarasah S, et al. Bone marrow contributes to renal parenchymal turnover and regeneration. J Pathol. 2001;195:229–35.
18. Witzgall R, Brown D, Schwarz C, Bonventre JV. Localization of proliferating cell nuclear antigen, vimentin, c-Fos, and clusterin in the postischemic kidney. Evidence for a heterogenous genetic response among nephron segments, and a large pool of mitotically active and dedifferentiated cells. J Clin Invest. 1994;93:2175–88.
19. Bonventre JV. Dedifferentiation and proliferation of surviving epithelial cells in acute renal failure. J Am Soc Nephrol. 2003;14(Suppl 1):S55–61.
20. Bussolati B, Bruno S, Grange C, Buttiglieri S, Deregibus MC, Cantino D, et al. Isolation of renal progenitor cells from adult human kidney. Am J Pathol. 2005;166:545–55.
21. Lazzeri E, Crescioli C, Ronconi E, Mazzinghi B, Sagrinati C, Netti GS, et al. Regenerative potential of embryonic renal multipotent progenitors in acute renal failure. J Am Soc Nephrol. 2007;18:3128–38.
22. Angelotti ML, Ronconi E, Ballerini L, Peired A, Mazzinghi B, Sagrinati C, et al. Characterization of renal progenitors committed toward tubular lineage and their regenerative potential in renal tubular injury. Stem Cells. 2012;30:1714–25.
23. Humphreys BD, Valerius MT, Kobayashi A, Mugford JW, Soeung S, Duffield JS, et al. Intrinsic epithelial cells repair the kidney after injury. Cell Stem Cell. 2008;2:284–91.
24. Sagrinati C, Netti GS, Mazzinghi B, Lazzeri E, Liotta F, Frosali F, et al. Isolation and characterization of multipotent progenitor cells from the Bowman's capsule of adult human kidneys. J Am Soc Nephrol. 2006;17:2443–56.
25. Romagnani P, Remuzzi G. Renal progenitors in non-diabetic and diabetic nephropathies. Trends Endocrinol Metab. 2013;24:13–20.
26. Grobstein C. Trans-filter induction of tubules in mouse metanephrogenic mesenchyme. Exp Cell Res. 1956;10:424–40.
27. Grobstein C. Inductive tissue interaction in development. Adv Cancer Res. 1956;4:187–236.
28. Kitamura S, Yamasaki Y, Kinomura M, Sugaya T, Sugiyama H, Maeshima Y, et al. Establishment and characterization of renal progenitor like cells from S3 segment of nephron in rat adult kidney. FASEB J. 2005;19:1789–97.
29. Lindgren D, Bostrom AK, Nilsson K, Hansson J, Sjolund J, Moller C, et al. Isolation and characteriza-

tion of progenitor-like cells from human renal proximal tubules. Am J Pathol. 2011;178:828–37.
30. Maeshima A, Yamashita S, Nojima Y. Identification of renal progenitor-like tubular cells that participate in the regeneration processes of the kidney. J Am Soc Nephrol. 2003;14:3138–46.
31. Bruno S, Camussi G. Isolation and characterization of resident mesenchymal stem cells in human glomeruli. Methods Mol Biol. 2012;879:367–80.
32. Oliver JA, Maarouf O, Cheema FH, Martens TP, Al-Awqati Q. The renal papilla is a niche for adult kidney stem cells. J Clin Invest. 2004;114:795–804.
33. Kobayashi A, Valerius MT, Mugford JW, Carroll TJ, Self M, Oliver G, et al. Six2 defines and regulates a multipotent self-renewing nephron progenitor population throughout mammalian kidney development. Cell Stem Cell. 2008;3:169–81.
34. Self M, Lagutin OV, Bowling B, Hendrix J, Cai Y, Dressler GR, et al. Six2 is required for suppression of nephrogenesis and progenitor renewal in the developing kidney. EMBO J. 2006;25:5214–28.
35. Guimaraes-Souza NK, Yamaleyeva LM, AbouShwareb T, Atala A, Yoo JJ. In vitro reconstitution of human kidney structures for renal cell therapy. Nephrol Dial Transplant. 2012;27:3082–90.
36. Lazzeri E, Ronconi E, Angelotti ML, Peired A, Mazzinghi B, Becherucci F, et al. Human urine-derived renal progenitors for personalized modeling of genetic kidney disorders. J Am Soc Nephrol. 2015;26:1961–74.
37. Weissman IL, Anderson DJ, Gage F. Stem and progenitor cells: origins, phenotypes, lineage commitments, and transdifferentiations. Annu Rev Cell Dev Biol. 2001;17:387–403.
38. Coskun V, Wu H, Blanchi B, Tsao S, Kim K, Zhao J, et al. CD133+ neural stem cells in the ependyma of mammalian postnatal forebrain. Proc Natl Acad Sci U S A. 2008;105:1026–31.
39. Ivanova L, Hiatt MJ, Yoder MC, Tarantal AF, Matsell DG. Ontogeny of CD24 in the human kidney. Kidney Int. 2010;77:1123–31.
40. Pleniceanu O, Harari-Steinberg O, Dekel B. Concise review: kidney stem/progenitor cells: differentiate, sort out, or reprogram? Stem Cells. 2010;28:1649–60.
41. Hendry CE, Little MH. Reprogramming the kidney: a novel approach for regeneration. Kidney Int. 2012;82:138–46.
42. Kusaba T, Lalli M, Kramann R, Kobayashi A, Humphreys BD. Differentiated kidney epithelial cells repair injured proximal tubule. Proc Natl Acad Sci U S A. 2014;111:1527–32.
43. Humphreys BD, Czerniak S, DiRocco DP, Hasnain W, Cheema R, Bonventre JV. Repair of injured proximal tubule does not involve specialized progenitors. Proc Natl Acad Sci U S A. 2011;108:9226–31.
44. Smeets B, Boor P, Dijkman H, Sharma SV, Jirak P, Mooren F, et al. Proximal tubular cells contain a phenotypically distinct, scattered cell population involved in tubular regeneration. J Pathol. 2013;229:645–59.
45. Berger K, Bangen JM, Hammerich L, Liedtke C, Floege J, Smeets B, et al. Origin of regenerating tubular cells after acute kidney injury. Proc Natl Acad Sci U S A. 2014;111:1533–8.
46. Dekel B, Shezen E, Even-Tov-Friedman S, Katchman H, Margalit R, Nagler A, et al. Transplantation of human hematopoietic stem cells into ischemic and growing kidneys suggests a role in vasculogenesis but not tubulogenesis. Stem Cells. 2006;24:1185–93.
47. Duffield JS, Park KM, Hsiao LL, Kelley VR, Scadden DT, Ichimura T, et al. Restoration of tubular epithelial cells during repair of the postischemic kidney occurs independently of bone marrow-derived stem cells. J Clin Invest. 2005;115:1743–55.
48. Friedenstein AJ, Latzinik NW, Grosheva AG, Gorskaya UF. Marrow microenvironment transfer by heterotopic transplantation of freshly isolated and cultured cells in porous sponges. Exp Hematol. 1982;10:217–27.
49. Bi B, Schmitt R, Israilova M, Nishio H, Cantley LG. Stromal cells protect against acute tubular injury via an endocrine effect. J Am Soc Nephrol. 2007;18:2486–96.
50. Imberti B, Morigi M, Tomasoni S, Rota C, Corna D, Longaretti L, et al. Insulin-like growth factor-1 sustains stem cell mediated renal repair. J Am Soc Nephrol. 2007;18:2921–8.
51. Togel F, Hu Z, Weiss K, Isaac J, Lange C, Westenfelder C. Administered mesenchymal stem cells protect against ischemic acute renal failure through differentiation-independent mechanisms. Am J Physiol Renal Physiol. 2005;289:F31–42.
52. Togel F, Cohen A, Zhang P, Yang Y, Hu Z, Westenfelder C. Autologous and allogeneic marrow stromal cells are safe and effective for the treatment of acute kidney injury. Stem Cells Dev. 2009;18:475–85.
53. Crisan M, Yap S, Casteilla L, Chen CW, Corselli M, Park TS, et al. A perivascular origin for mesenchymal stem cells in multiple human organs. Cell Stem Cell. 2008;3:301–13.
54. da Silva ML, Chagastelles PC, Nardi NB. Mesenchymal stem cells reside in virtually all post-natal organs and tissues. J Cell Sci. 2006;119:2204–13.
55. Dominici M, Le Blanc K, Mueller I, Slaper-Cortenbach I, Marini F, Krause D, et al. Minimal criteria for defining multipotent mesenchymal stromal cells. The International Society for Cellular Therapy position statement. Cytotherapy. 2006;8:315–7.
56. Horwitz EM, Le Blanc K, Dominici M, Mueller I, Slaper-Cortenbach I, Marini FC, et al. Clarification of the nomenclature for MSC: The International Society for Cellular Therapy position statement. Cytotherapy. 2005;7:393–5.
57. Nombela-Arrieta C, Ritz J, Silberstein LE. The elusive nature and function of mesenchymal stem cells. Nat Rev Mol Cell Biol. 2011;12:126–31.
58. Pittenger MF, Mackay AM, Beck SC, Jaiswal RK, Douglas R, Mosca JD, et al. Multilineage potential of adult human mesenchymal stem cells. Science. 1999;284:143–7.

59. Salem HK, Thiemermann C. Mesenchymal stromal cells: current understanding and clinical status. Stem Cells. 2010;28:585–96.
60. Pelekanos RA, Li J, Gongora M, Chandrakanthan V, Scown J, Suhaimi N, et al. Comprehensive transcriptome and immunophenotype analysis of renal and cardiac MSC-like populations supports strong congruence with bone marrow MSC despite maintenance of distinct identities. Stem Cell Res. 2012;8:58–73.
61. Matsumoto K, Mizuno S, Nakamura T. Hepatocyte growth factor in renal regeneration, renal disease and potential therapeutics. Curr Opin Nephrol Hypertens. 2000;9:395–402.
62. Humes HD, Cieslinski DA, Coimbra TM, Messana JM, Galvao C. Epidermal growth factor enhances renal tubule cell regeneration and repair and accelerates the recovery of renal function in postischemic acute renal failure. J Clin Invest. 1989;84:1757–61.
63. Sakai M, Zhang M, Homma T, Garrick B, Abraham JA, McKanna JA, et al. Production of heparin binding epidermal growth factor-like growth factor in the early phase of regeneration after acute renal injury. Isolation and localization of bioactive molecules. J Clin Invest. 1997;99:2128–38.
64. Homma T, Sakai M, Cheng HF, Yasuda T, Coffey RJ Jr, Harris RC. Induction of heparin-binding epidermal growth factor-like growth factor mRNA in rat kidney after acute injury. J Clin Invest. 1995;96:1018–25.
65. Zeisberg M, Hanai J, Sugimoto H, Mammoto T, Charytan D, Strutz F, et al. BMP-7 counteracts TGF-beta1-induced epithelial-to-mesenchymal transition and reverses chronic renal injury. Nat Med. 2003;9:964–8.
66. Vukicevic S, Basic V, Rogic D, Basic N, Shih MS, Shepard A, et al. Osteogenic protein-1 (bone morphogenetic protein-7) reduces severity of injury after ischemic acute renal failure in rat. J Clin Invest. 1998;102:202–14.
67. Ding H, Kopple JD, Cohen A, Hirschberg R. Recombinant human insulin-like growth factor-I accelerates recovery and reduces catabolism in rats with ischemic acute renal failure. J Clin Invest. 1993;91:2281–7.
68. Miller SB, Martin DR, Kissane J, Hammerman MR. Insulin-like growth factor I accelerates recovery from ischemic acute tubular necrosis in the rat. Proc Natl Acad Sci U S A. 1992;89:11876–80.
69. Nakagawa T, Sasahara M, Haneda M, Kataoka H, Nakagawa H, Yagi M, et al. Role of PDGF B-chain and PDGF receptors in rat tubular regeneration after acute injury. Am J Pathol. 1999;155:1689–99.
70. Yanagita M, Okuda T, Endo S, Tanaka M, Takahashi K, Sugiyama F, et al. Uterine sensitization-associated gene-1 (USAG-1), a novel BMP antagonist expressed in the kidney, accelerates tubular injury. J Clin Invest. 2006;116:70–9.
71. Imgrund M, Grone E, Grone HJ, Kretzler M, Holzman L, Schlondorff D, et al. Re-expression of the developmental gene Pax-2 during experimental acute tubular necrosis in mice 1. Kidney Int. 1999;56:1423–31.
72. Yoshino J, Monkawa T, Tsuji M, Hayashi M, Saruta T. Leukemia inhibitory factor is involved in tubular regeneration after experimental acute renal failure. J Am Soc Nephrol. 2003;14:3090–101.
73. Terada Y, Tanaka H, Okado T, Shimamura H, Inoshita S, Kuwahara M, et al. Expression and function of the developmental gene Wnt-4 during experimental acute renal failure in rats. J Am Soc Nephrol. 2003;14:1223–33.
74. Chen J, Chen JK, Harris RC. Deletion of the epidermal growth factor receptor in renal proximal tubule epithelial cells delays recovery from acute kidney injury. Kidney Int. 2012;82:45–52.
75. Sugimoto H, LeBleu VS, Bosukonda D, Keck P, Taduri G, Bechtel W, et al. Activin-like kinase 3 is important for kidney regeneration and reversal of fibrosis. Nat Med. 2012;18:396–404.
76. Zhou D, Tan RJ, Lin L, Zhou L, Liu Y. Activation of hepatocyte growth factor receptor, c-met, in renal tubules is required for renoprotection after acute kidney injury. Kidney Int. 2013;84:509–20.
77. Maeshima A, Miya M, Mishima K, Yamashita S, Kojima I, Nojima Y. Activin A: autocrine regulator of kidney development and repair. Endocr J. 2008;55:1–9.
78. Maeshima A, Nojima Y, Kojima I. The role of the activin-follistatin system in the developmental and regeneration processes of the kidney. Cytokine Growth Factor Rev. 2001;12:289–98.
79. Maeshima A, Sakurai H, Choi Y, Kitamura S, Vaughn DA, Tee JB, et al. Glial cell-derived neurotrophic factor independent ureteric bud outgrowth from the Wolffian duct. J Am Soc Nephrol. 2007;18:3147–55.
80. Maeshima A, Vaughn DA, Choi Y, Nigam SK. Activin A is an endogenous inhibitor of ureteric bud outgrowth from the Wolffian duct. Dev Biol. 2006;295:473–85.
81. Maeshima A, Yamashita S, Maeshima K, Kojima I, Nojima Y. Activin a produced by ureteric bud is a differentiation factor for metanephric mesenchyme. J Am Soc Nephrol. 2003;14:1523–34.
82. Maeshima A, Zhang YQ, Furukawa M, Naruse T, Kojima I. Hepatocyte growth factor induces branching tubulogenesis in MDCK cells by modulating the activin-follistatin system. Kidney Int. 2000;58:1511–22.
83. Ritvos O, Tuuri T, Eramaa M, Sainio K, Hilden K, Saxen L, et al. Activin disrupts epithelial branching morphogenesis in developing glandular organs of the mouse. Mech Dev. 1995;50:229–45.
84. Maeshima A, Zhang YQ, Nojima Y, Naruse T, Kojima I. Involvement of the activin-follistatin system in tubular regeneration after renal ischemia in rats. J Am Soc Nephrol. 2001;12:1685–95.
85. Gordon S, Taylor PR. Monocyte and macrophage heterogeneity. Nat Rev Immunol. 2005;5:953–64.
86. Ricardo SD, van Goor H, Eddy AA. Macrophage diversity in renal injury and repair. J Clin Invest. 2008;118:3522–30.
87. Wyburn KR, Jose MD, Wu H, Atkins RC, Chadban SJ. The role of macrophages in allograft rejection. Transplantation. 2005;80:1641–7.

88. Kluth DC, Erwig LP, Rees AJ. Multiple facets of macrophages in renal injury. Kidney Int. 2004;66:542–57.
89. Sean Eardley K, Cockwell P. Macrophages and progressive tubulointerstitial disease. Kidney Int. 2005;68:437–55.
90. Gordon S. Alternative activation of macrophages. Nat Rev Immunol. 2003;3:23–35.
91. Sica A, Mantovani A. Macrophage plasticity and polarization: in vivo veritas. J Clin Invest. 2012;122:787–95.
92. Ikezumi Y, Atkins RC, Nikolic-Paterson DJ. Interferon-gamma augments acute macrophage-mediated renal injury via a glucocorticoid-sensitive mechanism. J Am Soc Nephrol. 2003;14:888–98.
93. Lee S, Huen S, Nishio H, Nishio S, Lee HK, Choi BS, et al. Distinct macrophage phenotypes contribute to kidney injury and repair. J Am Soc Nephrol. 2011;22:317–26.
94. Nishida M, Fujinaka H, Matsusaka T, Price J, Kon V, Fogo AB, et al. Absence of angiotensin II type 1 receptor in bone marrow-derived cells is detrimental in the evolution of renal fibrosis. J Clin Invest. 2002;110:1859–68.
95. Lopez-Guisa JM, Cai X, Collins SJ, Yamaguchi I, Okamura DM, Bugge TH, et al. Mannose receptor 2 attenuates renal fibrosis. J Am Soc Nephrol. 2012;23:236–51.
96. Menke J, Iwata Y, Rabacal WA, Basu R, Yeung YG, Humphreys BD, et al. CSF-1 signals directly to renal tubular epithelial cells to mediate repair in mice. J Clin Invest. 2009;119:2330–42.
97. Alikhan MA, Jones CV, Williams TM, Beckhouse AG, Fletcher AL, Kett MM, et al. Colony-stimulating factor-1 promotes kidney growth and repair via alteration of macrophage responses. Am J Pathol. 2011;179:1243–56.
98. Cao Q, Wang Y, Zheng D, Sun Y, Lee VW, Zheng G, et al. IL-10/TGF-beta-modified macrophages induce regulatory T cells and protect against adriamycin nephrosis. J Am Soc Nephrol. 2010;21:933–42.
99. Kluth DC, Ainslie CV, Pearce WP, Finlay S, Clarke D, Anegon I, et al. Macrophages transfected with adenovirus to express IL-4 reduce inflammation in experimental glomerulonephritis. J Immunol. 2001;166:4728–36.
100. Lu J, Cao Q, Zheng D, Sun Y, Wang C, Yu X, et al. Discrete functions of M2a and M2c macrophage subsets determine their relative efficacy in treating chronic kidney disease. Kidney Int. 2013;84:745–55.
101. Riquelme P, Tomiuk S, Kammler A, Fandrich F, Schlitt HJ, Geissler EK, et al. IFN-gamma-induced iNOS expression in mouse regulatory macrophages prolongs allograft survival in fully immunocompetent recipients. Mol Ther. 2013;21:409–22.
102. Wilson HM, Chettibi S, Jobin C, Walbaum D, Rees AJ, Kluth DC. Inhibition of macrophage nuclear factor-kappaB leads to a dominant anti-inflammatory phenotype that attenuates glomerular inflammation in vivo. Am J Pathol. 2005;167:27–37.
103. Wilson HM, Stewart KN, Brown PA, Anegon I, Chettibi S, Rees AJ, et al. Bone-marrow-derived macrophages genetically modified to produce IL-10 reduce injury in experimental glomerulonephritis. Mol Ther. 2002;6:710–7.
104. Yamagishi H, Yokoo T, Imasawa T, Mitarai T, Kawamura T, Utsunomiya Y. Genetically modified bone marrow-derived vehicle cells site specifically deliver an anti-inflammatory cytokine to inflamed interstitium of obstructive nephropathy. J Immunol. 2001;166:609–16.
105. Yokoo T, Ohashi T, Utsunomiya Y, Kojima H, Imasawa T, Kogure T, et al. Prophylaxis of antibody-induced acute glomerulonephritis with genetically modified bone marrow-derived vehicle cells. Hum Gene Ther. 1999;10:2673–8.
106. Ferenbach DA, Ramdas V, Spencer N, Marson L, Anegon I, Hughes J, et al. Macrophages expressing heme oxygenase-1 improve renal function in ischemia/reperfusion injury. Mol Ther. 2010; 18:1706–13.
107. Nigam eS, Lieberthal W. Acute renal failure. III. The role of growth factors in the process of renal regeneration and repair. Am J Physiol Renal Physiol. 2000;279:F3–F11.
108. Lieberthal W, Fuhro R, Andry CC, Rennke H, Abernathy VE, Koh JS, et al. Rapamycin impairs recovery from acute renal failure: role of cell-cycle arrest and apoptosis of tubular cells. Am J Physiol Renal Physiol. 2001;281:F693–706.
109. Fresno Vara JA, Casado E, de Castro J, Cejas P, Belda-Iniesta C, Gonzalez-Baron M. PI3K/Akt signalling pathway and cancer. Cancer Treat Rev. 2004;30:193–204.
110. Angers S, Moon RT. Proximal events in Wnt signal transduction. Nat Rev Mol Cell Biol. 2009; 10:468–77.
111. Grimes CA, Jope RS. The multifaceted roles of glycogen synthase kinase 3beta in cellular signaling. Prog Neurobiol. 2001;65:391–426.
112. Lin SL, Li B, Rao S, Yeo EJ, Hudson TE, Nowlin BT, et al. Macrophage Wnt7b is critical for kidney repair and regeneration. Proc Natl Acad Sci U S A. 2010;107:4194–9.
113. Doble BW, Woodgett JR. GSK-3: tricks of the trade for a multi-tasking kinase. J Cell Sci. 2003;116:1175–86.
114. Jope RS, Johnson GV. The glamour and gloom of glycogen synthase kinase-3. Trends Biochem Sci. 2004;29:95–102.
115. Wang Z, Havasi A, Gall J, Bonegio R, Li Z, Mao H, et al. GSK3beta promotes apoptosis after renal ischemic injury. J Am Soc Nephrol. 2010;21:284–94.
116. Zhou D, Li Y, Lin L, Zhou L, Igarashi P, Liu Y. Tubule-specific ablation of endogenous beta-catenin aggravates acute kidney injury in mice. Kidney Int. 2012;82:537–47.
117. Bao H, Ge Y, Zhuang S, Dworkin LD, Liu Z, Gong R. Inhibition of glycogen synthase kinase-3beta prevents NSAID-induced acute kidney injury. Kidney Int. 2012;81:662–73.
118. Salahudeen AK, Haider N, Jenkins J, Joshi M, Patel H, Huang H, et al. Antiapoptotic properties

of erythropoiesis-stimulating proteins in models of cisplatin-induced acute kidney injury. Am J Physiol Renal Physiol. 2008;294:F1354–65.
119. Ucero AC, Berzal S, Ocana-Salceda C, Sancho M, Orzaez M, Messeguer A, et al. A polymeric nanomedicine diminishes inflammatory events in renal tubular cells. PLoS One. 2013;8:e51992.
120. Lewis TS, Shapiro PS, Ahn NG. Signal transduction through MAP kinase cascades. Adv Cancer Res. 1998;74:49–139.
121. Kyriakis JM, Avruch J. Mammalian mitogen-activated protein kinase signal transduction pathways activated by stress and inflammation. Physiol Rev. 2001;81:807–69.
122. di Mari JF, Davis R, Safirstein RL. MAPK activation determines renal epithelial cell survival during oxidative injury. Am J Physiol. 1999;277:F195–203.
123. Arany I, Megyesi JK, Nelkin BD, Safirstein RL. STAT3 attenuates EGFR-mediated ERK activation and cell survival during oxidant stress in mouse proximal tubular cells. Kidney Int. 2006;70:669–74.
124. Ishizuka S, Yano T, Hagiwara K, Sone M, Nihei H, Ozasa H, et al. Extracellular signal-regulated kinase mediates renal regeneration in rats with myoglobinuric acute renal injury. Biochem Biophys Res Commun. 1999;254:88–92.
125. Ikarashi K, Li B, Suwa M, Kawamura K, Morioka T, Yao J, et al. Bone marrow cells contribute to regeneration of damaged glomerular endothelial cells. Kidney Int. 2005;67:1925–33.
126. Lin F, Moran A, Igarashi P. Intrarenal cells, not bone marrow-derived cells, are the major source for regeneration in postischemic kidney. J Clin Invest. 2005;115:1756–64.
127. Prodromidi EI, Poulsom R, Jeffery R, Roufosse CA, Pollard PJ, Pusey CD, et al. Bone marrow-derived cells contribute to podocyte regeneration and amelioration of renal disease in a mouse model of Alport syndrome. Stem Cells. 2006;24:2448–55.
128. Bruno S, Grange C, Deregibus MC, Calogero RA, Saviozzi S, Collino F, et al. Mesenchymal stem cell-derived microvesicles protect against acute tubular injury. J Am Soc Nephrol. 2009;20:1053–67.
129. Caplan AI, Dennis JE. Mesenchymal stem cells as trophic mediators. J Cell Biochem. 2006;98:1076–84.
130. Tomasoni S, Longaretti L, Rota C, Morigi M, Conti S, Gotti E, et al. Transfer of growth factor receptor mRNA via exosomes unravels the regenerative effect of mesenchymal stem cells. Stem Cells Dev. 2013;22:772–80.
131. Wang Y, He J, Pei X, Zhao W. Systematic review and meta-analysis of mesenchymal stem/stromal cells therapy for impaired renal function in small animal models. Nephrology (Carlton). 2013;18:201–8.
132. Xinaris C, Morigi M, Benedetti V, Imberti B, Fabricio AS, Squarcina E, et al. A novel strategy to enhance mesenchymal stem cell migration capacity and promote tissue repair in an injury specific fashion. Cell Transplant. 2013;22:423–36.
133. Duffy MM, Ritter T, Ceredig R, Griffin MD. Mesenchymal stem cell effects on T-cell effector pathways. Stem Cell Res Ther. 2011;2:34.
134. Camussi G, Deregibus MC, Tetta C. Paracrine/endocrine mechanism of stem cells on kidney repair: role of microvesicle-mediated transfer of genetic information. Curr Opin Nephrol Hypertens. 2010;19:7–12.
135. Ke YH, He JW, Fu WZ, Zhang ZL. Identification of two novel mutations in the OCRL1 gene in two Chinese families with Lowe syndrome. Nephrology (Carlton). 2012;17:20–5.
136. Phinney DG, Prockop DJ. Concise review: mesenchymal stem/multipotent stromal cells: the state of transdifferentiation and modes of tissue repair--current views. Stem Cells. 2007;25:2896–902.
137. Burdon TJ, Paul A, Noiseux N, Prakash S, Shum-Tim D. Bone marrow stem cell derived paracrine factors for regenerative medicine: current perspectives and therapeutic potential. Bone Marrow Res. 2011;2011:207326.
138. Li J, Deane JA, Campanale NV, Bertram JF, Ricardo SD. The contribution of bone marrow-derived cells to the development of renal interstitial fibrosis. Stem Cells. 2007;25:697–706.
139. Thirabanjasak D, Tantiwongse K, Thorner PS. Angiomyeloproliferative lesions following autologous stem cell therapy. J Am Soc Nephrol. 2010;21:1218–22.
140. Donizetti-Oliveira C, Semedo P, Burgos-Silva M, Cenedeze MA, Malheiros DM, Reis MA, et al. Adipose tissue-derived stem cell treatment prevents renal disease progression. Cell Transplant. 2012;21:1727–41.
141. Eirin A, Zhu XY, Krier JD, Tang H, Jordan KL, Grande JP, et al. Adipose tissue-derived mesenchymal stem cells improve revascularization outcomes to restore renal function in swine atherosclerotic renal artery stenosis. Stem Cells. 2012;30:1030–41.
142. Kim JH, Park DJ, Yun JC, Jung MH, Yeo HD, Kim HJ, et al. Human adipose tissue-derived mesenchymal stem cells protect kidneys from cisplatin nephrotoxicity in rats. Am J Physiol Renal Physiol. 2012;302:F1141–50.
143. Zhu XY, Urbieta-Caceres V, Krier JD, Textor SC, Lerman A, Lerman LO. Mesenchymal stem cells and endothelial progenitor cells decrease renal injury in experimental swine renal artery stenosis through different mechanisms. Stem Cells. 2013;31:117–25.
144. de Almeida DC, Donizetti-Oliveira C, Barbosa-Costa P, Origassa CS, Camara NO. In search of mechanisms associated with mesenchymal stem cell-based therapies for acute kidney injury. Clin Biochem Rev. 2013;34:131–44.
145. Chen YT, Sun CK, Lin YC, Chang LT, Chen YL, Tsai TH, et al. Adipose-derived mesenchymal stem cell protects kidneys against ischemia-reperfusion injury through suppressing oxidative stress and inflammatory reaction. J Transl Med. 2011;9:51.

146. Rota C, Imberti B, Pozzobon M, Piccoli M, De Coppi P, Atala A, et al. Human amniotic fluid stem cell preconditioning improves their regenerative potential. Stem Cells Dev. 2012;21:1911–23.
147. Song B, Niclis JC, Alikhan MA, Sakkal S, Sylvain A, Kerr PG, et al. Generation of induced pluripotent stem cells from human kidney mesangial cells. J Am Soc Nephrol. 2011;22:1213–20.
148. Song B, Smink AM, Jones CV, Callaghan JM, Firth SD, Bernard CA, et al. The directed differentiation of human iPS cells into kidney podocytes. PLoS One. 2012;7:e46453.
149. Zhou T, Benda C, Duzinger S, Huang Y, Li X, Li Y, et al. Generation of induced pluripotent stem cells from urine. J Am Soc Nephrol. 2011;22:1221–8.
150. Polo JM, Liu S, Figueroa ME, Kulalert W, Eminli S, Tan KY, et al. Cell type of origin influences the molecular and functional properties of mouse induced pluripotent stem cells. Nat Biotechnol. 2010;28:848–55.
151. Lee PY, Chien Y, Chiou GY, Lin CH, Chiou CH, Tarng DC. Induced pluripotent stem cells without c-Myc attenuate acute kidney injury via downregulating the signaling of oxidative stress and inflammation in ischemia-reperfusion rats. Cell Transplant. 2012;21:2569–85.
152. Imberti B, Tomasoni S, Ciampi O, Pezzotta A, Derosas M, Xinaris C, et al. Renal progenitors derived from human iPSCs engraft and restore function in a mouse model of acute kidney injury. Sci Rep. 2015;5:8826.
153. Zhao T, Zhang ZN, Rong Z, Xu Y. Immunogenicity of induced pluripotent stem cells. Nature. 2011;474:212–5.
154. Harari-Steinberg O, Metsuyanim S, Omer D, Gnatek Y, Gershon R, Pri-Chen S, et al. Identification of human nephron progenitors capable of generation of kidney structures and functional repair of chronic renal disease. EMBO Mol Med. 2013;5:1556–68.
155. Woolf AS, Palmer SJ, Snow ML, Fine LG. Creation of a functioning chimeric mammalian kidney. Kidney Int. 1990;38:991–7.
156. Rogers SA, Lowell JA, Hammerman NA, Hammerman MR. Transplantation of developing metanephroi into adult rats. Kidney Int. 1998;54:27–37.
157. Statter MB, Fahrner KJ, Barksdale EM Jr, Parks DE, Flavell RA, Donahoe PK. Correlation of fetal kidney and testis congenic graft survival with reduced major histocompatibility complex burden. Transplantation. 1989;47:651–60.
158. Abrahamson DR, St John PL, Pillion DJ, Tucker DC. Glomerular development in intraocular and intrarenal grafts of fetal kidneys. Lab Invest. 1991;64:629–39.
159. Mikos AG, Herring SW, Ochareon P, Elisseeff J, Lu HH, Kandel R, et al. Engineering complex tissues. Tissue Eng. 2006;12:3307–39.
160. Basu J, Ludlow JW. Developmental engineering the kidney: leveraging principles of morphogenesis for renal regeneration. Birth Defects Res C Embryo Today. 2012;96:30–8.
161. Cuppage FE, Tate A. Repair of the nephron following injury with mercuric chloride. Am J Pathol. 1967;51:405–29.
162. Oliver J. Correlations of structure and function and mechanisms of recovery in acute tubular necrosis. Am J Med. 1953;15:535–57.
163. Bissell MJ, Hall HG, Parry G. How does the extracellular matrix direct gene expression? J Theor Biol. 1982;99:31–68.
164. Vorotnikova E, McIntosh D, Dewilde A, Zhang J, Reing JE, Zhang L, et al. Extracellular matrix-derived products modulate endothelial and progenitor cell migration and proliferation in vitro and stimulate regenerative healing in vivo. Matrix Biol. 2010;29:690–700.
165. Bornstein P, Sage EH. Matricellular proteins: extracellular modulators of cell function. Curr Opin Cell Biol. 2002;14:608–16.
166. Lelongt B, Ronco P. Role of extracellular matrix in kidney development and repair. Pediatr Nephrol. 2003;18:731–42.
167. Lanza RP, Chung HY, Yoo JJ, Wettstein PJ, Blackwell C, Borson N, et al. Generation of histocompatible tissues using nuclear transplantation. Nat Biotechnol. 2002;20:689–96.
168. Bergsma EJ, Rozema FR, Bos RR, de Bruijn WC. Foreign body reactions to resorbable poly(L-lactide) bone plates and screws used for the fixation of unstable zygomatic fractures. J Oral Maxillofac Surg. 1993;51:666–70.
169. Lo H, Kadiyala S, Guggino SE, Leong KW. Poly(L-lactic acid) foams with cell seeding and controlled-release capacity. J Biomed Mater Res. 1996;30:475–84.
170. Martin C, Winet H, Bao JY. Acidity near eroding polylactide-polyglycolide in vitro and in vivo in rabbit tibial bone chambers. Biomaterials. 1996;17:2373–80.
171. Zhang R, Ma PX. Poly(alpha-hydroxyl acids)/hydroxyapatite porous composites for bone-tissue engineering. I. Preparation and morphology. J Biomed Mater Res. 1999;44:446–55.
172. Nam YS, Yoon JJ, Park TG. A novel fabrication method of macroporous biodegradable polymer scaffolds using gas foaming salt as a porogen additive. J Biomed Mater Res. 2000;53:1–7.
173. Orlando G, Baptista P, Birchall M, De Coppi P, Farney A, Guimaraes-Souza NK, et al. Regenerative medicine as applied to solid organ transplantation: current status and future challenges. Transpl Int. 2011;24:223–32.
174. Orlando G, Wood KJ, De Coppi P, Baptista PM, Binder KW, Bitar KN, et al. Regenerative medicine as applied to general surgery. Ann Surg. 2012;255:867–80.
175. Orlando G, Wood KJ, Stratta RJ, Yoo JJ, Atala A, Soker S. Regenerative medicine and organ trans-

plantation: past, present, and future. Transplantation. 2011;91:1310–7.
176. Badylak SF, Taylor D, Uygun K. Whole-organ tissue engineering: decellularization and recellularization of three-dimensional matrix scaffolds. Annu Rev Biomed Eng. 2011;13:27–53.
177. Crapo PM, Medberry CJ, Reing JE, Tottey S, van der Merwe Y, Jones KE, et al. Biologic scaffolds composed of central nervous system extracellular matrix. Biomaterials. 2012;33:3539–47.
178. Badylak SF, Gilbert TW. Immune response to biologic scaffold materials. Semin Immunol. 2008;20:109–16.
179. Brown BN, Valentin JE, Stewart-Akers AM, McCabe GP, Badylak SF. Macrophage phenotype and remodeling outcomes in response to biologic scaffolds with and without a cellular component. Biomaterials. 2009;30:1482–91.
180. Brown B, Lindberg K, Reing J, Stolz DB, Badylak SF. The basement membrane component of biologic scaffolds derived from extracellular matrix. Tissue Eng. 2006;12:519–26.
181. Reing JE, Brown BN, Daly KA, Freund JM, Gilbert TW, Hsiong SX, et al. The effects of processing methods upon mechanical and biologic properties of porcine dermal extracellular matrix scaffolds. Biomaterials. 2010;31:8626–33.
182. Ott HC, Matthiesen TS, Goh SK, Black LD, Kren SM, Netoff TI, et al. Perfusion-decellularized matrix: using nature's platform to engineer a bioartificial heart. Nat Med. 2008;14:213–21.
183. Baptista PM, Siddiqui MM, Lozier G, Rodriguez SR, Atala A, Soker S. The use of whole organ decellularization for the generation of a vascularized liver organoid. Hepatology. 2011;53:604–17.
184. Song JJ, Kim SS, Liu Z, Madsen JC, Mathisen DJ, Vacanti JP, et al. Enhanced in vivo function of bioartificial lungs in rats. J Biomed Mater Res A. 2011;92:998–1005. discussion-6
185. Loai Y, Yeger H, Coz C, Antoon R, Islam SS, Moore K, et al. Bladder tissue engineering: tissue regeneration and neovascularization of HA-VEGF-incorporated bladder acellular constructs in mouse and porcine animal models. J Biomed Mater Res A. 2010;94:1205–15.
186. Ross EA, Williams MJ, Hamazaki T, Terada N, Clapp WL, Adin C, et al. Embryonic stem cells proliferate and differentiate when seeded into kidney scaffolds. J Am Soc Nephrol. 2009;20:2338–47.
187. Orlando G, Farney AC, Iskandar SS, Mirmalek-Sani SH, Sullivan DC, Moran E, et al. Production and implantation of renal extracellular matrix scaffolds from porcine kidneys as a platform for renal bioengineering investigations. Ann Surg. 2012;256:363–70.
188. Nakayama KH, Batchelder CA, Lee CI, Tarantal AF. Decellularized rhesus monkey kidney as a three-dimensional scaffold for renal tissue engineering. Tissue Eng Part A. 2010;16:2207–16.
189. Sullivan DC, Mirmalek-Sani SH, Deegan DB, Baptista PM, Aboushwareb T, Atala A, et al. Decellularization methods of porcine kidneys for whole organ engineering using a high-throughput system. Biomaterials. 2012;33:7756–64.
190. Oliver JA, Klinakis A, Cheema FH, Friedlander J, Sampogna RV, Martens TP, et al. Proliferation and migration of label-retaining cells of the kidney papilla. J Am Soc Nephrol. 2009;20:2315–27.
191. Reule S, Gupta S. Kidney regeneration and resident stem cells. Organogenesis. 2011;7:135–9.
192. Maeshima A, Sakurai H, Nigam SK. Adult kidney tubular cell population showing phenotypic plasticity, tubulogenic capacity, and integration capability into developing kidney. J Am Soc Nephrol. 2006;17:188–98.
193. Humphreys BD, Bonventre JV. The contribution of adult stem cells to renal repair. Nephrol Ther. 2007;3:3–10.
194. Vogetseder A, Karadeniz A, Kaissling B, Le Hir M. Tubular cell proliferation in the healthy rat kidney. Histochem Cell Biol. 2005;124:97–104.
195. Challen GA, Little MH. A side order of stem cells: the SP phenotype. Stem Cells. 2006;24:3–12.
196. Iwatani H, Ito T, Imai E, Matsuzaki Y, Suzuki A, Yamato M, et al. Hematopoietic and nonhematopoietic potentials of Hoechst(low)/side population cells isolated from adult rat kidney. Kidney Int. 2004;65:1604–14.
197. Hishikawa K, Marumo T, Miura S, Nakanishi A, Matsuzaki Y, Shibata K, et al. Musculin/MyoR is expressed in kidney side population cells and can regulate their function. J Cell Biol. 2005;169:921–8.
198. Hristov M, Erl W, Weber PC. Endothelial progenitor cells: mobilization, differentiation, and homing. Arterioscler Thromb Vasc Biol. 2003;23:1185–9.
199. Bidlingmaier S, Zhu X, Liu B. The utility and limitations of glycosylated human CD133 epitopes in defining cancer stem cells. J Mol Med (Berl). 2008;86:1025–32.
200. Dekel B, Zangi L, Shezen E, Reich-Zeliger S, Eventov-Friedman S, Katchman H, et al. Isolation and characterization of nontubular sca-1+lin- multipotent stem/progenitor cells from adult mouse kidney. J Am Soc Nephrol. 2006;17:3300–14.
201. Bertoncello I, Williams B. Hematopoietic stem cell characterization by Hoechst 33342 and rhodamine 123 staining. Methods Mol Biol. 2004;263:181–200.
202. Gupta S, Verfaillie C, Chmielewski D, Kren S, Eidman K, Connaire J, et al. Isolation and characterization of kidney-derived stem cells. J Am Soc Nephrol. 2006;17:3028–40.
203. Lasagni L, Romagnani P. Glomerular epithelial stem cells: the good, the bad, and the ugly. J Am Soc Nephrol. 2010;21:1612–9.
204. Ronconi E, Sagrinati C, Angelotti ML, Lazzeri E, Mazzinghi B, Ballerini L, et al. Regeneration of glomerular podocytes by human renal progenitors. J Am Soc Nephrol. 2009;20:322–32.
205. Appel D, Kershaw DB, Smeets B, Yuan G, Fuss A, Frye B, et al. Recruitment of podocytes from glomerular parietal epithelial cells. J Am Soc Nephrol. 2009;20:333–43.

206. Eirin A, Lerman LO. Mesenchymal stem cell treatment for chronic renal failure. Stem Cell Res Ther. 2014;5:83.
207. Rosenberg ME. Cell-based therapies in kidney disease. Kidney Int Suppl. 2011;3:364–7.
208. Herzlinger D, Koseki C, Mikawa T, al-Awqati Q. Metanephric mesenchyme contains multipotent stem cells whose fate is restricted after induction. Development. 1992;114:565–72.
209. Hartman HA, Lai HL, Patterson LT. Cessation of renal morphogenesis in mice. Dev Biol. 2007;310:379–87.
210. Strutz F, Zeisberg M, Renziehausen A, Raschke B, Becker V, van Kooten C, et al. TGF-beta 1 induces proliferation in human renal fibroblasts via induction of basic fibroblast growth factor (FGF-2). Kidney Int. 2001;59:579–92.
211. Phillips AO, Steadman R. Diabetic nephropathy: the central role of renal proximal tubular cells in tubulointerstitial injury. Histol Histopathol. 2002;17:247–52.
212. Helbert MJ, Dauwe S, De Broe ME. Flow cytometric immunodissection of the human nephron in vivo and in vitro. Exp Nephrol. 1999;7:360–76.
213. Cummings BS, Lasker JM, Lash LH. Expression of glutathione-dependent enzymes and cytochrome P450s in freshly isolated and primary cultures of proximal tubular cells from human kidney. J Pharmacol Exp Ther. 2000;293:677–85.
214. Qi W, Johnson DW, Vesey DA, Pollock CA, Chen X. Isolation, propagation and characterization of primary tubule cell culture from human kidney. Nephrology (Carlton). 2007;12:155–9.
215. Presnell SC, Bruce AT, Wallace SM, Choudhury S, Genheimer CW, Cox B, et al. Isolation, characterization, and expansion methods for defined primary renal cell populations from rodent, canine, and human normal and diseased kidneys. Tissue Eng Part C Methods. 2011;17:261–73.
216. Baer PC, Geiger H. Human renal cells from the thick ascending limb and early distal tubule: characterization of primary isolated and cultured cells by reverse transcription polymerase chain reaction. Nephrology (Carlton). 2008;13:316–21.
217. Aggarwal S, Moggio A, Bussolati B. Concise review: stem/progenitor cells for renal tissue repair: current knowledge and perspectives. Stem Cells Transl Med. 2013;2:1011–9.
218. Kitamura S, Sakurai H, Makino H. Single adult kidney stem/progenitor cells reconstitute three-dimensional nephron structures in vitro. Stem Cells. 2014;33:774–84.
219. Bruno S, Bussolati B, Grange C, Collino F, di Cantogno LV, Herrera MB, et al. Isolation and characterization of resident mesenchymal stem cells in human glomeruli. Stem Cells Dev. 2009;18:867–80.
220. Wang H, Gomez JA, Klein S, Zhang Z, Seidler B, Yang Y, et al. Adult renal mesenchymal stem cell-like cells contribute to juxtaglomerular cell recruitment. J Am Soc Nephrol. 2013;24:1263–73.
221. Simerman AA, Dumesic DA, Chazenbalk GD. Pluripotent muse cells derived from human adipose tissue: a new perspective on regenerative medicine and cell therapy. Clin Transl Med. 2014;3:12.
222. Jiang Y, Jahagirdar BN, Reinhardt RL, Schwartz RE, Keene CD, Ortiz-Gonzalez XR, et al. Pluripotency of mesenchymal stem cells derived from adult marrow. Nature. 2002;418:41–9.
223. Hayakawa M, Ishizaki M, Hayakawa J, Migita M, Murakami M, Shimada T, et al. Role of bone marrow cells in the healing process of mouse experimental glomerulonephritis. Pediatr Res. 2005;58:323–8.
224. Kale S, Karihaloo A, Clark PR, Kashgarian M, Krause DS, Cantley LG. Bone marrow stem cells contribute to repair of the ischemically injured renal tubule. J Clin Invest. 2003;112:42–9.
225. Rookmaaker MB, Smits AM, Tolboom H, Van't Wout K, Martens AC, Goldschmeding R, et al. Bone-marrow-derived cells contribute to glomerular endothelial repair in experimental glomerulonephritis. Am J Pathol. 2003;163:553–62.
226. Sugimoto H, Mundel TM, Sund M, Xie L, Cosgrove D, Kalluri R. Bone-marrow-derived stem cells repair basement membrane collagen defects and reverse genetic kidney disease. Proc Natl Acad Sci U S A. 2006;103:7321–6.
227. Herrera MB, Bussolati B, Bruno S, Fonsato V, Romanazzi GM, Camussi G. Mesenchymal stem cells contribute to the renal repair of acute tubular epithelial injury. Int J Mol Med. 2004;14:1035–41.
228. Morigi M, Imberti B, Zoja C, Corna D, Tomasoni S, Abbate M, et al. Mesenchymal stem cells are renotropic, helping to repair the kidney and improve function in acute renal failure. J Am Soc Nephrol. 2004;15:1794–804.
229. Wong CY, Cheong SK, Mok PL, Leong CF. Differentiation of human mesenchymal stem cells into mesangial cells in post-glomerular injury murine model. Pathology. 2008;40:52–7.
230. Togel F, Weiss K, Yang Y, Hu Z, Zhang P, Westenfelder C. Vasculotropic, paracrine actions of infused mesenchymal stem cells are important to the recovery from acute kidney injury. Am J Physiol Renal Physiol. 2007;292:F1626–35.
231. Chou YH, Pan SY, Yang CH, Lin SL. Stem cells and kidney regeneration. J Formos Med Assoc. 2014;113:201–9.
232. Terada N, Hamazaki T, Oka M, Hoki M, Mastalerz DM, Nakano Y, et al. Bone marrow cells adopt the phenotype of other cells by spontaneous cell fusion. Nature. 2002;416:542–5.
233. Alvarez-Dolado M, Pardal R, Garcia-Verdugo JM, Fike JR, Lee HO, Pfeffer K, et al. Fusion of bone-marrow-derived cells with Purkinje neurons, cardiomyocytes and hepatocytes. Nature. 2003;425:968–73.
234. Togel FE, Westenfelder C. Mesenchymal stem cells: a new therapeutic tool for AKI. Nat Rev Nephrol. 2010;6:179–83.
235. Perico N, Casiraghi F, Introna M, Gotti E, Todeschini M, Cavinato RA, et al. Autologous mesenchymal

stromal cells and kidney transplantation: a pilot study of safety and clinical feasibility. Clin J Am Soc Nephrol. 2011;6:412–22.
236. Tan J, Wu W, Xu X, Liao L, Zheng F, Messinger S, et al. Induction therapy with autologous mesenchymal stem cells in living-related kidney transplants: a randomized controlled trial. JAMA. 2012;307:1169–77.
237. Casiraghi F, Perico N, Remuzzi G. Mesenchymal stromal cells to promote solid organ transplantation tolerance. Curr Opin Organ Transplant. 2013;18:51–8.
238. Reinders ME, de Fijter JW, Roelofs H, Bajema IM, de Vries DK, Schaapherder AF, et al. Autologous bone marrow-derived mesenchymal stromal cells for the treatment of allograft rejection after renal transplantation: results of a phase I study. Stem Cells Transl Med. 2013;2:107–11.
239. Lee RH, Pulin AA, Seo MJ, Kota DJ, Ylostalo J, Larson BL, et al. Intravenous hMSCs improve myocardial infarction in mice because cells embolized in lung are activated to secrete the anti-inflammatory protein TSG-6. Cell Stem Cell. 2009;5:54–63.
240. Ebrahimi B, Eirin A, Li Z, Zhu XY, Zhang X, Lerman A, et al. Mesenchymal stem cells improve medullary inflammation and fibrosis after revascularization of swine atherosclerotic renal artery stenosis. PLoS One. 2013;8:e67474.
241. Katsuno T, Ozaki T, Saka Y, Furuhashi K, Kim H, Yasuda K, et al. Low serum cultured adipose tissue-derived stromal cells ameliorate acute kidney injury in rats. Cell Transplant. 2013;22:287–97.
242. Furuhashi K, Tsuboi N, Shimizu A, Katsuno T, Kim H, Saka Y, et al. Serum-starved adipose-derived stromal cells ameliorate crescentic GN by promoting immunoregulatory macrophages. J Am Soc Nephrol. 2013;24:587–603.
243. Morigi M, Rota C, Montemurro T, Montelatici E, Lo Cicero V, Imberti B, et al. Life-sparing effect of human cord blood-mesenchymal stem cells in experimental acute kidney injury. Stem Cells. 2010;28:513–22.
244. Shalaby RH, Rashed LA, Ismaail AE, Madkour NK, Elwakeel SH. Hematopoietic stem cells derived from human umbilical cord ameliorate cisplatin-induced acute renal failure in rats. Am J Stem Cells. 2014;3:83–96.
245. Panepucci RA, Siufi JL, Silva WA Jr, Proto-Siquiera R, Neder L, Orellana M, et al. Comparison of gene expression of umbilical cord vein and bone marrow-derived mesenchymal stem cells. Stem Cells. 2004;22:1263–78.
246. De Coppi P, Bartsch G Jr, Siddiqui MM, Xu T, Santos CC, Perin L, et al. Isolation of amniotic stem cell lines with potential for therapy. Nat Biotechnol. 2007;25:100–6.
247. Perin L, Giuliani S, Jin D, Sedrakyan S, Carraro G, Habibian R, et al. Renal differentiation of amniotic fluid stem cells. Cell Prolif. 2007;40:936–48.
248. Hauser PV, De Fazio R, Bruno S, Sdei S, Grange C, Bussolati B, et al. Stem cells derived from human amniotic fluid contribute to acute kidney injury recovery. Am J Pathol. 2010;177:2011–21.
249. Perin L, Sedrakyan S, Giuliani S, Da Sacco S, Carraro G, Shiri L, et al. Protective effect of human amniotic fluid stem cells in an immunodeficient mouse model of acute tubular necrosis. PLoS One. 2010;5:e9357.
250. Sedrakyan S, Da Sacco S, Milanesi A, Shiri L, Petrosyan A, Varimezova R, et al. Injection of amniotic fluid stem cells delays progression of renal fibrosis. J Am Soc Nephrol. 2012;23:661–73.
251. Sun D, Bu L, Liu C, Yin Z, Zhou X, Li X, et al. Therapeutic effects of human amniotic fluid-derived stem cells on renal interstitial fibrosis in a murine model of unilateral ureteral obstruction. PLoS One. 2013;8:e65042.
252. Da Sacco S, Lemley KV, Sedrakyan S, Zanusso I, Petrosyan A, Peti-Peterdi J, et al. A novel source of cultured podocytes. PLoS One. 2013;8:e81812.
253. Siegel N, Rosner M, Unbekandt M, Fuchs C, Slabina N, Dolznig H, et al. Contribution of human amniotic fluid stem cells to renal tissue formation depends on mTOR. Hum Mol Genet. 2010;19:3320–31.
254. Patschan D, Krupincza K, Patschan S, Zhang Z, Hamby C, Goligorsky MS. Dynamics of mobilization and homing of endothelial progenitor cells after acute renal ischemia: modulation by ischemic preconditioning. Am J Physiol Renal Physiol. 2006;291:F176–85.
255. Patschan D, Patschan S, Gobe GG, Chintala S, Goligorsky MS. Uric acid heralds ischemic tissue injury to mobilize endothelial progenitor cells. J Am Soc Nephrol. 2007;18:1516–24.
256. Asahara T, Murohara T, Sullivan A, Silver M, van der Zee R, Li T, et al. Isolation of putative progenitor endothelial cells for angiogenesis. Science. 1997;275:964–7.
257. Chade AR, Zhu X, Lavi R, Krier JD, Pislaru S, Simari RD, et al. Endothelial progenitor cells restore renal function in chronic experimental renovascular disease. Circulation. 2009;119:547–57.
258. Chade AR, Zhu XY, Krier JD, Jordan KL, Textor SC, Grande JP, et al. Endothelial progenitor cells homing and renal repair in experimental renovascular disease. Stem Cells. 2010;28:1039–47.
259. Jie KE, Zaikova MA, Bergevoet MW, Westerweel PE, Rastmanesh M, Blankestijn PJ, et al. Progenitor cells and vascular function are impaired in patients with chronic kidney disease. Nephrol Dial Transplant. 2010;25:1875–82.
260. Evans MJ, Kaufman MH. Establishment in culture of pluripotential cells from mouse embryos. Nature. 1981;292:154–6.
261. Takahashi K, Yamanaka S. Induction of pluripotent stem cells from mouse embryonic and adult fibroblast cultures by defined factors. Cell. 2006;126:663–76.
262. Takahashi K, Tanabe K, Ohnuki M, Narita M, Ichisaka T, Tomoda K, et al. Induction of pluripotent

stem cells from adult human fibroblasts by defined factors. Cell. 2007;131:861–72.
263. Basma H, Soto-Gutierrez A, Yannam GR, Liu L, Ito R, Yamamoto T, et al. Differentiation and transplantation of human embryonic stem cell-derived hepatocytes. Gastroenterology. 2009;136:990–9.
264. Chambers SM, Fasano CA, Papapetrou EP, Tomishima M, Sadelain M, Studer L. Highly efficient neural conversion of human ES and iPS cells by dual inhibition of SMAD signaling. Nat Biotechnol. 2009;27:275–80.
265. D'Amour KA, Bang AG, Eliazer S, Kelly OG, Agulnick AD, Smart NG, et al. Production of pancreatic hormone-expressing endocrine cells from human embryonic stem cells. Nat Biotechnol. 2006;24:1392–401.
266. Dhara SK, Stice SL. Neural differentiation of human embryonic stem cells. J Cell Biochem. 2008; 105:633–40.
267. Hay DC, Zhao D, Fletcher J, Hewitt ZA, McLean D, Urruticoechea-Uriguen A, et al. Efficient differentiation of hepatocytes from human embryonic stem cells exhibiting markers recapitulating liver development in vivo. Stem Cells. 2008;26:894–902.
268. Ledran MH, Krassowska A, Armstrong L, Dimmick I, Renstrom J, Lang R, et al. Efficient hematopoietic differentiation of human embryonic stem cells on stromal cells derived from hematopoietic niches. Cell Stem Cell. 2008;3:85–98.
269. Takahashi T, Lord B, Schulze PC, Fryer RM, Sarang SS, Gullans SR, et al. Ascorbic acid enhances differentiation of embryonic stem cells into cardiac myocytes. Circulation. 2003;107:1912–6.
270. Zhang D, Jiang W, Liu M, Sui X, Yin X, Chen S, et al. Highly efficient differentiation of human ES cells and iPS cells into mature pancreatic insulin-producing cells. Cell Res. 2009;19:429–38.
271. Spence JR, Mayhew CN, Rankin SA, Kuhar MF, Vallance JE, Tolle K, et al. Directed differentiation of human pluripotent stem cells into intestinal tissue in vitro. Nature. 2011;470:105–9.
272. Gadue P, Huber TL, Paddison PJ, Keller GM. Wnt and TGF-beta signaling are required for the induction of an in vitro model of primitive streak formation using embryonic stem cells. Proc Natl Acad Sci U S A. 2006;103:16806–11.
273. Lam AQ, Freedman BS, Morizane R, Lerou PH, Valerius MT, Bonventre JV. Rapid and efficient differentiation of human pluripotent stem cells into intermediate mesoderm that forms tubules expressing kidney proximal tubular markers. J Am Soc Nephrol. 2014;25:1211–25.
274. Mae S, Shono A, Shiota F, Yasuno T, Kajiwara M, Gotoda-Nishimura N, et al. Monitoring and robust induction of nephrogenic intermediate mesoderm from human pluripotent stem cells. Nat Commun. 2013;4:1367.
275. Takasato M, Er PX, Becroft M, Vanslambrouck JM, Stanley EG, Elefanty AG, et al. Directing human embryonic stem cell differentiation towards a renal lineage generates a self-organizing kidney. Nat Cell Biol. 2014;16:118–26.
276. Xia Y, Nivet E, Sancho-Martinez I, Gallegos T, Suzuki K, Okamura D, et al. Directed differentiation of human pluripotent cells to ureteric bud kidney progenitor-like cells. Nat Cell Biol. 2013;15:1507–15.
277. Taguchi A, Kaku Y, Ohmori T, Sharmin S, Ogawa M, Sasaki H, et al. Redefining the in vivo origin of metanephric nephron progenitors enables generation of complex kidney structures from pluripotent stem cells. Cell Stem Cell. 2014;14:53–67.
278. Cai J, Zhao Y, Liu Y, Ye F, Song Z, Qin H, et al. Directed differentiation of human embryonic stem cells into functional hepatic cells. Hepatology. 2007;45:1229–39.
279. Wichterle H, Lieberam I, Porter JA, Jessell TM. Directed differentiation of embryonic stem cells into motor neurons. Cell. 2002;110:385–97.
280. Laflamme MA, Chen KY, Naumova AV, Muskheli V, Fugate JA, Dupras SK, et al. Cardiomyocytes derived from human embryonic stem cells in pro-survival factors enhance function of infarcted rat hearts. Nat Biotechnol. 2007;25:1015–24.
281. Martin GR. Isolation of a pluripotent cell line from early mouse embryos cultured in medium conditioned by teratocarcinoma stem cells. Proc Natl Acad Sci U S A. 1981;78:7634–8.
282. Osafune K. In vitro regeneration of kidney from pluripotent stem cells. Exp Cell Res. 2010;316:2571–7.
283. Thomson JA, Itskovitz-Eldor J, Shapiro SS, Waknitz MA, Swiergiel JJ, Marshall VS, et al. Embryonic stem cell lines derived from human blastocysts. Science. 1998;282:1145–7.
284. Reubinoff BE, Pera MF, Fong CY, Trounson A, Bongso A. Embryonic stem cell lines from human blastocysts: somatic differentiation in vitro. Nat Biotechnol. 2000;18:399–404.
285. Schuldiner M, Yanuka O, Itskovitz-Eldor J, Melton DA, Benvenisty N. Effects of eight growth factors on the differentiation of cells derived from human embryonic stem cells. Proc Natl Acad Sci U S A. 2000;97:11307–12.
286. Kobayashi T, Tanaka H, Kuwana H, Inoshita S, Teraoka H, Sasaki S, et al. Wnt4-transformed mouse embryonic stem cells differentiate into renal tubular cells. Biochem Biophys Res Commun. 2005;336:585–95.
287. Mae S, Shirasawa S, Yoshie S, Sato F, Kanoh Y, Ichikawa H, et al. Combination of small molecules enhances differentiation of mouse embryonic stem cells into intermediate mesoderm through BMP7-positive cells. Biochem Biophys Res Commun. 2010;393:877–82.
288. Steenhard BM, Isom KS, Cazcarro P, Dunmore JH, Godwin AR, St John PL, et al. Integration of embryonic stem cells in metanephric kidney organ culture. J Am Soc Nephrol. 2005;16:1623–31.

289. Kim D, Dressler GR. Nephrogenic factors promote differentiation of mouse embryonic stem cells into renal epithelia. J Am Soc Nephrol. 2005;16:3527–34.
290. Yamamoto M, Cui L, Johkura K, Asanuma K, Okouchi Y, Ogiwara N, et al. Branching ducts similar to mesonephric ducts or ureteric buds in teratomas originating from mouse embryonic stem cells. Am J Physiol Renal Physiol. 2006;290:F52–60.
291. Esteban MA, Xu J, Yang J, Peng M, Qin D, Li W, et al. Generation of induced pluripotent stem cell lines from Tibetan miniature pig. J Biol Chem. 2009;284:17634–40.
292. Ezashi T, Telugu BP, Alexenko AP, Sachdev S, Sinha S, Roberts RM. Derivation of induced pluripotent stem cells from pig somatic cells. Proc Natl Acad Sci U S A. 2009;106:10993–8.
293. Honda A, Hirose M, Hatori M, Matoba S, Miyoshi H, Inoue K, et al. Generation of induced pluripotent stem cells in rabbits: potential experimental models for human regenerative medicine. J Biol Chem. 2010;285:31362–9.
294. Li W, Wei W, Zhu S, Zhu J, Shi Y, Lin T, et al. Generation of rat and human induced pluripotent stem cells by combining genetic reprogramming and chemical inhibitors. Cell Stem Cell. 2009;4:16–9.
295. Liao J, Cui C, Chen S, Ren J, Chen J, Gao Y, et al. Generation of induced pluripotent stem cell lines from adult rat cells. Cell Stem Cell. 2009;4:11–5.
296. Liu H, Zhu F, Yong J, Zhang P, Hou P, Li H, et al. Generation of induced pluripotent stem cells from adult rhesus monkey fibroblasts. Cell Stem Cell. 2008;3:587–90.
297. Montserrat N, Ramirez-Bajo MJ, Xia Y, Sancho-Martinez I, Moya-Rull D, Miquel-Serra L, et al. Generation of induced pluripotent stem cells from human renal proximal tubular cells with only two transcription factors, OCT4 and SOX2. J Biol Chem. 2012;287:24131–8.
298. Kim K, Doi A, Wen B, Ng K, Zhao R, Cahan P, et al. Epigenetic memory in induced pluripotent stem cells. Nature. 2010;467:285–90.
299. Zhou T, Benda C, Dunzinger S, Huang Y, Ho JC, Yang J, et al. Generation of human induced pluripotent stem cells from urine samples. Nat Protoc. 2012;7:2080–9.
300. Okita K, Nagata N, Yamanaka S. Immunogenicity of induced pluripotent stem cells. Circ Res. 2011;109:720–1.
301. Eiraku M, Takata N, Ishibashi H, Kawada M, Sakakura E, Okuda S, et al. Self-organizing optic-cup morphogenesis in three-dimensional culture. Nature. 2011;472:51–6.
302. Eiraku M, Watanabe K, Matsuo-Takasaki M, Kawada M, Yonemura S, Matsumura M, et al. Self-organized formation of polarized cortical tissues from ESCs and its active manipulation by extrinsic signals. Cell Stem Cell. 2008;3:519–32.
303. Suga H, Kadoshima T, Minaguchi M, Ohgushi M, Soen M, Nakano T, et al. Self-formation of functional adenohypophysis in three-dimensional culture. Nature. 2011;480:57–62.
304. Nakano T, Ando S, Takata N, Kawada M, Muguruma K, Sekiguchi K, et al. Self-formation of optic cups and storable stratified neural retina from human ESCs. Cell Stem Cell. 2012;10:771–85.
305. Osafune K, Takasato M, Kispert A, Asashima M, Nishinakamura R. Identification of multipotent progenitors in the embryonic mouse kidney by a novel colony-forming assay. Development. 2006;133:151–61.
306. Unbekandt M, Davies JA. Dissociation of embryonic kidneys followed by reaggregation allows the formation of renal tissues. Kidney Int. 2010;77:407–16.
307. Chen J, Lansford R, Stewart V, Young F, Alt FW. RAG-2-deficient blastocyst complementation: an assay of gene function in lymphocyte development. Proc Natl Acad Sci U S A. 1993;90:4528–32.
308. Espejel S, Roll GR, McLaughlin KJ, Lee AY, Zhang JY, Laird DJ, et al. Induced pluripotent stem cell-derived hepatocytes have the functional and proliferative capabilities needed for liver regeneration in mice. J Clin Invest. 2010;120:3120–6.
309. Fraidenraich D, Stillwell E, Romero E, Wilkes D, Manova K, Basson CT, et al. Rescue of cardiac defects in id knockout embryos by injection of embryonic stem cells. Science. 2004;306:247–52.
310. Kobayashi T, Yamaguchi T, Hamanaka S, Kato-Itoh M, Yamazaki Y, Ibata M, et al. Generation of rat pancreas in mouse by interspecific blastocyst injection of pluripotent stem cells. Cell. 2010;142:787–99.
311. Matsunari H, Nagashima H, Watanabe M, Umeyama K, Nakano K, Nagaya M, et al. Blastocyst complementation generates exogenic pancreas in vivo in apancreatic cloned pigs. Proc Natl Acad Sci U S A. 2013;110:4557–62.
312. Muller SM, Terszowski G, Blum C, Haller C, Anquez V, Kuschert S, et al. Gene targeting of VEGF-A in thymus epithelium disrupts thymus blood vessel architecture. Proc Natl Acad Sci U S A. 2005;102:10587–92.
313. Ueno H, Turnbull BB, Weissman IL. Two-step oligoclonal development of male germ cells. Proc Natl Acad Sci U S A. 2009;106:175–80.
314. Ueno H, Weissman IL. Clonal analysis of mouse development reveals a polyclonal origin for yolk sac blood islands. Dev Cell. 2006;11:519–33.
315. Usui J, Kobayashi T, Yamaguchi T, Knisely AS, Nishinakamura R, Nakauchi H. Generation of kidney from pluripotent stem cells via blastocyst complementation. Am J Pathol. 2012;180:2417–26.
316. Dekel B, Burakova T, Arditti FD, Reich-Zeliger S, Milstein O, Aviel-Ronen S, et al. Human and porcine early kidney precursors as a new source for transplantation. Nat Med. 2003;9:53–60.
317. Hammerman MR. Renal organogenesis from transplanted metanephric primordia. J Am Soc Nephrol. 2004;15:1126–32.
318. Matsumoto K, Yokoo T, Matsunari H, Iwai S, Yokote S, Teratani T, et al. Xenotransplanted embryonic kidney provides a niche for endogenous mesenchymal stem cell differentiation into erythropoietin-producing tissue. Stem Cells. 2012;30:1228–35.

319. Matsumoto K, Yokoo T, Yokote S, Utsunomiya Y, Ohashi T, Hosoya T. Functional development of a transplanted embryonic kidney: effect of transplantation site. J Nephrol. 2012;25:50–5.
320. Rogers SA, Talcott M, Hammerman MR. Transplantation of pig metanephroi. ASAIO J. 2003;49:48–52.
321. Sariola H, Ekblom P, Lehtonen E, Saxen L. Differentiation and vascularization of the metanephric kidney grafted on the chorioallantoic membrane. Dev Biol. 1983;96:427–35.
322. Xinaris C, Benedetti V, Rizzo P, Abbate M, Corna D, Azzollini N, et al. In vivo maturation of functional renal organoids formed from embryonic cell suspensions. J Am Soc Nephrol. 2012;23:1857–68.
323. Yokote S, Yokoo T, Matsumoto K, Ohkido I, Utsunomiya Y, Kawamura T, et al. Metanephros transplantation inhibits the progression of vascular calcification in rats with adenine-induced renal failure. Nephron Exp Nephrol. 2012;120:e32–40.
324. Yokote S, Yokoo T, Matsumoto K, Utsunomiya Y, Kawamura T, Hosoya T. The effect of metanephros transplantation on blood pressure in anephric rats with induced acute hypotension. Nephrol Dial Transplant. 2012;27:3449–55.
325. Dekel B, Marcus H, Herzel BH, Bocher WO, Passwell JH, Reisner Y. In vivo modulation of the allogeneic immune response by human fetal kidneys: the role of cytokines, chemokines, and cytolytic effector molecules. Transplantation. 2000;69:1470–8.
326. Steer DL, Nigam SK. Developmental approaches to kidney tissue engineering. Am J Physiol Renal Physiol. 2004;286:F1–7.
327. Sakurai H, Barros EJ, Tsukamoto T, Barasch J, Nigam SK. An in vitro tubulogenesis system using cell lines derived from the embryonic kidney shows dependence on multiple soluble growth factors. Proc Natl Acad Sci U S A. 1997;94:6279–84.
328. Machiguchi T, Nakamura T. Cellular interactions via conditioned media induce in vivo nephron generation from tubular epithelial cells or mesenchymal stem cells. Biochem Biophys Res Commun. 2013;435:327–33.
329. Auerbach R, Grobstein C. Inductive interaction of embryonic tissues after dissociation and reaggregation. Exp Cell Res. 1958;15:384–97.
330. Ganeva V, Unbekandt M, Davies JA. An improved kidney dissociation and reaggregation culture system results in nephrons arranged organotypically around a single collecting duct system. Organogenesis. 2011;7:83–7.
331. D'Agati VD. Growing new kidneys from embryonic cell suspensions: fantasy or reality? J Am Soc Nephrol. 2012;23:1763–6.
332. Ogle B, Cascalho M, Platt JL. Fusion of approaches to the treatment of organ failure. Am J Transplant. 2004;4(Suppl 6):74–7.
333. Cascalho M, Ogle BM, Platt JL. Xenotransplantation and the future of renal replacement. J Am Soc Nephrol. 2004;15:1106–12.
334. Yokoo T, Fukui A, Ohashi T, Miyazaki Y, Utsunomiya Y, Kawamura T, et al. Xenobiotic kidney organogenesis from human mesenchymal stem cells using a growing rodent embryo. J Am Soc Nephrol. 2006;17:1026–34.
335. Yokoo T, Ohashi T, Shen JS, Sakurai K, Miyazaki Y, Utsunomiya Y, et al. Human mesenchymal stem cells in rodent whole-embryo culture are reprogrammed to contribute to kidney tissues. Proc Natl Acad Sci U S A. 2005;102:3296–300.
336. Yokoo T, Fukui A, Matsumoto K, Ohashi T, Sado Y, Suzuki H, et al. Generation of a transplantable erythropoietin-producer derived from human mesenchymal stem cells. Transplantation. 2008;85:1654–8.
337. Davies JA, Fisher CE. Genes and proteins in renal development. Exp Nephrol. 2002;10:102–13.
338. Lipschutz JH. Molecular development of the kidney: a review of the results of gene disruption studies. Am J Kidney Dis. 1998;31:383–97.
339. Gheisari Y, Yokoo T, Matsumoto K, Fukui A, Sugimoto N, Ohashi T, et al. A thermoreversible polymer mediates controlled release of glial cell line-derived neurotrophic factor to enhance kidney regeneration. Artif Organs. 2010;34:642–7.
340. Fukui A, Yokoo T, Matsumoto K, Kawamura T, Hosoya T, Okabe M. Integration of human mesenchymal stem cells into the Wolffian duct in chicken embryos. Biochem Biophys Res Commun. 2009;385:330–5.
341. Pino CJ, Humes HD. Stem cell technology for the treatment of acute and chronic renal failure. Transl Res. 2010;156:161–8.
342. Fissell WH, Fleischman AJ, Humes HD, Roy S. Development of continuous implantable renal replacement: past and future. Transl Res. 2007;150:327–36.
343. Tasnim F, Deng R, Hu M, Liour S, Li Y, Ni M, et al. Achievements and challenges in bioartificial kidney development. Fibrogenesis Tissue Repair. 2010;3:14.
344. Aebischer P, Ip TK, Panol G, Galletti PM. The bioartificial kidney: progress towards an ultrafiltration device with renal epithelial cells processing. Life Support Syst. 1987;5:159–68.
345. Humes HD, Weitzel WF, Bartlett RH, Swaniker FC, Paganini EP, Luderer JR, et al. Initial clinical results of the bioartificial kidney containing human cells in ICU patients with acute renal failure. Kidney Int. 2004;66:1578–88.
346. MacKay SM, Funke AJ, Buffington DA, Humes HD. Tissue engineering of a bioartificial renal tubule. ASAIO J. 1998;44:179–83.
347. Nikolovski J, Gulari E, Humes HD. Design engineering of a bioartificial renal tubule cell therapy device. Cell Transplant. 1999;8:351–64.
348. Humes HD, Cieslinski DA. Interaction between growth factors and retinoic acid in the induction of kidney tubulogenesis in tissue culture. Exp Cell Res. 1992;201:8–15.
349. Humes HD, Krauss JC, Cieslinski DA, Funke AJ. Tubulogenesis from isolated single cells of adult

mammalian kidney: clonal analysis with a recombinant retrovirus. Am J Physiol. 1996;271:F42–9.
350. O'Neil JJ, Stegemann JP, Nicholson DT, Mullon CJ, Maki T, Monaco AP, et al. Immunoprotection provided by the bioartificial pancreas in a xenogeneic host. Transplant Proc. 1997;29:2116–7.
351. Humes HD, Buffington DA, MacKay SM, Funke AJ, Weitzel WF. Replacement of renal function in uremic animals with a tissue-engineered kidney. Nat Biotechnol. 1999;17:451–5.
352. Fissell WH, Dyke DB, Weitzel WF, Buffington DA, Westover AJ, MacKay SM, et al. Bioartificial kidney alters cytokine response and hemodynamics in endotoxin-challenged uremic animals. Blood Purif. 2002;20:55–60.
353. Humes HD, Buffington DA, Lou L, Abrishami S, Wang M, Xia J, et al. Cell therapy with a tissue-engineered kidney reduces the multiple-organ consequences of septic shock. Crit Care Med. 2003;31:2421–8.
354. Humes HD, Sobota JT, Ding F, Song JH. A selective cytopheretic inhibitory device to treat the immunological dysregulation of acute and chronic renal failure. Blood Purif. 2010;29:183–90.
355. Tumlin JA, Chawla L, Tolwani AJ, Mehta R, Dillon J, Finkel KW, et al. The effect of the selective cytopheretic device on acute kidney injury outcomes in the intensive care unit: a multicenter pilot study. Semin Dial. 2012;26:616–23.
356. Pino CJ, Yevzlin AS, Lee K, Westover AJ, Smith PL, Buffington DA, et al. Cell-based approaches for the treatment of systemic inflammation. Nephrol Dial Transplant. 2013;28:296–302.
357. Buffington DA, Pino CJ, Chen L, Westover AJ, Hageman G, Humes HD. Bioartificial renal epithelial cell system (BRECS): a compact, cryopreservable extracorporeal renal replacement device. Cell Med. 2012;4:33–43.
358. Gura V, Beizai M, Ezon C, Polaschegg HD. Continuous renal replacement therapy for end-stage renal disease. The wearable artificial kidney (WAK). Contrib Nephrol. 2005;149:325–33.
359. Roberts M, Ash SR, Lee DB. Innovative peritoneal dialysis: flow-thru and dialysate regeneration. ASAIO J. 1999;45:372–8.
360. Ash SR, Janle EM. Continuous flow-through peritoneal dialysis (CFPD): comparison of efficiency to IPD, TPD, and CAPD in an animal model. Perit Dial Int. 1997;17:365–72.
361. Fissell WH, Dubnisheva A, Eldridge AN, Fleischman AJ, Zydney AL, Roy S. High-performance silicon nanopore hemofiltration membranes. J Memb Sci. 2009;326:58–63.
362. Fissell WH, Roy S. The implantable artificial kidney. Semin Dial. 2009;22:665–70.
363. Roy S, Goldman K, Marchant R, Zydney A, Brown D, Fleischman A, et al. Implanted renal replacement for end-stage renal disease. Panminerva Med. 2011;53:155–66.
364. Conlisk AT, Datta S, Fissell WH, Roy S. Biomolecular transport through hemofiltration membranes. Ann Biomed Eng. 2009;37:722–36.
365. Kanani DM, Fissell WH, Roy S, Dubnisheva A, Fleischman A, Zydney AL. Permeability – selectivity analysis for ultrafiltration: effect of pore geometry. J Memb Sci. 2010;349:405.
366. Muthusubramaniam L, Lowe R, Fissell WH, Li L, Marchant RE, Desai TA, et al. Hemocompatibility of silicon-based substrates for biomedical implant applications. Ann Biomed Eng. 2011;39:1296–305.

8 Bladder

Yun-Sok Ha and Tae Gyun Kwon

8.1 Introduction

The bladder is susceptible to a variety of congenital anomalies, injuries, and disorders, such as cancer, trauma, infection, and chronic inflammation [1]. In cases of muscle-invasive and refractory non-muscle-invasive bladder cancer and end-stage (congenital) bladder disease, the current standard treatment is radical cystectomy in combination with urinary diversion or bladder augmentation, which serves to restore bladder capacity and compliance and prevent vesicoureteral reflux and renal damage [2, 3]. The method of diversion depends in part on the nature of defect and the patient's needs and wishes. Bladder reconstruction is one of the greatest surgical challenges in the field of urology. To repair or replace the bladder, gastrointestinal segments are commonly used. However, gastrointestinal tissues are designed to absorb specific solutes, whereas bladder tissues are designed to excrete solutes. Conventional bladder reconstruction using gastrointestinal tissue (enterocystoplasty) is associated with numerous complications, including mucus production, bacterial colonization, electrolyte imbalance, anastomotic leakage, enteric fistulae, bowel obstruction, and prolonged episodes of ileus, intestinal failure, or malignancy, as well as significant morbidity and functional alterations [4]. To overcome these problems, numerous alternative reconstructive procedures have been introduced including autoaugmentation and ureterocystoplasty [5–7], as well as tissue expansion [8, 9]. However, almost all of them resulted in failure due to technical difficulty or complication. Tissue engineering techniques, transplantation of biomaterials seeded with cells, have been developed as a promising alternative that circumvent many of the limitations and complications associated with enterocystoplasty; therefore, it is considered a potentially promising option for bladder reconstruction.

Regenerative medicine aims to regenerate tissues and organs by creating biological equivalents through supplementation of scaffolding materials with bioactive components, cells, or a combination thereof [10]. Various attempts have been made to reconstruct the bladder in both animal and human studies using regenerative medicine techniques [11]. In 2006, the first clinical application of regenerative medicine for bladder reconstruction was published by Atala et al. [12] In this study, they implanted a collagen/polyglycolic acid composite scaffold with urothelial and smooth muscle cells (SMCs) into the bladder dome area after partial cystectomy.

The fields of tissue engineering and regenerative medicine have achieved substantial progression over the previous two decades. It

Y.-S. Ha (✉) • T.G. Kwon
Department of Urology, School of Medicine,
Kyungpook National University, Daegu, South Korea
e-mail: yunsokha@gmail.com; tgkwon@knu.ac.kr

B.W. Kim (ed.), *Clinical Regenerative Medicine in Urology*, DOI 10.1007/978-981-10-2723-9_8

encompasses the cell biology, transplantation, material science, and biomedical engineering, toward identifying alternatives that can reestablish and preserve the regular function of damaged tissues and organs [13]. Studies for bladder tissue engineering strategies have been orientated in two directions: firstly, to identify the most appropriate type of cells for regeneration and to proficiently incorporate it into bladder cells and, secondly, to determine the most appropriate biomaterial and technique of embedding cells into engineered grafts [14, 15]. The selected grafts must exhibit all the qualities of the native tissue, acting ultimately as microenvironments for the implanted cells to proliferate and differentiate. Numerous studies have been tried to identify ideal materials for bladder regeneration including acellular tissue matrices and synthetic matrices [16]. For example, bladder submucosa matrix (BSM), obtained by decellularization of bladder submucosa, has been shown to be an effective material for clinical applications because it maintains the extracellular matrix of the bladder as well as the biological activity of growth factors [17].

On the other hand, efforts have been exerted to find ideal cell source for bladder regeneration. Autologous bladder cells are the gold standard source for bladder regeneration in cell-based tissue engineering. To provide functional cells for tissue reconstruction and remodeling, cultured cells at early passage (before passage 5) provide optimal results. In addition, cells should be cultured on the scaffold for a short time (less than 2 weeks) since SMCs lose their phenotype after long-term static culture [18]. Mechanical preconditioning of muscle cells might be an alternative culture method that would maintain the phenotype and functional characteristics of cells before implantation [19]. Immortalized muscle cell lines are attractive since they may generate a large number of functional cells. However, these types of cell lines have limited clinical application, as immortalized cells carry the risk of tumor formation [20]. For patients with bladder cancer, normal bladder cells are not available; thus, an alternative cell source needs to be found, and several cell types are being investigated for this purpose, including bone marrow stem cells, skeletal myoblasts, and adipose progenitor cells [21–24].

In the 1950s, efforts to find alternatives to intestinal cystoplasty included the use of alloplastic materials [25–31]. Although these early research efforts met with very limited success owing to foreign body complications, they provided the foundation for regenerative medicine and led to a focus on biodegradable, collagen-rich tissues for use as a scaffold with and without cell seeding [32–38]. Atala et al. [12] reported the first data regarding the use of a seeded biodegradable construct for augmentation of the neurogenic cell population. This work led to an industry-supported, prospective, multi-institution investigation using a neo-bladder construct. Unfortunately, compliance improved in only four patients after 12 months and in five after 36 months but in none to a clinically or statistically significant degree. Capacity also did not improve [39]. Using autologous tissue from a neuropathic source may lower the yield of usable cells or could potentiate the growth of abnormal tissue [40]. An alternate cell source for regenerative tissue, such as stem cells, might circumvent these concerns. However, the use of stem cells carries its own set of obstacles [41–44]. Further investigation into the clinical applicability of regenerative tissue engineering is required, but this technology remains poised to alter reconstructive bladder surgery.

This chapter is focused on current status of regenerative medicine including achievements and perspectives on the use of biomaterials and stem cells in the field of bladder reconstruction.

8.2 The Use of Matrices for Bladder Regeneration

The bladder is considered as a good candidate for regenerative medicine-based approach because it is one of the simple structured organs (not so many different kinds of cells, minimal metabolic activity, and less interaction with adjacent organs). Therefore, many researchers have studied bladder reconstruction with various methods.

One of the approaches is implantation of biomaterials into the bladder. In this approach, selection of the appropriate bioscaffold for large bladder tissue replacement is a critical component. There are distinct benefits to using biocompatible material in regenerative medicine for the purpose of cell delivery vehicles and for bearing the physical maintenance required for tissue replacement [45]. Scaffolds are constructs that are designed to direct tissue development and the growth of cells during the process of healing [46]. Bladder replacements should therefore provide provisional mechanical support, adequate to endure forces exerted from neighboring structures, while maintaining a potential zone for tissue development. Biomaterials used for bladder replacements should possess the ability to be easily manipulated into a hollow, spherical configuration. Furthermore, the biomaterials should possess the ability to biodegrade for complete tissue development, without causing inflammation. An ideal scaffold for bladder tissue reconstruction should have an adjustable 3-D porous structure, high mechanical strength, uniformity, and appropriate degradation rate, and it should be easily fabricated. Over the last few decades, several bladder wall substitutes, composed of both synthetic and organic materials, have been investigated extensively. Biocompatible synthetic materials are considered potential substitutes for bladder tissue. Biomaterials can be divided into four main categories: (1) naturally derived matrices, including collagen; (2) synthetic matrices, including polylactic-co-glycolic acid (PLGA); (3) acellular tissue matrices, including bladder submucosa; and (4) hybrid or composite scaffold [47].

8.2.1 Naturally Derived Matrices

Collagen, a naturally derived matrix, is considered to be the most ubiquitous protein in the human body, and it is often used alongside alginate as a natural matrix. It is useful in tissue engineering, as it possesses the ability to be easily manipulated and does not provoke an immune response [48]. Through the use of innovative inkjet technology, it has been possible to use bioprinting to create a naturally derived 3-D construct, with a precise arrangement of growth factors and other cellular components, into a patient-specific scaffold [16].

8.2.2 Synthetic Matrices

Synthetic polymers are promising materials because of their reproducibility, large-scale production, mechanical properties, ease of fabrication, and manipulable strength, degradation rate, and microstructure [49]. The typical synthetic biomaterials utilized for urinary tissue regeneration are polyglycolic acid (PGA) and polylactic-co-glycolic acid (PLGA), which are biocompatible, biodegradable, and FDA-approved for human applications [1]. However, the brittleness of these synthetic polymers and the acidic by-products produced during their degradation have been linked to inflammation and immune dysfunction [50–54]. Furthermore, studies have shown that stiff synthetic materials can cause mechanical failure and urinary stone formation, and over time, the degradation products cause fibroblast deposition, scarring, and graft contracture and reduce reservoir volume [25, 55, 56]. To avoid these side effects, interest in polycaprolactone (PCL) has increased because it is flexible, biocompatible, stable, and resistant to resorption; its degradation products show low toxicity; and it could help protect the urothelium [57, 58]. However, it is well known that the use of scaffolds with low hydrophilicity (hydrophobic), like PCL, can lead to low initial cell seeding efficacy and heterogeneous/slow cell growth due to inadequate diffusion of the cell culture medium into the scaffold and the lack of specific interaction sites with cells [59]. Besides abovementioned biomaterials, numerous synthetic materials including polyvinyl sponge, Teflon, collagen matrices, Vicryl PGA matrices, and silicone have been studied in both experimental and clinical settings. However, most of these materials failed to show possibilities of clinical application due to mechanical, structural, functional, or biocompatibility problems. Permanent synthetic materials often succumb to mechanical failure and urinary stone formation, while degradable materials lead to fibroblast deposition, scarring, graft contracture, and a reduced reservoir volume [60, 61].

8.2.3 Acellular Tissue Matrices

Decellularized matrices are the most commonly used naturally derived urological matrices. They are usually harvested from autologous, allogenic, or xenogeneic tissue [55, 62]. Chemical or mechanical processing decellularizes the matrix, removing all cellular components and leaving a natural platform for tissue development [63]. The most common origin of decellularized matrices is the tissue harvested from the bladder or small intestinal mucosa.

Bladder submucosa matrix (BSM), obtained by decellularization of bladder submucosa, has been shown to be an effective material for clinical applications because it maintains the extracellular matrix of the bladder as well as the biological activity of growth factors [17]. Non-seeded allogeneic acellular bladder matrices have been used as scaffolds for the ingrowth of host bladder wall components. These matrices are prepared by mechanically and chemically removing all cellular components from bladder tissue. Probst et al. reported a method of bladder augmentation that avoided the complications encountered with the use of bowel segments using a newly developed acellular biomaterial, the bladder acellular matrix graft (BAMG), as a homologous graft [64]. Thirty-four rats underwent a partial cystectomy (40–50%) and grafting with a BAMG of equal size. After initial bladder enlargement, the graft was progressively infiltrated by host vessels and SMCs, and the mucosal lining was complete within 10 days. After 4 weeks, histological analysis showed that all bladder wall components were present in the graft. Ingrowth was complete after 8 weeks, except for neural regeneration, which was only partial. At 12 weeks, the bladder wall muscle structure in the graft was so well developed that it was difficult to delineate the junction between the host bladder and BAMG. Neural regeneration continued to improve, and normal bladder capacity was maintained throughout the study. They concluded that the BAMG appeared to serve, without rejection, as a collagen and elastin framework for the ingrowth of all bladder wall components.

Small intestinal submucosa (SIS), a biodegradable, acellular, xenogeneic collagen-based tissue matrix graft, was first described in the early 1960s as an acellular matrix for tissue replacement in the vascular field [65]. The matrix is derived from pig small intestine in which the mucosa is mechanically removed from the inner surface and the serosa and muscular layer are removed from the outer surface. Animal studies have shown that, when used for bladder augmentation, non-seeded SIS matrix is able to regenerate [66]. In this study, 15-month small intestinal submucosa-regenerated canine bladder strips in vitro were analyzed by muscle bath compliance, contractility testing, and immunohistochemical staining, and the results showed that small intestinal submucosa-regenerated canine bladder. Compliance studies demonstrated no significant difference between small intestinal submucosa-regenerated and control bladders, which were 30-fold more compliant than a native small intestinal submucosal graft. Contractility studies demonstrated that the contractile responses and innervation were similar to those of normal canine bladder. Afferent nerves were observed by immunohistochemical techniques. Histologically, the transitional layer was the same as that of the native bladder tissue; however, as with other non-seeded collagen matrices, the muscle layer was not fully developed. In vitro contractility studies performed on SIS-regenerated dog bladders showed a 50% decrease in maximal contractile response compared to that of normal bladder tissues.

Bladder augmentation with porcine acellular bowel tissue matrix, human placental membranes, or porcine SIS was performed using laparoscopic techniques in minipigs. At 12 weeks post-op, the grafts had contracted to 60% of their original size, and histologically, the grafts showed predominantly mucosal regeneration [67]. The human placental membranes, acellular tissue matrices, and SIS grafts persisted as well-vascularized fibrous bands, without evidence of significant inflammatory responses. These results suggest that a laparoscopic technique for partial bladder wall replacement using a cell-free graft is feasible. At 1 year, the hemicystectomy and bladder replacement with SIS group showed muscle tissue at the graft periphery and center; however, it consisted of small fused bundles with significant fibrosis. Compared with primary bladder closure after hemicystectomy, there was no improvement in bladder capacity or compliance [68]. In this study, 12 minipigs underwent laparoscopic hemicystectomy. Then, six pigs underwent bladder reconstruction with SIS and ipsilateral ureteral reimplantation. In the SIS group, four out of five surviving pigs had unobstructed reimplanted ureters without evidence of hydroureteronephrosis, while one had a high-grade obstruction at the reimplantation site. Histopathologic studies 1 year after implantation revealed muscle at the graft periphery and center; however, it consisted of small fused bundles with significant fibrosis. Although nerves were present at the graft periphery and center, they were decreased in number. Compared to primary bladder closure, no increase in bladder capacity or compliance was observed.

Studies of acellular matrices that may provide the necessary environment to promote cell migration, growth, and differentiation have been conducted. Chin et al. identified and characterized the bioactive factors in decellularized BSM by using ELISA, Western blotting, and immunohistochemistry for its effective utilization in regenerative medicine [17]. At least ten growth factors, including VEGF, BMP4, PDGF-BB, KGF, TGF-beta 1, IGF, bFGF, EGF, and TGF-alpha, were detected in the decellularized BSM. The presence of collagen (types 1, 2, 3, 4), laminin, and elastin within the matrix was also demonstrated. Supplementation with soluble BSM extracts showed an obvious effect on cell proliferation. These findings demonstrate that the growth factors and extracellular matrix components in the BSM maintain biological activity even after decellularization and extraction, supporting the wide applicability of BSM in tissue regeneration. The identification and characterization of the growth factors and extracellular matrix components in BSM are a prerequisite for understanding tissue regeneration using this scaffold. A study by Kikuno et al. showed that grafts of acellular collagen matrices can be enhanced by adding growth factors to improve bladder regeneration [69]. They evaluated the combined effects of nerve growth factor (NGF) and vascular endothelial growth factor (VEGF) on regeneration of the BAMG in spinal cord injury (SCI)-mediated neurogenic bladder in rats. Bladder capacity and compliance were significantly increased in all BAMG groups 8 weeks after surgery compared with that before bladder replacement surgery. However, bladder capacity and compliance were much higher in the VEGF and NGF group than in the control, NGF alone, and VEGF alone groups. NGF had a significant synergistic effect on the development, differentiation, and functional restoration of the BAMG when administered with VEGF in neurogenic bladders. Therefore, NGF may be a useful cytokine for enhancing the regeneration of a functional bladder following acellular matrix grafting in a neurogenic rat model.

8.2.4 Hybrid or Composite Scaffolds

While it is difficult to find a single material that satisfies the numerous requirements for bladder regeneration, composite materials might have the necessary characteristics for use as scaffolds [70]. Tensile strength and elastic modulus are important properties for urine storage and expulsion, and changes in the viscoelastic properties of the bladder wall can lead to voiding dysfunction [71]. Therefore, composite scaffolds which have many benefits, including

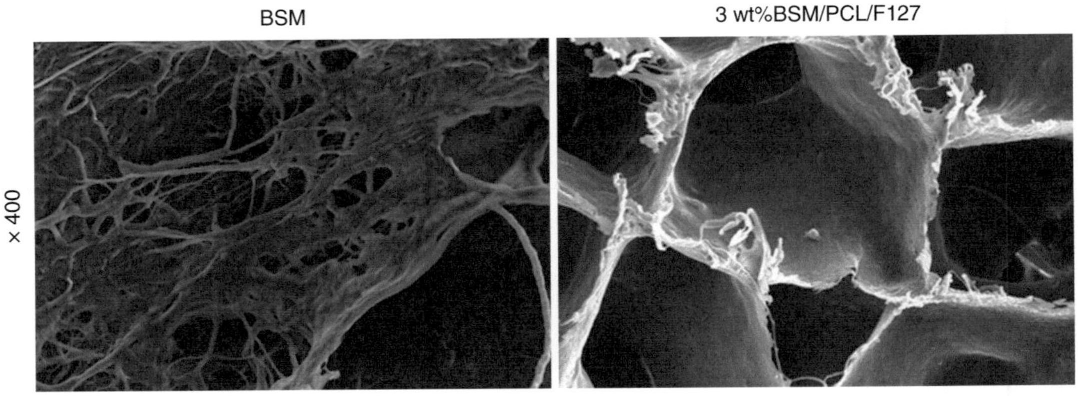

Fig. 8.1 Field emission scanning electron microscopy images of cell morphology on a 3 wt% BSM composite scaffold. For the analyses, upper urinary tract-derived urine stem cells were used. The BSM scaffold was used as the control. *PCL* polycaprolactone, *F127* Pluronic F127 *BSM* bladder submucosa matrix

reproducibility, large-scale production, and appropriate mechanical properties, as well as adjustable strength, degradation rate, and microstructure, may be useful synthetic polymers, because the desirable configuration is easily achieved. Pluronic F127 (F127) does not alter the properties of the scaffold, but it improves hydrophilicity and the microenvironment for cell attachment [72]. PCL was investigated as a potential substitute for bladder tissue reconstruction because of its flexibility, biocompatibility, stability, and resistance to resorption [57]. Jang et al. [1] fabricated a blend of PCL with F127 and BSM, which has been shown to be a nonimmunogenic, noncytotoxic, collagen-rich membrane that can be rapidly replaced by native tissues. In their study, the artificial polymer PCL/F127 was chosen for its natural elastic characteristics and physiochemical properties, which enable fabrication of a reservoir of sufficient volume, and naturally occurring BSM was used to improve its biocompatibility by increasing the proliferation of the urothelium and SMCs [1]. Testing for the proper PCL/F127/BSM ratio for the scaffold showed that a PCL/F127/3 wt% BSM composite scaffold exhibited significantly enhanced hydrophilicity, the surface was easily immobilized, and there was no evidence of teratoma formation in vivo. The cell morphology on the 3 wt% BSM composite scaffold was assessed by field emission scanning electron microscopy (FE-SEM; Fig. 8.1). The BSM scaffold was used as the control. The primary upper urinary tract-derived urine stem cells (uUSCs) on the 3 wt% BSM/F127 composite scaffold were flattened, with cell processes that extended to the surface of the scaffold. A similar morphology was seen in the control. FE-SEM analysis showed that the uUSCs were tightly attached to the surface of the scaffold and had extended into the pores on day 8, indicating that the BSM/PCL/F127 composite scaffold provided a suitable microenvironment. Evaluation of the in vivo tumorigenicity of the scaffold with or without uUSCs 8 weeks after implantation showed that the 3 wt% BSM composite scaffold did not cause teratomas. The tumorigenicity of the 3 wt% BSM/F127 composite scaffolds was evaluated by implanting the scaffold into the subcapsular space of the kidneys in ICR mice (Fig. 8.2). BSM was used as the control. Scaffolds seeded with uUSCs were also tested. After 8 weeks, H&E staining showed no signs of tumor formation in any scaffold, with or without cells. Evaluation of the in vivo tumorigenicity of the scaffold with or without uUSCs, 8 weeks after implantation, showed that the 3 wt% BSM composite scaffold did not cause teratomas. With respect to tumorigenicity, this result supports the safety of the scaffold with or without cells for therapeutic applications.

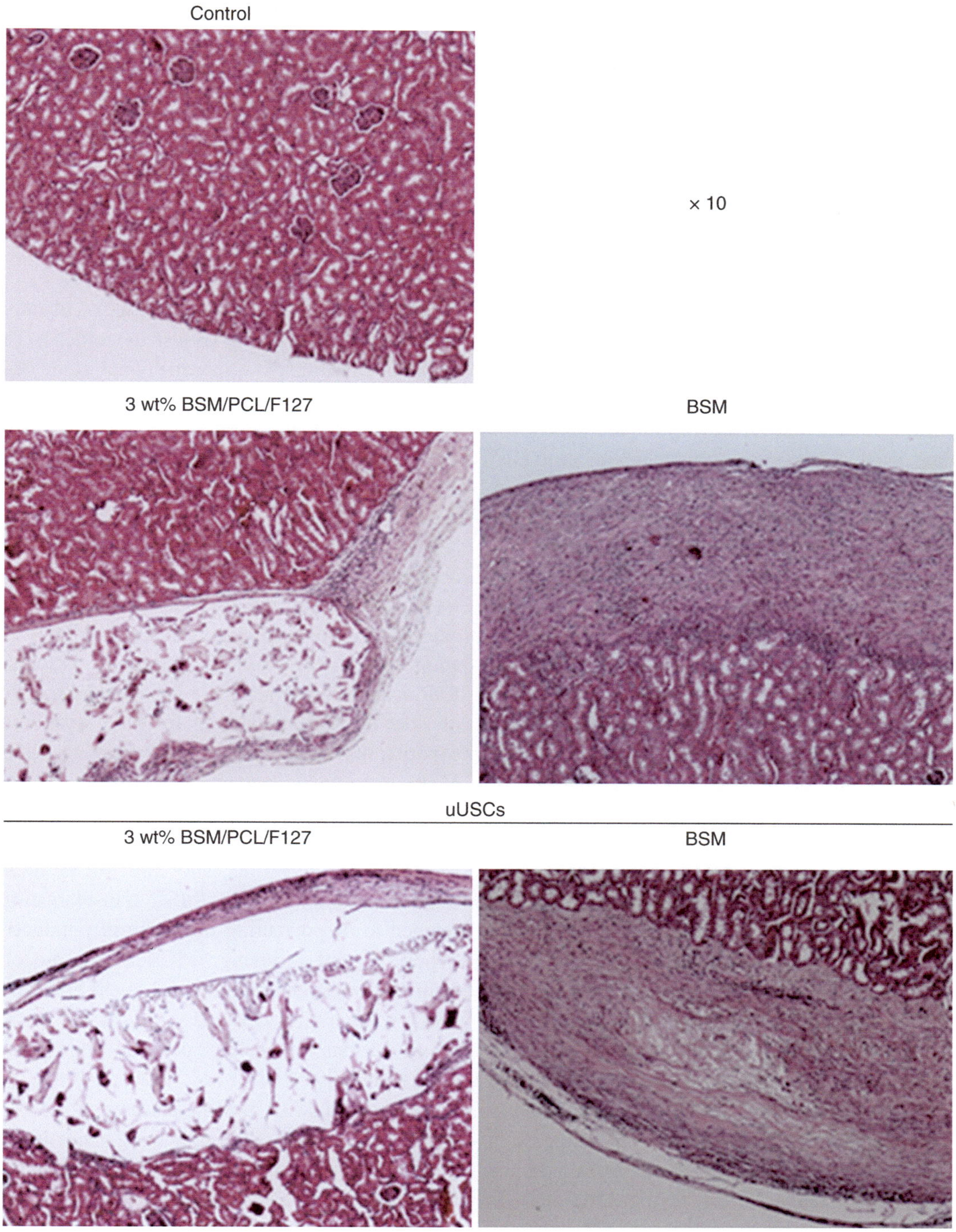

Fig. 8.2 Tumorigenicity analysis of the 3 wt% BSM composite scaffold. For this analysis, uUSCs were used. The BSM scaffold and a sham-operated kidney were used as controls. Control, non-scaffold-treated kidney

8.3 Bladder Regeneration Using Cell Transplantation

Regenerative medicine with selective cell transplantation may provide a means to create new functional bladder segments. The success of cell transplantation strategies for bladder reconstruction depends on the ability to efficiently use donor tissue and provide the right conditions for long-term survival, differentiation, and growth. Various cell sources have been explored for bladder regeneration. Cilento et al. [73] showed that native cells are preferable because they can be used without rejection. A simple method for harvesting bladder cell types from surgical specimens was used to generate normal human urothelial cell lines that could be reproducibly cultivated, passaged, and extensively expanded in serum-free medium. Immunostaining of the bladder epithelial cells with broadly reacting anti-cytokeratin antibodies and an anti-cytokeratin antibody specific to cytokeratin 7, a transitional cell marker, indicated that they expressed a stable epithelial phenotype during serial passaging. Low levels of immunostaining for E-cadherin and E-cadherin messenger ribonucleic acid by Northern blot analysis and strongly positive immunostaining for vimentin indicated that the uroepithelial cells express a non-barrier-forming phenotype under these culture conditions. However, when the urothelial cells were implanted subcutaneously into athymic mice on biodegradable synthetic polymers, they formed multilayered structures, suggesting that they retain the capability to differentiate in a living host. The urothelial cells proliferated in an epidermal growth factor-independent manner and expressed high levels of transforming growth factor-alpha and amphiregulin messenger ribonucleic acids, suggesting autocrine regulation of growth by epidermal growth factor-like factors. Cytogenetic analysis indicated that urothelial cells cultured for six passages possessed a normal chromosomal complement. These results demonstrate that primary cultures of autologous human bladder epithelial cells can be extensively expanded in vitro and, consequently, might be useful in cell transplantation strategies for genitourinary reconstruction. Amniotic fluid- and bone marrow-derived stem cells can also be used in an autologous manner and have the potential to differentiate into bladder muscle and urothelium [74–76]. Embryonic stem cells and uUSCs also have the potential to differentiate into bladder tissue [77, 78].

8.3.1 Mesenchymal Stem Cell

De Coppi et al. [74] set up a model of bladder acute necrotizing injury to test the efficacy of adult rat bone marrow mesenchymal stem cell (MSC) vs. fetal rat amniotic fluid MSC transplantation for the treatment of the impaired detrusor muscle contractility that develops when a cryoinjury is applied to the bladder wall. Impaired detrusor muscle contractility is part of the wound-healing process, which is characterized by SMC depletion, followed by SMC hyperplasia and tissue remodeling in the surviving bladder. At 30 days after transplantation, only a few fetal or adult MSCs gave rise to enteric or vascular SMCs, whereas most MSCs appeared incapable of specific differentiation. In vitro coculture experiments using SMCs with fetal or adult MSCs selectively labeled with distinct fluorochromes showed the presence of hybrid cells, suggesting that some MSCs can undergo cell fusion. Surprisingly, the major effect of rat bone marrow or amniotic fluid MSC transplantation seemed to be preventing the cryoinjury-induced hypertrophy of surviving SMCs. In this model, stem cell transplantation had a limited effect on SMC regeneration. Instead, it regulated post-injury bladder remodeling, possibly via a paracrine mechanism.

The potential risk of using diseased tissue as a source of cells for tissue engineering necessitates the investigation of other possible cell sources. One promising source is the bone marrow, which is a known source of stem cells for hematopoietic and mesenchymal lineages. Bone marrow-derived MSCs mobilize from the marrow in response to tissue damage and contribute to the normal regeneration process, and MSCs have been shown to localize in urinary bladders undergoing tissue regeneration after acellular bladder augmentation in rats [79]. Shukla et al. [67] characterized bone marrow-derived MSCs from pigs and demonstrated their ability to differentiate into SMCs and utility for autologous augmentation cystoplasty. MSCs were isolated from pigs and analyzed for common markers of MSCs by flow cytometry, and SMC differentiation was assessed by immunoblotting. MSCs were isolated, genetically labeled, expanded in vitro, seeded onto SIS, and used for autologous bladder augmentation. Porcine MSCs are morphologically and immunophenotypically similar to human MSCs. Culturing MSCs at low density enhances their proliferation rates, whereas maintenance at confluence consistently induces differentiation into mature SMCs. Labeled MSCs grew on SIS for 1 week in vitro and survived a 2-week implantation as an autologous bladder augment in vivo. Some SMC label-positive cells with typical SMC morphology were detected; however, most cells were SMC label negative. Notably, many cells with a urothelial morphology stained positively for SMC markers. Porcine MSCs have properties similar to those of MSCs from other species and consistently undergo differentiation into mature SMCs in vitro under specific culture conditions. The addition of MSCs to SIS may enhance tissue regeneration in augmentation cystoplasty; however, they may not be significantly incorporated into smooth muscle bundles.

Anumanthan et al. [76] reported directed differentiation of bone marrow-derived MSCs into bladder urothelium for use as a source of pluripotent or multipotent progenitor cells. The epithelium was separated from the mesenchymal shells of embryonic day 14 rat bladders. MSCs were isolated from mouse femoral and tibial bone marrow, and heterospecific recombinant xenografts were created by combining embryonic rat bladder mesenchymal shells with the MSCs and grafting them into the renal subcapsular space of athymic nude mice. Grafts were harvested at time points of up to 42 days and stained for urothelial and stromal differentiation. Histological examination of xenografts comprising mouse MSCs and rat embryonic rat bladder mesenchyma yielded mature bladder structures with normal microscopic architecture and expression of proteins confirming functional characteristics. Specifically, the induced urothelium expressed uroplakin, a marker of urothelial differentiation. These differentiated bladder structures also showed appropriate alpha-smooth muscle actin staining. Finally, Hoechst staining of the xenografts revealed a nuclear architecture consistent with a mouse mesenchymal stem cell origin in the urothelium, supporting the differentiation of these cells. In the appropriate signaling environment, bone marrow-derived MSCs can undergo directed differentiation toward endodermal-derived urothelium and develop into mature bladder tissue within a tissue recombination model. This model serves as an important tool for the study of bladder development with a long-term goal of cell replacement therapy applications.

8.3.2 Embryonic Stem Cell

In 1981, embryonic stem (ES) cells were isolated from mice for the first time [80]. This major breakthrough revolutionized the field of developmental biology. ES cells are capable of prolonged self-renewal and differentiation, providing a tool to investigate the molecular mechanisms occurring during differentiation from the embryo to adult. ES cells are considered to be pluripotent and can differentiate into almost all cell types that arise from the three embryonic germ layers [81]. In vitro, these cells can differentiate into multiple embryonic and adult cell types but rarely cells of endodermal lineage [82]. In contrast, differentiation of ES cells in an in vivo environment shows their full developmental potential. Whether stem and/or progenitor cells exist within the bladder is unknown, but hypothetically, they should exist. Identification of stem cells within the vast population of cells in the bladder would be challenging, especially without bladder-specific stem/progenitor cell markers. Oottamasathien et al. [77] determined the specific mesenchymal to ES cell ratios necessary to promote organ-specific differentiation while completely suppressing teratomatous growth. The embryonic mesenchyme is well established as an inductive tissue that dictates organ-specific programming of epithelial tissues, and this study showed that embryonic bladder mesenchyme can also drive ES cell differentiation toward endodermal-derived urothelium. These approaches allow us to capture specific stages of stem cell differentiation and better define stem cell hierarchies.

8.3.3 Urine-Derived Stem Cell

Urine-derived stem cells (USCs) consistently expressed MSC/pericyte markers and some key cell surface markers but no hematopoietic stem cell markers (except for MHC-1), endothelial markers (CD31), or human leukocyte antigen (locus) DR (HLA-DR) [83]. Compared to other MSCs, USCs have several advantages: (1) they can be collected using a simple, low-cost, safe, noninvasive procedure; (2) they display telomerase activity, and, thus, they are able to generate more cells; and (3) they can efficiently differentiate into SMCs, UCs, and endothelial cells. Chun and Kim et al. [70] investigated whether cells isolated from the upper urinary tract (UTCs) possess stem cell characteristics and could be used as an alternative cell source for patients with bladder cancer. Current tissue engineering approaches for urologic tissue regeneration require invasive tissue biopsies to obtain autologous cells, and these procedures are associated with various potential complications, such as donor site morbidity. Recently, cells isolated from voided urine (VUCs) have been proposed as an alternative cell source for urologic tissue engineering. However, VUCs should not be used in patients with bladder cancer, because the voided urine sample could contain malignant cells. In the study, urine samples were collected from the upper urinary tract of four male patients with bladder cancer using a ureteral catheter. The samples were centrifuged, and the cell pellets were plated for primary culture. The cells were analyzed for the number of colony-forming units, proliferation rate, cytogenetics, stem cell characteristics, and tumorigenicity, and the results were compared to those of VUCs collected from three healthy men. The UTCs were able to form colonies, had a greater proliferation rate than the VUCs, and had a normal karyotype. The UTCs possessed stem cell characteristics (expression of CD44+, CD73+, CD90+, CD105+, and SSEA4+) and expressed several markers of the urothelial, smooth muscle, and endothelial cell lineages. The UTCs did not form teratomas when implanted into the subcapsular space of a mouse kidney. Since the UTCs possessed stem cell characteristics, they could potentially be an alternative cell source for urologic tissue regeneration in patients with bladder cancer.

8.3.4 Induced Pluripotent Stem Cell

Induced pluripotent stem cells (iPSCs) are naturally programmed to divide continuously and remain undifferentiated. Although these cells can give rise to ectodermal, mesodermal, or endodermal cell lineages, a significant risk of teratoma exists. Any undifferentiated iPSCs placed in the body might continue to divide in an uncontrolled manner, forming tumors. In addition, it takes a long time (4 months) to derive and characterize iPSCs from an individual. Furthermore, the low efficiency of differentiation, genetic abnormalities, and high cost prohibit their clinical applicability. Despite this, a few studies of ESCs or iPSCs for bladder tissue engineering have been reported. Frimberger et al. [76] reported that human embryoid body-derived stem cells showed improved migration in the presence of mature human bladder SMCs and urothelial cells. In addition, Moad et al. [77] reported the generation of human iPSCs derived from normal and aging human urinary tract tissue. These iPSCs underwent bladder differentiation more efficiently than skin-derived iPSCs, as shown by the expression of urothelial-specific markers (uroplakins, claudins, and cytokeratin) and stromal smooth muscle markers (alpha-smooth muscle actin, calponin, and desmin), indicating the importance of organ-specific iPSCs for tissue-specific studies. Immobilized cell lines are not suitable for bladder regeneration due to safety concerns. Therefore, multipotent adult stem cells are currently used in bladder repair and reconstruction. Of particular interest is the paper published by Xue et al. [84], in which they describe a practical method to generate human iPSCs from USCs under feeder-free, virus-free, serum-free conditions without the c-MYC oncogene. The authors showed that this approach could be applied in a large population with different genetic backgrounds. USCs are easily accessible and exhibit high reprogramming efficiency, offering several advantages over other cell types used for iPSC generation. Using the approach described in this study, the authors generated 93 iPSC lines from 20 donors with diverse genetic backgrounds. The nonviral iPSC bank containing these cell lines is a valuable resource for iPSC research, facilitating future applications of human iPSCs. Table 8.1 shows a comparison of the various stem cell types used in bladder repair studies.

Table 8.1 Comparison of the various stem cell types used for bladder repair

Cell type/parameter	BMSCs	ASCs	USCs	ESC/iPSCs	Bladder SMCs and UCs
Self-renewal and expansion capability	Limited, PD ~30		High, PD 60–70	Very high, PD >200	Limited, PD <30
Multi-lineage differentiation capability	Multipotent but mainly limited to mesodermal cell lineages	Similar to BMSCs	Multipotent differentiation potential	Pluripotent (can generate all lineages)	None
Urothelial and endothelial differentiation capability	Low (<10%)	Low (10%)	High (60–85%)	Low	
Telomerase activity (TA)/ telomere length	Cannot be detected	Cannot be detected	Up to 75% of USC clones possess TA and relatively long telomeres	Possess TA and long telomeres	None
Harvesting approach	Invasive	Invasive	Noninvasive, simple, low cost, safe	Invasive to harvest somatic cells to generate iPSCs	Invasive
Pure stem cell isolation	Difficult	Difficult	Very easy	Easy	None
Number of stem cells harvested	MSC/10^4 bone marrow stromal cells in newborns, 1MSC/10^6		100–140 USC clones/24 h urine from adults		Unknown
Angiogenic trophic factors	Yes	Yes	Yes	Unknown	Moderate
Immunomodulatory properties	Yes	Yes	Yes	Unknown	Unknown
Rejection after implantation	No rejection as allogenous or xenogenous cells (e.g., human BMSCs or USCs) when implanted in rodent, rabbit, or canine models			Likely to be rejected	No rejection as autogenous cells
Oncogenic potential	No	No	No	Yes	None
Clinical trial utility	Potential	Potential	Potential	Safety concern	Yes

ASC adipose-derived stem cell, *BMSC* bone marrow-derived mesenchymal stromal cell, *ESC* embryonic stem cell, *iPSC* induced pluripotent stem cell, *MSC* mesenchymal stem cell, *PD* population doubling, *SMC* smooth muscle cell *UC* urothelial cell, *USC* urine-derived stem cell

8.4 Tissue Engineering Approach for Bladder Regeneration

Even in multiple studies, implantation of biomaterials without cells into the bladder has shown some promising results, especially the urothelial layer which was able to regenerate normally; however the regeneration of muscle layer was not fully developed [55, 62, 64, 66, 85]. Therefore, many investigators preferred tissue engineering approach (grafting biomaterials seeded with cells) for bladder tissue regeneration. In the early stage of investigation, synthetic polymer seeded with autologous cells was the most commonly used approach. The autologous urothelial and muscle cells can be expanded in vitro, seeded onto polymer scaffolds, and allowed to attach and form sheets of cells. The resulting cell-polymer scaffold can then be implanted in vivo. Histological analysis showed that viable cells were able to self-assemble back into their respective tissue types and retain their native phenotypes [86]. Synthetic polymer fibers of polyglycolic acid can serve as both a scaffold and delivery vehicle for the implantation of rabbit uroepithelial cells into athymic host animals. The polymers, which slowly degrade in vivo, allow the urothelial cells to survive at the implant site. In the abovementioned study [79], the authors demonstrated that polyglycolic acid polymers support the proliferation of rabbit urothelial cells in situ and can serve as a malleable substrate for the creation of new urological structures that replace the degrading polymer fibers. They also showed that when implanted on polyglycolic acid fibers, human urothelial and bladder muscle cells form new urological structures composed of both cell types. Human cell-polymer xenografts can be recovered from host animals at extended time points after implantation. These data suggest the feasibility of using polyglycolic acid polymers as substrates for the creation of human urothelial and muscle grafts for genitourinary reconstruction. These experiments demonstrated, for the first time, that composite-layered tissue-engineered structures could be created de novo.

8.4.1 Animal Models

Jayo et al. compared the in situ cellular responses to two biopolymer implants, a polylactic-co-glycolic acid-based biodegradable mesh scaffold with autologous urothelial cells and SMCs (construct) and a PLGA-based biodegradable mesh scaffold without cells (scaffold), in a canine model of augmentation cystoplasty [87]. Healing events were correlated with urodynamic assessments. Construct implants regenerated baseline urodynamics as early as 4 months after implantation. In contrast, following scaffold implantation, urodynamics failed to return to baseline by study termination (9 months). The functional differences elicited by the construct and scaffold implants were correlated with structural differences in the neo-tissues. The construct stroma had greater vascularization, with gently folded, interwoven connective tissue elements. Conversely, the scaffold stroma was dense, with haphazardly organized connective tissue. The urothelium was regenerated in response to both construct and scaffold implantation. However, only the construct urothelium had normal stroma, well-developed detrusor, and abundant alpha-smooth muscle actin cell staining at early time points, leading to a structurally and functionally complete bladder wall at 9 months. They concluded that early cellular and stromal events distinguished the healing processes that led to bladder wall regeneration or repair. Construct implants containing cells elicited early healing processes that culminated with the regeneration of complete mucosal and muscular components. In contrast, the response to scaffold implantation was consistent with reparative healing, i.e., mucosal growth, but incomplete tissue layer development. These independent studies demonstrate that cells are necessary to improve bladder function when a large bladder tissue implant is required.

The utility of allogenic bladder submucosa seeded with cells as a biomaterial for bladder augmentation was investigated by Yoo et al. in 1998 [62]. Partial cystectomies were performed in ten beagle dogs. Both urothelial and SMCs were harvested from five animals and expanded separately. Allogenic bladder submucosa obtained from sacrificed dogs was seeded with muscle cells on one side and urothelial cells on the opposite side. All beagles underwent cruciate cystotomies on the bladder dome. Augmentation cystoplasty was performed with the cell-seeded allogenic bladder submucosa in five animals and unseeded allogenic bladder submucosa in five animals. The augmented bladders were retrieved 2 and 3 months after augmentation. Bladders augmented with the cell-seeded allogenic bladder submucosa showed a 99% increase in capacity compared to bladders augmented with the cell-free allogenic bladder submucosa, which showed only a 30% increase in capacity. All dogs showed normal bladder compliance, as evidenced by urodynamic studies. Histologically, all retrieved bladders contained a normal cellular organization consisting of a urothelial-lined lumen surrounded by submucosal tissue and smooth muscle. Immunocytochemical analyses confirmed the urothelial and muscle cell phenotypes and showed the presence of nerve fibers. In summary, these matrices can function as vehicles for partial bladder regeneration, and no relevant antigenicity is evident.

It has been known for decades that the bladder can regenerate extensively over free grafts, as the urothelium has a high reparative capacity [88]. In their study, de Boer et al. investigated the spatiotemporal changes in the RNA and protein expression of growth factors and their receptors by in situ hybridization and immunocytochemistry during regeneration after acute injury of mouse urothelium. These expression data were well correlated with the changes in cell morphology and proliferation. Except for enhanced muscular transforming growth factor-beta 1 (TGF-beta 1) and TGF-beta type 2 receptor expression, the changes in the expression patterns of growth factors or receptors were confined to the urothelium. Increased mucosal RNA expression of insulin-like growth factor-2 (IGF-2) and particularly type 1 IGF receptor, as well as fibroblast growth factor-1 (FGF-1) but not FGF-2, coincided with reepithelialization and urothelial proliferation. High levels of urothelial TGF-beta 1 RNA and protein expression were associated with reepithelialization and differentiation. In addition, TGF-beta type 2 receptor protein expression was enhanced in the urothelium. Platelet-derived growth factor-A (PDGF-A) RNA was constitutively expressed in the mucosa, but expression decreased in the reepithelialization phase. These data are consistent with the notion that urothelial regeneration can be achieved through paracrine or autocrine mechanisms via urothelium-derived growth factors. The observation of analogous growth factor RNA expression patterns in regenerating skin epidermis suggests a more general growth factor-regulated mechanism for epithelial regeneration.

Bladder muscle tissue is less likely to regenerate normally. Both urothelial and muscle ingrowth are believed to be initiated from the edges of the normal bladder toward the region of the graft [89]. Regeneration of smooth muscle appears to take place within the fibrous tissue characteristically found when biodegradable collagen/Vicryl prosthesis is used to repair full-thickness defects in the rabbit urinary bladder. The question of whether the central smooth muscle was generated via myoblastic differentiation within the fibrous tissue or arose from healthy preexisting detrusor muscle was addressed by serial sectioning and specific staining. Only in situ transmutation, or differentiation, explains the observed morphology, and the results strongly suggest that the central smooth muscle was regenerated from within the repair area.

However, contracture or resorption of the graft is usually evident. The inflammatory response to the matrix may contribute to resorption of the free graft. It was hypothesized that building 3-D constructs in vitro prior to implantation might facilitate the eventual terminal differentiation of the cells after implantation while minimizing the inflammatory response toward the matrix, thus avoiding graft contracture and shrinkage. A study in dogs demonstrated a major difference between matrices with autologous cells (tissue-engineered matrices) and those without cells [62]. Matrices implanted with cells retained most of their implanted diameter, whereas matrices implanted without cells showed graft contraction and shrinkage. The histomorphology demonstrated a marked paucity of muscle cells and a more aggressive inflammatory reaction in the matrices implanted without cells.

To better address the functional parameters of tissue-engineered bladders, a canine animal model was designed that required a subtotal cystectomy and subsequent replacement with a tissue-engineered organ [90]. Cystectomy-only and non-seeded controls maintained average capacities of 22% and 46% of the preoperative values, respectively. In contrast, an average bladder capacity of 95% of the original precystectomy volume was achieved in the cell-seeded tissue-engineered bladder replacements. These findings were confirmed radiographically. Histologically, the non-seeded scaffold bladders presented a pattern of normal urothelial cells with a thickened fibrotic submucosa and a thin layer of muscle fibers. The retrieved tissue-engineered bladders showed a normal cellular organization, consisting of a trilayer of urothelium, submucosa, and muscle. These studies, performed with PGA-based scaffolds, have been repeated by other investigators, and similar results in long-term studies of large numbers of animals have been reported [85, 87]. Jayo et al. [85] evaluated bladder regeneration following 80% cystectomy and augmentation using a synthetic biopolymer with autologous urothelial and SMCs (autologous neo-bladder augmentation construct [construct]) or autotransplantation of native bladder (reimplanted native urinary bladder [reimplant]) in canines. Voiding function, urodynamic assessment, and neo-organ capacity-to-body-weight ratio (C/BW) were assessed longitudinally for 24 months following trigone-sparing augmentation cystoplasty in juvenile canines. Within 30 days post-implantation, hematology and urinalysis returned to baseline. Both the constructs and reimplants yielded neo-organs with statistically equivalent urodynamics and histology. Linear regression analysis of C/BW showed that the constructs regained baseline slope and continued to adapt with animal growth. Constructs and reimplants regained and maintained native bladder histology by 3 months, capacity at 3–6 months, and compliance by 12–24 months. Furthermore, the construct C/BW demonstrated the ability of the regenerated bladder to respond to growth regulation. However, not all scaffolds perform well when used to replace a large portion of the bladder. In a study using SIS for subtotal bladder replacement in dogs, both the unseeded and cell-seeded experimental groups showed graft shrinkage and poor results [91]. In the study, 22 male dogs had a 90% partial cystectomy and were divided into three groups. At 1 month after cystectomy, dogs in the unseeded ($n = 6$) and seeded ($n = 6$) groups received a bladder augmentation with a corresponding SIS graft. The dogs in the surgical control group ($n = 10$) received no further surgery. All dogs were evaluated before and after surgery with blood chemistry, urine culture, intravenous urography, cystogram, and cystometrogram. After surgery (at 1, 5, and 9 months), the bladders were examined by routine histology and immunohistochemistry. All 22 dogs survived the subtotal cystectomy, and 18 survived to the end of their intended survival period. One dog in the seeded group died 1 month after augmentation due to a bladder perforation caused by a large piece of incompletely absorbed SIS. Three other dogs (two in the unseeded group and one in the seeded group) died within 2 months after augmentation due to bladder obstruction by stones. Unseeded and seeded SIS grafts showed moderate to heavy adhesion and graft shrinkage, and some had bone and calcification at the graft site. In both groups, histology showed limited bladder regeneration. Interestingly, dogs in the control group at 1 month after cystectomy (when the seeded and unseeded groups received their augmentations) had severely shrunken bladders and histologically showed severe inflammation, fibroblast infiltration, and muscle hypertrophy. These results verify the subtotal cystectomy model. The use of seeded or unseeded SIS in a subtotal cystectomy model does not yield the same quality and quantity of bladder regeneration observed in the 40% non-inflammatory cystectomy model. This study provides important insights into the process of regeneration in a severely damaged bladder. These results led us to reevaluate the critical elements required for complete bladder replacement using tissue engineering.

The type of scaffold used is critical for the success of tissue engineering-based bladder replacement. The use of bioreactors, wherein mechanical stimulation is initiated at organ production, has also been proposed as an important parameter for success [92, 93]. Farhat and Yegar [92] reported that mechanical stimulation may have a role in urinary bladder tissue engineering. Currently, tissue engineering of the urinary bladder relies on biocompatible scaffolds that deliver biological and physical functionality with negligible immunogenic or tumorigenic risks, and recent research suggests that autologous cells propagated in culture and seeded on scaffolds prior to implantation improve clinical outcomes. In addition, as normal urinary bladder development in utero requires regular filling and emptying, current research suggests that bladders constructed in vitro may also benefit from regular mechanical stimulation. Such stimulation appears to induce favorable cellular changes, proliferation, and the production of structurally suitable extracellular matrix (ECM) components that are essential for the normal function of hollow dynamic organs. To mimic in vivo urinary bladder dynamics, tissue bioreactors that imitate the filling and emptying of a normal bladder have been devised. A "urinary bladder tissue bioreactor" that is able to recapitulate these dynamics while providing a cellular environment that facilitates the normal cell-cell and cell-matrix interactions may be necessary to successfully engineer bladder tissue. Validation of a urinary bladder tissue bioreactor that permits careful control of physiological conditions will generate broad interest from researchers in urinary bladder physiology and tissue engineering. A similar study was conducted by Bouhout et al. [86], showing a bladder substitute that was reconstructed in a physiological pressure environment. Bladder reconstruction by enterocystoplasty or with bioengineered substitutes is still associated with complications, which led us to develop an autologous vesical equivalent (VE). This model has already proven its structural conformity. The current challenge is to reconstruct our model in a more physiological environment, with the use of a bioreactor that mimics the dynamics of bladder filling and emptying, to acquire the proper physiological properties. In our model, fibroblasts and urothelial cells were evolved in a 3-D culture to obtain a reconstructed VE. This was then cultured in our bioreactor, which delivers a cyclic pressure increase up to 15 cm H_2O, followed by a rapid decrease, to achieve a dynamically cultured VE (dcVE). The dcVE was characterized by histology and immunofluorescence and compared to the characteristics of statically cultured VE. Mechanical resistance was evaluated by uniaxial tensile tests, and permeability was measured with 14C-urea. Compared to our static model, the dynamic model led to a urothelium profile similar to that of native bladder. Permeability analysis showed a profile comparable to that of native bladder, coinciding with the basal cell organization in the dcVE, and appropriate resistance for suturing and handling was also shown. This new alternative method offers a promising avenue for regenerative medicine. It is distinguished by its autologous character and efficiency as a urea barrier. These properties could significantly reduce inflammation, necrosis, and possibly rejection.

8.4.2 Human Models

Clinical trials of engineered bladder tissue for cystoplasty reconstruction began in 1998. The first was a small pilot study of seven patients using either a cell-seeded collagen scaffold (with or without omentum coverage) or a combined PGA-collagen cell-seeded scaffold with omental coverage. The patients who underwent reconstruction with the engineered bladder tissue created with the PGA-collagen cell-seeded scaffolds with omental coverage showed increased compliance, decreased end-filling pressure, increased capacity, and longer dry periods over time [12]. In this study, the outcomes were measured by serial urodynamics, cystograms, ultrasounds, bladder biopsies, and serum analyses, and the average follow-up was 46 months (range, 22–61 months). Postoperatively, the mean bladder leak point pressure decreases at capacity, and the greatest increase in volume and compliance was observed in the composite engineered bladders with an omental wrap (56%, 1.58-fold and 2.79-fold, respectively). Bowel function returned promptly after surgery. No metabolic consequences were noted, urinary calculi did not form, mucus production was normal, and renal function was preserved. The engineered bladder biopsies showed an adequate structural architecture and phenotype. Based on these results, engineered bladder tissues created with autologous cells seeded on collagen-polyglycolic acid scaffolds and wrapped in omentum after implantation can be used in patients who require cystoplasty. Although these results are promising since they show that engineered tissues can be implanted safely, it is just a start in terms of accomplishing the goal of engineering fully functional bladders. This was a limited clinical experience, and the technology is not yet ready for wide dissemination; further experimental and clinical studies are required, and phase 2 studies have been completed.

8.4.3 Neo-urinary Conduits

From the aforementioned and the urethral studies, it is evident that the use of cell-seeded matrices is superior to non-seeded matrices for the creation of engineered bladder tissues. Although advances have been made in bladder tissue engineering, many challenges remain. Much of the current research is aimed at the development of biologically active, "smart" biomaterials that may improve tissue regeneration. Similar engineering techniques are now being used in patients with bladder cancer who are having engineered urinary conduits implanted after cystectomy [94]. Muscle-invasive and recurrent non-muscle-invasive bladder cancers have been traditionally treated with a radical cystectomy and urinary diversion. The urinary diversion is generally accomplished through the creation of an incontinent ileal conduit, continent catheterizable reservoir, or orthotopic neo-bladder utilizing the small or large intestine. While radical extirpation of the bladder is often successful from an oncological perspective, there is significant morbidity associated with enteric interposition within the genitourinary tract. Therefore, there is a great opportunity to decrease the morbidity associated with the current surgical management of bladder cancer by utilizing novel technologies to create a urinary diversion without the intestine. Clinical trials using neo-urinary conduits (NUC) seeded with autologous SMCs are currently in progress and may offer a significant surgical advance by eliminating the complications associated with the use of gastrointestinal segments in urinary reconstruction, simplifying the surgical procedure, and greatly facilitating recovery from cystectomy. A conduit from the ureters to the skin surface addresses the current standard of care while simplifying the surgical procedure, and it may also improve patient outcomes. The NUC created by Tengion serves as a template to catalyze the regeneration of native-like urinary tissue that can connect the ureters to the skin surface. To ensure native urinary tissue regeneration, a biocompatible, biodegradable scaffold with an extended history of safety and clinical utility is necessary. The broadly used PLGA scaffold can enhance tissue regeneration and promote neo-tissue integration when properly seeded with SMCs. The NUC construct has two principal components. The first is the biomaterials. The NUC scaffold is composed of a PGA polymer mesh fashioned into the required tubular shape and coated with a 50/50 blend of PLGA copolymer. The specific structural parameters of the construct can be modified during the surgical procedure according to the patient's needs. The choice of well-established, synthetic, degradable biopolymers reflects the same requirements for reliability and reproducibility inherent in the choice of these polymers for applications in other bladder-related neo-organs. The second component is the cells. Autologous SMCs sourced from bladder or non-bladder tissue may be applied for NUC construction. Based on the successful outcomes in a porcine cystectomy model, Tengion has initiated phase I clinical trials of NUC constructs in human patients requiring urinary diversion. This phase I study, "Incontinent Urinary Diversion Using an

Autologous Neo-Urinary Conduit" (http://www.clinicaltrials.gov/ct2/show/NCT01087697), is currently recruiting patients, with the objective of implanting up to ten patients by the end of 2012. The objective of the study is to evaluate if NUC constructs made using autologous adipose-derived SMCs in combination with defined degradable biomaterial scaffolds can form a functional conduit to safely facilitate passage of urine from the kidneys subsequent to radical cystectomy. Primary outcome indices over a 12-month postimplantation period include structural integrity and conduit patency. CT scans will be used to demonstrate that urine flows safely through the NUC construct. Additional measures of primary outcomes up to 12-month postimplantation include evaluation of any product- or procedure-related adverse events. Similarly, secondary outcome indices will include analysis of NUC structural integrity and patency over a 12–60-month postimplantation period. CT scan and renal ultrasound will be applied to demonstrate that urine flows safely through the NUC construct up to 60 months after implantation. Procedural- and product-related adverse events will also be monitored up to 60 months after implantation. Finally, the overall safety of the NUC construct will be assessed by evaluation of nonproduct-/procedural-related adverse events and patient vital signs.

Stem cells derived from fat can be differentiated into smooth muscle for use in the conduits, thus avoiding the use of bladder cells from bladder cancer patients [95]. Basu et al. [88] described the isolation and characterization of SMCs from porcine adipose tissue and peripheral blood that are phenotypically and functionally indistinguishable from bladder-derived SMCs. In a preclinical Good Laboratory Practice study, we demonstrated that autologous adipose- and peripheral blood-derived SMCs can be used to seed synthetic, biodegradable, tubular scaffold structures and that implantation of these seeded scaffolds into a porcine cystectomy model leads to successful de novo regeneration of a tubular neo-organ composed of urinary-like neo-tissue that is histologically identical to native bladder. The ability to create urologic structures de novo from scaffolds seeded with autologous adipose- or peripheral blood-derived SMCs will greatly facilitate the translation of urologic tissue engineering technologies into clinical practice.

8.4.4 Kyungpook National University Experiences

In a very important study, Lee et al. [96] investigated the synergistic effect of human USCs and a surface-modified composite scaffold for bladder reconstruction in a rat model. The composite scaffold (PCL/F127/3 wt% BSM) was fabricated using an immersion precipitation method, and heparin was immobilized on the surface via covalent conjugation. A PCL pellet/Pluronic F127 powder mixture (95/5 [w/w]) was dissolved in tetraglycol (12 wt%), and BSM powder was evenly mixed with the polymer solution. The mixed solution was poured into a polytetrafluoroethylene mold (70 × 70 × 0.4 μL) and then immersed in water for 1 h at room temperature. After additional washing and vacuum drying, the PCL/F127/BSM composite scaffold was sterilized with ethanol. The PCL/F127/3 wt% BSM composite scaffold exhibited significantly enhanced hydrophilicity, the surface was easily immobilized, and there was no evidence of teratoma formation in vivo [1]. Basic fibroblast growth factor (bFGF) was loaded onto the heparin-immobilized scaffold by a simple dipping method. bFGF has been shown to stimulate the proliferation and survival of both SMCs and urothelial cells [97], suggesting the benefit of bFGF-loaded scaffolds for urological tissue engineering applications. To fabricate a scaffold for bFGF delivery, the authors covalently conjugated heparin to the surface of a scaffold to form a heparin-immobilized scaffold, which was then loaded with bFGF (scaffold$^{\text{heparin-bFGF}}$) [98]. Urine samples from the upper urinary tract were obtained from a 52-year-old female patient. The urine samples (100 mL each) were centrifuged, and the cell pellets were washed with PBS. The cells were cultured in a mixture of keratinocyte serum-free medium and progenitor cell medium (Gibco-Invitrogen, Grand Island, NY, USA) in a

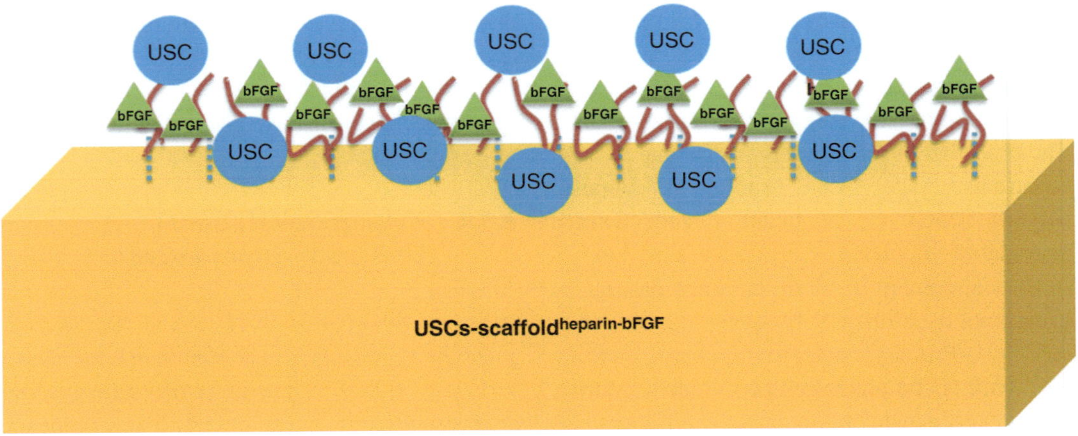

Fig. 8.3 Schematic diagram of the USC-seeded composite scaffold$^{heparin-bFGF}$ graft. *USC* urine-derived stem cell, *bFGF* basic fibroblast growth factor

1:1 ratio. USCs have been proposed as an alternative stem cell source for urological tissue reconstruction since they have mesenchymal stem cell characteristics and have the capacity to differentiate into a variety of urological cell lineages [78]. Figure 8.3 shows a schematic diagram of the USC-seeded composite scaffold$^{heparin-bFGF}$ graft. Twenty-five rats were divided into five groups: (1) control (sham operated), (2) partial cystectomy group (an ~40% defect was created in the dome of the bladder wall), (3) scaffold (an unmodified scaffold was attached after partial cystectomy), (4) scaffold$^{heparin-bFGF}$ (the heparin-immobilized bFGF-loaded scaffold was attached after partial cystectomy), and (5) USC-scaffold$^{heparin-bFGF}$ (scaffold$^{heparin-bFGF}$ combined with 1×10^4 USCs was attached after partial cystectomy) groups. The single-layer scaffold (disk shaped, 6 mm diameter) was sutured as a patch onto the bladder defect with 7-0 Vicryl sutures. The omentum was loosely wrapped over the graft and fixed with fibrin glue (Greenplast; Green Cross, Seoul, Korea) (Fig. 8.4). In maximal bladder capacity and compliance analyses at 8 weeks postoperation, the USC-scaffold$^{heparin-bFGF}$ group showed significant functional improvement (2.34 ± 0.25 mL and 55.09 ± 11.81 μL/cm H_2O) compared to the other groups (2.60 ± 0.23 mL and 56.14 ± 9.00 μL/cm H_2O, control group; 1.46 ± 0.18 mL and 34.27 ± 4.42 μL/cm H_2O, partial cystectomy group; 1.76 ± 0.22 mL and 35.62 ± 6.69 μL/cm H_2O, scaffold group; and 1.92 ± 0.29 mL and 40.74 ± 7.88 μL/cm H_2O, scaffold$^{heparin-bFGF}$ group; Table 8.2). In the histological and immunohistochemical analyses, the USC-scaffold$^{heparin-bFGF}$ group showed pronounced, well-differentiated, and organized smooth muscle bundle formation, a multilayered and pan-cytokeratin-positive urothelium, and high condensation of the submucosal area. The implanted scaffolds were not visible in the tissue sections because they were dissolved by xylene during specimen processing. The histological features of the implanted grafts showed that the regenerated portions of the bladders in the USC-scaffold$^{heparin-bFGF}$ group exhibited pronounced smooth muscle bundles, a multilayered urothelium, condensed submucosa layer formation, and restored bladder volume. These anatomical reconstructions are essential for functional compliance. However, the other scaffold groups showed only weak SMC bundles, a thin urothelium, and loose submucosa regeneration at the graft. These results suggest that the seeded USCs contributed to tissue regeneration. While the seeded cells were only expected to survive for 2 weeks in vivo [99], these exogenous cells were more effective for bladder regeneration than the cells recruited from the surrounding host tissues or circulating blood flow. Although the USC regenerative mechanism was not determined in this study, based on previous reports, regeneration is presumed to result from

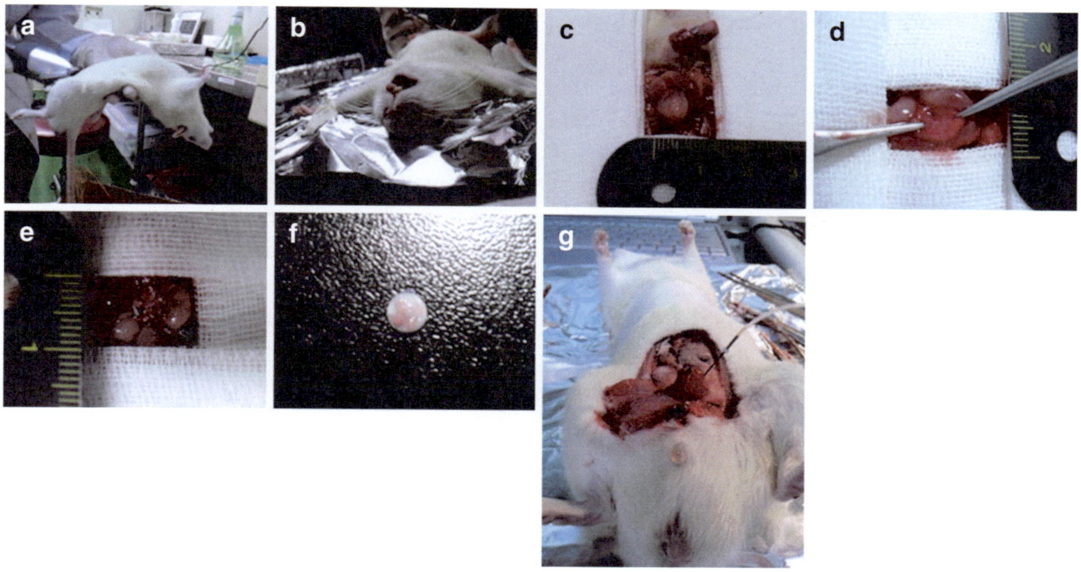

Fig. 8.4 Bladder reconstruction procedure using the USC-scaffold$^{heparin\text{-}bFGF}$. (**a**) Rats were anesthetized with an intraperitoneal injection of ketamine (40 mg/kg) and Rompun (10 mg/kg), with periodic supplementation as needed. (**b**) The rats were placed on the operating board in a supine position in laminar flow. (**c**) After shaving and sterilization with povidone iodine scrub over the suprapubic region, the abdomen was opened through a 4-cm longitudinal midline incision, and the urinary bladder was exposed. (**d**) The upper half of the bladder (dome and upper portion) was transected and removed with scissors. (**e**) The single-layer scaffold (disk form; diameter, 6 mm) was sutured as a patch onto the defect with 7-0 Vicryl sutures. (**f**) Representative picture of the scaffold and USC-scaffold$^{heparin\text{-}bFGF}$. (**g**) A urodynamic study (filling cystometry). The bladder was filled with PBS, and maximal capacity was defined as the minimum infusion volume that triggered urine leakage. Compliance was defined as the maximal capacity/(pressure that triggered leakage)—(baseline pressure). *USC* urine-derived stem cell, *bFGF* basic fibroblast growth factor

Table 8.2 Measurement of cross-sectional area and urodynamics, including maximal bladder capacity and compliance

	Control	Partial cystectomy	Scaffold	Scaffold$^{heparin\text{-}bFGF}$	USC-scaffold$^{heparin\text{-}bFGF}$
Cross-sectional area, mm^2 (mean ± SD)	16.85 ± 1.21	7.87 ± 1.37	9.67 ± 0.87	11.19 ± 0.87	15.71 ± 1.34
Maximal bladder capacity, mL (mean ± SD)	2.60 ± 0.23	1.46 ± 0.18	1.76 ± 0.22	1.92 ± 0.29	2.34 ± 0.25
Compliance, μL/cm H_2O (mean ± SD)	56.14 ± 9.00	34.27 ± 4.42	35.62 ± 6.69	40.74 ± 7.88	55.09 ± 11.81

USC urine-derived stem cell, *bFGF* basic fibroblast growth factor

the paracrine effects of trophic factors secreted from the USCs [99]. Thus, the transplanted stem cells influenced the surrounding host cells, resulting in improved cell migration and differentiation into the target cell type. The USC-seeded scaffold$^{heparin\text{-}bFGF}$ exhibited significantly increased bladder capacity, compliance, regeneration of smooth muscle tissue, multilayered urothelium, and condensed submucosa layers in the study. Based on these results, the USC-scaffold$^{heparin\text{-}bFGF}$ would be ideal for bladder reconstruction.

8.4.5 Whole Bladder Reconstruction

An interesting study was published by Hoogenkamp et al [100]. They showed that scaffolds made from molecularly defined biomaterials are instrumental in the regeneration of tissues but are generally confined to small flat patches and do not comprise the whole organ. In this study, a simple, one-step casting method was developed to produce a seamless, large, hollow, collagen-based scaffold, mimicking the shape of the whole bladder with integrated anastomotic sites for ureters and the urethra. This hollow bladder scaffold is highly standardized with uniform wall thickness and a unidirectional pore structure to facilitate cell infiltration in vivo. Human and porcine bladder urothelial cells and SMCs were able to attach to the scaffold and maintained their phenotypes in vitro. The closed luminal side and the porous outside of the scaffold facilitated the formation of a urothelial lining and infiltration of SMCs, respectively. The cells aligned according to the scaffold template. The technology used is highly adjustable in terms of shape, size, and materials and could be used as a starting point for research aimed at an off-the-shelf medical device for neo-bladders.

Tissue engineering of the bladder may become a reality in the future. It is important to design a preclinical study using the best predictive model with an adequate disease background to mimic the clinical situation. For bladder tissue engineering, large animals with diseased bladders appear to represent the best experimental model, which should aid in the development of clinically applicable tissue-engineered bladder augmentation or replacement with satisfactory long-term outcomes.

8.5 Conclusions and Perspectives

To prevent the problematic outcomes of bowel-based bladder reconstructive surgery, such as prosthetic tumors and systemic metabolic complications, the research focus has switched from regenerating and strengthening the failing organ or building an organ replacement in the 1990s to regenerative medicine and tissue engineering. Various types of acellular matrices, naturally derived materials, and synthetic polymers have been used for either unseeded (cell-free) or autologous cell-seeded tissue engineering scaffolds. Different categories of cell sources, from autologous differentiated urothelial cells and SMCs to natural or laboratory-derived stem cells, have been tested to obtain suitable "cell-seeded" templates. The current clinically validated bladder tissue engineering approaches essentially consist of augmentation cystoplasty in patients suffering from poorly compliant neuropathic bladder. There have been no clinical applications of whole tissue-engineered neo-bladder for radical-reconstructive surgical treatment of bladder malignancies or chronic inflammation due to vesical coarctation. The reasons why bladder tissue engineering has not yet been clinically applied include the risk of graft ischemia and subsequent fibrous contraction and perforation. The generation of a graft vascular network (vasculogenesis), together with the promotion of surrounding vessel sprouting (angiogenesis), could allow an effective graft blood supply and avoid serious, ischemia-related complications.

Current research suggests that the use of biomaterial-based, bladder-shaped scaffolds seeded with autologous urothelial cells and SMCs is the best option for bladder tissue engineering. Further studies to develop novel biomaterials and identify cell sources, as well as the information gained from research in developmental biology, signal transduction, and wound healing, would be beneficial.

References

1. Jang YJ, Chun SY, Kim GN, Kim JR, Oh SH, Lee JH, et al. Characterization of a novel composite scaffold consisting of acellular bladder submucosa matrix, polycaprolactone and Pluronic F127 as a substance for bladder reconstruction. Acta Biomater. 2014;10:3117–25.
2. Hautmann RE, Hautmann SH, Hautmann O. Complications associated with urinary diversion. Nat Rev Urol. 2011;8:667–77.
3. Nieuwenhuijzen JA, de Vries RR, Bex A, van der Poel HG, Meinhardt W, Antonini N, et al. Urinary diversions

after cystectomy: the association of clinical factors, complications and functional results of four different diversions. Eur Urol. 2008;53:834–42. discussion 42-4
4. Jednak R. The evolution of bladder augmentation: from creating a reservoir to reconstituting an organ. Front Pediatr. 2014;2:10.
5. Cartwright PC, Snow BW. Bladder autoaugmentation: early clinical experience. J Urol. 1989;142:505–8. discussion 20-1
6. Cartwright PC, Snow BW. Bladder autoaugmentation: partial detrusor excision to augment the bladder without use of bowel. J Urol. 1989;142:1050–3.
7. Adams MC, Brock JW 3rd, Pope JC, Rink RC. Ureterocystoplasty: is it necessary to detubularize the distal ureter? J Urol. 1998;160:851–3.
8. Lailas NG, Cilento B, Atala A. Progressive ureteral dilation for subsequent ureterocystoplasty. J Urol. 1996;156:1151–3.
9. Satar N, Yoo JJ, Atala A. Progressive dilation for bladder tissue expansion. J Urol. 1999;162:829–31.
10. Langer R, Vacanti JP. Tissue engineering. Science. 1993;260:920–6.
11. Sloff M, Simaioforidis V, de Vries R, Oosterwijk E, Feitz W. Tissue engineering of the bladder–reality or myth? A systematic review. J Urol. 2014;192:1035–42.
12. Atala A, Bauer SB, Soker S, Yoo JJ, Retik AB. Tissue-engineered autologous bladders for patients needing cystoplasty. Lancet. 2006;367:1241–6.
13. Aboushwareb T, McKenzie P, Wezel F, Southgate J, Badlani G. Is tissue engineering and biomaterials the future for lower urinary tract dysfunction (LUTD)/pelvic organ prolapse (POP)? Neurourol Urodyn. 2011;30:775–82.
14. Petrovic V, Stankovic J, Stefanovic V. Tissue engineering of the urinary bladder: current concepts and future perspectives. ScientificWorldJournal. 2011;11:1479–88.
15. Shokeir AA, Harraz AM, El-Din AB. Tissue engineering and stem cells: basic principles and applications in urology. Int J Urol. 2010;17:964–73.
16. El-Taji OM, Khattak AQ, Hussain SA. Bladder reconstruction: The past, present and future. Oncol Lett. 2015;10:3–10.
17. Chun SY, Lim GJ, Kwon TG, Kwak EK, Kim BW, Atala A, et al. Identification and characterization of bioactive factors in bladder submucosa matrix. Biomaterials. 2007;28:4251–6.
18. Chamley-Campbell J, Campbell GR, Ross R. The smooth muscle cell in culture. Physiol Rev. 1979;59:1–61.
19. Akhyari P, Fedak PW, Weisel RD, Lee TY, Verma S, Mickle DA, et al. Mechanical stretch regimen enhances the formation of bioengineered autologous cardiac muscle grafts. Circulation. 2002;106:I137–42.
20. Koh GY, Klug MG, Soonpaa MH, Field LJ. Differentiation and long-term survival of C2C12 myoblast grafts in heart. J Clin Invest. 1993;92:1548–54.
21. Chung SY, Krivorov NP, Rausei V, Thomas L, Frantzen M, Landsittel D, et al. Bladder reconstitution with bone marrow derived stem cells seeded on small intestinal submucosa improves morphological and molecular composition. J Urol. 2005;174:353–9.
22. Zhang Y, Lin HK, Frimberger D, Epstein RB, Kropp BP. Growth of bone marrow stromal cells on small intestinal submucosa: an alternative cell source for tissue engineered bladder. BJU Int. 2005;96:1120–5.
23. Lu SH, Sacks MS, Chung SY, Gloeckner DC, Pruchnic R, Huard J, et al. Biaxial mechanical properties of muscle-derived cell seeded small intestinal submucosa for bladder wall reconstitution. Biomaterials. 2005;26:443–9.
24. Jack GS, Almeida FG, Zhang R, Alfonso ZC, Zuk PA, Rodriguez LV. Processed lipoaspirate cells for tissue engineering of the lower urinary tract: implications for the treatment of stress urinary incontinence and bladder reconstruction. J Urol. 2005;174:2041–5.
25. Gleeson MJ, Griffith DP. The use of alloplastic biomaterials in bladder substitution. J Urol. 1992;148:1377–82.
26. Kropp BP, Eppley BL, Prevel CD, Rippy MK, Harruff RC, Badylak SF, et al. Experimental assessment of small intestinal submucosa as a bladder wall substitute. Urology. 1995;46:396–400.
27. Kropp BP, Cheng EY, Lin HK, Zhang Y. Reliable and reproducible bladder regeneration using unseeded distal small intestinal submucosa. J Urol. 2004;172:1710–3.
28. Kanematsu A, Yamamoto S, Ogawa O. Changing concepts of bladder regeneration. Int J Urol. 2007;14:673–8.
29. Lewis JM, Cheng EY. Non-traditional management of the neurogenic bladder: tissue engineering and neuromodulation. ScientificWorldJournal. 2007;7:1230–41.
30. Yamzon J, Perin L, Koh CJ. Current status of tissue engineering in pediatric urology. Curr Opin Urol. 2008;18:404–7.
31. Roth CC, Mondalek FG, Kibar Y, Ashley RA, Bell CH, Califano JA, et al. Bladder regeneration in a canine model using hyaluronic acid-poly(lactic-co-glycolic-acid) nanoparticle modified porcine small intestinal submucosa. BJU Int. 2011;108:148–55.
32. Kelami A. Lyophilized human dura as a bladder wall substitute: experimental and clinical results. J Urol. 1971;105:518–22.
33. Fishman IJ, Flores FN, Scott FB, Spjut HJ, Morrow B. Use of fresh placental membranes for bladder reconstruction. J Urol. 1987;138:1291–4.
34. Atala A, Vacanti JP, Peters CA, Mandell J, Retik AB, Freeman MR. Formation of urothelial structures in vivo from dissociated cells attached to biodegradable polymer scaffolds in vitro. J Urol. 1992;148:658–62.
35. Kambic H, Kay R, Chen JF, Matsushita M, Harasaki H, Zilber S. Biodegradable pericardial implants for bladder augmentation: a 2.5-year study in dogs. J Urol. 1992;148:539–43.

36. Kropp BP, Badylak S, Thor KB. Regenerative bladder augmentation: a review of the initial preclinical studies with porcine small intestinal submucosa. Adv Exp Med Biol. 1995;385:229–35.
37. Zhang Y, Kropp BP, Lin HK, Cowan R, Cheng EY. Bladder regeneration with cell-seeded small intestinal submucosa. Tissue Eng. 2004;10:181–7.
38. Harrington DA, Sharma AK, Erickson BA, Cheng EY. Bladder tissue engineering through nanotechnology. World J Urol. 2008;26:315–22.
39. Joseph DB, Borer JG, De Filippo RE, Hodges SJ, McLorie GA. Autologous cell seeded biodegradable scaffold for augmentation cystoplasty: phase II study in children and adolescents with spina bifida. J Urol. 2014;191:1389–95.
40. Subramaniam R, Hinley J, Stahlschmidt J, Southgate J. Tissue engineering potential of urothelial cells from diseased bladders. J Urol. 2011;186:2014–20.
41. Aboushwareb T, Atala A. Stem cells in urology. Nat Clin Pract Urol. 2008;5:621–31.
42. Aitken KJ, Bagli DJ. The bladder extracellular matrix. Part II: regenerative applications. Nat Rev Urol. 2009;6:612–21.
43. Soler R, Fullhase C, Atala A. Regenerative medicine strategies for treatment of neurogenic bladder. Therapy. 2009;6:177–84.
44. Chen BS, Xie H, Zhang SL, Geng HQ, Zhou JM, Pan J, et al. Tissue engineering of bladder using vascular endothelial growth factor gene-modified endothelial progenitor cells. Int J Artif Organs. 2011;34:1137–46.
45. Furth ME, Atala A, Van Dyke ME. Smart biomaterials design for tissue engineering and regenerative medicine. Biomaterials. 2007;28:5068–73.
46. Chan BP, Leong KW. Scaffolding in tissue engineering: general approaches and tissue-specific considerations. Eur Spine J. 2008;17(Suppl 4):467–79.
47. Atala A. Tissue engineering of human bladder. Br Med Bull. 2011;97:81–104.
48. Dahms SE, Piechota HJ, Dahiya R, Lue TF, Tanagho EA. Composition and biomechanical properties of the bladder acellular matrix graft: comparative analysis in rat, pig and human. Br J Urol. 1998;82:411–9.
49. Shakhssalim N, Dehghan MM, Moghadasali R, Soltani MH, Shabani I, Soleimani M. Bladder tissue engineering using biocompatible nanofibrous electrospun constructs: feasibility and safety investigation. Urol J. 2012;9:410–9.
50. Bergsma EJ, Rozema FR, Bos RR, de Bruijn WC. Foreign body reactions to resorbable poly(L-lactide) bone plates and screws used for the fixation of unstable zygomatic fractures. J Oral Maxillofac Surg. 1993;51:666–70.
51. Martin C, Winet H, Bao JY. Acidity near eroding polylactide-polyglycolide in vitro and in vivo in rabbit tibial bone chambers. Biomaterials. 1996;17:2373–80.
52. Zhang R, Ma PX. Poly(alpha-hydroxyl acids)/hydroxyapatite porous composites for bone-tissue engineering. I. Preparation and morphology. J Biomed Mater Res. 1999;44:446–55.
53. Lo H, Kadiyala S, Guggino SE, Leong KW. Poly(L-lactic acid) foams with cell seeding and controlled-release capacity. J Biomed Mater Res. 1996;30:475–84.
54. Nam YS, Yoon JJ, Park TG. A novel fabrication method of macroporous biodegradable polymer scaffolds using gas foaming salt as a porogen additive. J Biomed Mater Res. 2000;53:1–7.
55. Sutherland RS, Baskin LS, Hayward SW, Cunha GR. Regeneration of bladder urothelium, smooth muscle, blood vessels and nerves into an acellular tissue matrix. J Urol. 1996;156:571–7.
56. Monsour MJ, Mohammed R, Gorham SD, French DA, Scott R. An assessment of a collagen/vicryl composite membrane to repair defects of the urinary bladder in rabbits. Urol Res. 1987;15:235–8.
57. Yu DS, Lee CF, Chen HI, Chang SY. Bladder wall grafting in rats using salt-modified and collagen-coated polycaprolactone scaffolds: preliminary report. Int J Urol. 2007;14:939–44.
58. Beiser IH, Kanat IO. Biodegradable internal fixation. A literature review. J Am Podiatr Med Assoc. 1990;80:72–5.
59. Oh SH, Lee JH. Hydrophilization of synthetic biodegradable polymer scaffolds for improved cell/tissue compatibility. Biomed Mater. 2013;8:014101.
60. Atala A. This month in investigative urology: Commentary on the replacement of urologic associated mucosa. J Urol. 1996;156:338–9.
61. Atala A. Autologous cell transplantation for urologic reconstruction. J Urol. 1998;159:2–3.
62. Yoo JJ, Meng J, Oberpenning F, Atala A. Bladder augmentation using allogenic bladder submucosa seeded with cells. Urology. 1998;51:221–5.
63. Crapo PM, Gilbert TW, Badylak SF. An overview of tissue and whole organ decellularization processes. Biomaterials. 2011;32:3233–43.
64. Probst M, Dahiya R, Carrier S, Tanagho EA. Reproduction of functional smooth muscle tissue and partial bladder replacement. Br J Urol. 1997;79:505–15.
65. Rotthoff G, Braun B, Bilgin I, Mahjoub M, Zimmermann U, Zinganell K. Aortic replacement with multi-layer submucosa prostheses made from heterologous small intestine. J Cardiovasc Surg (Torino). 1969;10:31–5.
66. Kropp BP, Sawyer BD, Shannon HE, Rippy MK, Badylak SF, Adams MC, et al. Characterization of small intestinal submucosa regenerated canine detrusor: assessment of reinnervation, in vitro compliance and contractility. J Urol. 1996;156:599–607.
67. Portis AJ, Elbahnasy AM, Shalhav AL, Brewer A, Humphrey P, McDougall EM, et al. Laparoscopic augmentation cystoplasty with different biodegradable grafts in an animal model. J Urol. 2000;164:1405–11.
68. Landman J, Olweny E, Sundaram CP, Andreoni C, Collyer WC, Rehman J, et al. Laparoscopic mid sagittal hemicystectomy and bladder reconstruction with small intestinal submucosa and reimplantation of ureter into small intestinal submucosa: 1-year followup. J Urol. 2004;171:2450–5.

69. Kikuno N, Kawamoto K, Hirata H, Vejdani K, Kawakami K, Fandel T, et al. Nerve growth factor combined with vascular endothelial growth factor enhances regeneration of bladder acellular matrix graft in spinal cord injury-induced neurogenic rat bladder. BJU Int. 2009;103:1424–8.
70. Wang C, Meng G, Zhang L, Xiong Z, Liu J. Physical properties and biocompatibility of a core-sheath structure composite scaffold for bone tissue engineering in vitro. J Biomed Biotechnol. 2012;2012:579141.
71. Ghoniem GM. The recalcitrant overactive bladder patient. Geriatrics. 2002;57(Suppl 1):23–9.
72. Khattak SF, Bhatia SR, Roberts SC. Pluronic F127 as a cell encapsulation material: utilization of membrane-stabilizing agents. Tissue Eng. 2005;11:974–83.
73. Cilento BG, Freeman MR, Schneck FX, Retik AB, Atala A. Phenotypic and cytogenetic characterization of human bladder urothelia expanded in vitro. J Urol. 1994;152:665–70.
74. De Coppi P, Callegari A, Chiavegato A, Gasparotto L, Piccoli M, Taiani J, et al. Amniotic fluid and bone marrow derived mesenchymal stem cells can be converted to smooth muscle cells in the cryo-injured rat bladder and prevent compensatory hypertrophy of surviving smooth muscle cells. J Urol. 2007;177:369–76.
75. Shukla D, Box GN, Edwards RA, Tyson DR. Bone marrow stem cells for urologic tissue engineering. World J Urol. 2008;26:341–9.
76. Anumanthan G, Makari JH, Honea L, Thomas JC, Wills ML, Bhowmick NA, et al. Directed differentiation of bone marrow derived mesenchymal stem cells into bladder urothelium. J Urol. 2008;180:1778–83.
77. Oottamasathien S, Wang Y, Williams K, Franco OE, Wills ML, Thomas JC, et al. Directed differentiation of embryonic stem cells into bladder tissue. Dev Biol. 2007;304:556–66.
78. Chun SY, Kim HT, Lee JS, Kim MJ, Kim BS, Kim BW, et al. Characterization of urine-derived cells from upper urinary tract in patients with bladder cancer. Urology. 2012;79:1186 e1-7.
79. Kanematsu A, Yamamoto S, Iwai-Kanai E, Kanatani I, Imamura M, Adam RM, et al. Induction of smooth muscle cell-like phenotype in marrow-derived cells among regenerating urinary bladder smooth muscle cells. Am J Pathol. 2005;166:565–73.
80. Martin GR. Isolation of a pluripotent cell line from early mouse embryos cultured in medium conditioned by teratocarcinoma stem cells. Proc Natl Acad Sci U S A. 1981;78:7634–8.
81. Alison MR, Poulsom R, Forbes S, Wright NA. An introduction to stem cells. J Pathol. 2002;197:419–23.
82. Trounson A. The production and directed differentiation of human embryonic stem cells. Endocr Rev. 2006;27:208–19.
83. Qin D, Long T, Deng J, Zhang Y. Urine-derived stem cells for potential use in bladder repair. Stem Cell Res Ther. 2014;5:69.
84. Xue Y, Cai X, Wang L, Liao B, Zhang H, Shan Y, et al. Generating a non-integrating human induced pluripotent stem cell bank from urine-derived cells. PLoS One. 2013;8:e70573.
85. Jayo MJ, Jain D, Ludlow JW, Payne R, Wagner BJ, McLorie G, et al. Long-term durability, tissue regeneration and neo-organ growth during skeletal maturation with a neo-bladder augmentation construct. Regen Med. 2008;3:671–82.
86. Atala A, Freeman MR, Vacanti JP, Shepard J, Retik AB. Implantation in vivo and retrieval of artificial structures consisting of rabbit and human urothelium and human bladder muscle. J Urol. 1993;150:608–12.
87. Jayo MJ, Jain D, Wagner BJ, Bertram TA. Early cellular and stromal responses in regeneration versus repair of a mammalian bladder using autologous cell and biodegradable scaffold technologies. J Urol. 2008;180:392–7.
88. de Boer WI, Schuller AG, Vermey M, van der Kwast TH. Expression of growth factors and receptors during specific phases in regenerating urothelium after acute injury in vivo. Am J Pathol. 1994;145:1199–207.
89. Gorham SD, French DA, Shivas AA, Scott R. Some observations on the regeneration of smooth muscle in the repaired urinary bladder of the rabbit. Eur Urol. 1989;16:440–3.
90. Oberpenning F, Meng J, Yoo JJ, Atala A. De novo reconstitution of a functional mammalian urinary bladder by tissue engineering. Nat Biotechnol. 1999;17:149–55.
91. Zhang Y, Frimberger D, Cheng EY, Lin HK, Kropp BP. Challenges in a larger bladder replacement with cell-seeded and unseeded small intestinal submucosa grafts in a subtotal cystectomy model. BJU Int. 2006;98:1100–5.
92. Farhat WA, Yeger H. Does mechanical stimulation have any role in urinary bladder tissue engineering? World J Urol. 2008;26:301–5.
93. Bouhout S, Gauvin R, Gibot L, Aube D, Bolduc S. Bladder substitute reconstructed in a physiological pressure environment. J Pediatr Urol. 2011;7:276–82.
94. Hyndman ME, Kaye D, Field NC, Lawson KA, Smith ND, Steinberg GD, et al. The use of regenerative medicine in the management of invasive bladder cancer. Adv Urol. 2012;2012:653652.
95. Basu J, Jayo MJ, Ilagan RM, Guthrie KI, Sangha N, Genheimer CW, et al. Regeneration of native-like neo-urinary tissue from nonbladder cell sources. Tissue Eng Part A. 2012;18:1025–34.
96. Lee JN, Chun SY, Lee HJ, Jang YJ, Choi SH, Kim DH, et al. Human Urine-derived Stem Cells Seeded Surface Modified Composite Scaffold Grafts for Bladder Reconstruction in a Rat Model. J Korean Med Sci. 2015;30:1754–63.
97. Lee M, Wu BM, Stelzner M, Reichardt HM, Dunn JC. Intestinal smooth muscle cell maintenance by basic fibroblast growth factor. Tissue Eng Part A. 2008;14:1395–402.

98. Yoon JJ, Chung HJ, Lee HJ, Park TG. Heparin-immobilized biodegradable scaffolds for local and sustained release of angiogenic growth factor. J Biomed Mater Res A. 2006;79:934–42.
99. Kim BS, Chun SY, Lee JK, Lim HJ, Bae JS, Chung HY, et al. Human amniotic fluid stem cell injection therapy for urethral sphincter regeneration in an animal model. BMC Med. 2012;10:94.
100. Hoogenkamp HR, Pot MW, Hafmans TG, Tiemessen DM, Sun Y, Oosterwijk E, et al. Scaffolds for whole organ tissue engineering: Construction and in vitro evaluation of a seamless, spherical and hollow collagen bladder construct with appendices. Acta Biomater. 2016;43:112–21.

9 Urethra

Yun-Sok Ha and Tae-Hwan Kim

9.1 Introduction

A variety of congenital and acquired urethral pathologies including hypospadias, strictures, fistulae, and straddle injuries can severely impair normal urethral function, necessitating surgical reconstruction [1–4]. No universally recognized algorithms for the surgical management of these conditions are in place, although some progress in this field has been made [5–7]. A urethral stricture is a narrowing of the urethral lumen, arising from iatrogenic (33%), idiopathic (33%), traumatic (19%), and inflammatory (15%) causes, that may lead to lower urinary tract obstruction [3, 8]. Such obstruction may significantly impair patients' quality of life by causing voiding difficulties; these, in turn, may negatively affect the upper urinary tract, resulting in deterioration of renal function [9]. The mechanism of urethral stricture development involves the processes of fibrosis and cicatrix formation in the urethral mucosa and surrounding connective tissues. Strictures can occur at any portion of the urethra between the meatus and bladder neck [10]. In the context of urethral stricture description, the term "urethra" typically refers to the anterior urethra including the bulbar and penile urethra [11]. However, posterior urethral strictures including those of the membranous and prostatic urethrae are not included in this definition; instead, these are typically termed "urethral contractures" or "urethral stenoses" [11]. It should be noted that urethroplasties are performed not only in patients with urethral strictures, but also in those with hypospadias and other urethral defects (e.g., urethral fistulae). In particular, repair of the anterior urethra is one of the most demanding surgical problems in urology. A multitude of techniques has been used to treat anterior urethral strictures using buccal mucosa or a penile skin flap for reconstruction of the long urethral defect [12–14]. However, in addition to potential donor site morbidity, there is an inadequate amount of graft in many cases. Urothelial cell based tissue engineering has shown promise as an alternative to urethral substitution [15–20]. In this chapter, we have evaluated currently available studies regarding the potential of tissue engineering in urethral reconstruction, in particular those describing the use of both acellular and recellularized tissue-engineered constructs in animal and human models. Possible future developments in this field are also discussed.

Y.-S. Ha (✉) • T.-H. Kim
Department of Urology, School of Medicine, Kyungpook National University, Daegu, South Korea
e-mail: yunsokha@gmail.com; doctork@knu.ac.kr

B.W. Kim (ed.), *Clinical Regenerative Medicine in Urology*, DOI 10.1007/978-981-10-2723-9_9

9.2 Tissue Engineering for Urethroplasty to Treat Urethral Lesions

To reconstruct a new tissue, a triad of components is needed: cells, biomaterials to be used as supports or scaffolds, and growth factors [21–23]. Various strategies have been proposed over the years for the regeneration of urethral tissue. A woven mesh of polyglycolic acid (PGA) both with and without cells was used to regenerate urethras in various animal models [24–26]. Naturally derived collagen-based materials such as bladder-derived acellular submucosa, acellular urethral submucosa, and collagen gels have also been tried experimentally in various animal models for urethral reconstruction [27–29]. Urethral reconstruction involves the use of different "patches" (which may be called matrices, scaffolds, or substrates) that have to be transplanted into the urethral wall, thus providing a patent lumen. Conventional substitution urethroplasty utilizes different tissues from the patient. Tissue engineered urethroplasty procedures may be performed with both naturally derived and synthetic materials [30, 31].

Many pediatric and adult patients with urethral disease have been successfully treated in an onlay manner using collagen-based matrices. One of the advantages of this method over nongenital tissue grafts used for urethroplasty is that the material is "off the shelf." This eliminates the necessity of additional surgical procedures for graft harvesting, which may decrease operative time as well as eliminate morbidity from the harvest procedure.

One of the fundamental tasks when developing tissue-engineered "patches" is forming a support matrix that can serve as a substrate for cells (cell-adhesion substrate) and control the configuration of the tissue-engineered structure and the direction of tissue regeneration. Each of the naturally derived and synthetic biocompatible materials utilized for tissue engineering has its advantages and disadvantages.

9.2.1 Natural Materials

Natural polymers may be divided into natural polymers and acellular matrices [32, 33]. In the quest for an ideal urethral substitute, acellular scaffolds are promising, as they have demonstrated the ability to induce tissue regeneration layer by layer. Following several experimental studies, the use of acellular matrices for urethral reconstruction has become a clinical reality over the last decade. Such biomaterials are prepared via a process leading to the removal of all cells and their components, i.e. decellularization. The remaining product is an acellular matrix, mostly containing collagen that (theoretically) should be non-immunogenic and should cause no allergic reactions. All the above-mentioned matrices are biodegradable and are gradually substituted by the host's own intercellular matrix.

The bladder submucosal matrix proved to be a suitable graft for repair of urethral defects in rabbits [27]. The neourethras exhibited a normal urothelial luminal lining and organized muscle bundles. These results were confirmed clinically in a series of patients with a history of failed hypospadias reconstruction wherein the urethral defects were repaired with human bladder acellular collagen matrices [34]. The neourethras were created by anastomosing the matrix in an onlay fashion to the urethral plate. The size of the created neourethra ranged from 5 to 15 cm. After a 3-year follow-up, three of the four patients had a successful outcome with regard to cosmetic appearance and function. The acellular collagen-based matrix eliminated the necessity for performing additional surgical procedures for graft harvesting, and both operative time and the potential morbidity from the harvest procedure were decreased. Similar results were obtained in pediatric and adult patients with primary urethral stricture disease using the same collagen matrices [35]. Another study by El-Kassaby et al. of 30 patients with recurrent stricture disease showed that a healthy urethral bed (two or fewer prior urethral surgeries) was needed for successful urethral reconstruction using acellular collagen-based grafts [36]. In their study, the length of the strictures ranged from 2 to 18 cm (average 6.9 cm). All grafts were implanted as ventral onlay patches. The patients were followed for a mean period of 25 months. Where the patients had one or no previous interventions, the success rate between both groups was similar (100% for buccal mucosa vs. 89% for acellular matrix).

Feng et al. compared the mechanical properties and biocompatibility of biomaterials, including bladder submucosa, small intestinal submucosa (SIS), acellular corpus spongiosum matrix, and PGA, to identify the optimal scaffold for urethral tissue engineering [37]. Cytotoxicity, tensile mechanical properties, and pore size were evaluated. Smooth muscle cells were seeded onto the biomaterials to evaluate differences in cell infiltration. They concluded that the acellular corpus spongiosum matrix had better overall performance, indicating that an organ specific matrix may be the most suitable type of scaffold for urethral reconstruction.

A clinical trial using tubularized nonseeded SIS for endoscopic urethroplasty was performed in eight evaluable patients [38]. Two patients with short inflammatory strictures maintained urethral patency. Stricture recurrence developed in the six patients within 3 months of surgery. Authors concluded that endoscopic urethroplasty using an unseeded SIS graft was unsuccessful. On the other hand, various authors usually reported favorable results from SIS urethroplasties in humans. Palminteri et al. observed an 85% success rate with short-term follow-up (mean 21 months) [39]. A dorsal inlay graft was performed in 14 cases, ventral onlay graft in 1, and dorsal inlay plus ventral onlay in 5. The three failures were penile repairs with long strictures (over 5 cm). This group updated their series in 2012, publishing long-term results (71 months) with an overall patency rate of 76% [40].

Sievert et al. performed single-stage urethral reconstruction with homologous acellular urethral matrix grafts [41]. All tissue components were observed in the grafted matrix after 3 months, with further improvement over time; however, the smooth muscle in the matrix was less extensive than in the normal rabbit urethra and was not well oriented.

The value of homologous dermal acellular matrix grafts for urethral reconstruction in humans was investigated by Lin et al. in 2005 [42]. The dermal acellular matrix graft was sutured to a tubular graft and it replaced the defect in the urethra. Following the operation, urethrography revealed the excellent caliber of the reconstructed urethra. Urethroscopic examination showed that the graft was covered by epithelial tissue and grew into the native tissue. They suggested that the homologous dermal acellular matrix graft may serve as an ideal replacement material for complex urethral strictures or defects, without the risk of rejection.

Fossum et al. surgically treated six boys aged 14–44 months with severe hypospadias with autologous urothelial cell constructs [43]. A two-staged procedure starting with repair of the chordee was performed in all cases. Urothelial cells were harvested via bladder lavage during the first operation and were seeded onto acellular dermis. The seeded scaffold was implanted in a second operation in an onlay fashion. These children were followed up for 3.5–5 years. One patient developed a partial stricture, which was treated conservatively, and another developed an obstruction in the proximal anastomosis that was managed successfully with internal urethrotomy. Two other children developed fistulae requiring surgical correction.

9.2.2 Synthetic Polymers

Synthetic polymers of naturally occurring α-hydroxy acids are widely used in the field of regenerative medicine. Some of these have been approved by the US Federal Drug Administration (FDA) for use in a number of applications, including sutures [44]. Of interest for urethral tissue engineering are biodegradable matrices (both synthetic and natural). The degradation products of synthetic biodegradable polymers based on α-hydroxy acids are carbon dioxide and water.

As these polymers are thermoplastics, they can easily be shaped into a three-dimensional (3D) scaffold with the required porosity and size.

Synthetic polymer scaffolds designed for cell transplantation were reproducibly made on a large scale and studied with respect to biocompatibility, structure, and biodegradation rate [45]. The scaffold induced chondrocyte differentiation with respect to morphology and phenotype and represented a model cell culture substrate that may be useful for a variety of tissue engineering applications [45]. Micos et al. suggested that biodegradable foams of hydrophobic polymers can be efficiently wet by a two-step immersion in ethanol and water, which overcomes the hindered entry of water into air-filled pores [46]. In their study, ethanol readily enters into the porous polymer, after which it is diluted and replaced by water. This method was evaluated for porous disks of poly(L-lactic acid) (PLLA) and poly(DL-lactic-co-glycolic acid) (PLGA) foams of copolymer ratios 85:15 and 50:50. Furthermore, water entry even after 1 h was very close to the plateau value for all pre-wet polymers tested.

Harris et al. established open pore biodegradable matrices, formed using a gas foaming method developed for fabricating matrices without the use of organic solvents and/or elevated temperatures [47]. Disks comprised of polymer and NaCl particles were compression molded at room temperature and subsequently allowed to equilibrate with high pressure CO_2 gas (800 psi). Creation of a thermodynamic instability led to the nucleation and growth of gas pores in the polymer particles, resulting in the expansion of the polymer particles. The polymer particles fused to form a continuous matrix with entrapped salt particles. The NaCl particles subsequently were leached to yield macropores within the polymer matrix. The overall porosity and extent of pore connectivity were regulated by the ratio of polymer/salt particles and the size of the salt particles. The utility of these matrices was demonstrated by engineering smooth muscle tissue in vitro. This process, a combination of high-pressure gas foaming and particulate leaching techniques, allows the fabrication of matrices with well-controlled porosity and pore structure.

The fabrication technique of electrospinning allows the fast production of high-porosity scaffolds with defined shapes and architectures. Han and Gouma developed electrospun bioscaffolds that mimic the topology of extracellular matrix (ECM) [48]. ECM is a natural scaffold for cell, tissue, and organ growth. Its topology plays an important role in cell differentiation. The design challenge is to fabricate biomaterials that mimic the ECM's three-dimensional (3D) structures with defined shapes and complex porous architecture. The urinary bladder matrix (UBM) is used in this work as the model system of the ECM architecture. Cellulose acetate (CA) is the biomaterial of choice for building UBM-mimicking scaffolds. Electrospinning is the fabrication method used to form complex, porous, 3D structures with specific designs in a single-step process. Lee et al. used thermal treatments to enhance the mechanical properties of electrospun poly-(epsilon-caprolactone) (PCL) scaffolds [49]. The biomechanical properties of the thermally bonded electrospun PCL scaffolds were significantly increased without any gross observable and ultrastructural changes when compared to untreated PCL scaffolds. They suggested that the introduction of thermal fiber bonding to electrospun PCL scaffolds improved the biomechanical properties of these scaffolds, making them more suitable for tissue engineering applications.

PGA polymers were used as tissue-engineered autologous urethras for patients who required reconstruction [50]. In this study, epithelial cells were expanded and seeded onto tubularized polyglycolic acid:poly(lactide-co-glycolide acid) scaffolds. Patients then underwent urethral reconstruction with the tissue-engineered tubularized urethras. Urethral biopsies showed that the engineered grafts had developed apparently normal architecture by 3 months after implantation. Tubularized urethras can be engineered and can remain functional in a clinical setting for up to 6 years. Kanatami et al. reported fabrication of an optimal urethral graft using collagen-sponge tubes reinforced with Copoly(L-lactide/epsilon-caprolactone) [P(LA/CL)] fabric [51]. The tubes were dipped in aqueous collagen solution and lyophilyzed to prepare the P(LA/CL)-collagen sponge graft. The grafts were applied to a 1.5 cm rabbit urethral defect ($n = 14$ for each condition), and tissue repair was evaluated using urethrographical, urethroscopical, and histological examination 1, 3, and 6 months after surgery. For the mesh style graft, all urethras were patent, without fistulae or stenoses, over the entire observation period. Histologically, urethral structure was disorganized for the stent style graft, whereas the urethral tissue on the mesh style graft was slightly fibrotic but completely epithelialized and supported by a regenerated smooth muscle layer at 6 months. These findings suggest that creation of a scaffold suitable for urethral tissue regeneration depends not only on biomaterial composition, but also on the fabrication technique.

9.2.3 Hybrid or Composite Scaffolds

Hybrid or composite scaffolds are constructed from a synthetic polymer combined with a corresponding natural matrix [52, 53]. Such scaffolds were developed by researchers in the quest for an ideal tissue-engineered material to be used in urology. Hybrid scaffolds are constructed by electrospinning PLGA microfibers onto the abluminal surface of a bladder acellular matrix [53].

There are two types of matrix used in tissue-engineered urethral substitution procedures: acellular matrices and autologous cell–seeded matrices. Acellular matrices have been utilized for a long period of time. An acellular matrix transfer procedure will be successful if: (a) the entire abluminal surface is infiltrated by host epithelial cells; (b) the defect being substituted is short; and (c) there is a good vascular urethral bed. Therefore, procedures using acellular matrices are expected to fail in patients with recurring strictures, marked spongiofibrosis, or long strictures. An autologous cell-seeded matrix is a tissue-engineered construct containing an acellular matrix populated (ex vivo) with autologous cells. A biopsy is obtained from the patient. In specialized sterile laboratories, cells of the desired type are harvested from the biopsy and grown in a culture. Depending on the type of cells and culture method used, this process may take 4–12 days to 3–6 weeks [50, 54, 55]. As soon as the required number of cells is obtained, they are seeded onto a matrix and, 1–7 days later, are implanted into a patient or animal. Frequently used autologous cells are as follows: urothelial cells from the bladder, urethra, or ureter; buccal mucosa epitheliocytes; keratinocytes; fibroblasts; and smooth muscle cells [50, 54, 56–58].

De Filippo et al. showed that acellular collagen matrices derived from bladder submucosa seeded with cells from normal urethral tissue can be used for tubularized replacement [56]. They suggested that the collagen matrices seeded with cells may offer a useful alternative in the future for patients requiring a tubularized urethral segment replacement. Li et al. investigated the feasibility of urethral reconstruction using oral keratinocyte (OK)-seeded bladder acellular matrix grafts [57]. In their study, OKs had good biocompatibility with bladder acellular matrix grafts, and urethral reconstruction with these grafts was better than that with bladder acellular matrix grafts alone. Fu et al. evaluated alterations in foreskin epidermal cells, which were used to reconstruct male rabbit anterior urethras in combination with acellular collagen matrices. They concluded that epidermal cells seeded onto acellular collagen matrices can be successfully used to reconstruct urethras that have defects, and are transformed to transitional epithelial cells [58]. Autologous tissue-engineered buccal mucosa was confirmed as a clinically useful autologous urethral replacement tissue in a group of patients with lichen sclerosis urethral stricture. However, it was not without complications, namely fibrosis and contraction [54]. Raya-Rivera et al. constructed engineered urethras with autologous cells and implanted them into patients with urethral defects. A tissue biopsy was taken from each patient, and the muscle and epithelial cells were expanded and seeded onto tubularized polyglycolic acid:poly (lactide-co-glycolide acid) scaffolds. Patients then underwent urethral reconstruction with the tissue-engineered tubularized urethras. Tubularized urethras can be engineered and remain functional in

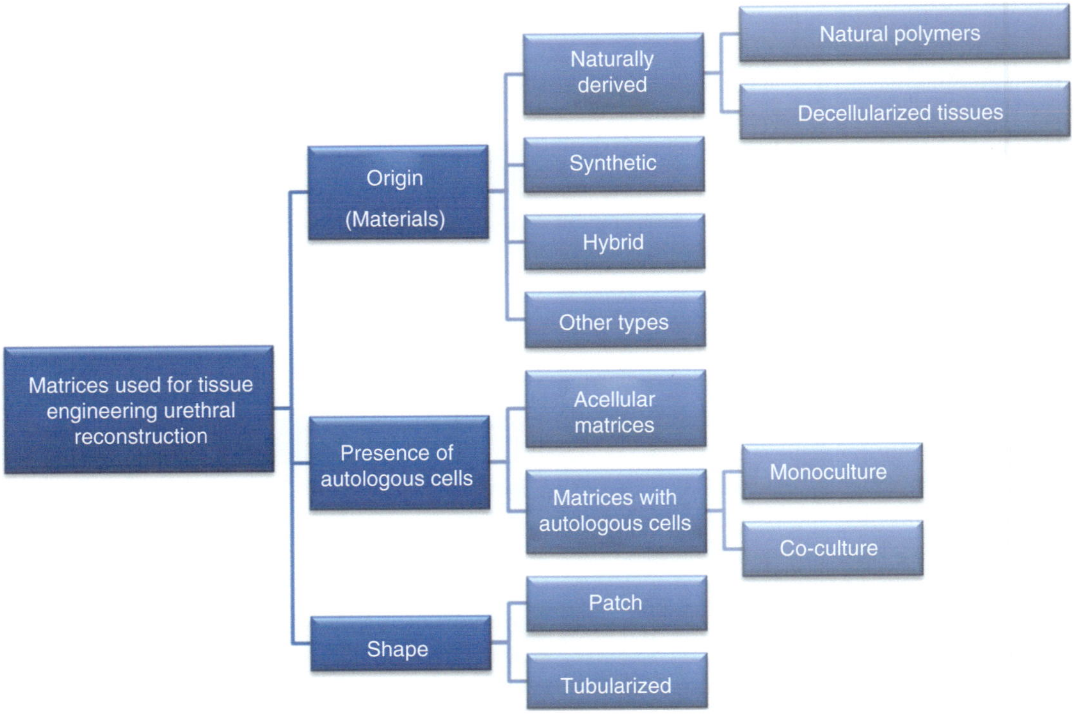

Fig. 9.1 Matrices and cells used for tissue-engineered urethral reconstruction

a clinical setting for up to 6 years. These engineered urethras can be used in patients who need complex urethral reconstruction [50].

Tissue-engineered autologous cells can be divided into two types: tissue-engineered constructs developed using a monoculture; and those using co-culture. To develop tissue-engineered constructs using a monoculture, autologous cells are seeded onto one of the surfaces of the support matrix and, as a rule, a certain type of epitheliocyte is used [55]. Constructs using a co-culture are more complex, as the abluminal surface is seeded with epitheliocytes, while the opposite side is seeded with either fibroblasts or smooth muscle cells [50, 59].

According to macroscopic parameters, tissue-engineered constructs can be divided into patch constructs and tubularized constructs. The stricture length to be reconstructed with patch constructs grafts is usually limited, as it is in cases of conventional substitution procedures. Tubularized grafts are usually developed using a co-culture and may be utilized to reconstruct entire urethras [50]. Figure 9.1 overviews matrices and cells used for tissue-engineered urethral reconstruction.

9.2.3.1 Acellular Matrices Utilized in Animal Models

Kropp et al. suggested using SIS as an acellular urethroplasty matrix [60]. In one of the control groups, full thickness preputial skin from the host rabbit was used. The success rates were the same in both groups (100%), yet preputial graft procedures resulted in the formation of urethral diverticulum in all eight animals.

Chen et al. utilized BAMG (obtained and processed from porcine bladder submucosa) for ventral onlay urethroplasties in 10 rabbits [27]. Complete graft epithelialization was achieved after 2 months; the migration of organized muscle bundles was detected 6 months after implantation. No strictures or complications were observed.

In a study conducted by Nuininga et al., rabbits were divided into four groups, with six animals in each, according to the type of biomatrix used: in group 1, the rabbits underwent partial urethral replacement, which was replaced with one layer of SIS; in group 2, four-layer SIS patch grafts were utilized; group 3 was treated with collagen-based matrices produced from bovine tendons; and in group 4, the animals underwent a sham control operation (urethras were incised ventrally and then sutured) [61]. In groups 1 and 3, complete epithelialization was observed at 1 month after implantation, whereas in group 2, it was achieved at 3 months. One rabbit from group 3 developed a stricture and one rabbit from group 2 developed a fistula near the operation site.

Yang et al. suggested using an acellular corpus spongiosum matrix substitute as a patch [62]. These kinds of matrices were obtained from rabbit urethras and transplanted to repair urethral defects of 10–15 mm in length. Complete epithelialization of the extracellular matrix was achieved at 3 weeks post operation. Well-formed smooth muscle cells were observed after 6 weeks. No strictures or complications were reported.

Huang et al. explored the potential of SIS for urethral reconstruction in rabbits [63]. Tubular SIS grafts were applied in group 1 ($n = 6$), while a ventral onlay graft was used as a patch in group 2 ($n = 6$). The epithelium covered the grafts fully after 6 weeks.

An interesting study was published by Kanatani et al., wherein two types of tubularized graft based on type I collagen sponge, reinforced with co-poly(Lactide/ε-caprolactone) (CLLC) fabric, were created [51]. In group 1, the P(LA/CL) fibers were knitted into a vascular stent style; in group 2, the P(LA/CL) fibers were weaved into a mesh style. Each of the grafts was applied to a 15-mm urethral defect. Numerous complications (stenoses, fistulae, or stone formation) were observed in group 1, whereas no complications were seen in group 2. The authors concluded that creation of a scaffold suitable for urethral tissue regeneration depends not only on the biomaterial composition, but also on the fabrication technique.

In a very important study, Dorin et al. determined the maximum potential distance of normal native tissue regeneration when using tubularized unseeded matrices [64]. Twelve rabbits were divided into four groups. In each group, urethral defects of varying length (0.5, 1.0, 2.0, and 3.0 cm) were created. At week 4, only the 0.5 cm group had a normal layer of epithelium surrounded by a layer of smooth muscle within the urethral lumen. In all groups with longer defects, strictures developed by 4 weeks. Therefore, this study proved that acellular matrices in tubularized grafts may be used successfully only to repair defects limited in size.

Villoldo et al. assessed the efficacy of onlay urethroplasty with SIS in rabbits [65]. A total of 30 animals included in the study had 15 mm of the ventral wall of the penile urethra excised. One month later, the rabbits were divided into two groups: group 1 was the control group; in group 2, where the created defect was patched with an SIS onlay graft, epithelialization was achieved after 15 days and smooth muscle bundles were detected at 6 months follow-up. The fact that urethral strictures were created 1 month before urethroplasties were performed brings this study a step closer to clinical reality.

Chung et al. studied the potential of using acellular matrices derived from silk fibroin for urethral repair [66]. The results of urethroplasty with silk fibroin matrices (SFMs) were compared to those with SIS. The follow-up period was 3 months, with no intermediate time points. The success rates of SFM and SIS were found to be the same. No strictures or complications were reported. However, the silk fibroin scaffolds showed reduced immunogenicity.

Kajbafzadeh et al. explored the role of preputial acellular matrix (PAM) for urethral reconstruction [67]. Prepuces were obtained from circumcised boys. The prepuces were decellularized and the produced matrices were used as patches in ventral onlay urethroplasty procedures: PAM in group 1; and PAM plus fibrin sealant in group 2. The effectiveness of repair was evaluated in both groups at a 9 month follow-up. The authors noted that satisfactory vascularity and smooth muscle layer formation were more significant in group 2 (PAM + fibrin sealant).

In all the above-mentioned studies, only male rabbits were used as animal models. In most studies, a longitudinal urethral defect was created, varying in length (10–20 mm); in some studies it was a tubular defect 5–30 mm long. SIS was used as a urethroplasty matrix in four of ten studies, with BAMG being the second most frequently utilized type of scaffold (applied in three of ten studies). One should note that no episodes of graft rejection were reported. Acellular matrices were found to be highly effective in animal models, yet their application range was limited by the size of urethral defect. Complete formation of the urothelial layer on the inner surface of the graft was achieved 4–12 weeks after implantation; either single smooth muscle cells or a regenerated smooth muscle layer was observed 2–12 months after implantation.

9.2.3.2 Acellular Matrices Utilized in Human Models

The feasibility of applying a bladder submucosal, collagen-based inert matrix as a free graft substitute for urethral repair in patients with hypospadias was explored [34]. All four patients had had a history of hypospadia repair procedures and required yet another repair. The neourethras were created from BAMGs in the size range 5–15 cm. Postoperatively, only one patient developed a fistula. Histological evaluation showed typical urethral stratified epithelium at the site of surgery.

Mantovani et al. were the first to utilize an SIS graft for substitution urethroplasty in a 72 year-old patient with a long stricture of the penile and bulbar urethra [68]. The dorsal onlay technique was used. At a 16 month follow-up, the maximum urine flow rate was 14 ml/s. No complications were observed intraoperatively or postoperatively. In the same year, El-Kassaby et al. used a BAMG matrix for urethral stricture repair [35]. A total of 28 patients with urethral strictures of varied length (1.5–16 cm) underwent reconstructive surgery wherein the ventral onlay technique was used. The success rate was 86%, as one patient developed a urethrocutaneous fistula that closed spontaneously 1 year after repair.

Lin et al. suggested using an acellular dermal matrix (ADM) graft for urethral reconstruction [42]. Homologous ADM was applied as a tubularized graft in 16 patients with urethral strictures and hypospadias. No episodes of rejection were observed. During the 46 month follow-up period, only four patients needed periodical urethral dilatation.

Le Roux et al. evaluated SIS as a substitute in endoscopic urethroplasty [38]. Nine patients were enrolled. Optical urethrotomy was performed prior to the SIS graft implantation. Subsequently, a tubularized SIS graft was implanted into the stricture site. Only two patients had patent lumen at 1 and 2 years follow-up, respectively; all the other patients developed recurring strictures. The authors concluded that endoscopic urethroplasty with unseeded SIS grafts is not to be recommended.

Donkov et al. assessed the success rates of SIS grafts used for dorsal onlay substitution urethroplasty [69]. The graft was fixed using a modified Barbagli technique. The success rate was 89% (eight of nine patients) at 18 months follow-up.

While Hauser et al. also chose SIS for dorsal onlay substitution urethroplasty, the success rate was much lower: only one of five patients did not have stricture recurrence at a 12 month follow-up [70]. The complications observed postoperatively were extravasation, severe urethritis, and urinary tract infection.

Palminteri et al. evaluated the role of SIS grafts for substitution urethroplasties [39]. Three techniques were employed: a dorsal inlay graft was performed in 14 patients, ventral onlay in one patient, and dorsal onlay plus ventral onlay in five patients; 21 month follow-up data demonstrated an average success rate of 85%. The average stricture length was 3 cm. No complications were noted. The three failures were in penile and penile–bulbar urethral strictures with an average length of 5.7 cm.

Fiala et al. used SIS grafts for urethroplasties in 50 patients [71]. The success rate averaged 80%, with a mean follow-up time of 31 months. Recurring strictures developed in one of 10 bulbar, five of 31 bulbopenile, and four of nine penile strictures; these occurred during the first 6 months postoperatively. Farahat et al. placed SIS grafts endoscopically [72]. Only two cases of 10 exhibited stricture recurrence at a mean follow-up time of 14 months; there were no complications.

El-Kassaby et al. conducted a comparative study between BAMGs and buccal mucosal grafts in anterior urethral strictures [36]. The patients were followed for 25 months on average. The success rate for BAMG was 53%, and that of the buccal mucosal graft was 100%. The authors divided patients from the two groups into subgroups: (a) those with a healthy urethral bed (less than two prior surgeries), where BAMG surgeries were successful in 89% of cases and buccal mucosa surgeries in 100%; and (b) those with an unhealthy urethral bed (more than two prior surgeries), where the success rates were 33% (BAMG) and 100% (buccal mucosa).

Palminteri et al. reported the longest (as of today) mean follow-up period (71 months), having worked with patients who had bulbar urethral strictures and underwent urethroplasties using SIS [40]. The success rate was 76%. It should be noted that the failure rate was 14% for strictures <4 cm and 100% for strictures >4 cm.

In the studies mentioned above, the most frequently used type of matrix was SIS, which was, in most cases, used as a material for patch grafts. Tubularized grafts were applied twice: in endoscopic urethroplasties (with a success rate of 22%) [38] and in conventional urethroplasties (with a success rate of 88%); however, when describing the cases of conventional urethroplastic surgery, the authors did not mention the lengths of the urethral stricture [42]. No episodes of graft rejection were reported. Biopsies taken in five studies out of 13 showed normal urethral tissue characteristics at the site of surgery. Almost all studies using acellular matrices reported success rates of 75% or more. Failures were more common in patients with long urethral strictures (>4 cm), in those who had had previous urethroplasties (unhealthy urethral bed, unsatisfactory vascularity), and in those with penile or penile–bulbar strictures. The two studies that used endoscopic approaches exhibited different success rates: 22% vs 80% [38, 72]. The key differences among the groups were in the way the SIS grafts were secured: Le Roux et al. used tubularized grafts [38], while Farahat et al. chose patches [72]. Additionally, the average stricture length in the former study was 2.2 cm, as opposed to 1.47 cm in the latter study. Apparently, the difference in the success rates can be attributed to these two factors.

In addition, Donkov et al. and Hauser et al. both applied SIS in a similar technique (dorsal onlay substitution), but the success rates were completely different (89% vs 20%, respectively) [69, 70]. The difference can be explained by the difference in the way SIS was secured: in the study by Donkov et al. [69], the patch was spread-fixed onto the tunica albuginea, while the mucosa was sutured to the graft margins; in the study by Hauser et al. [70], SIS was anastomosed only to the incised urethra, without being fixed onto the tunica albuginea. Moreover, the strictures in the former study were shorter than those in the latter (4–6 cm vs 3.5–10 cm).

9.2.3.3 Cellularized Matrices Utilized in Animal Models (Monoculture)

De Filippo et al. were the first to use a tubularized tissue-engineered construct derived from bladder submucosa and seeded with autologous urothelial bladder cells for urethroplasty in rabbits [56]. The cells were obtained via an open biopsy. Urethroplasties were performed to repair 10 mm-long defects. In the control group, similar defects were repaired with unseeded tubularized BAMGs. After 6 months of follow-up, there were no strictures in the former group, whereas the control group exhibited strictures at the site of surgery.

A similar study was conducted by Li et al., although instead of urothelial bladder cells they used OKs [57]. The ventral onlay graft procedure was used to repair 20 mm-long defects. Rabbits in the experimental group (BAMG plus OKs) developed no strictures or complications; in the control group, two rabbits died of infection, two had fistulae, and the remaining eight developed strictures.

Fu et al. used tubularized BAMGs seeded with autologous preputial keratinocytes to repair 15 mm-long defects in three rabbits; there was no control group in the study [58]. At a 2 month follow-up, urethrography showed a wide urethral caliber in all three animals, and histological evaluation showed complete epithelialization at the site of surgery.

Of particular interest is the paper published by Gu et al. [73] They implanted silastic tubes into the peritoneal cavities of nine rabbits. After 2 weeks, the authors harvested the tubes covered by tissue, histological analysis of which revealed myofibroblasts embedded in collagen bundles covered by an outer layer of mesothelium. The tissue was everted and used as a tubularized graft to repair 15 mm-long urethral defects. At 1 month post operation, histological and immunohistochemical analyses showed normal urethral architecture.

In their study, Micol et al. used high-density collagen gel tubes as grafts that were seeded with autologous bladder smooth muscle cells [29]. It took the authors approximately 24 h to produce each tubularized graft. They did not place catheters postoperatively. Seven of sixteen rabbits developed fistulae. In both groups (with eight animals in each), equal numbers of smooth muscle cells were observed after 1 month.

Gu et al. isolated mesothelial cells via omentum biopsy and seeded them onto BAMGs [74]. Fifteen mm-long defects were substituted by tubularized matrices seeded with cells (in nine rabbits) and by those without cells (in nine rabbits from the control group). In the latter, all animals developed strictures, while no stricture formation was observed in the experimental group. At 6 months after implantation, the neo-urethras could not be histologically distinguished from the native urethras.

Xie et al. prepared electrospun silk fibroin matrices and seeded them with urothelial bladder cells that had been isolated and expanded [18]. Dorsal urethral defects 30 mm long were created in female dogs. At 6 months after implantation, the neo-urethras could not be histologically distinguished from the native urethras. It should be noted that the artificial defects involved only the mucosa, not the smooth muscle layer.

Li et al. seeded BAMGs with either epithelial differentiated or undifferentiated rabbit adipose-derived stem cells [75]. BAMGs were labelled with 5-bromo-2′-deoxyuridine. Thirty-six rabbits were divided into three groups (12 animals/group); 20 mm ventral urethral defects were created. In group A, defects were substituted only with BAMGs; in group B with BAMGs plus TGFβ1 siRNA-transfected fibroblasts (Und-rASCs); and in group C with BAMGs plus epithelial-differentiated rabbit adipose-derived stem cells (Epith-rASCs). Only one stricture was observed in group C, whereas almost all animals developed strictures in the first two groups. Complete epithelialization was achieved only in group C.

Wang et al. used a denuded human amniotic scaffold (dHAS); 5 × 10 mm defects were created on the ventral wall of the urethra and were repaired with either dHAS alone (group 1, $n = 6$) or dHAS + urethral urothelial cells (group 2, $n = 6$). In group 1, one animal developed an infection and one had a fistula; group 2 exhibited no complications or strictures (the efficacy rate was 100%). Mild inflammatory infiltration was observed in cell-seeded dHAS grafts, as revealed by less pronounced accumulation of CD4+ and CD8+ cells (or neutrophils or other immune cells). Histopathological analysis identified a mild immune response in the cell-seeded dHAS group ($p < 0.05$). Urethral defects in group 2 were completely resolved in 1 month. At 3 months after surgery, the formation of a smooth muscular layer and rich blood vessels was apparent [76].

In a very recently published Korean study by Chun and Kim et al., they evaluated the combined effect of acellular bladder submucosa matrix (BSM) and autologous urethral tissue for the treatment of long segment urethral stricture in a rabbit model [20]. To prepare the BSM, porcine bladder submucosa was processed, decellularized, configured into a sheet-like shape, and sterilized (Fig. 9.2). Twenty rabbits were randomly divided into groups: normal control, urethral stricture, urethroplasty using BSM only, or BSM/autologous urethral tissue ($n = 5$ per group). Brief urethroplasty methods are represented in Fig. 9.3. The width of the penile urethra was measured by postoperative

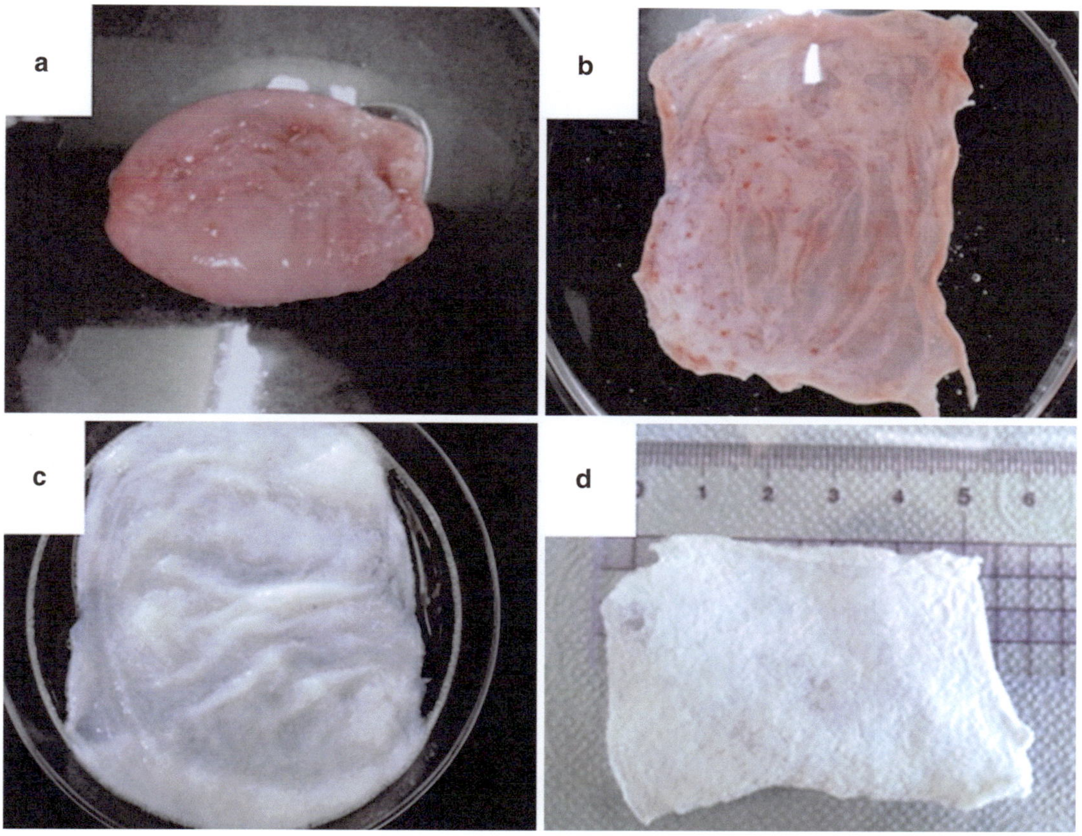

Fig. 9.2 Preparation of bladder submucosa matrix (BSM). (**a**) Bladder extraction from pig. (**b**) Separation of submucosa. (**c**) Gross image of acellular BSM, characterized by a thin sheet structure, acellular composition, multidirectional tensile strength, and lack of chemical cross-links. (**d**) Lyophilized acellular BSM sheet

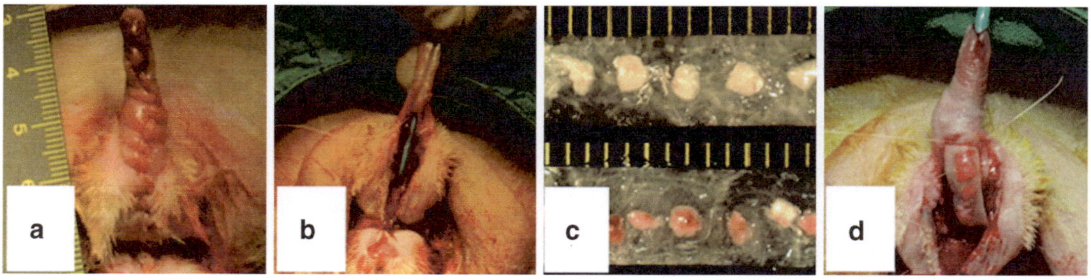

Fig. 9.3 Process of nontransected ventral urethral stricture model generation and urethroplasty. (**a**) Urethral defect created by urethrectomy and suturing. (**b**) Excision of healthy urethral muscle and endothelial tissue from normal urethra. (**c**) Embedded urethral tissue on the BSM with fibrin glue (shown below the magnified figure). Prepared combined graft with BSM and autologous urethral tissue. (**d**) Completed onlay graft. Scale bar, 10 μm

urethrograms at 4, 8, and 12 weeks. The control urethrogram showed a wide urethral caliber. Both graft groups showed a similar width to those of the normal group, and the stricture group revealed stenosis. The mean urethral width of the control, stricture, BSM, and BSM/autologous urethral tissue groups at week 12 was 10.3 ± 0.80, 3.8 ± 1.35, 8.8 ± 0.84, and 9.1 ± 1.14 mm, respectively. Although the difference in the width of the BSM and BSM/tissue graft urethroplasty groups was not statistically significant, the BSM/tissue graft group diameter

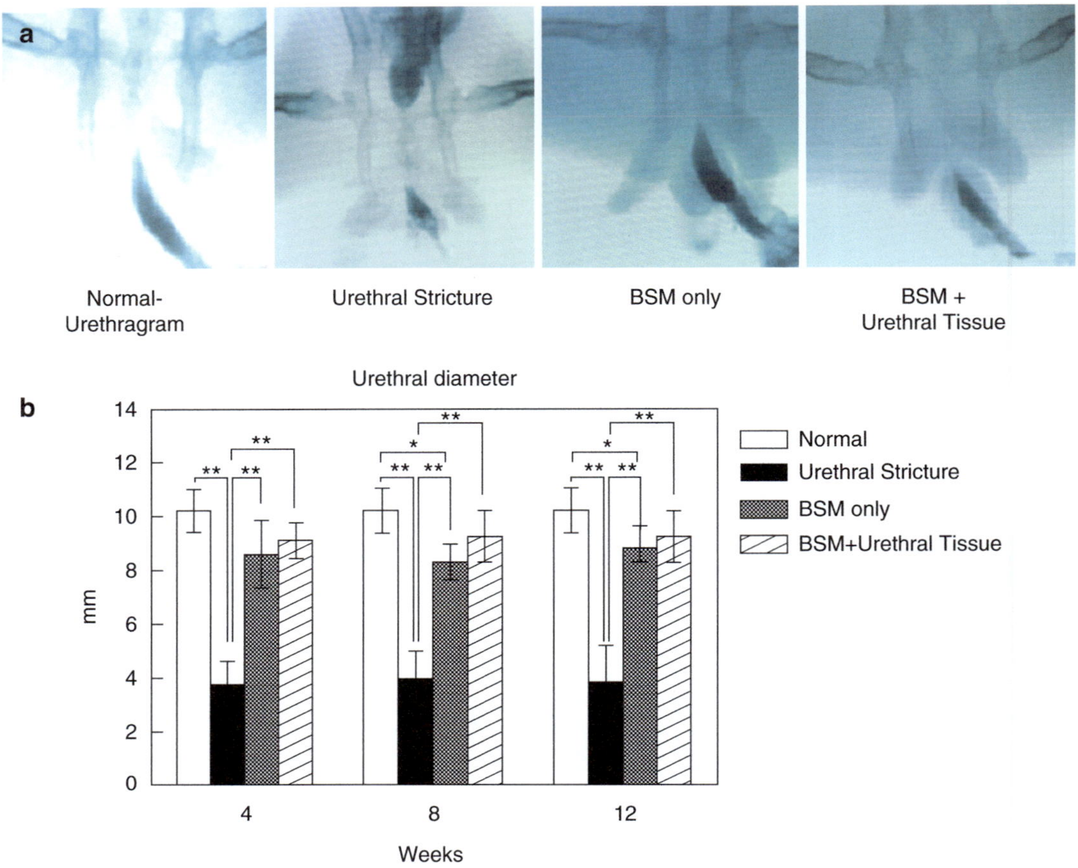

Fig. 9.4 Width of the penile urethra. (**a**) Retrograde urethrography (representative images taken at week 12) shows complete tubularization of both grafted urethras similar to the control group throughout the study period. The stricture group shows stenosis. (**b**) Urethral diameter measured at 4, 8, and 12 weeks post operation. The *symbols* on the top of the bars indicate significant differences at $P < 0.05$. P values were obtained with analysis of variance and Tukey's test. BSM, urethrotomy, and onlay urethroplasty with an acellular BSM scaffold graft; BSM/tissue, urethrotomy and onlay urethroplasty with a graft composed of autologous urethral tissue and an acellular BSM scaffold

was approximately 0.567 mm wider (Fig. 9.4). The histopathology study revealed that the BSM/autologous urethral tissue graft had a normal urethral lumen area, compact muscular layers, complete epithelialization, and progressive infiltration by vessels in the regenerated urethra. In contrast, the BSM grafts revealed keratinized epithelium, abundant collagenized fibrous connective tissue, and were devoid of bundles of circular smooth muscle (Fig. 9.5). According to these results, the BSM/tissue graft had well-organized integration with the recipient tissues, and the cells within embedded autologous tissue and BSM can synchronistically act as biological activators for incorporation, promoting ingrowth of surrounding urethral tissue into the scaffold. This indicates that autologous urethral tissue at the time of implantation is effective for urethral reconstruction. A BSM graft embedded with autologous urethral tissue can be applied for clinical treatment of strictures requiring partially resected urethra replacements.

Male rabbits were used in ten of the eleven studies reviewed above. Female dogs were used as animal models in one study. Despite the fact that preclinical and clinical studies demonstrated better success rates with SIS, six of the eleven reviewed studies utilized BAMGs. SIS

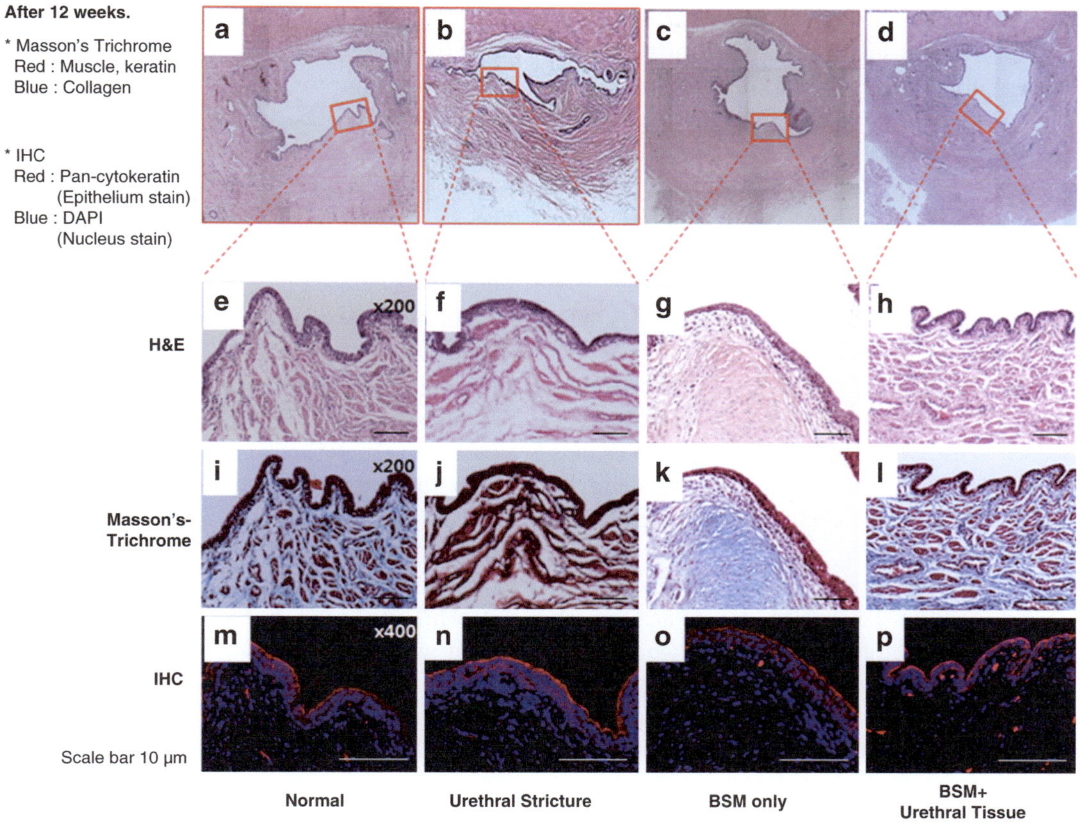

Fig. 9.5 Histological, Masson's trichrome, and immunohistochemical analysis of harvested urethra. (**a–d**) Low-magnification images of H&E-stained sections (40×); both graft groups show thick circular smooth muscle and a similar area of the urethral lumen as observed in the control group. The stricture group shows loose muscular layers and narrow lumen. (**e–h**) High-magnification images of H&E-stained sections (200×); the BSM/tissue grafts show compressed columnar epithelium, newly formed capillaries, and abundant circular bundles of smooth muscle, similar to controls. BSM grafts show fibrosis-like morphology and a simple smooth muscle layer as observed in the stricture group. (**i–l**) BSM/tissue grafts reveal scattered, variably sized circular bundles of smooth muscle. BSM grafts show few muscle fibers, extensive collagen deposition, and keratinized squamous cell epithelium. (**m–p**) BSM/tissue grafts express stratified columnar urothelium and stratified squamous cells. BSM grafts have a thin and irregular urothelial layer (400×)

grafts were not applied at all. In most studies, matrices were seeded with epithelial cells. Also, mesothelial cells, bladder smooth muscle cells, bone marrow mesenchymal stem cells, and rabbit adipose-derived stem cells were used. The authors did not mention any problems with expanding the cells or seeding them onto different matrices; 50% of the surgeries were performed with tubularized constructs; and 8 of 11 studies had control groups, where similar matrices without cells were used. All 11 studies showed significantly better results with cellularized matrices. Epithelial cells and smooth muscle cells at the site of surgery were observed at a mean follow-up time of 4 weeks.

9.2.3.4 Cellularized Matrices Utilized in Human Models (Monoculture)

Fossum et al. obtained urothelial cells via bladder washing [55]. The cells were cultured and seeded onto decellularized dermal matrices that were used to surgically treat six boys with scrotal or perineal hypospadias. The mean follow-up time was 87 months. Only one boy had to undergo an optical urethrotomy to improve voiding.

9.2.3.5 Recellularized Matrices Utilized in Animal Models (Co-Culture)

Feng et al. seeded two types of autologous cells (lingual keratinocytes and corporal smooth muscle cells) onto both sides of porcine acellular corpus spongiosum matrices (ACSMs) [59]. The best results were achieved in the group in which urethral defects were repaired with matrices containing both types of cells. Animals in the other two groups (ACSMs alone, and ACSMs plus OKs) developed strictures.

De Filippo et al. substituted 30 mm defects with either tubularized BAMGs alone, or BAMGs seeded with two types of cells (BAMGs plus BUCs plus BSMCs) [77]. The success rate in the group wherein cell-seeded matrices were utilized was 100%.

Mikami et al. formed tubularized tissue engineered urethras comprised of collagen mesh matrices and autologous cells of two types—keratinocytes and buccal mucosa smooth muscle cells [78]. These grafts were used to repair urethral defects in the experimental group (10 dogs) and were not applied to defects in the control group (10 dogs). The authors reported statistically significant differences in the number of complications between the two groups. The control dogs developed fistulae and strictures.

Orabi et al. used the same types of matrix in male dogs; 6 cm perineal urethra segments were removed and substituted with either tubularized matrices with two types of cells (BAMGs plus BUCs and BSMCs) or BAMGs alone (control group) [19]. In the control group, 100% of animals developed strictures and fistulae, whereas in the experimental group there were no strictures at the 12 month follow-up.

Li et al. hypothesized that TGFβ1 plays a very important role in the formation of fibrosis. Using OK- and TGFβ1 siRNA transfected fibroblast-seeded BAMGs, the authors were able to minimize the activity of TGFβ1 and avoid the formation of fibrosis [79]. In three groups of rabbits (nine animals/group) the urethral defects were repaired with: (a) BAMGs alone; (b) BAMGs plus OKs; and (c) BAMGs plus OKs plus TGF-β1-siRNA transfected fibroblasts. Rabbits in group (a) developed strictures, while those in groups (b) and (c) had none. However, in group (c) the formation of capillaries in the epithelial lower layer was observed at 6 months after implantation.

Xie et al. seeded autologous OKs and fibroblasts onto silk fibroin matrices; 5 cm urethral mucosal defects were repaired using matrices alone and matrices with two cell types (in five female dogs in each case) [80]. No strictures or complications developed in the group with cell-seeded matrices.

Tissue-engineered constructs with different cell types in a co-culture system are structurally more similar to the native urethra than other constructs. In three of the six studies reviewed above, BAMGs were utilized. The inner surface of the matrix was seeded with OKs (in four of six studies) or bladder urothelial cells (in two of six studies), while the outer surface of the matrix was seeded with smooth muscle cells or fibroblasts (in two of six studies). Each study included a control group in which matrices without cells were used. The authors did not mention any problems regarding culturing autologous cells or creating tissue-engineered constructs with two types of cells. The animal models were male rabbits (in three of six studies), male dogs (in two of six studies), and female dogs (in one study). Tubularized constructs were applied in 50% of studies. The authors used tissue-engineered constructs containing different cell types with 100% success in five of six studies and with 70% success in one study.

9.2.3.6 Recellularized Matrices Utilized in Human Models (Co-Culture)

Bhargava et al. were the first to use a tissue-engineered construct for urethral repair in clinical patients [54]. Sterilized donor de-epidermized dermis was utilized as a matrix, onto which buccal mucosa keratinocytes and fibroblasts were seeded. Five patients with urethral stricture secondary to lichen sclerosis participated in the study. Three patients underwent a two-stage procedure and two patients a one-stage procedure. Postoperatively, one patient required complete graft excision because of marked fibrosis, and one had partial graft excision due to hyperproliferation of tissue. Three patients required some form of instrumentation because of recurring strictures. At present, the mean follow-up time is 111.8 months [81]. All patients who underwent urethroplasties are able to void independently. All four patients with tissue-engineered buccal mucosa (TEBM) void spontaneously. Three of five patients used intermittent self-calibration.

Raya-Rivera et al. used a PGA matrix, of which the inner surface was seeded with expanded bladder epithelial cells and the outer surface with bladder smooth muscle cells [50]. All patients had post-traumatic obliteration of membranous urethra. They had scar tissues excised and tissue-engineered urethras implanted. The success rate was 100% at a 71 month follow-up. Biopsies showed that the grafts had developed a normal-appearing architecture by 3 months after implantation.

To sum up, the authors applied tissue-engineered constructs with two types of cells to manage very complicated cases of urethral stricture disease: lichen sclerosis in Bhargava et al. and post-traumatic membranous urethral obliteration in Raya-Rivera et al. [50, 54]. In the former, additional interventions were needed, while in the latter, the use of tubularized synthetic matrices seeded with urothelial and smooth muscle cells obtained from bladder demonstrated 100% success at long-term follow-up. Histological evaluation showed that the engineered grafts had developed apparently normal architecture by 3 months after implantation. All adverse fibrotic reactions were observed within the first year after implantation. The results have been durable for three out of five patients for 9 years, although all of them required re-interventions [81].

9.3 Current Challenges and Future Perspectives

Urethral tissue engineering provides fertile ground for further research, with many problems remaining unresolved. One of the questions to be answered is, "What are the potential indications for substitution urethroplasty procedures using tissue-engineered constructs?" At present, clinicians are successfully utilizing oral (buccal) mucosa to repair complex urethral strictures and hypospadias. Its use has yielded good results, not only in patients with primary urethral strictures but also in those with recurring strictures and lichen sclerosis, with most authors reporting low donor-site morbidity (sometimes extremely low, especially when mucosa is harvested from the cheek or the ventral surface of the tongue). Moreover, the procedure of harvesting is simple, reproducible, and well tolerated by the majority of patients. The view that oral mucosa may not provide enough tissue for urethral repair seems to be groundless. Mucosa from both cheeks as well as from the ventral surface of the tongue may provide enough material to create a 16 cm graft.

While buccal mucosa possesses many advantages, its use is associated with a number of problems that have to be addressed. First, any graft harvesting procedure entails some degree of donor site morbidity, which, no matter how minimal, still has a negative impact on the patients. The larger the amount of tissue removed from the donor site, the higher the negative impact. Second, long strictures may require harvesting of long grafts from several donor sites. This increases surgical duration as well as intraoperative and postoperative complication rates [82]. Third, in cases of recurrent strictures where a secondary urethroplasty is needed, the lack of oral mucosa may drive the need to find other substitution tissues. A similar problem may arise during the first stage of a two-stage reconstruction that often requires additional epithelialization of the urethral bed. Thus, selecting and designing an ideal substitution material remains a great challenge. Tissue-engineering options may help address this challenge. The development of a perfect tissue-engineered construct (matrix + cells obtained using a non-invasive technique, possibly from urine) will eliminate the problems of donor site morbidity and limited amount of available substitution tissue [83, 84], even in patients with long strictures and in those who have had previous urethroplasties. It will also decrease operative time and the risks of intraoperative and postoperative complications.

Among the problems our analysis of the reviewed preclinical studies brings to light is the problem of developing animal models for urethral repair. Only one preclinical study reviewed above created a model of urethral stricture: its authors excised a segment of the urethra 1 month before urethroplasties were performed, thus simulating the conditions under which grafts normally take place [65]. In most of the reviewed preclinical trials, a urethral defect was created intraoperatively, to be immediately repaired with a matrix. However, the process of graft take in an injured urethra may substantially differ from that in a normal urethral bed. What is more, only four out of twenty-eight preclinical studies used large animals (dogs) [18, 19, 78, 80], and only two of them used male dogs [19, 78]. While we are aware that, anatomically, the female and male urethras are different, and that spongiofibrosis, being the major cause of strictures in males, never occurs in females, we still decided to include the studies conducted on female animals in our review. The reason for doing so was that the authors of the studies introduced a novel technique to prepare matrices by electrospinning, and suggested using silk fibroin matrices [18, 80]. As for developing animal models for urethral repair, Sievert et al. introduced three methods of stricture induction in large animals (minipigs): urethrotomy, ligation, and thermocoagulation [85]. Stricture severity was found to be higher in the model that used thermocoagulation. From these results, we may assume that it is best to conduct experiments on large animals and develop a model of urethral stricture to mimic the conditions of graft take in humans.

It should be noted that SIS grafts were the most frequently applied matrices in the preclinical and clinical studies reviewed above. They were used in five of eleven preclinical studies and in nine of thirteen clinical studies. Synthetic matrices were not used at all, with the exception of one study that utilized a hybrid matrix [CLLC plus collagen sponge tubes (CST)]. All the matrices took well and caused no immunogenic and/or proinflammatory reactions. All of them (except CLLC plus CST, and SFM) were homogeneous or heterogeneous acellular grafts that contained mostly collagen and were devoid of any native cellular components, thus having reduced immunogenicity and bioactivity. The success of acellular matrices is known to be contingent on the vascularization of the graft and regeneration of native mucosa at the implantation site. Moreover, an ideal matrix must be biodegradable, its breakdown products must be non-toxic, and it must be substituted by the host's own intercellular matrix [86, 87]. Also, an ideal matrix should provide an environment in which adhesion, proliferation, migration, and differentiation of cells can occur to enable functional tissue formation [30, 31]. Therefore, when an acellular matrix is used for urethral reconstruction, the following processes should take place: (a) the matrix becomes covered with the native urothelium; (b) it completely degrades; and (c) it becomes substituted by the host's extracellular matrix (ideally, with smooth muscle fibers being formed).

Another issue to be touched upon is the efficacy of acellular matrices in both preclinical and clinical studies. As was shown by Dorin et al. [64], urothelial regeneration has some limitations, which is why acellular matrices are less successful in patients with long urethral strictures and in those who have had previous repairs (unsatisfactory vascularity). Hence, it is not advisable to use acellular matrices to repair strictures longer than 2 cm. As for short strictures (<2 cm), which, in theory, can be reconstructed with acellular matrices, they are successfully managed using either primary anastomosis or a non-transecting anastomotic technique [7, 88]. In cases of longer strictures, where urethroplasty is indicated [6], the use of acellular matrices does not seem to be expedient. Patients with an unhealthy urethral bed (severe spongiofibrosis, recurring strictures, or previous urethroplasties) have higher risks of poor graft vascularization, graft contraction, and stricture recurrence, as was shown in several clinical trials [21, 39, 40, 70]. To sum up, the applications for acellular matrices in the treatment of urethral strictures are limited.

Problems with acellular matrices made researchers look for other options. This is why they turned to designing tissue-engineered constructs, containing not only a matrix but also autologous cells. Interestingly, the cells seeded onto the surface of a matrix (when cultured in vitro and immediately after implantation onto a patient) survive by molecular diffusion. Later, the seeded cells enter a hypoxic state, thus stimulating the endogenous release of angiogenic growth factors [89, 90]. If adequate angiogenesis is achieved, the cells expand and new tissue is formed. In the absence of adequate angiogenesis, the cells die.

It is noteworthy that in the experimental studies that compared the efficacy of acellular and cellularized tissue engineered constructs, failure rates when using acellular matrices were significantly higher than were those in the studies that utilized only acellular matrices. This may be explained either by some degree of bias or by differences in study designs—the lengths of the urethral defects in the studies presented in the animal models using acellular matrices were 5–20 mm, whereas the lengths of the urethral defects in the studies presented in the animal models using cellularized matrices varied from 10 to 30 mm.

While SIS has been successfully utilized as a material for grafts in earlier trials, it was not used in the preclinical and clinical studies reviewed here. Most preclinical studies applied BAMGs, while two clinical studies used MukoCell® matrices and ADMs. Furthermore, the authors seeded matrices with different types of autologous epithelial cells, smooth muscle cells, and stem cells. Preclinical studies showed better success with cellularized matrices than with acellular ones. Fossum et al. obtained urothelial cells via bladder washing [55]. Urethroplasties with this type of cell were rather successful (83%) at a mean follow-up time of 87 months. We believe bladder washing is a promising technique, as it is minimally invasive and simple. Recently, some authors have reported that stem cells can also be obtained from urine [83, 84, 91], which makes this approach even more attractive.

Since the native urethra is comprised of the inner epithelial lining and corpus spongiosum that contain endothelial and smooth muscle cells, the next step taken by some researchers was to create a tissue-engineered construct that would contain a matrix and two types of cell. As a rule, in such cases, the inner surface is seeded with epithelial cells, while the outer surface is seeded with smooth muscle cells. Preclinical studies mostly applied BAMGs, whereas two clinical studies used ADMs and PGA matrices (thus, in the reviewed studies, a synthetic matrix was used only once). The inner surface was seeded either with bladder epithelial cells or with buccal mucosa cells. The use of bladder urothelial cells seems to be preferable; however, the main drawback is that they are obtained via an open bladder biopsy. Therefore, in this respect, buccal mucosa cells appear to be a better choice, as the technique employed to obtain them is less invasive.

Another important question regards graft placement. Ventral graft placement is preferred in patients with urethral strictures located in the proximal bulbar urethra [14, 92, 93]. This is because, in the proximal bulbar urethra, the lumen is dorsally eccentric, and the ventral aspect of the surrounding tissues of the corpus spongiosum is rather thick. This thick portion of the bulbar urethra has sufficient blood supply to vascularize the graft. Moving distally, the urethral lumen becomes more centrally placed and the corpus spongiosum becomes thinner, which is why the use of ventral graft location may be associated with higher risks of inadequate vascularization and stricture recurrence. Thus, in patients with strictures located in the distal bulbar urethra, one should opt for dorsal (onlay or inlay) graft placement. This is why we believe dorsal graft placement should be preferred in clinical studies assessing the use of tissue-engineered constructs.

Overall, the evidence presented in the reviewed studies appears to support the view that tissue-engineered constructs with different cell types should be used as grafts of choice in substitution urethroplasty procedures. Tissue-engineered constructs should be utilized in the following groups of patients: in those with long strictures (>2 cm); in those with multiple strictures; in those with subtotal and total strictures; in those with an unhealthy urethral bed (severe spongiofibrosis, recurring strictures after multiple urethrotomies and dilations, previous urethroplasties); and in those with soft tissue deficit at the site of surgery (hypospadias, failed hypospadia repair). The use of "off-the-shelf" tissue engineered constructs can minimize (or, in cases where non-invasive techniques to obtain autologous cells are employed, even eliminate) donor site morbidity, decrease operative time, and reduce the risks of intraoperative and postoperative complications. However, further investigations are needed to prove this.

Interestingly, despite the fact that there has been much international debate regarding naturally derived vs synthetic scaffolds, almost all the studies reviewed here utilized natural matrices. Only two studies applied a hybrid matrix and a synthetic one. Therefore, more attention should be paid to synthetic matrices and their possible applications in substitution urethroplasty procedures.

Urethral tissue engineering is, without any doubt, a promising field of regenerative medicine. Yet creating a tissue-engineered construct with different types of autologous cells is a complex biotechnological process, requiring clean room laboratories and highly specialized personnel. A lack of the mentioned facilities and specialists, as well as a lack of clinical studies, limit the wide application of tissue engineering to replace urethras. Urethral tissue engineering provides effective treatment options for patients with strictures and congenital anomalies. However, with the development of biotechnologies in this field, we will harness new research results that will enable us to tackle other problems as well, such as various lesions of the bladder and ureter, chordee, or erectile dysfunction.

Conclusions

The studies reviewed in this chapter have described different approaches to the development of matrices, harvesting and expanding autologous cells, and producing tissue-engineered constructs of various complexities (patch grafts and tubularized grafts). More clinical studies are needed to explore these and other problems on a larger scale.

References

1. Bhatnagar A, Upadhyaya VD, Kumar B. Congenital urethrocutaneous fistula: case report with review of literature. Indian J Plast Surg. 2012;45:563–5.
2. Elgammal MA. Straddle injuries to the bulbar urethra: management and outcome in 53 patients. Int Braz J Urol. 2009;35:450–8.
3. Kim BS, Kwon TG. Urethral reconstruction using autologous vein grafts for the management of urethral strictures. Curr Urol Rep. 2015;16:467.
4. Macedo A Jr, Rondon A, Ortiz V. Hypospadias. Curr Opin Urol. 2012;22:447–52.
5. Chapple C. Anterior urethral surgery: current concepts and future directions. Eur Urol. 2010;58:42–5.
6. Chapple C, Andrich D, Atala A, Barbagli G, Cavalcanti A, Kulkarni S, et al. SIU/ICUD consultation on urethral strictures: the management of anterior urethral stricture disease using substitution urethroplasty. Urology. 2014;83:S31–47.
7. Morey AF, Watkin N, Shenfeld O, Eltahawy E, Giudice C. SIU/ICUD consultation on urethral strictures: anterior urethra--primary anastomosis. Urology. 2014;83:S23–6.
8. Fenton AS, Morey AF, Aviles R, Garcia CR. Anterior urethral strictures: etiology and characteristics. Urology. 2005;65:1055–8.
9. Tritschler S, Roosen A, Fullhase C, Stief CG, Rubben H. Urethral stricture: etiology, investigation and treatments. Dtsch Arztebl Int. 2013;110:220–6.
10. Lumen N, Hoebeke P, Willemsen P, De Troyer B, Pieters R, Oosterlinck W. Etiology of urethral stricture disease in the 21st century. J Urol. 2009;182:983–7.
11. Chapple C, Barbagli G, Jordan G, Mundy AR, Rodrigues-Netto N, Pansadoro V, et al. Consensus statement on urethral trauma. BJU Int. 2004;93:1195–202.
12. Bhandari M, Dubey D, Verma BS. Dorsal or ventral placement of the preputial/penile skin onlay flap for anterior urethral strictures:does it make a difference? BJU Int. 2001;88:39–43.
13. Kozinn SI, Harty NJ, Zinman L, Buckley JC. Management of complex anterior urethral strictures with multistage buccal mucosa graft reconstruction. Urology. 2013;82:718–22.
14. Barbagli G, Lazzeri M. Penile urethral stricture reconstruction--flap or graft? Graft. J Urol. 2011;186:375–6.
15. Venn SN, Mundy AR. Urethroplasty for balanitis xerotica obliterans. Br J Urol. 1998;81:735–7.
16. Xu YM, Xu QK, Fu Q, Sa YL, Zhang J, Song LJ, et al. Oral complications after lingual mucosal graft harvesting for urethroplasty in 110 cases. BJU Int. 2011;108:140–5.
17. Cattan V, Bernard G, Rousseau A, Bouhout S, Chabaud S, Auger FA, et al. Mechanical stimuli-induced urothelial differentiation in a human tissue-engineered tubular genitourinary graft. Eur Urol. 2011;60:1291–8.

18. Xie M, Song L, Wang J, Fan S, Zhang Y, Xu Y. Evaluation of stretched electrospun silk fibroin matrices seeded with urothelial cells for urethra reconstruction. J Surg Res. 2013;184:774–81.
19. Orabi H, AbouShwareb T, Zhang Y, Yoo JJ, Atala A. Cell-seeded tubularized scaffolds for reconstruction of long urethral defects: a preclinical study. Eur Urol. 2013;63:531–8.
20. Chun SY, Kim BS, Kwon SY, Park SI, Song PH, Yoo ES, et al. Urethroplasty using autologous urethral tissue-embedded acellular porcine bladder submucosa matrix grafts for the management of long-segment urethral stricture in a rabbit model. J Korean Med Sci. 2015;30:301–7.
21. Ahn HH, Kim KS, Lee JH, Lee MS, Song IB, Cho MH, et al. Porcine small intestinal submucosa sheets as a scaffold for human bone marrow stem cells. Int J Biol Macromol. 2007;41:590–6.
22. Narayanan K, Leck KJ, Gao S, Wan AC. Three-dimensional reconstituted extracellular matrix scaffolds for tissue engineering. Biomaterials. 2009;30:4309–17.
23. Ng SL, Narayanan K, Gao S, Wan AC. Lineage restricted progenitors for the repopulation of decellularized heart. Biomaterials. 2011;32:7571–80.
24. Bazeed MA, Thuroff JW, Schmidt RA, Tanagho EA. New treatment for urethral strictures. Urology. 1983;21:53–7.
25. Olsen L, Bowald S, Busch C, Carlsten J, Eriksson I. Urethral reconstruction with a new synthetic absorbable device. An experimental study. Scand J Urol Nephrol. 1992;26:323–6.
26. Atala A, Vacanti JP, Peters CA, Mandell J, Retik AB, Freeman MR. Formation of urothelial structures in vivo from dissociated cells attached to biodegradable polymer scaffolds in vitro. J Urol. 1992;148:658–62.
27. Chen F, Yoo JJ, Atala A. Acellular collagen matrix as a possible "off the shelf" biomaterial for urethral repair. Urology. 1999;54:407–10.
28. Sievert KD, Tanagho EA. Organ-specific acellular matrix for reconstruction of the urinary tract. World J Urol. 2000;18:19–25.
29. Micol LA, Arenas da Silva LF, Geutjes PJ, Oosterwijk E, Hubbell JA, Feitz WF, et al. In-vivo performance of high-density collagen gel tubes for urethral regeneration in a rabbit model. Biomaterials. 2012;33:7447–55.
30. Atala A. Regenerative medicine strategies. J Pediatr Surg. 2012;47:17–28.
31. Atala A. Tissue engineering of reproductive tissues and organs. Fertil Steril. 2012;98:21–9.
32. Micol LA, Ananta M, Engelhardt EM, Mudera VC, Brown RA, Hubbell JA, et al. High-density collagen gel tubes as a matrix for primary human bladder smooth muscle cells. Biomaterials. 2011;32:1543–8.
33. Ribeiro-Filho LA, Sievert KD. Acellular matrix in urethral reconstruction. Adv Drug Deliv Rev. 2015;82-83:38–46.
34. Atala A, Guzman L, Retik AB. A novel inert collagen matrix for hypospadias repair. J Urol. 1999;162:1148–51.
35. El-Kassaby AW, Retik AB, Yoo JJ, Atala A. Urethral stricture repair with an off-the-shelf collagen matrix. J Urol. 2003;169:170–3. discussion 3
36. el-Kassaby A, AbouShwareb T, Atala A. Randomized comparative study between buccal mucosal and acellular bladder matrix grafts in complex anterior urethral strictures. J Urol. 2008;179:1432–6.
37. Feng C, Xu YM, Fu Q, Zhu WD, Cui L, Chen J. Evaluation of the biocompatibility and mechanical properties of naturally derived and synthetic scaffolds for urethral reconstruction. J Biomed Mater Res A. 2010;94:317–25.
38. le Roux PJ. Endoscopic urethroplasty with unseeded small intestinal submucosa collagen matrix grafts: a pilot study. J Urol. 2005;173:140–3.
39. Palminteri E, Berdondini E, Colombo F, Austoni E. Small intestinal submucosa (SIS) graft urethroplasty: short-term results. Eur Urol. 2007;51:1695–701. discussion 701
40. Palminteri E, Berdondini E, Fusco F, De Nunzio C, Salonia A. Long-term results of small intestinal submucosa graft in bulbar urethral reconstruction. Urology. 2012;79:695–701.
41. Sievert KD, Bakircioglu ME, Nunes L, Tu R, Dahiya R, Tanagho EA. Homologous acellular matrix graft for urethral reconstruction in the rabbit: histological and functional evaluation. J Urol. 2000;163:1958–65.
42. Lin J, Hao JR, Jin J, Deng SM, Hu J, Na YQ. Homologous dermal acellular matrix graft for urethral reconstruction in man (report of 16 cases). Zhonghua Yi Xue Za Zhi. 2005;85:1057–9.
43. Fossum M, Svensson J, Kratz G, Nordenskjold A. Autologous in vitro cultured urothelium in hypospadias repair. J Pediatr Urol. 2007;3:10–8.
44. Williams DF. Biomaterials and biocompatibility. Med Prog Technol. 1976;4:31–42.
45. Freed LE, Vunjak-Novakovic G, Biron RJ, Eagles DB, Lesnoy DC, Barlow SK, et al. Biodegradable polymer scaffolds for tissue engineering. Biotechnology (N Y). 1994;12:689–93.
46. Mikos AG, Lyman MD, Freed LE, Langer R. Wetting of poly(L-lactic acid) and poly(DL-lactic-co-glycolic acid) foams for tissue culture. Biomaterials. 1994;15:55–8.
47. Harris LD, Kim BS, Mooney DJ. Open pore biodegradable matrices formed with gas foaming. J Biomed Mater Res. 1998;42:396–402.
48. Han D, Gouma PI. Electrospun bioscaffolds that mimic the topology of extracellular matrix. Nanomedicine. 2006;2:37–41.
49. Lee SJ, Oh SH, Liu J, Soker S, Atala A, Yoo JJ. The use of thermal treatments to enhance the mechanical properties of electrospun poly(epsilon-caprolactone) scaffolds. Biomaterials. 2008;29:1422–30.
50. Raya-Rivera A, Esquiliano DR, Yoo JJ, Lopez-Bayghen E, Soker S, Atala A. Tissue-engineered

autologous urethras for patients who need reconstruction: an observational study. Lancet. 2011; 377:1175–82.
51. Kanatani I, Kanematsu A, Inatsugu Y, Imamura M, Negoro H, Ito N, et al. Fabrication of an optimal urethral graft using collagen-sponge tubes reinforced with Copoly(L-lactide/epsilon-caprolactone) fabric. Tissue Eng. 2007;13:2933–40.
52. Fu WJ, Xu YD, Wang ZX, Li G, Shi JG, Cui FZ, et al. New ureteral scaffold constructed with composite poly(L-lactic acid)-collagen and urothelial cells by new centrifugal seeding system. J Biomed Mater Res A. 2012;100:1725–33.
53. Horst M, Madduri S, Milleret V, Sulser T, Gobet R, Eberli D. A bilayered hybrid microfibrous PLGA--acellular matrix scaffold for hollow organ tissue engineering. Biomaterials. 2013;34:1537–45.
54. Bhargava S, Patterson JM, Inman RD, MacNeil S, Chapple CR. Tissue-engineered buccal mucosa urethroplasty-clinical outcomes. Eur Urol. 2008;53: 1263–9.
55. Fossum M, Skikuniene J, Orrego A, Nordenskjold A. Prepubertal follow-up after hypospadias repair with autologous in vitro cultured urothelial cells. Acta Paediatr. 2012;101:755–60.
56. De Filippo RE, Yoo JJ, Atala A. Urethral replacement using cell seeded tubularized collagen matrices. J Urol. 2002;168:1789–92. discussion 92-3
57. Li C, Xu YM, Song LJ, Fu Q, Cui L, Yin S. Urethral reconstruction using oral keratinocyte seeded bladder acellular matrix grafts. J Urol. 2008;180:1538–42.
58. Fu Q, Deng CL, Song XF, Xu YM. Long-term study of male rabbit urethral mucosa reconstruction using epidermal cell. Asian J Androl. 2008;10:719–22.
59. Feng C, Xu YM, Fu Q, Zhu WD, Cui L. Reconstruction of three-dimensional neourethra using lingual keratinocytes and corporal smooth muscle cells seeded acellular corporal spongiosum. Tissue Eng Part A. 2011;17:3011–9.
60. Kropp BP, Ludlow JK, Spicer D, Rippy MK, Badylak SF, Adams MC, et al. Rabbit urethral regeneration using small intestinal submucosa onlay grafts. Urology. 1998;52:138–42.
61. Nuininga JE, van Moerkerk H, Hanssen A, Hulsbergen CA, Oosterwijk-Wakka J, Oosterwijk E, et al. Rabbit urethra replacement with a defined biomatrix or small intestinal submucosa. Eur Urol. 2003;44:266–71.
62. Yang SX, Yao Y, Hu YF, Song C, Wang LL, Jin HM. Reconstruction of rabbit urethra using urethral extracellular matrix. Chin Med J. 2004;117:1786–90.
63. Huang X, Luo J, Liao Y, Qu Y, Yang Z. Study on small intestinal submucosa as repair materials in urethral reconstruction. Zhongguo Xiu Fu Chong Jian Wai Ke Za Zhi. 2006;20:206–9.
64. Dorin RP, Pohl HG, De Filippo RE, Yoo JJ, Atala A. Tubularized urethral replacement with unseeded matrices: what is the maximum distance for normal tissue regeneration? World J Urol. 2008;26:323–6.
65. Villoldo GM, Loresi M, Giudice C, Damia O, Moldes JM, DeBadiola F, et al. Histologic changes after urethroplasty using small intestinal submucosa unseeded with cells in rabbits with injured urethra. Urology. 2013;81:1380 e1-5.
66. Chung YG, Tu D, Franck D, Gil ES, Algarrahi K, Adam RM, et al. Acellular bi-layer silk fibroin scaffolds support tissue regeneration in a rabbit model of onlay urethroplasty. PLoS One. 2014;9:e91592.
67. Kajbafzadeh AM, Sabetkish S, Tourchi A, Amirizadeh N, Afshar K, Abolghasemi H, et al. The application of tissue-engineered preputial matrix and fibrin sealant for urethral reconstruction in rabbit model. Int Urol Nephrol. 2014;46:1573–80.
68. Mantovani F, Trinchieri A, Castelnuovo C, Romano AL, Pisani E. Reconstructive urethroplasty using porcine acellular matrix. Eur Urol. 2003;44:600–2.
69. Donkov II, Bashir A, Elenkov CH, Panchev PK. Dorsal onlay augmentation urethroplasty with small intestinal submucosa: modified Barbagli technique for strictures of the bulbar urethra. Int J Urol. 2006;13:1415–7.
70. Hauser S, Bastian PJ, Fechner G, Muller SC. Small intestine submucosa in urethral stricture repair in a consecutive series. Urology. 2006;68:263–6.
71. Fiala R, Vidlar A, Vrtal R, Belej K, Student V. Porcine small intestinal submucosa graft for repair of anterior urethral strictures. Eur Urol. 2007;51:1702–8. discussion 8
72. Farahat YA, Elbahnasy AM, El-Gamal OM, Ramadan AR, El-Abd SA, Taha MR. Endoscopic urethroplasty using small intestinal submucosal patch in cases of recurrent urethral stricture: a preliminary study. J Endourol. 2009;23:2001–5.
73. Gu GL, Zhu YJ, Xia SJ, Zhang J, Jiang JT, Hong Y, et al. Peritoneal cavity as bioreactor to grow autologous tubular urethral grafts in a rabbit model. World J Urol. 2010;28:227–32.
74. Gu GL, Xia SJ, Zhang J, Liu GH, Yan L, Xu ZH, et al. Tubularized urethral replacement using tissue-engineered peritoneum-like tissue in a rabbit model. Urol Int. 2012;89:358–64.
75. Li H, Xu Y, Xie H, Li C, Song L, Feng C, et al. Epithelial-differentiated adipose-derived stem cells seeded bladder acellular matrix grafts for urethral reconstruction: an animal model. Tissue Eng Part A. 2014;20:774–84.
76. Wang F, Liu T, Yang L, Zhang G, Liu H, Yi X, et al. Urethral reconstruction with tissue-engineered human amniotic scaffold in rabbit urethral injury models. Med Sci Monit. 2014;20:2430–8.
77. De Filippo RE, Kornitzer BS, Yoo JJ, Atala A. Penile urethra replacement with autologous cell-seeded tubularized collagen matrices. J Tissue Eng Regen Med. 2015;9:257–64.
78. Mikami H, Kuwahara G, Nakamura N, Yamato M, Tanaka M, Kodama S. Two-layer tissue engineered urethra using oral epithelial and muscle derived cells. J Urol. 2012;187:1882–9.

79. Li C, Xu YM, Liu ZS, Li HB. Urethral reconstruction with tissue engineering and RNA interference techniques in rabbits. Urology. 2013;81:1075–80.
80. Xie M, Xu Y, Song L, Wang J, Lv X, Zhang Y. Tissue-engineered buccal mucosa using silk fibroin matrices for urethral reconstruction in a canine model. J Surg Res. 2014;188:1–7.
81. Osman NI, Patterson JM, MacNeil S, Chapple CR. Long-term follow-up after tissue-engineered buccal mucosa urethroplasty. Eur Urol. 2014;66:790–1.
82. Kim BD, Ver Halen JP, Grant DW, Kim JY. Anesthesia duration as an independent risk factor for postoperative complications in free flap surgery: a review of 1,305 surgical cases. J Reconstr Microsurg. 2014;30:217–26.
83. Zhang Y, McNeill E, Tian H, Soker S, Andersson KE, Yoo JJ, et al. Urine derived cells are a potential source for urological tissue reconstruction. J Urol. 2008;180:2226–33.
84. Bharadwaj S, Liu G, Shi Y, Markert C, Andersson KE, Atala A, et al. Characterization of urine-derived stem cells obtained from upper urinary tract for use in cell-based urological tissue engineering. Tissue Eng Part A. 2011;17:2123–32.
85. Sievert KD, Selent-Stier C, Wiedemann J, Greiner TO, Amend B, Stenzl A, et al. Introducing a large animal model to create urethral stricture similar to human stricture disease: a comparative experimental microscopic study. J Urol. 2012;187:1101–9.
86. Kim BS, Mooney DJ. Development of biocompatible synthetic extracellular matrices for tissue engineering. Trends Biotechnol. 1998;16:224–30.
87. Silver FH, Pins G. Cell growth on collagen: a review of tissue engineering using scaffolds containing extracellular matrix. J Long-Term Eff Med Implants. 1992;2:67–80.
88. Andrich DE, Mundy AR. Non-transecting anastomotic bulbar urethroplasty: a preliminary report. BJU Int. 2012;109:1090–4.
89. Kannan RY, Salacinski HJ, Sales K, Butler P, Seifalian AM. The roles of tissue engineering and vascularisation in the development of micro-vascular networks: a review. Biomaterials. 2005;26:1857–75.
90. Rouwkema J, Rivron NC, van Blitterswijk CA. Vascularization in tissue engineering. Trends Biotechnol. 2008;26:434–41.
91. Wu S, Wang Z, Bharadwaj S, Hodges SJ, Atala A, Zhang Y. Implantation of autologous urine derived stem cells expressing vascular endothelial growth factor for potential use in genitourinary reconstruction. J Urol. 2011;186:640–7.
92. Kulkarni S, Barbagli G, Sansalone S, Lazzeri M. One-sided anterior urethroplasty: a new dorsal onlay graft technique. BJU Int. 2009;104:1150–5.
93. Barbagli G, Sansalone S, Djinovic R, Romano G, Lazzeri M. Current controversies in reconstructive surgery of the anterior urethra: a clinical overview. Int Braz J Urol. 2012;38:307–16. discussion 16

Urethral Sphincter: Stress Urinary Incontinence

10

Eun Sang Yoo and Jun Nyung Lee

10.1 Introduction

10.1.1 Stress Urinary Incontinence

10.1.1.1 Introduction of Stress Urinary Incontinence Including Symptoms

Urinary incontinence is described as the involuntary loss of urine and is a common condition in middle-aged and elderly women and men [1]. Urinary incontinence can be generally classified into the following three: (1) stress urinary incontinence, (2) urge urinary incontinence, and (3) a mixed form of (1) and (2) [1]. Stress urinary incontinence occurs when increased intra-abdominal pressure causes bladder pressure to exceed urethral pressure, resulting in involuntary leakage of urine. In stress urinary incontinence, urine leakage can be observed during coughing, sneezing, laughing, lifting, and exercising. Classically stress urinary incontinence often relies on distinguishing between intrinsic sphincter deficiency and urethral malposition or hypermobility [2]. However, this is controversial as each component may contribute in varying proportion to the occurrence of stress urinary incontinence.

E.S. Yoo (✉) • J.N. Lee
Department of Urology, School of Medicine, Kyungpook National University, Daegu, South Korea
e-mail: uroyoo@knu.ac.kr; ljnlover@gmail.com

10.1.1.2 Definition of Stress Urinary Incontinence

The International Continence Society defined stress urinary incontinence as the involuntary leakage of urine on effort, exertion, sneezing, or coughing. It is estimated that one third of the procedures that are performed to treat stress urinary incontinence are performed on patients with recurrent disease. Currently accepted procedures for stress urinary incontinence are usually based on compensatory and nonphysiological mechanisms. Although the complete success of long-term management of any condition is based on an understanding of its pathophysiology, the pathophysiology of stress urinary incontinence is not well defined. Therefore, further investigations including the development of animal models to properly understand the condition and alternative managements based on pathophysiological changes are necessary.

10.1.1.3 Prevalence of Stress Urinary Incontinence

Inappropriate micturition accounts for a large portion of the affected patients. Especially, urinary incontinence is a major health problem causing a significant social and economic impact affecting more than 200 million people worldwide [3]. The prevalence rises with age, and therefore the magnitude of this problem is expected to substantially increase as the life expectancy and aging population continue to rise in the world [3]. The incidence of incontinence is

B.W. Kim (ed.), *Clinical Regenerative Medicine in Urology*, DOI 10.1007/978-981-10-2723-9_10

increasing by more than one third of middle-aged females report leakage at least weekly with 10% reporting daily or severe leakage [4].

10.1.1.4 Etiology of Stress Urinary Incontinence

Voiding dysfunction represents a diverse spectrum of urologic conditions including stress urinary incontinence and overactive bladder: two of the most common and challenging conditions faced by urologists today. Several of these disorders are precipitated by an acute injury such as vaginal delivery in female stress urinary incontinence or radical prostatectomy in male stress urinary incontinence. The underlying pathophysiology comprises a complex interplay of injuries or altered function at the tissue, nerve, and vascular level that is then exacerbated by underlying comorbidities, including obesity, increasing age, and diabetes mellitus.

Male

The most common cause in the development of male stress urinary incontinence is iatrogenic damage due to radical prostatectomy, transurethral resection of the prostate, and radiation therapy. Postprostatectomy stress urinary incontinence is of particular concern because data shows that anywhere from 2% to 66% of males are affected depending on the chosen group of patients and methods of measurement [5]. Radical prostatectomy may cause direct injury to the urethral sphincter via transection, penetration, a combination of both, or indirect neuromuscular damage.

Female

In women, the most influential factors for stress urinary incontinence are vaginal delivery and increased age [6]. The exact mechanism of injury is not well understood but is likely multifactorial. Vaginal childbirth produces mechanical and neurovascular injury to the pelvic floor, and aging plays a negative role in the structure and function of the pelvic floor. Combined with other potential risk factors such as parity, obesity, and menopause, vaginal childbirth contributes to a decrease in the number and diameter of the periurethral striated and smooth muscle fibers. As a result, patients often do not become symptomatic until years after the initial trauma of childbirth due to a cascade of events which continues to occur with age. Therefore, relevant animal models are also essential in understanding these events and in the development of preventative interventions.

10.1.1.5 Mechanism of Stress Urinary Incontinence

Maintaining urinary continence involves several aspects:

1. A stable bladder with adequate capacity and accommodation,
2. An anatomically normal and functionally competent continence mechanism (consisting of a bladder neck, urethra, urethral sphincter [itself consisting of striated and smooth muscle layers, neuronal innervations, a vascular plexus, submucosa, and epithelium], endopelvic fascia, arcus tendineus, and pelvic muscle support)
3. The correct integrity of somatic and autonomic innervation of the structures involved

The muscular structures are controlled by three sets of major nerves: (1) parasympathetic sacral nerves (pelvic nerves), (2) sympathetic thoracolumbar nerves innervating the periurethral smooth muscle, and (3) somatic sacral nerves (pudendal nerves) innervating the urethral striated muscles including the external periurethral sphincter and the pelvic floor muscles. Assuming proper bladder function, damage to the urethral sphincter muscle or its innervation can be produced by various ways in both men and women, resulting in urethral sphincter damage or deficit and ultimately stress urinary incontinence. Thus, the mechanism of human stress urinary incontinence is a complex and usually multifactorial process sometimes combining denervation, muscle degeneration and apoptosis, chronic muscle atrophy, fibrosis, and connective tissue disorders, among others, which makes the generation of an in vivo model that integrates all pathophysiological aspects of stress urinary incontinence difficult, if not impossible.

The pathophysiology of stress urinary incontinence has not been fully understood. Several hypotheses have been proposed to explain the mechanism of urethral closure [7]. One school of thought considers the rhabdosphincter (a horseshoe-shaped muscle which sits on the dorsal part of the midurethra), along with the urethral smooth muscle and the vascular elements within the submucosa, as the principal structures for continence. The second school of thought considers that stress urinary incontinence results from the weakening or loss of the structures which support the bladder neck and proximal urethra (ligaments, fascias, and muscles) [8] and that the rhabdosphincter plays a minor role in closure [8]. Probably the majority of women with stress urinary incontinence have varying degrees of both intrinsic sphincter weakness and loss of the normal anatomic support to the bladder neck and proximal urethra resulting in hypermobility [9]. The primary cause of male stress urinary incontinence is damage to the urethral sphincter during surgical procedures, mostly radical prostatectomy.

In females, urinary continence relies on an intact urethral sphincteric mechanism. Multiple factors contribute to urethral pressure including bladder neck position, urethral sphincter musculature, sphincter innervation, and surrounding vascular supply and tissue support. Pregnancy and childbirth are well-recognized risk factors for stress urinary incontinence, and four related major mechanisms of injury have been identified: (1) injury to connective tissue support during vaginal delivery, (2) vascular damage due to fetal compression of surrounding pelvic structures, (3) traumatic injury to pelvic nerves and musculature, and (4) direct injury to the lower urinary tract during childbirth.

In postprostatectomy stress urinary incontinence, contemporary series report improved continence rates following prostatectomy although wide variation continues to exist due to discrepancies in data collection methodology and patient-versus surgeon-reported outcomes. Nonetheless, while treatment options have improved in recent years, the burden of disease remains high due to the increasing numbers of radical prostatectomies performed annually. Postprostatectomy incontinence results from a failure to store urine secondary to inadequate resistance of the outlet sphincter. Surgical damage to the urethral sphincter occurs due to direct injury to the sphincter itself, as well as to surrounding nerves and supportive tissue [10].

10.1.1.6 Current Clinical Management of Stress Urinary Incontinence

Male

Over time diverse treatments for stress urinary incontinence have been used. As in females, surgical therapies are the mainstay of therapy and include transurethral bulking agents, bulbar urethral slings, and artificial urinary sphincters. These treatments aim to augment a deficient urinary sphincter by allowing urethral coaptation during periods of increased abdominal pressure without causing outlet obstruction during voiding. Currently, the gold standard in treatment for moderate male stress urinary incontinence is the use of the male sling or, in the case of severe incontinence, implantation of an artificial sphincter. However, 5 years post-artificial urinary sphincter implantation, up to 24% of men still report severe incontinence, and 28% will have undergone an artificial urinary sphincter revision [11]. In addition, there are many complications associated with these procedures. Although treatment for male stress urinary incontinence due to sphincter damage has evolved in the past 40 years, this disease will continue to be an unresolved social health problem. The development of new treatments focusing on cellular sphincter regeneration could play a very important role in the years to come.

Female

Clinical treatment of stress urinary incontinence is complex and includes conservative techniques, pharmacologic therapy, and surgical procedures. Patients with stress urinary incontinence can benefit from initial conservative measures, including pelvic floor physiotherapy, biofeedback, and electrical stimulation. The goal of surgical procedures is anatomical and functional restoration of the pelvic structures. Urethral slings and suspensions aim to correct stress urinary incontinence by correcting hypermobility and augmenting intrinsic sphincter deficiency by allowing urethral compression during periods of increased intra-abdominal pressure without causing obstruction during voiding. As of now, more than 100 surgical methods have been performed for treatment of stress urinary incontinence, with various success rates. These techniques may be broadly subcategorized into four major types: periurethral bulking agent injection, retropubic suspension, sling and tape procedures, and artificial urinary sphincter devices [12]. In contemporary practice, the use of midurethral slings has shown cost-effective objective results and subjective success [13]. Currently, the gold standard for surgical treatment of all types of stress urinary incontinence in women is midurethral slings. The surgical options remain the mainstay for cases nonresponsive to conservative measures [14, 15], with a success rate for different sling operations of up to 80–90% and with observation times up to 17 years [16]. In cases of severe incontinence such as intrinsic sphincter deficiency, artificial urinary sphincter implantation is an appropriate treatment option [17] since other treatments such as placement of suburethral bands or injection of bulking agents have not proved effective enough for this condition [13]. Although the use of artificial sphincters is highly effective in severe incontinence cases, the overall complication rates are not negligible, and recurrence of stress urinary incontinence can occur.

10.1.1.7 Limitation of Current Therapy

Although the tension-free vaginal tape is accepted as gold standard procedure for surgical treatment of stress urinary incontinence in women, midurethral slings are associated with complications including overactive bladder/urgency; urinary retention; bladder, urethral, or vaginal perforation; erosion; infections; persistent groin/suprapubic pain; and dyspareunia [18, 19]. Incidences of complications after midurethral slings are erosion rates in up to 23% of cases, permanent retention in up to 5% of cases, as well as wound complications, bladder perforation, persistent groin/ suprapubic pain, and dyspareunia in up to 15% of cases. Because of not sustainable efficacy and complications of standard therapies of stress urinary incontinence, there is a need to find a new, effective, and minimally invasive method of treatment. Although injection of a bulking agent is attempted as a minimal alternative of stress urinary incontinence treatment, the long-term cure rates are poor [20].

10.1.2 Regenerative Medicine for Stress Urinary Incontinence

10.1.2.1 Purpose of Regenerative Medicine

Despite significant developments in the underlying pathophysiology of stress urinary incontinence, current gold standard therapies remain lacking in their ability to target these fundamental mechanisms of injury or etiology of disease. To date, none of the standard therapies focus on halting disease progression or reversing underlying injury. Regenerative medicine concepts have emerged as an exciting means of fulfilling this therapeutic void by restoring and maintaining normal function through direct effects on injured or dysfunctional tissues. Over the past decade, the use of stem cells has shown promise for a host of urologic disorders including applications in lower urinary tract dysfunction, ureteral and bladder trauma, erectile dysfunction, and renal disease.

Stem cells are classically thought to improve tissue repair through multilineage differentiation and self-renewal. Stem cells may also exert a therapeutic effect through the secretion of bioactive factors that have antiapoptotic, antiscarring, neovascularization, and immunomodulatory effects on innate tissues and can direct innate stem and progenitor cells to the area of injury. Multiple treatment avenues using stem cells for voiding dysfunction, especially stress urinary incontinence, have been evaluated with preclinical animal models and clinical trials demonstrating their potential to restore function via direct effects on the underlying mechanisms that lead to incontinence or voiding dysfunction. Currently, many challenges remain to translate these promising results to clinical practice.

10.1.2.2 Application of Regenerative Medicine for Stress Urinary Incontinence

Currently, the most promising method of stress urinary incontinence treatment is cell implantation which is based on regeneration of the damaged rhabdosphincter. The main concept of cell-based therapies for stress urinary incontinence involves the implantation of cells into the damaged parts of the urethral sphincter complex, which can restore its proper function through the induction of muscle and nerve regeneration. This is an active, complex process that involves matrix remodeling necessary for restoration of functional cells responsible for the proper function of the sphincter complex and ultimately continence. This effect is achieved by stem cell differentiation into new striated muscle cells or neurons that replace injured parts or by induced regeneration of host tissue, triggered by trophic factors released by implanted cells. Therefore, a cell-based regenerative therapy could be established as a novel treatment option for future urethral sphincter muscle regeneration.

10.1.2.3 Purpose of This Chapter

In this chapter, we provide the current state of knowledge regarding cell therapy for stress urinary incontinence including stem cell sources, the potential mechanisms of stem cell therapy, experimental trials, and clinical trials, as well as new approaches of using regenerative pharmacology as an adjunct or replacement for cell therapy.

10.2 Body

10.2.1 Regenerative Therapy of Stress Urinary Incontinence As Alternative Approach

10.2.1.1 Overview of Cell Therapy for Stress Urinary Incontinence

The field of stem cell research began with the discovery of mouse embryonic stem cells in the early 1970s and with the description of human embryonic stem cells in 1998 [21, 22]. Stem cells comprise a unique population of cells with three defining characteristics: (1) the ability to self-renewal, (2) multipotent differentiation potential or the ability to differentiate into a number of different cell types, and (3) clonogenicity or the ability to form clonal cell populations derived from a single stem cell. It is these unique abilities for differentiation and self-renewal that give these cells the potential for restoration of function in multiple tissue types.

Regenerative medicine may offer such an alternative by using a cell-based therapeutic approach to regenerate the sphincter muscle. This approach has the advantage of treating the cause and not only the symptoms [23–25] and may improve the external (striated muscle) and internal (smooth muscle) urethral sphincter muscle function, the neuromuscular transmission, and blood supply [24, 26]. Since the majority of women with stress urinary incontinence may have hypermobility rather than intrinsic sphincter deficiency, this would limit the cell-based approach to patients with intrinsic sphincter deficiency. Clearly, some pathological processes place a woman at risk of both sphincter deficiency and urethral hypermobility/malposition (e.g., childbirth injuries) and an age-dependent loss of both striated muscle thickness and contractility [27].

Many recent reviews have discussed different aspect of stem cell treatment of stress urinary incontinence [14, 23–25], and most of these reviews emphasize the potential of this approach, but also that additional research is required.

10.2.1.2 Source of Stem Cells for Stress Urinary Incontinence Therapy

Currently, four broad categories describe the diversity of stem cells being investigated in regenerative medicine: embryonic stem cells, stem cells derived from placenta or amniotic fluid, induced pluripotent stem cells, and adult stem cells [21, 22, 28–30]. Embryonic stem cells are stem cells isolated from an early-stage embryo [22]. They represent a pluripotent cell source and can differentiate into all adult cell types. Although embryonic stem cells have great therapeutic potential, their use is limited by ethical factors. In addition, their clinical application is limited because they represent an allogeneic cell source with the potential to provoke an immune response and because of concerns regarding potential tumorigenicity.

Embryonic Stem Cells

Embryonic stem cells are currently being investigated for application to type 1 diabetes mellitus and cardiomyopathy [31, 32]. However, embryonic stem cells are not being investigated as a treatment for voiding dysfunction or urinary incontinence because of several limitations stated above.

Amniotic Fluid Stem Cells

Amniotic fluid stem cells are a new class of stem cells with properties intermediate to those of embryonic stem cells and adult stem cells [33]. They represent a heterogeneous stem cell population derived from the amniotic fluid and placental membrane of the developing fetus. Cells found in these tissues include mesenchymal stem cells as well as multipotent amniotic fluid stem cells that possess extensive self-renewal capacity. In addition, amniotic fluid stem cells can be induced to differentiate into cells of all three germ cell layers including cells of adipogenic, osteogenic, myogenic, endothelial, neural, and hepatic lineages. Amniotic fluid stem cells are currently being investigated for a variety of applications including treatment of urinary incontinence, acute tubular necrosis, cardiac valve regeneration for the early correction of congenital cardiac malformations, and as a source of immunomodulatory cells for a variety of immunotherapies [34].

Induced Pluripotent Stem Cells

Induced pluripotent stem cells are a unique class of stem cells recently discovered by Takahashi and Yamanaka who demonstrated that specific transcription factors could be used to reprogram differentiated cells to a pluripotent state [35]. Induced pluripotent stem cells have been shown to possess capacity for multipotent differentiation and self-renewal. In addition, they can be used autologously, and their derivation from adult cells obviates concerns regarding ethical issues with embryonic stem cells. A disadvantage of induced pluripotent stem cells is the time involved in resetting the cells to a pluripotent state followed by the additional time required to induce the cells to differentiate into the desired lineage [36]. Furthermore, there remain concerns that full transition to this new desired lineage may not occur [36]. To date, IPSCs have not nearly been used for urologic applications.

Adult Stem Cells

Adult stem cells are the most well-understood cell type in the field of stem cell therapy and are the most studied stem cell type in urologic fields [28–30, 37]. Over the past decade, adult stem cells have been identified throughout the body and are thought to act as tissue-specific progenitors that repair damage and restore function locally. Mesenchymal stem cells, also known as multipotent adult progenitor cells, are a unique subset of adult stem cells and are described for the first time by Friedenstein et al. in the 1970s [38]. Classically, mesenchymal stem cells were isolated from bone marrow stroma although more contemporary studies have demonstrated that they may also be found in other well-vascularized tissues including adipose, muscle, endometrium, and kidney [39]. Unlike tissue-specific progenitor cells, they can be induced to differentiate into multiple cell lineages including bone, neuronal, adipose, muscle, liver, lungs, spleen, and gastrointestinal tissues [39].

Mesenchymal Stem Cells

Currently, mesenchymal stem cells are the primary source of stem cells tested for therapeutic benefit in urologic applications. However, this cell population is relatively rare in the bone marrow (approximately 1 per 10,000 cells), and traditional bone marrow procurement is a painful and underproductive procedure and requires general or spinal anesthesia [40]. Alternative cell sources that have been investigated for urologic application include both muscle-derived stem cells and adipose-derived stem cells which can be obtained via less invasive biopsies and in larger quantities under local anesthesia [28]. Investigators continue to identify novel, less invasive cell sources for urologic application including cells derived from hair follicles, menstrual fluid, and urine [41–43].

Urine-Derived Stem Cells

Urine-derived stem cells are of particular interest for potential urologic applications [43–48]. In recent studies, Zhang et al. have reported the successful isolation and expansion of stem cells from voided human urine [43]. These cells, thought to be mesenchymal stem cells or pericytes from the kidney [45], can be obtained noninvasively from human urine specimens and demonstrate stem cell-like features including clonogenicity, self-renewal, and multipotent differentiation capacity [47]. Single clones of urine-derived stem cells have been shown to have the capacity to expand to yield a large population. Comprehensive characterization studies of these urine-derived stem cells demonstrate their ability to differentiate into multiple cell lineages at the gene/protein expression, cellular, and tissue levels [43, 47]. In vitro studies show that urine-derived stem cells derived from a single clone were able to differentiate into smooth muscle and urothelial lineages as evidenced by histologic examination and gene/protein marker assays [46]. Urine-derived stem cells can also be induced to differentiate into osteogenic, adipogenic, and chondrogenic lineages using different differentiation protocols [46]. In vivo studies using BMP-2- and BMP-9-transduced urine-derived stem cells also demonstrated their ability to form bone, fat, and cartilage tissue in immunodeficient female mice [49]. Urine-derived stem cells hold several advantages as a cell source for incontinence and voiding dysfunction [43]. They are easily harvested and do not require invasive surgical techniques to obtain. Furthermore, they are easily isolated with a significant cost advantage. Finally, they can be used autologously, obviating potential ethical issues or the potential of adverse immune reactions.

10.2.1.3 Stem Cell Procurement

The procurement and isolation of stem cells for therapeutic use typically occurs in three stages. The patient initially undergoes a procedure to harvest the cells. Following procurement, the tissue specimen is then transported to a regulated facility for isolation from other cell types. Here, the cells are grown via a process of ex vivo stem cell expansion involving multiple cycles of differentiation and senescence until adequate numbers are reached. Finally, cell sorting may or may not be used to isolate stem cells prior to therapeutic application, as this process is often labor-intensive and expensive. Thus, the cells that are ultimately administered to the patient may comprise a heterogeneous combination of stem cells and differentiated cells. Although still investigational, it is feasible during this process to manipulate the cells for a variety of therapeutic applications. For example, cells can be predifferentiated toward a specific cell lineage, preexposed to an environment similar to the in vivo posttransplant environment, or transfected to produce specific cytokines or growth factors.

10.2.1.4 Stem Cell Homing

The process of innate systemic stem cell delivery to the site of injury is termed “homing” and can be taken advantage in delivering cells systemically rather than locally. In contrast to tissue-specific progenitor cells, mesenchymal stem cells derived from the bone marrow traverse the circulatory system with access to all tissues in the body but then migrate to specific locations such as areas of acute injury following chemokine gradients where they can engraft and facilitate healing and regeneration [50]. Much of the insight regarding stem cell homing derives from literature regarding leukocyte migration into injured tissue, metastatic cancer cells, and hematopoietic stem cells [50]. Similar to leukocytes, mesenchymal stem cells express cell surface receptors and adhesion molecules responsible for directing cellular migration and homing to particular tissues, including the chemokine receptor CXCR4 and its binding partner CXCL12 [51] as well as the chemokine ligands: CCR1, CCR4, CCR7, CCR10, CCR9, CXCR5, and CXCR6 [52]. Mesenchymal stem cells are hypothesized to migrate to target tissues via a process similar to that of leukocyte migration: initial localization by means of chemoattraction, adhesion to vascular endothelial cells at the target site, and, finally, transmigration across the endothelium to the site of injury. Integrins and selectins are other classes of cell surface molecules that direct migration and adhesion of a variety of cells, including mesenchymal stem cells [53]. While the role of these molecules in leukocyte-endothelial adhesion is well established, their exact role in facilitating mesenchymal stem cells interaction with endothelium is less well characterized. Ruster and colleagues found that binding and rolling of mesenchymal stem cells were mediated by P-selectin, while migration involved binding of the integrin VLA-4 on mesenchymal stem cells with VCAM-1 on endothelial cells [54].

A large body of literature utilizing animal models has demonstrated the ability of mesenchymal stem cells to home to injured tissues in several disease models including cardiac injury, renal failure, and skin wounds. In addition, recent work investigating mesenchymal stem cells for stress urinary incontinence has shown that chemokine ligand 7 (CCL7), a homing factor for mesenchymal stem cells, is upregulated in both the urethra and vagina after vaginal distension, suggesting that intravenously administered mesenchymal stem cells could have the potential to home to sites of injury in stress urinary incontinence [55–57]. Subsequent literature examining intravenous injection of mesenchymal stem cells in a vaginal distension rat model of simulated childbirth injury showed that mesenchymal stem cells home to the urethra and vagina and facilitate recovery of continence as measured by leak point pressure [58, 59]. Recent work by Lenis and colleagues used a rat model of childbirth injury with both virgin rats that had undergone vaginal distension and postpartum rats to further investigate the expression of chemokines and receptors involved in stem cell homing and tissue repair [57]. They showed that vaginal distension in virgin and postpartum rats resulted in upregulation of urethral CCL7 expression. In addition, pregnancy and delivery was found to upregulate the chemokine receptor CD191 but to decrease the expression of hypoxia inducible factor 1α and vascular endothelial growth factor. Similarly, in a mouse model of overactive bladder via bladder outlet obstruction, Woo and colleagues showed that intravenously injected mesenchymal stem cells following bladder injury were associated with increased chemokine ligand 2 (CCL2) expression in the affected tissue [60].

In therapeutic applications, stem cell homing is affected by a variety of factors including age and passage number of the cells, culture conditions, and the delivery method [50]. Rombouts and Ploemacher demonstrated that with increasing age and passage number, the efficiency of mesenchymal stem cells engraftment decreases, possibly due to rapid aging of the cells with increased in vitro multiplication [61]. Reproduction of the innate mesenchymal stem cells microenvironment in vitro is challenging and plays a significant role in stem cell homing potential. For example, in vitro expression of matrix metalloproteinases (MMPs), key factors to stem cell migration, is affected by cell culture factors such as hypoxia and culture confluence and may be upregulated with the addition of certain inflammatory cytokines, such as transforming growth factor beta 1 (TGFβ1), interleukin (IL)-1β, and tumor necrosis factor (TNF)-α [62]. Despite the appeal of delivering cells systemically rather than locally, systemically infused mesenchymal stem cells often suffer from a first-pass effect whereby these larger cells become trapped in capillary beds of various tissues, especially the lungs, liver, and spleen, decreasing their therapeutic bioavailability and functionality [63]. To overcome this, recent investigations have utilized an intraperitoneal delivery route for mesenchymal stem cells and have obtained higher yield in target tissues [64].

10.2.1.5 Stem Cell Differentiation

The mechanism of action of stem cells was initially thought to primarily derive from their ability to differentiate into multiple cell types and regenerate damaged tissues. To date, treatment avenues for voiding dysfunction and urinary incontinence have focused on this potential to restore function via differentiation to replace injured or diseased tissues such as smooth or striated muscle for urethral sphincter regeneration and urothelial tissue for bladder, urethral, and upper urinary tract reconstruction. It is thought that mesenchymal stem cells restore function in stress urinary incontinence primarily by their ability to differentiate into multiple cell lineages, with animal studies demonstrating increases in urethral muscle, nerves, and connective tissue following mesenchymal stem cells injection [65–67]. Mesenchymal stem cells have also been induced to differentiate into a smooth muscle phenotype when exposed to conditioned media from smooth muscle cell cultures or when induced with specific myogenic growth factors for application to bladder reconstruction [68]. However, more recent researches suggest a more complex role of stem cells in functional recovery, with some studies demonstrating that paracrine secretions of mesenchymal stem cells play an important role in the regeneration process [59, 69].

10.2.1.6 Bioactive Effects of Stem Cells

Recent research has demonstrated that stem cells likely exert a therapeutic effect via the secretion of bioactive factors since few mesenchymal stem cells engraft and remain long term in target tissues, in contrast to their large therapeutic effect [59, 69]. Mesenchymal stem cells also activate and direct endogenous stem and progenitor cells to areas of injury via the secretion of cytokines and chemokines [70]. In addition, their secretions have antiapoptotic, antiscarring, and neovascularization effects, as well as immunomodulatory properties [71]. The investigation of this protein milieu or "secretome" is a subject of growing interest with the increasing recognition of the paracrine/autocrine role of cell secretions in the regulation of many physiological processes and their potential for therapeutic application. The investigation of these cell-specific proteins often begins in cell culture. While in vitro studies cannot fully capture and test the totality of mesenchymal stem cells secretions in the in vivo microenvironment, researchers seek to replicate the effects of the mesenchymal stem cells secretome via the use of media conditioned by the mesenchymal stem cells and containing their secretions [72].

Mesenchymal stem cells also possess immunomodulatory and immunological tolerance inducing characteristics. These cells typically express MHC-I but lack expression of MHC-II, CD40, CD80, and CD86. Owing to the lack of co-stimulatory cell surface molecules, mesenchymal stem cells fail to induce an immune response by the transplant host [73]. Mesenchymal stem cells have also been shown to play a role in suppressing immune responses by modulation of T cell activation and proliferation through both direct cell to cell interaction and via the action of soluble factors [73]. These immunomodulating properties are currently being investigated in myriad applications including prevention of graft versus host disease following allogeneic transplantation and Crohn's disease [74]. Clinical trials in cardiology have taken advantage of these properties of mesenchymal stem cells to investigate the efficacy of non-type-matched allogeneic MSC transplantation [75, 76]. In a study by Hare et al. 1 year after intravenous administration of allogeneic human mesenchymal stem cells in reperfused myocardial infarction patients, recovery, as measured by global symptom score, ejection fraction, ambulatory electrocardiogram monitoring, and pulmonary function testing, was significantly improved in treated patients compared with those that received a placebo [75]. In addition, no signs of rejection were observed and adverse event rates were comparable between treated and placebo arms.

A recent study by Dissaranan et al. demonstrates that the secretome of mesenchymal stem cells can facilitate recovery of continence as measured by leak point pressure following vaginal distension in a rat model. Moreover, rats treated locally with mesenchymal stem cells secretome, as contained in conditioned media, exhibited an increase in elastin fibers and urethral smooth muscle, which may have contributed to the restoration of continence [59]. A study of hamsters with heart failure suggests that mesenchymal stem cells act systemically as well as locally [77]. In this study mesenchymal stem cells injected into the hamstrings of affected animals were unable to migrate from the injection site; however, the authors found that treated animals still benefited from stem cell injection based on histologic and functional analysis of the myocardium. More recent work by Timmers et al. utilizes human mesenchymal stem cells secretions collected as conditioned medium in a pig model of myocardial infarction [78]. Pigs underwent left circumflex coronary artery and were then administered intravenous conditioned media for 7 days. At 3 weeks following initial cardiac injury and treatment, pigs treated with mesenchymal stem cells-conditioned media were found not only to have increased myocardial vascular density but also reduced infarction size and more preserved cardiac function.

These findings have been borne out in other urologic investigations as well. Lin and colleagues investigated rats subjected to vaginal distension and ovariectomy that were subsequently treated with intraurethral injection of adipose-derived stem cells; despite limited cell engraftment on histologic analysis, these rats demonstrated significant functional recovery [67]. Similarly, a study by Song and colleagues utilizing bladder wall injections of mesenchymal stem cells demonstrated functional recovery in a rat model of overactive bladder from partial bladder outlet obstruction despite limited engraftment 4 weeks posttreatment [79]. Further investigations into the use of stem cell bioactive factors could someday obviate the need for cellular injections in future stem cell therapies.

10.2.2 Experimental Trials

10.2.2.1 Animal Model

In the last 45 years, several animal models for urethral dysfunction have emerged (Table 10.1). These models include vaginal distension, pudendal nerve crush, urethrolysis, periurethral cauteriza-

Table 10.1 Regenerative therapy targeting the urethral sphincter animal models

Cell type	Biomaterial	Animal model	SUI model	Continence assessment
BMSC [66, 108]	–	Rodent	PNT	LPP
BMSC [92]	–	Rodent	Urethrolysis/cardiotoxin	ALPP
BMSC [109]	Alginate/Calcium	Rodent	PNT	LPP
ADSC [67, 113]	–	Rodent	VD	Bladder capacity, LPP
ADSC [106]	PLGA ± FNG	Rodent	PNT	Bladder capacity, LPP, RUPP
ADSC [65]	–	Rodent	VD	Bladder capacity, LPP
DFAT [114]	–	Rodent	VD	LPP
MDC [116]	–	Rodent	Sciatic nerve resection and T9 spinal cord transection	LPP, EFS
MDSC [105]	Fibrin gel	Rodent	PNT	LPP
MPC [98]	–	Rodent	Sphincterotomy	LPP, CP
MPC [117]	–	Swine	None	UPP
MPC [96]	–	Rodent	Electrocoagulation	Cystometry
MPC [152]	–	Canine	Sphincterotomy	DLP, UPP, USP, cystourethrogram, electrical stimulation
Human MDC [119]	–	Rodent	Sciatic nerve resection and T9 spinal cord transection	LPP
Human UCB [94]	–	Rodent	Electrocauterization and T9 spinal cord transection	LPP
Human AFSC [120]	–	Rodent	Bilateral PNT and T9 spinal cord transection	LPP, CP
Human myoblasts [121]	–	Rodent	Botulinum-A toxin	Urine volume
Human AFSC [110, 120]	–	Rodent	Bilateral PNT and T9 spinal cord transection	LPP, CP

ADSC adipose-derived stem cells, *AFSC* amniotic fluid stem cells, *ALPP* abdominal leak point pressure, *ASMA* alpha smooth muscle actin, *BMSC* bone marrow mesenchymal stems cells, *CP* closure urethral pressure, *DFAT* mature adipocyte-derived dedifferentiated fat, *DLP* lowest bladder pressure, *EFS* electrical field stimulation, *FNG* nerve growth factor, *LPP* leak point pressure, *MDC* muscle-derived cells, *MPC* muscle precursor cells, *MSCs* mesenchymal stem cells, *PLGA* polylactic-co-glycolic acid, *RUPP* retrograde urethral perfusion pressure, *UCB* umbilical cord blood, *UPP* urethral pressure curve, *USP* urethral pressure curve with abdominal compression

tion, urethral sphincterotomy, pudendal nerve transection, and toxins injected into the sphincter. Some of these models were based on pathophysiological theories of urethral sphincter dysfunction. For instance, the vaginal distention model was developed to simulate maternal childbirth trauma and its related direct sheering, ischemic, or neurogenic effects on the function of the urethral sphincter. The pudendal nerve crush model is a more specific model of neurogenic urethral dysfunction and isolates the injuries to the pudendal nerve. Both models are unique for the characteristic of reversibility where the dysfunction in the urethral function spontaneously resolves. This characteristic is ideal for researchers who are primarily interested in understanding mechanisms of injury and repair in the pathophysiology of urethral dysfunction. Researchers interested in generating a durable model of dysfunction have developed non-pathophysiologic-based models such as urethrolysis, cauterization, and pudendal nerve transection. A durable model is generally defined as stress urinary incontinence lasting 3 months or longer. Therefore what is advantageous to one group may be disadvantageous to the other group. The necessity of animal models is essential to determine the best cell therapeutic approach to treatment of stress urinary incontinence. However, animal models are limited in that we do not have their cooperation, and thus we assess continence through various surrogate outcome measurements that also constitute a source of interlaboratory variability. In this section, the different models of stress urinary incontinence urethral dysfunction are introduced.

10.2.2.2 Pathophysiological Animal Models of Reversible Incontinence

Vaginal Distension Model

Murine Vaginal Distension Model

One of the first animal models to study the pathophysiological mechanisms of stress urinary incontinence associated with childbirth was developed 15 years ago by Lin et al. [80] and performed in rats using a vaginal distension method that introduced an intravaginal balloon filled with 2 mLl of saline using a transurethral 12 F catheter left in situ for 4 h. This murine model showed a significant decrease in the periurethral striated and smooth muscle and c-Fos immunostaining neurons in the L6 to S1 spinal cord segments, indicating that irritation causes nerve and muscle injury. This finding paved the way for the next studies using vaginal distension methods for animal models of stress urinary incontinence and recovery after childbirth [81]. Since then, many authors have modified the vaginal distension method in rats or mice using different sized catheters or balloons or by changing the distension time. However, the results have always been similar with short-term effects including urethral dysfunction, a decrease in urethral resistance, neuromuscular damage of structures responsible for continence with edema in the pelvic musculature, and overexpression of hypoxia-inducible factors and cellular markers of innate repair. The main limitation of this model is the short durability of the functional, structural, and biomechanical defects due to vaginal distension, with a recovery within 10 days [82] and up to 6 weeks [81], depending on the vaginal distension time. This limitation makes the vaginal distension method useful for mechanistic studies of childbirth-induced tissue injury and recovery, but it is challenging for testing the efficacy of stress urinary incontinence treatments beyond 6 weeks. The strength of this model is that it recapitulates childbirth injury which is one of the strongest known risk factors for the development of stress urinary incontinence. However, to the best of our knowledge, the long-term effects of simulated childbirth injury have not been investigated in animals. Of the women who have full remission of stress urinary incontinence after delivery, almost half redevelop stress urinary incontinence 5 years later, and many more develop stress urinary incontinence decades later in menopause. For this reason it is important to test the effect of age in vaginally distended rodents, since they should be expected to redevelop stress urinary incontinence as a consequence of their altered function or support.

Cell Therapy in Vaginal Distension Murine Model

Vaginal distension rat models have been used to study stem cell regeneration therapies. For example, autologous mesenchymal stem cells were shown to home to the urethra and vagina after tail vein injection [59]. Although this study showed improvement in continence as measured by leak point pressure 1 week posttreatment, the external urethral sphincter function as measured by electromyography was not improved at that time. Therefore, regenerative treatment investigations may be limited by the short durability exhibited in the vaginal distension models. Other authors have combined vaginal distension with bilateral ovariectomy in rats [67] to study tail vein or urethral injection of adipose-derived stem cells and showed that 66.7% of the animals had normal voiding function after 1 month of treatment as measured by four-channel conscious cystometry.

Cell Therapy in Vaginal Distension Larger Animal Model

The vaginal distension technique has been established in larger animals as well. Burdzin'ska et al. developed the vaginal distension method in pigs and showed a decrease in maximal urethral closure pressure (50%) and functional urethral length (52%) 28 days post-vaginal distension [83]. The advantage of this animal model is that the external urethral sphincter of pigs and humans consists of a high percentage of slow-twitch myofibers and urodynamic evaluations can be carried out using the same procedure used for human patients, whereas this is not the case in rodents or dogs [83]. The disadvantage of larger animal models is the rate of animal growth during the study and the maintenance costs of larger animals compared to murine models. However, it is important to have both small and large animal models available since regulatory requirements that apply to novel therapies usually involve completion of a full nonclinical development program.

Pudendal Nerve Crush

Pudendal Nerve Crush Model Introduction

The pudendal nerve that controls external urethral sphincter activity can be injured during vaginal delivery. Among other authors, Kerns et al. developed a murine model to study external urethral sphincter dysfunction associated with stress urinary incontinence due to pudendal nerve damage during childbirth [84]. The pudendal nerve was accessed dorsally and then crushed bilaterally in the ischiorectal fossa. The electromyograms showed motor units undergoing typical denervation changes followed by regeneration and recovery. Pan et al. used a similar model [85] and demonstrated greater alterations in histological muscle atrophy in diabetic rats, as well as a reversible reduction in leak point pressure versus the control group. Recently, Castiglione et al. [86] used a rat model for pudendal neuropraxia to test betamethasone as a treatment for bladder dysfunction and showed that pudendal nerve crush leads to stress urinary incontinence with irregular micturition by involuntary muscle contractions.

Fecal Incontinence

This model has been used to test stem cells as a therapy in a rat model of fecal incontinence [87]. However, no significant differences in either resting or peak pressure of the anal sphincter following treatment at 10 days postinjection with mesenchymal stem cells were observed. Although this model is useful for investigating mechanisms of neuromuscular recovery and for testing neuroregenerative agents and can be used in conjunction with other pathological animal models to assess mechanisms of injury and recovery in populations at increased risk for stress urinary incontinence, its short-term durability could represent a handicap in cell-based therapy investigations.

Vaginal Distension Combined with Pudendal Nerve Crush Model

Other animal models combine the two methodologies vaginal distension and pudendal nerve crush to mimic stress urinary incontinence after childbirth. Song et al. developed a murine model combining vaginal distension and pudendal nerve crush to study the long-term effects of damage associated with childbirth [88]. Six weeks after surgery, there was no difference in the leak point pressure data between the vaginal distension/pudendal nerve crush and the sham group. However, after 9 weeks distal nerve regions were diffuse and were innervated by tortuous and multiple axons, demonstrating that reinnervation of the area was still in progress and could explain female recurrent stress urinary incontinence after the recovery following the first childbirth.

Conclusion of Reversible Incontinence Model

In conclusion, reversible stress urinary incontinence models such as vaginal distension, pudendal nerve crush, or a combination of both may be useful to evaluate pathophysiological mechanisms associated with stress urinary incontinence after childbirth, urethral damage and recovery, and neuromuscular recovery to test potentially neuroregenerative treatments and to identify favoring situations for recurrence of stress urinary incontinence. Cell therapy investigations using the vaginal distension model would be interesting, but due to their reversibility in the absence of treatment, they must be combined with other stress urinary incontinence models to ensure structural and functional viability of the injected cells.

10.2.2.3 Pathophysiological Animal Models of Durable Incontinence

Durable Animal Model

Another strategy to generate stress urinary incontinence in animal models consists of simulation of periurethral sphincter damage through direct or indirect mechanisms that produce a longer-lasting form of stress urinary incontinence. Among other direct mechanisms, it is important to mention sphincter damage by cauterization and periurethral damage by sphincterotomy. Indirect mechanisms include bilateral pudendal nerve transection.

Urethrolysis

First Study of Urethrolysis and Method Including Durability

In 1999, Kato et al. developed a canine model to evaluate the effects of urethrolysis and periurethral nerve resection on the capacity and compliance of the bladder [89]. Their experiments laid the groundwork for others, such as Rodriguez et al., to establish a murine model of stable and durable urethral dysfunction by transabdominal urethrolysis that resulted in neuromuscular and functional differences between incontinent and sham-operated rats [90]. Urethrolysis was performed by making a lower abdominal midline incision. After the bladder and the urethra were identified, the proximal urethra was detached circumferentially by incising the endopelvic fascia, and the remaining urethra was detached from the anterior vagina and the pubic bone. They observed a significant decrease in retrograde urethral perfusion pressure. The mean leak point pressure and retrograde urethral perfusion pressure decreased to 9.8 cm H2O and 11.2 cm H2O, respectively, at 1 week post-urethrolysis, and these changes were maintained for up to 24 weeks. After urethrolysis, histological analysis showed a 65% reduction in urethral smooth muscle with a high correlation between retrograde urethral perfusion pressure and muscular atrophy. The number of apoptotic cells in the urethra and bladder was also significantly higher, with the most obvious apoptosis rate in the submucosa and muscle layers. Since then, others have improved this technique showing the reproducibility of the urethrolysis mechanism as an animal model for lasting stress urinary incontinence with decreased urethral resistance and connective tissue damage for at least 8 weeks and up to 12 weeks [91]. Recently, Skaff et al. developed a durable animal model of stress urinary incontinence in rabbits by using urethrolysis which resulted in a significant decrease in the leak point pressure (from 33 to 12 cm H2O) and a decrease in 22% of the smooth muscle density for up to 12 weeks according to histological analysis [91].

Limitation of Urethrolysis

Although the urethrolysis model induces a severe type of incontinence that is not entirely representative of the complex etiologic factors associated with stress urinary incontinence, and it is not known if full physiological recovery to the normal preoperative status could occur in such a destructive model, these studies show how the alteration of anatomical structural components of the continence mechanism plays a key role in the development of stress urinary incontinence. The urethrolysis model could represent severe forms of incontinence associated with non-sparing prostatectomy during cystectomy or urethral intrinsic sphincter deficiency or apoptosis of the muscle that correlates with stress urinary incontinence and age. However this model may or may not be reflective of stress urinary incontinence in women over a lifetime.

Enhanced Model of Urethrolysis (Urethrolysis with Cardiotoxin Injection)

Kinebuchi et al. used this methodology in a rat model of stress urinary incontinence which combined urethrolysis with cardiotoxin injection and tested the use of bone-marrow-derived mesenchymal stem cells as a cell-based therapy [92]. In this model cultured autologous bone-marrow-derived mesenchymal stem cells were injected into the periurethral tissues, and leak point pressure was measured before injury and during the follow-up for up to 13 weeks postinjection. Despite that the authors view that the bone-marrow-derived mesenchymal stem cells group transplanted cells survived and differentiated into striated muscle cells and peripheral nerve cells and that there was a clear trend toward recovery of leak point pressure in bone-marrow-derived mesenchymal stem cells-transplanted urethras, no significant effect was detected. Although this study did not result in improved continence, the significant reduction in smooth muscle and correlation with urethral pressure observed in this model [90, 91] suggests that it may be useful for investigators interested in muscular regeneration of the internal urethral sphincter composed of smooth muscle. This may be especially important since it has been suggested that the density of the circular smooth muscle decreases up to 50% with age which may account for age-related declines in urethral closure pressure, a predominant factor associated with stress urinary incontinence and aging in women more often than loss of urethral support.

Electrocauterization

Method of Electrocauterization and Durability

Like urethrolysis, electrocauterization achieves a similar outcome of stress urinary incontinence but via a different method. The most representative animal model of periurethral electrocauterization was developed by Chermansky et al. in which cauterization of the periurethral region 1 cm from the bladder neck to the upper face of the symphysis of the pubis for 30 s at an elevated temperature (up to 1204°C) was performed in rats [93]. Sphincter function was assessed by measuring the leak point pressure through a suprapubic bladder catheter. Before the measurement, the spinal cord was transected at the T9–T10 level to eliminate the micturition reflex with increased intravesical pressure. Six weeks after surgery, cauterized rats with sphincter damage demonstrated a significant decrease in leak point pressure versus the sham-operated group, and differences remained for 16 weeks. Histological evaluation of the midurethra at 6 and 16 weeks post-electrocauterization showed an alteration in neuromuscular periurethral striated fibers, which was even more evident in the later evaluation. Although to date self-regeneration of striated muscles after electrocauterization has not been documented and therefore one may think that this is a good model to test cell-based therapies, we should note that this model mainly reproduces damage of the urethral sphincter that can occur accidentally during a surgical procedure, such as during prostatectomy in men or any procedure approaching the urethra and anterior vagina in women. Therefore, this type of stress urinary incontinence could represent a conventional model.

Limitation of Electrocauterization

However, it is not fully representative of the complex etiologic factors associated with stress urinary incontinence, and like urethrolysis, it is unclear if full physiological recovery to the preoperative state occurs. Having said that, regenerative therapies have been tested in rat models of stress urinary incontinence induced by electrocauterization.

Electrocauterization Studies

For example, Lim et al. used mononuclear cells isolated from human umbilical cord blood as a source of stem cells [94]. Four weeks after injection, leak point pressure values were significantly higher in the experimental group, and the external urethral sphincter of the experimental group remained well organized and intact, while the saline-treated control group showed severe degeneration and disruption of the sphincter muscle layers. However, the mechanism as to how the injected human umbilical cord blood mononuclear cells acted on the urethral sphincter is not explained in this work. Although they found human cells in the lamina propria and in the striated sphincter layers at 2 weeks, they could not find human cells at 4 weeks. In a previous study, intraurethral injection of muscle-derived cells improved urethral sphincter function in an electrocauterized rat model of intrinsic sphincter deficiency [95]. Despite that they showed improvement in sphincter function after 4 weeks and that the muscle-derived cells had integrated within the striated muscle layer of the cauterized tissue, they could not answer the question whether muscle-derived cells were bulking the urethra or promoting the functional recovery of the injured sphincter. In another study the use of muscle precursor cells isolated from limb myofibers of rats were used to regenerate the sphincter in an electrocauterized rat model of stress urinary incontinence and showed partial muscle regeneration in the context of denervation and fibrosis as demonstrated by histologic staining and immunofluorescence detection of beta-galactosidase expressed in the injected muscle precursor cells [96].

Urethral Sphincterotomy

Method of Sphincterotomy and Durability

Similar models of direct sphincter damage by muscle resection have been developed as a model not only for stress urinary incontinence but also for fecal incontinence [97]. In 2007 Praud et al. developed three mouse models of stress urinary incontinence (freezing, longitudinal sphincterotomy, and periurethral injection) and observed that only sphincterotomy caused stress urinary incontinence as measured by a decrease in urethral closure pressure 21 days after surgery [98]. Eberli et al. described a durable canine model of sphincter deficiency by microsurgical abdominal sectioning of 25% of the periurethral sphincter (smooth and skeletal) muscle that resulted in a decrease in urethral profiles up to 7 months post-surgery and an absence of muscle cells [99]. Furthermore, in vivo pudendal nerve stimulation confirmed the loss of sphincter tissue function. Therefore, if this model [99] is reproducible in other animals and results in long-lasting stress urinary incontinence, it may represent a tool for the evaluation of methods reproducing sphincter function and intending to regenerate smooth and striated muscle of the internal and external urethral sphincter, respectively.

Limitation of Sphincterotomy

Although this model does not represent the multifaceted etiology of stress urinary incontinence, one needs to consider that a major postoperative complication of radical prostatectomy is damage to the muscle-nerve-blood vessel units around the urethra. Vaginal delivery also causes varying degrees of neuromuscular damage. Such a microsurgical approach to remove the sphincter muscle offers the advantage that it produces durable damage and that the surgical techniques are relatively simple and reproducible. In addition, it can generate a direct muscle injury without pronounced nerve damage therefore potentially allowing a better direct assessment of a muscle regenerative therapy with easier electromyographic measurements. However, like the other destructive models, it is not known if full physiological recovery to the normal preoperative status can occur.

Pubourethral Ligament and Pudendal Nerve Transection

Durability of Pubourethral Ligament and Pudendal Nerve Transection

Animal models of incontinence by transection of the pubourethral ligament, which removes the structural support of the urethra, and/or transection of the pudendal nerve generated stress urinary incontinence for up to 1 month. Kefer et al. compared the two methods and transected the pubourethral ligament between the middle urethra and posterior symphysis of the rat and the pudendal nerve via a dorsal incision and the bilateral ischiorectal opening [100, 101]. After 28 days, a decrease in leak point pressure was observed in the rats with pubourethral ligament or pudendal nerve transection compared to the sham-operated group. Following pubourethral ligament transection, histologic studies demonstrated an absent pubourethral ligament for up to 4 weeks post-injury. Therefore, the long-term durability of this model is unclear at this point. Applications directed at treating hypermobility of the urethra could benefit from the use of this model. Pubourethral ligament could have similar effects to the urethrolysis model, since transectioning the ligaments may damage vascular, nerve, and possibly muscle structures, and could also be combined with other models to investigate cell therapies, such as transection of the pudendal nerve, favoring the neurogenic atrophy and decreased periurethral neurofilaments.

Bilateral Pudendal Nerve Transection (Transection of the Pudendal Nerve)

Durability of Bilateral Transection of the Pudendal Nerve

Bilateral pudendal nerve transection has been used in several animal models as a mechanism of a stress urinary incontinence model due to indirect periurethral sphincter damage. In 2003 Kamo et al. demonstrated the importance of the pudendal nerve in the continence mechanism in a murine model by demonstrating an 80% decrease in the muscle response at the midurethral level when transectioning the bilateral pudendal nerves and the nerves to the iliococcygeus and pubococcygeus muscles, but not by transection of the visceral branches of the pelvic nerves and hypogastric nerves [101]. Their data also showed that in stress conditions such as sneezing, the contribution of pudendal nerve-mediated striated muscle activity in the external urethral sphincter to the continence mechanism may be greater than that of iliococcygeus and pubococcygeus muscle activity. Peng et al. investigated the effect of the unilateral versus the bilateral transection of the pudendal nerve in rats [102]. After 6 weeks they observed a significant decrease in leak point pressure and striated muscle atrophy in rats that underwent unilateral transection of the pudendal nerve, which was even more prominent in the bilateral transection of the pudendal nerve group based on voiding efficiency. Other non-rodent animal models of stress urinary incontinence have been developed as well, including cats, dogs, and more recently nonhuman primates. Badra et al. developed a model for stress urinary incontinence in premenopausal female primates by abdominal bilateral transection of the pudendal nerve and cauterization of the pudendal nerve [103]. Urodynamic investigations were performed before injury and 3, 6, and 12 months after injury. Electromyography prior to necropsy showed decreased levels in the transection of the pudendal nerve group versus the sham group. Histological and inmunohistochemical analysis showed a decrease in smooth and striated muscle fibers, increased collagen, and decreased vascularization in injured animals. Cystogram results were consistent with those results showing greater amplitude and incompetence of the bladder outlet and the external urethral sphincter.

Advantage of Transection of the Pudendal Nerve

This model offers the advantage that it may be representative of neurogenic damage that occurs in radical prostate surgery or urethral denervation and/or pudendal nerve and muscular damage that probably occurs during vaginal delivery [102]. These studies demonstrate the feasibility of using this model in cell therapy approaches investigating regeneration of neuromuscular tissue. It also demonstrates that transection of the pudendal nerve affects the structural and functional anatomical capacity of the periurethral urethral sphincter in a human-like animal model.

Limitation of Transection of the Pudendal Nerve

Despite the success of such stress urinary incontinence animal models, most if not all cell therapy regenerative assessments have been performed to date in murine models (rats or mice) using different cell types such as human or allogeneic skeletal muscle stem cells [104, 105], autologous or allogeneic adipose-derived stem cells [106, 107], or allogeneic bone-marrow-derived mesenchymal stem cells [66, 108]. These aspects of regenerative cell therapies will be discussed in the next section.

10.2.2.4 Regeneration of the Urethral Sphincter Using Cell Therapy in Animal Models

The feasibility and efficacy of different strategies of cell therapy, at times combined with bioengineering techniques, have been tested in various animal models of sphincter regeneration. Given the legal and ethical issues on the use of embryonic stem cells and their higher tumor potential, the strategy in the last decade has been based largely on the use of adult stem cells derived from muscle, adipose tissue, and bone marrow.

Mesenchymal Stem Cell-Based Study

Mesenchymal Stem Cell Homing Study

Therapies using mesenchymal stem cells have been used in stress urinary incontinence rodent models of vaginal distension and various strategies of bilateral transection of the pudendal nerve. Dissaranan used bone-marrow-derived mesenchymal stem cells in a vaginal distension rat model to investigate homing of mesenchymal stem cells to pelvic organs after i.v. injection and recovery from simulated childbirth injury [59]. However, a limitation of this study was that their assessments were short term (after 1 week postinjection) and mesenchymal stem cells from passage 16 were used. Interestingly though, they did show that the vaginal distension model disrupted the smooth and striated muscle of the urethral sphincter and that mesenchymal stem cells preferentially engrafted in the smooth muscle of the urethra and vagina. Although there was an improvement in leak point pressure, the external urethral sphincter function was not improved. While this study suggests that mesenchymal stem cells can home to the damaged tissue and may aid in some level of recovery in continence, other methods of injection such as direct injection into the sphincter may produce different results.

Mesenchymal Stem Cells + Periurethral Injection

Corcos et al. and Kim et al. tested periurethral injection of mesenchymal stem cells in rats that underwent bilateral transection of the pudendal nerve and reported an improvement in urinary continence (Valsalva leak point pressure) as measured by cystometry at 4 weeks [66, 108]. Mesenchymal stem cells survived and were viable in the injected area based on immunohistochemistry analysis. Histopathologic regeneration of striated muscle was observed by desmin and myosin staining, respectively. However this approach is far from optimal since it is well known that fluorescent dyes coming out of transplanted cells can be transferred to surrounding cells and can mistakenly be taken as evidence for differentiation. Other authors have evaluated the use of biomaterials combined with mesenchymal stem cells for sphincter regeneration. Du et al. evaluated submucosal injection at the level of the urethra and bladder neck of an alginate/calcium gel alone or in combination with mesenchymal stem cells or muscle-like predifferentiated cells in vitro with 5-azacytidine in a rat stress urinary incontinence model that used a bilateral transection of the pudendal nerve model combined with transectioning of the nerves innervating the iliococcygeus/pubococcygeus muscle [109]. The three experimental groups showed an improvement in leak point pressure at 1, 4, and 8 weeks posttreatment compared to the control group, and immunohistochemical analysis resulted in desmin- and alpha-SMA-positive cells in all three groups, without observing clear and obvious specific muscle regeneration. In an analysis performed at 8 weeks after injection, the gel still maintained a certain volume, and therefore it cannot be ascertained whether the improvement in leak point pressure was due to a bulking effect since there was no group injected with cells alone. A remarkable fact is that addition of the calcium alginate composite gel seemed to stimulate angiogenesis with a higher density of newly formed microvessels in the cell plus gel group, which was not observed in the gel group alone.

Amniotic Fluid Stem Cells

In a combined urethrolysis and cardiotoxin injection stress urinary incontinence female rat model, bone-marrow-derived mesenchymal stem cells differentiated into striated muscle cells and peripheral nerve cells in vivo, but only resulted in a small improvement in urodynamics

[92]. A recent study tested different combinations of what was defined by the authors as "early" muscle, neuron, and endothelial cells that were differentiated from human amniotic fluid stem cells for 7–21 days in vitro prior to injection in a transection of the pudendal nerve female mouse model [110]. Histological and immunohistochemical analysis revealed that regeneration of the sphincter muscle was accelerated in the muscle group through fusion with damaged cells, as well as activation of the local muscle progenitor cells, but the enhancement was less pronounced and the regenerated muscle quality was lower compared to the muscle/neuron/endothelial cell group. The muscle/neuron and muscle/neuron/endothelial cell groups showed increased muscle bundles in the urethral sphincter, while the triple combination group showed areas of connective tissue masses at the urethral sphincter and higher nerve and endothelial cell marker expression. Real-time PCR results also demonstrated higher levels of myogenic, neuronal, and endothelial marker expression in the muscle/neuron/endothelial cell group. At 4 weeks, the leak point pressure and closure pressure was significantly higher for each of the three cell injection groups versus the control which the authors propose may be due to tissue edema and increased outflow resistance due to over injection. This suggests that in addition to the effect on myogenesis, mesenchymal stem cells may have a synergistic effect on neurogenesis, the formation of neuromuscular junctions and angiogenesis during regeneration.

Limitation

The limitation of these studies was a short-term follow-up (4–8 weeks) for functional analyses. Another consideration is the injury models used. For example, if a denervated urethral sphincter (e.g., after bilateral transection of the pudendal nerve) is treated with a cell-based therapy, can the regenerated tissue become innervated in a physiological manner? Reinnervation of muscle tissue after injury in such a model is necessary for functional recovery to occur. Mesenchymal stem cells are known to generate neurotrophic factors and can promote endogenous neuronal growth [111, 112], and the use of biomaterials or delivery products may enhance such an effect, but are they enough to reinnvervate the tissue?

Adipose-Derived Stem Cell-Based Study

Adipose-Derived Stem Cell Introduction (Advantage and Disadvantage)

The use of adipose-derived stem cells or derivatives has been evaluated in different murine models of stress urinary incontinence using vaginal distension or transection of the pudendal nerve models [67, 106, 113, 114]. In most of the cases, the functional assessment after adipose-derived stem cells injection was evaluated by cystometric parameters (bladder capacity, leak point pressure, and retrograde urethral perfusion pressure) and reported a significant increase in these values in the cell-injected groups compared to control groups. However, the muscle regenerative capacity or the in vivo mechanisms of these cell sources to achieve such results are not well defined. The incontinence models most commonly used with this cellular source were investigated using the vaginal distension model as shown by Obinata et al. [114]. Although their group demonstrated a recovery in muscle atrophy starting at 2 weeks, which was fully evident at day 28, it is difficult to assess to what extent muscle tissue regeneration of the damaged tissue was achieved since muscle regeneration markers included collagen I/III, elastic fibers, and alpha-SMA which do not guarantee the regeneration of a quality and morpho-functionally competent smooth and/or striated muscle. What is evident in these models is the cell viability and the paracrine capacity of adipose-derived stem cells at the injection site.

Adipose-Derived Stem Cell Several Studies

Zhao et al. [106] and Obinata et al. [114] reported cell survival and viability of adipose-derived stem cells in the injection area. In a vaginal distension stress urinary incontinence model, Zhao et al. [106] demonstrated how the use of a biomaterial-derived hydrogel (polylactic-co-glycolic acid, PLGA), in conjunction with cell therapy (mature adipocyte-derived dedifferentiated fat cells), can be used to generate microparticles (growth factor delivery "vehicles") that allow the introduction of certain factors such as nerve growth factor to improve the microenvironment and generate neurofilaments. However, it is not clearly defined if the improvement in bladder capacity of this group (mature adipocyte-derived dedifferentiated fat cells + PLGA/nerve growth factor) versus control is due to the self-improvement of the cellular microenvironment in vivo or an increased bulking effect. Li et al. [67, 113] observed that some adipose-derived stem cells may exhibit a vascular endothelial growth factor paracrine activity. Moreover this group showed increased vascular density and increased expression of p-ERK1/2, which could induce favorable changes in the extracellular matrix and cause trophic effects of immunomodulation and cell survival, thus improving the microenvironment for regeneration. Mature adipocyte-derived dedifferentiated fat cells alone have also promoted functional recovery from spinal cord injury-induced motor dysfunction in rats [115], suggesting reinnervation may occur with adipose-derived stem cells or mature adipocyte-derived dedifferentiated fat cell treatment.

Innervation Comment

However, the same question arises as with mesenchymal stem cells; can cell-based therapy lead to innervation in a physiological manner after bilateral transection of the pudendal nerve or pudendal nerve crush (which may in fact be more representative of postpartum stress urinary incontinence)? Some groups have suggested that factors such as brain-derived neurotrophic factor may need to be delivered with the stem cells to lead to such an effect.

Muscle-Derived Cells and Muscle-Derived Stem Cell-Based Study

MDS Researches

Cell therapy with muscle-derived cells and muscle-derived stem cells has been tested in murine models of stress urinary incontinence by pudendal or sciatic denervation [105, 116]. Kwon et al. showed that injection of muscle-derived cells, fibroblasts, or both can produce a functional improvement of continence in leak point pressure [116]. However, electromyographic activity demonstrated that only the muscle-derived cells had the ability to differentiate into myotubes and myofibrils and integrated into the sphincter tissue. This model also suggests that a transient bulking effect, but not morphological and functional regeneration, could occur by injection of fibroblasts since the improvement reported was dose dependent and a linear increase in leak point pressure at increasing fibroblast doses occurred, except at high doses of fibroblast injection which caused urinary retention. In order to improve the disadvantages of using muscle-derived stem cells including migration and absorption, Xu et al. studied the effect of adding a biodegradable fibrin glue to the periurethral muscle-derived stem cells injection therapy [105]. They observed that although both groups injected with muscle-derived stem cells or muscle-derived stem cells + fibrin glue showed a significant improvement in the leak point pressure, only the muscle-derived stem cells implanted in combination with the fibrin glue appeared to integrate into the host tissue and result in an increase in survival and a tendency to form multinucleated myofibrils. These results demonstrated that the addition of biomaterials such as fibrin may not be necessary to generate an early functional improvement based on the analysis of the leak point pressure, but may promote the implantation and differentiation of muscle-derived stem cells into striated muscle.

Muscular Precursor Cell-Based Study

Myoblast Researches

Other strategies have included the isolation and implantation of muscular precursor cells from muscle biopsies. Some results suggest that the use of myoblasts could be a potential treatment for damaged urethral sphincter muscle regeneration. Praud et al. performed periurethral injections of skeletal myoblasts in a stress urinary incontinence murine model of sphincterotomy and reported a significant improvement in urethral closure pressure [98]. However, there was only fusing of myoblasts and regeneration of new myofibers when the cells were grafted into normal striated skeletal muscle, which was not observed in the incontinent animals. Due to a large similarity between human and porcine rhabdosphincter muscle cells, Mitterberger et al. investigated myoblast injection in a non-incontinent pig model [117]. They showed cell dose-dependent functional effects. Thus, injection of high cell numbers (7.8×10^7) led to a clear increase in urethral pressure values at 3 weeks compared with low cell dose groups (4.4×10^7). However, the mechanisms underlying these effects were not explained. In the histological analysis, integration and differentiation of myoblasts was observed, as well as an increase in the formation of new myofibrils in the high cell dose injected groups. Likewise, Yiou et al. demonstrated that periurethral injection of muscle precursor cells in a murine stress urinary incontinence model caused by electrocauterization promoted the formation of myotubes and neuromuscular junctions [118]. Muscle precursor cell injection resulted in formation of myotubes and 40% restoration of the sphincter function 1 month after the injection.

Muscle Precursor Cell Researches

While most studies on cell therapy using myoblasts injected into the bladder and urethra were performed in murine models, recently Eberli et al. investigated the use of muscle precursor cells in a sphincterotomy canine model of stress urinary incontinence [99]. Their group showed that injection of autologous muscle precursor cells, with a low concentration of collagen used as a cell carrier to provide a microenvironment for the muscle precursor cells, significantly decreased urethral pressure values and was structurally and functionally viable after 6 months. Histological analysis revealed the capacity of not only skeletal muscle development but also nerve development, demonstrating that the injected muscle precursor cells are able to survive in vivo for an extended period of time. Thus, they showed histologically how injected muscle precursor cells grow in clusters and form new myofibrils starting 1 month after the treatment and at 3 and 6 months new muscle fiber formation occurred, although thinner and less organized than native muscle tissue. Moreover, staining with S100b showed that nerve fibers of different sizes were present in between the newly formed muscle tissue in the cell-treated group. Urodynamic evaluations showed an improvement in static, stress, and detrusor leak point pressure, electromyographic parameters demonstrated an improvement in pressure during stimulation of the pudendal nerve, and anatomical parameters (cystourethrogram) showed a urethrovesical morphology in the muscle precursor cell-injected group comparable to normal anatomy at 1, 3, and 6 months postinjection. Again, this highlights the importance in survival and integration of the cells using a cell carrier in the context of cell therapy.

Human Cell-Based Study

Finally, other groups have gone a step further and instead of using allogeneic or autologous animal cells in animal models, they used human cells implanted in animal models of incontinence.

Human Muscle Precursor Cell Studies

Kim et al. evaluated the feasibility of using human muscle precursor cells that were injected periurethrally in a sciatic sectioning murine model of incontinence [119]. They showed homing of the muscle precursor cells to the sphincter by immunofluorescence using human-specific antibodies. An improvement in leak point pressure at 4 weeks posttreatment was reported.

Human-Human Umbilical Cord Blood Studies

Lim et al. used human umbilical cord blood as a treatment for stress urinary incontinence [94]. A significant improvement in leak point pressure 4 weeks after cell injection was reported, as well as homing of DiI-labeled injected cells to the lamina propria and the muscular urethral sphincter at 2 weeks postinjection. However, 4 weeks after injection human cells could not be detected in the urethral tissue, perhaps because the iron oxide label of the injected cells was already metabolized at that point or because the injected cells were phagocytized and lost. Histologically, desmin staining in the control group showed a markedly disrupted sphincter muscle with atrophic muscle layers and collagen deposition, while the sphincter muscle was restored without damage in the cell-treated group. Despite a functional improvement in the human umbilical cord blood group, the mechanism as to how the injected human umbilical cord blood mononuclear cells act on the urethral sphincter is not explained in this work.

Human Amniotic Fluid Stem Cell Studies

Recently, human amniotic fluid stem cells have been proposed as a stem cell source for various cell therapies and tissue engineering. Thus, Kim et al. used amniotic fluid stem cells in a bilateral transection of the pudendal nerve murine model of stress urinary incontinence and demonstrated an improvement in the leak point pressure and closure pressure (using the vertical tilt/intravesical pressure clamp model) in the cell-transplanted group after 4 weeks of injury [120]. In this work, FACS analysis showed a lower expression of HLADR in the human amniotic fluid stem cells, and their injection did not stimulate CD8 T cell infiltration into the injected area or tumor formation after 8 weeks postinjection, suggesting that these cells are immunologically tolerated and/or result in an immunosuppression effect and have low tumorigenic potential. By immunohistochemistry using a human-specific antihuman nuclei antibody, they confirmed that the injected cells were able to survive; however, at days 3–7 cell migration from the injection site to the periurethral area was reported, and at day 14 a loss of transplanted cells was observed. This is especially interesting since in vivo cell tracking was performed and showed homing of the injected cells and cell clustering at the injection site during the first 10 days postinjection. Moreover, they reported muscle regeneration in the periurethral region in the experimental group by H&E staining and confirmed this result by MyoD antibody staining and real-time PCR of early and late striated muscle cell marker expression. Alpha-bungarotoxin for the acetylcholine receptor expression was reported to be similar between the experimental and control group. The experimental group also showed higher neurogenic gene expression (nestin, vimentin, neurofilament, microtubule-associated protein 2, β III tubulin, glial fibrillary acidic protein) than the control group. Subsequently in 2014, the same authors published the use of a triple combination treatment of muscle, neuron, and endothelial cells derived from human amniotic fluid stem cells as a cell therapy in the same mouse model of stress urinary incontinence described above [110]. This time they established four experimental groups: (1) amniotic fluid stem cells, (2) a single-cell group containing early muscle progenitor differentiated cells, (3) a double-cell group containing muscle/endothelial or muscle/neuron progenitor cells, and (4) a triple-cell-combination group with muscle/neuron/endothelial progenitor cells. Functional evaluation of stress urinary incontinence as evaluated by leak point pressure and closure pressure reported a significant improvement in all experimental groups compared to the control group, with the triple-cell-injected group showing superlative improvement compared to other experimental groups. Again CD8 T cell evaluations demonstrated a low immunoreaction in the experimental groups versus the control group. The cell tracking method used in the previous work showed the best result in terms of homing for the triple-cell-combination injection. Using the same techniques as in the previous work, muscle regeneration and increased formation of blood vessels were reported in all groups but to a higher extent in the triple-cell-injected group. This work reinforces their previous work and adds the option of combining human progenitor cells obtained from human amniotic fluid stem cells, suggesting that mesenchymal stem cells may have a synergic effect on myogenesis, neuromuscular junction formation, and angiogenesis during the regeneration of the damaged urethral sphincter.

Human Myoblast Studies

Finally, Bandyopadhyay et al. reported the effectiveness of human myoblasts injection in a botulinum-A toxin injection murine model of stress urinary incontinence [121]. Significant improvement in the recovery of urination volume and muscle atrophy was observed. Following myoblast injection, GFP-positive myoblast cells and urethral staining for human desmin demonstrated homing of the cells to the urethra. They also demonstrated formation of myotubes and an increase in the thickness of the periurethral muscle in the treatment group compared to the control group.

10.2.3 Clinical Trials

10.2.3.1 Overview of Clinical Trials

Several clinical trials in human subjects with stress urinary incontinence were conducted using cell-based therapy in the last decade. Generally, the therapy was concluded to be safe and effective in the short term. Disregarding several studies, where doubts have been raised about the reliability of the results [14], reports on the efficacy of cellular therapy in humans have shown cure rates between 40% and 75%. However, these figures are based on small, non-controlled studies using different methodologies, different cell types, and varying terminology, and a reliable overall assessment of the efficacy of treatment is not possible [122]. In addition, the cell sources, methods of cell processing, cell number, and implantation techniques differed considerably between studies, so a comparison was very difficult. Multiple cell types were evaluated for treatment of stress urinary incontinence. In several studies, no distinction has been made to establish the mechanism of stress urinary incontinence at baseline, i.e., the patients have not been categorized as having ISD of hypermobility.

Cell Type on Clinical Trials

Different types of stem cells have been used for stress urinary incontinence treatment in humans as well as in animal models including skeletal muscle derived (myoblasts, satellite cells, muscle progenitor cells, and muscle-derived stem cells), bone marrow stem cells (bone-marrow-derived mesenchymal stem cells), human umbilical mononuclear cells, adipose-derived stem cells, and modified/sorted stem cells which show various efficacy [23]. Skeletal muscle-derived cells, including myoblasts, muscle-derived stem cells, and fibroblasts, were the most commonly used in recent clinical trials [123–133]. Adipose-derived regenerative cells [134, 135], cord blood stem cells [136], and total nucleated cells and platelets [137] were tested less frequently. Almost authors have not reported any significant advantage from the use of any particular type of cells thus far.

Cell Status (Number, Injection Routes, Duration) on Clinical Trials

The number of cells used for the treatment of stress urinary incontinence differed considerably between the studies and ranged from 0.6 mln to 1020 mln cells. However, it is still unclear by which route the implantation of cells should take place: peri- or transurethral, injected into the sphincter or submucosa, or both? Should injections be performed in single or repeated doses? How long after the first series of injections should the following one be performed? Which are better to implant: freshly isolated or expanded in vitro cells? These questions remain unanswered.

Overall Outcomes Including Efficacy and Safety on Clinical Trials

Recent review article [138] showed that the clinical outcomes of cell therapy in stress urinary incontinence based on 616 patients indicated that mean cure and improvement rates over a 12 month follow-up were 37.2 ± 29.7% and 33.1 ± 14.3% (patients), respectively (Table 10.2). A comparative analysis of cross-sex outcomes indicated that the cure rate was significantly higher in female (41 ± 30.7%) compared to male patients (28.7 ± 31.5%) ($p < 0.01$).

No serious adverse events or major complications related to the treatment were reported. The most common adverse events were worsening of incontinence, urinary retention (<24 h), pain at the injection site, cystitis, and urinary tract infection.

Regarding Usage of Bulking Agents

When the outcomes were analyzed separately for studies where only cells were implanted from those together with a bulking agent (collagen, adipose tissue), the cure rate over the 12 month follow-up was significantly lower ($p < 0.001$) for the cell-only group (21.7 ± 8.9%) compared to the cell-bulking agent combination (60.8 ± 36%). This trend was observed in both male and female patients. So in fact, the results of cell therapy in stress urinary incontinence are comparable or even worse to those observed in conventional urethral bulking [20].

Table 10.2 Clinical trials of regenerative therapy on stress urinary incontinence

Study	Cell type	Bulking agent	Female, *n*	Male, *n*	SUI type	Follow up months	Cu%	I%	Assessment
Mitterberger et al. [117, 123]	M, F	C	123	0	II or III	12	79	13	Objective
Mittenberger et al. [124]	M, F	C	0	63	Iatrogenic	12	65	27	Objective
Mittenberger et al. [125]	M, F	C	20	0	III	24	89	11	Objective
Carr et al. [126]	MDSCs	None	8	0	Not classified	12	20	40	Objective (PT)
Lee et al. [136]	CBSCs	None	39	0	II, III, or II and III	12	36	36	Subjective
Sebe et al./Cornu et al. [127, 128]	M	None	12	0	III	72	17	0	Objective (PT)
Blaganje et al. [129, 140]	M	None	38	0	Not classified	6	24	53	Subjective
Gerullis et al. [130]	M	None	0	222	Iatrogenic	12	12	42	Subjective
Yamamoto et al. [134]	ADSCs	AT	0	3	Iatrogenic	6	33	67	Objective (PT)
Shirvan et al. [137]	TNCs-Ps	None	9	0	II or III	6	89	11	Objective (PT)
Carr et al. [132]	MDSCs	None	38	0	Not classified	18	20	40	Objective (PT)
Stangel-Wojcikiewicz et al. [131]	MDSCs	None	16	0	Not classified	24	50	25	Subjective
Gotoh et al. [135]	ADSCs	AT	0	11	Iatrogenic	12	9	55	Objective (PT)
Peters et al. [133]	MDSCs	None	80	0	Not classified	12	24	24	Objective (PT)

Female SUI was classified into three basic types: type I defined as urine loss occurring in the absence of urethral hypermobility, type II defined as urine loss occurring due to the urethral hypermobility, and type III defined as urine leakage occurring from an intrinsic sphincter deficiency (ISD)

ADSCs adipose-derived stem cells, *AT* adipose tissue, *CBSCs* cord blood stem cells, *C* collagen, *Cu* cure, *F* fibroblasts, *I* improvement, *M* myoblasts, *MDCs* muscle-derived cells, *MDSCs* muscle-derived stem cells, *PT* pad test, *TNCs-Ps* total nucleated cells platelets

10.2.3.2 First Clinical Trial Using Myoblast and Fibroblast

The first series of clinical trials utilizing cell therapy for stress urinary incontinence were conducted by a group of investigators from Austria [123–125]. However, their resultant two publications were retracted due to ethical and legal irregularities. The first reported an approach to stress urinary incontinence treatment involving performing the muscle biopsy from the patients' arm, followed by myoblast and fibroblast isolation, expansion, and implantation. The myoblasts were implanted into the urethral rhabdosphincter, while the fibroblasts were mixed with collagen and implanted into the urethral submucosa. This innovative strategy resulted in a 65–90% cure rate over a 12-month follow-up, but their results raised several questions [123–125]. A 1 year follow-up seems to be too short to indisputably prove the superior efficacy of injecting autologous fibroblasts and myoblasts into the urethral submucosa and the rhabdosphincter versus delivering a bulking agent as therapy for stress urinary incontinence. The efficiency of bulking agents is estimated at 8–12 months, and long-term continence is dependent on multiple booster injections. It is highly probable that the bulking effect is partially responsible for the reported improvement of continence. Periurethral delivery of a solution of mixed stem cells and collagen might lead to gradual coaptation of the urethra, caused by spontaneous proliferation of the implanted cells and inflammatory reaction. Although it has been shown that collagen is absorbed 3–6 months after injection, and since the 1 year cure rates are poor, the addition of a high amount of fibroblast could affect collagen degradation and activate an intrinsic fibrosis reaction similar to scar formation [139].

10.2.3.3 Muscle-Derived Stem Cells and Muscle-Derived Cell Clinical Trial

Muscle-Derived Stem Cell Trials

In 2008 Carr et al. published the first American clinical trial of muscle-derived stem cell therapy in eight female patients [126]. Muscle-derived stem cells were isolated from the lateral thigh, expanded and implanted by periurethral or transurethral routes. The three patients were withdrawn from the study after 1 month follow-up. Partial improvement in symptoms was observed in the remaining five patients, but only one patient achieved total continence. The authors concluded from their results that both periurethral and transurethral routes of cell implantation could be successfully used for stress urinary incontinence treatment [126]. Numerous variables, including cell implantation routes (periurethral or transurethral), injection positions (3, 9 or 3, 6, 9, 12 o'clock positions), and number of cells (with or without reinjection 3 months following the first injection), make the results of this small study very difficult to compare and discuss. Furthermore, as the muscle-derived stem cells were shipped frozen to the clinic, the eventual number of viable cells implanted is unknown.

The same group of investigators later confirmed the safety and efficacy of this method of treatment in a larger dose-ranging study [132]. A total of 38 women with stress urinary incontinence underwent intrasphincter injection with low (1, 2, 4, 8, or 16 mln) or high (32, 64, or 128 mln) doses of muscle-derived stem cells. Of these, 32 patients selected a second treatment of the same dose after a 3 month follow-up. Patients who received two muscle-derived stem cell implantations exhibited a significant reduction in pad weight (61.5% and 88.9% in the low- and high-dose group, respectively) and stress leaks (53.3% and 77.8% in the low- and high-dose group, respectively) at 18 months. Only four patients (two low dose and two high dose) reported urinary continence at 18 month follow-up. All four patients who received only a single muscle-derived stem cell implantation reported a 50% or greater reduction in the incidence of stress leakage, including one patient who was dry at 12 months [132].

Limitation of Muscle-Derived Stem Cell Trial

The major weakness of the study design concerns the number of patients in the subgroups (three to four patients), which was insufficient to achieve statistical significance after data acquisition. The authors did not present their results for the various subgroups, namely, low (1, 2, 4, 8, or 16 mln) and high (32, 64, or 128 mln) muscle-derived stem cell dosage groups, respectively. This led to an oversimplification of their results and the inaccurate conclusion that increased effectiveness of stress urinary incontinence treatment results from usage of high doses of muscle-derived stem cells compared with low doses of muscle-derived stem cells. Determining the minimal number of cells providing therapeutic effect is fundamental to establishing any state-of-the-art method in the field of cell-based therapies for stress urinary incontinence. Despite the analyses of study results, it remains unclear what constitutes the minimal number of implanted cells allowing improvement of outcomes in stress urinary incontinence treatment—32, 64, 128, or 256 mln? How incredibly large would the muscle biopsy have to be to obtain 256 mln muscle-derived stem cells? As the authors did not present data concerning the phenotype analysis of the isolated cells, it is unknown whether muscle-derived stem cells or a co-culture of other cell types were used in this study. Furthermore, as muscle-derived stem cells were shipped frozen to their clinic, again the number of viable cells implanted is unknown. Besides, distribution of muscle-derived stem cells into the muscle tissue following their injection into only two areas of sphincter was certainly very difficult, especially when such a high dose of cells was implanted. The most important observation from this study was that no serious adverse events or major complications were related to the treatment [132].

Muscle-Derived Cell Trial and Limitation

More recently, the same group assessed the safety and efficacy of muscle-derived cells for urinary sphincter repair in 80 women with stress urinary incontinence [133]. Patients received intrasphincter injections of 10, 50, 100, or 200 mln autologous muscle-derived cells. However, as the muscle-derived cells were shipped frozen to the clinic, the exact number of viable cells implanted into the patients is unknown. Although the differences in efficacy between the groups were not statistically significant, the higher doses (100 and 200 mln muscle-derived cells) tended to be associated with a greater percentage of patients with at least a 50% reduction in stress leaks and pad weight at 12-month follow-up. No serious adverse events were reported in this study [133].

Myoblast Trial

Another dose-ranging study utilizing myoblasts for treatment of stress urinary incontinence was performed by Sebe et al. [127]. Twelve female patients randomly divided into three dose-related groups ($n = 4$ each) underwent myoblast implantation into the urethral sphincter. Five of 12 patients improved (>50% of improvement) after 12 month follow-up, but only three patients were totally continent. Interestingly, there was no correlation either with cell dosage (10, 25, 50 mln) or urinary incontinence severity (serve, moderate, and mild) [127]. These results appear to contradict the studies by Carr et al. [132, 133]. However, the power of this study is reduced by the small number of patients in the experimental groups, non-confirmed phenotype of isolated cells, cell freezing, and only two injection positions (3 and 9 o'clock). Surprisingly, a long-term 6 year follow-up study indicated that two patients categorized as cured at 1 year evaluation were still dry at last follow-up. However, of the five patients considered improved at 1 year, all worsened during follow-up [128].

Similar studies utilizing myoblasts or muscle-derived stem cells in stress urinary incontinence treatment were conducted by Blaganje et al. [129, 140], Gerullius et al. [130], and Stangel-Wojcikiewicz [131]. All of them confirmed that cell therapy provides encouraging results in stress urinary incontinence treatment. The delayed improvement suggests rather detrusor regeneration than bulking effect. Unfortunately, the number of cells utilized for stress urinary incontinence treatment differs considerably in each of the studies, e.g., 1–50 mln [129, 140], 1.2–19.2 mln [130], and 0.6–25 mln [131]. This makes the results very difficult to compare and interpret. The time of cell culture was very long, either 5–14 weeks [129, 140] or 2–17 weeks [130], which suggests that the methods of cell isolation and culture were not standardized. The regeneration potential of cells between second and 17th weeks can differ considerably.

Blaganje et al. combined ultrasound-guided injection of autologous myoblasts into the external urethral sphincter followed by electrical stimulation of the sphincter in 38 women with stress urinary incontinence (those with severe hypermobility were excluded) [140]. The rationale for performing the electrical stimulation was a previous report that electrical stimulation accelerated the maturation and organization of myoblasts into myotubes. After 6 weeks five patients were considered cured and 29 (78.4%) reported improvement. The short observation time and the lack of control group reduce the impact of these results.

Minced Skeletal Muscle Tissue (MSMT)

Freshly harvested, minced autologous skeletal muscle tissue with its inherent content of regenerative cells were injected intraurethrally in 20 and 15 women with uncomplicated and complicated stress urinary incontinence, respectively [141]. They were followed for 1 year. Significant reductions were observed in each group in the mean number of leakages ($p < 0.01$). In the uncomplicated group, cure or improvement was observed in 25% and 63% of patients and in the complicated group 7% and 57%. No voiding dysfunction developed, and only minor adverse events were noted.

10.2.3.4 Adipose-Derived Stem Cell Trial

Another source of cells used for stress urinary incontinence treatment is adipose tissue. Yamamoto et al. reported in 2010 two initial cases of adipose-derived stem cell implantation for treatment of stress urinary incontinence following radical prostatectomy [134]. This paper was initially retracted from the published journal due to ethical irregularities, but eventually published 2 years later in the same journal with minor changes [134]. Freshly isolated adipose-derived stem cells from the abdominal adipose tissue were implanted immediately without in vitro expansion into the urethral rhabdosphincter. Additionally, probably 2.2–3.8 mln of adipose-derived stem cells in 4 mL of vehicle were combined with intact adipose tissue and subsequently injected into the urethral submucosa. Interestingly, urinary incontinence improved significantly a few days after injection, deteriorated subsequently, and progressively improved up to 6 months. This could indicate a loss of adipose tissue mass in the initial period and regeneration of sphincter muscle by implanted adipose-derived stem cells at a later stage. However, MRI showed no change in the volume of injected adipose tissue between 4 days and 12 weeks following implantation, proving, in fact, that the bulking effect was the dominant factor responsible for success. Also, the rapid improvement is a characteristic for bulking therapy. The regeneration of the sphincter complex induced by implanted cells needs time to occur. The acute phase of inflammation and tissue remodeling lasts at least 4 weeks, and during this time the regeneration process continues. Both long-term follow-up and a larger group of patients are necessary to evaluate the efficacy of this method of treatment [134].

In addition, Kuismanen et al. [142] injected five women with autologous adipose-derived stem cells combined with bovine collagen gel and saline. Patients were evaluated, 3, 6, and 12 months after the injections. The primary endpoint was a cough test. At 6 months one patient was continent and two more at 12 months. No adverse effects of treatment were reported. This pilot study is limited by the low number of patients, but may support that the treatment was safe and possibly that there is a therapeutic potential for cells other than muscle precursor cells.

10.2.3.5 Umbilical Cord Blood Mononuclear Cell Trial

In another trial, allogenic umbilical cord blood mononuclear cells were injected into the urethral submucosa of 39 women with stress urinary incontinence. Lee et al. [136] used umbilical cord stem cells in 39 women with stress urinary incontinence. The injections were performed by one operator at a single hospital. Twenty-two of the patients had urethral hypermobility, eight had intrinsic sphincter deficiency, and nine had mixed stress urinary incontinence. After 12 months, there was a more than 50% subjective improvement in 72% of the patients evaluated by a patient satisfaction test. The investigators noted a progressive improvement over time with more than 90% improvement at 12 months. No adverse effects were reported. The study supports that the approach used may be safe. However, the evaluation of efficacy data is not convincing, and the results did not allow an assessment of efficacy differences between the different types of stress urinary incontinence.

10.2.3.6 Total Nucleated Cells with Platelet-Rich Plasma Trial

In a recently published trial, autologous total nucleated cells with platelet-rich plasma were used in the treatment of nine women with stress urinary incontinence [137]. Total nucleated cells isolated from peripheral blood were implanted into the rhabdosphincter, while platelets were implanted into the submucosa. There was a significant improvement at 6 month follow-up, with eight patients attaining full continence. However, it is not known which of the agents, total nucleated cells or platelet-rich plasma, were responsible for the improvement. Future research is needed to distinguish their effects. Moreover, larger trial and long-term follow-up are needed to confirm the safety and efficacy of this method of treatment.

10.2.3.7 Analysis of Clinical Outcomes and Adverse Events of Cell-Based Therapy in Urinary Incontinence

Summary and Limitation of Clinical Trials

In review of clinical trials, cell-based therapy, as practiced today, is a safe but ineffective method for stress urinary incontinence treatment. The present article provides an indication of new management for stress urinary incontinence patients, accurate diagnosis of specific anatomical defects, and application of targeted therapy: cell implantation to restore the smooth muscle and nerve cells and tissue engineering to restore the collagen, the key structural component of the pubourethral ligament. There are other issues, such as identifying the most effective sources of cells; optimizing the methods of their isolation, culture, and implantation; and, finally, randomized blinded controlled trials with long-term follow-up comparing any new cell therapy with a bulking agent and/or a midurethral sling. This technology has some way to go before it becomes a viable clinical option.

Discussion/Approach of Clinical Trials

Cell-based therapy as practiced today is safe but only moderately effective, comparable to conventional urethral bulking. A fundamental question needs to be addressed before any cell therapy is performed: what is the etiology of stress urinary incontinence? The huge success of the midurethral sling operation in all types of stress urinary incontinence indicates that it is the pubourethral ligament, which should be the focus of attention, not the rhabdosphincter, deemed to have only a minor role in continence control [8]. In some clinical trials evaluating cell-based therapy in stress urinary incontinence, the cells were implanted into the urethral rhabdosphincter to treat incontinence caused by urethral hypermobility [123, 136, 137]. In order to evaluate the effectiveness of cell implantation, those patients with combined stress urinary incontinence (urethral hypermobility and ISD) should not be included in the study groups. Urethral hypermobility as a cause of stress urinary incontinence is beyond the scope of treatment with cell injection within the sphincter complex. At least some cell stress urinary incontinence operations should be focused on the reconstruction of the structures, which support the bladder neck and proximal urethra. The treatment of stress urinary incontinence should be based on targeted therapy. The patients with stress urinary incontinence type II, caused by urethral hypermobility, should be treated by implantation of tissue-engineered pubourethral ligament. The treatment of stress urinary incontinence type III caused by urethral sphincter insufficiency or iatrogenic sphincter injury should focus on cell implantation into the urethral sphincter. The mixed etiology stress urinary incontinence (types II and III), however, should be treated using both methods of ligament and sphincter regeneration.

Limitations of (Pre-) Clinical Trials

What has delayed further development of cell treatment of stress urinary incontinence? Undoubtedly, the withdrawal of several initial studies for various reasons has created a doubtful attitude to the approach among clinicians, researchers, and the authorities. Importantly, the published reports, although preliminary, have shown varying effects, and there is no apparent consensus on how studies should be designed, what cells should be injected, and how effects should be evaluated. Furthermore, very few preclinical studies simulate the patient condition which is chronic and not an acute injury. Nonetheless, there seems to be a general agreement that to accelerate clinical progress more predictive preclinical research should be performed.

This, in turn, focuses interest on what predictive animal models and cell types/fractions should be used.

Limitations of Mesenchymal Stem Cells

As mentioned, a number of different cell types have been used in the treatment of stress urinary incontinence both clinically and experimentally [24–26, 143]. Mesenchymal stem cells which can be isolated from many different tissues are defined by their ability to indefinitely self-renew, differentiate into multiple cell lineages, and form clonal cell populations [72, 144–146]. The application of mesenchymal stem cells as part of a therapeutic strategy for functional regeneration of the urinary sphincter complex opens two ways of action, by application of mesenchymal stem cells capable to differentiate and functionally replace defect striated and smooth muscular cells of the urethral sphincter complex and by taking advantage of the growth factors released by the mesenchymal stem cells injected in or deposited around the urethral wall. However, based on the current knowledge, a differentiation of adult mesenchymal stem cells to yield functional striated or smooth muscle cells is very unlikely to occur at reasonable efficacy.

10.3 Conclusion

10.3.1 Summary

Although cell therapy will continue to be explored, a regenerative pharmacologic approach to treatment of stress urinary incontinence holds the promise of bypassing the lengthy and expensive process of cell isolation and increase availability of treatment in many clinical settings. However, this approach will require careful preclinical/clinical modeling and attention to its health benefit/risk ratio. The clinical trials have included small numbers of patients and results vary depending on the patient cohorts and the cells used. Results of preclinical studies also vary, but report a more favorable outcome. This difference is most likely explained by animal modeling not being directly translatable to the human condition. However, preclinical studies have identified an exciting new approach to the regeneration of the urinary sphincter by using the components of cells (secretomes) or chemokines that home reparative cells to the sites of injury.

10.3.2 Future Perspectives

Preclinical research in animal models of intrinsic sphincter deficiency/stress urinary incontinence and studies in humans have shown the feasibility of cellular therapy as a treatment alternative. However, in view of the promising, but rather, preliminary clinical results, it seems reasonable that before starting new randomized controlled trials based on existing information, novel approaches should be evaluated in optimized preclinical models, using well-defined cells or secretomes. Such endeavors may aid in the development of clinically applicable cellular therapy with satisfactory long-term outcome.

10.3.2.1 Optimization of In Vivo Animal Models

Most of the animal experiments for the evaluation of cell therapy have been performed in murine models with a more destructive or permanent form of stress urinary incontinence. However, other groups have used reversible incontinence models. One of the best predictors of later development of stress urinary incontinence is postpartum incontinence, whether it resolves or not. Therefore, investigating therapies for postpartum incontinence should have relevance to the development of later incontinence as well. The mechanism of human stress urinary incontinence is a complex and usually multifactorial process. Moreover, the pathological changes associated with these mechanisms are variable and complex, combining denervation, muscle degeneration and apoptosis, chronic muscle atrophy, fibrosis, and connective tissue disorders, among others. Hence, integrating all conditions of stress urinary incontinence into a single model is very difficult, if not impossible. Therefore, the specific animal models should take into account the potential patient to be treated and the target tissue to be regenerated. Besides that, more efforts must be done in order to be able to perform long-term functional, clinical, urodynamic, and electromyogram assessments in such animal models to demonstrate cell survival, accurate homing of the cells, viability, and cell regeneration capacities by use of more specific cell lineage markers. A special effort must be done to incorporate human cell therapy studies in animal models as previous steps to clinical trials. Key phenomena such as neurogenesis and angiogenesis should be evaluated in the animal models, as well as proper muscular regeneration.

Outline of Animal Models for Cell Therapy

Many animal models of stress urinary incontinence have been described and discussed in detail [23, 147–150]. Most of the studies, investigating the effects of cell therapy in these models, have been performed in small animals such as rodents and rabbits. However, large animals such as pigs, dogs, and nonhuman primates have also been used. The applicability of the different models is dependent on what is to be studied: effects of the intervention on acute injury, effects on recovery/regeneration, or pathophysiological mechanisms. The durability of the model is of major importance from a translational point of view. In an acute situation such as traumatic childbirth, cellular therapy may be expected to have a prophylactic effect, but can hardly be considered an option clinically. To mimic a clinical situation, the model should be in a stable chronic state before intervention, i.e., a condition where spontaneous recovery is not expected. If the intervention shows improvement under these conditions, the results may have a clinical impact.

Various Animal Models and Limitation of Small or Large Animal Model

A number of animal models of urethral dysfunction have emerged, including vaginal distension, pudendal nerve crush, urethrolysis, periurethral cauterization, urethral sphincterotomy, pudendal nerve transection, and toxins. In particular, the shortcomings and reliability issues with measuring techniques should be emphasized [147, 150]. Sloff et al. [151] performed a systematic review of preclinical results on the use of stem cells in the tissue engineering of the urinary bladder and concluded that the translational predictability of models, particularly those based on small animals, might be questioned. This may be the case also for cellular therapy for stress urinary incontinence. It seems reasonable to suggest that new approaches should be evaluated in large animal models of stress urinary incontinence before being applied clinically. As expected, rodent models do not always recapitulate the characteristics of stress urinary incontinence seen in patients.

A useful large animal model that replicates the irreversible loss of external sphincter function found in humans should fulfill several criteria. However, the disadvantages of large animal models are the rate of animal growth during the study and the specialized facility and husbandry required. An important issue and a limiting factor are the maintenance costs of larger animals compared to rodent models. However, it is important to have both small and large animal models available since regulatory requirements that apply to novel therapies usually involve completion of a full nonclinical development program.

Canine Model

Eberli et al. [99] described a canine model of irreversible (12 month duration) urethral sphincter insufficiency. Subgroups of the dogs were subjected standard urodynamic and radiographic studies before and at 1, 2, 3, 4, and 7 months after surgery. The urodynamic studies, which included a static urethral pressure profile, stress urethral profile, and detrusor leak point pressure, showed a significant and sustained decrease in sphincter function, indicating that the model was a reliable and reproducible model. Using this model, Eberli et al. [152] injected autologous skeletal muscle precursor cells into the damaged sphincter muscles of 12 animals. The animals were followed up for up to 6 months after injection, and urodynamic studies, functional organ bath studies, and ultrastructural and histological examinations were performed. Animals receiving cell injections had sphincter pressures of approximately

80% of normal values, while the pressures in the control animals without cells dropped and remained at 20% of normal values. Histological analysis indicated that the implanted cells survived and formed tissue, including new innervated muscle fibers, within the injected region of the sphincter. These results would indicate that muscle precursor cells may be able to restore otherwise irreversibly damaged urinary sphincter function clinically. Further study revealed that injection of cells resulted in no significant local or systemic pathologic features within the dose range that improved sphincter pressures [153].

Pig Model

Burdzinska et al. [83] developed a model of intrinsic sphincter deficiency in female pigs by balloon distension of the urethra. The urethral pressure profile was evaluated before injury, immediately post-injury, and at day 28 post-injury in the experimental group. The maximal urethral closure pressure, the functional urethral length, and the area under curve of the urethral pressure profile were measured. At 28 days post-urethral distension, there was a decrease in maximal urethral closure pressure (50%) and functional urethral length (52%). As pointed out by the authors, an advantage of this animal model is that the urinary sphincter of pigs and humans consists of a high percentage of slow-twitch myofibers and urodynamic evaluations can be carried out using the same procedure as used for human patients, whereas this is not the case in rodents or dogs. Unfortunately, after injection of autologous muscle-derived cells into the urethral sphincter using a urethrocystoscope, urethral pressure profilometry demonstrated no significant differences between urethral closure pressure in the transplanted group and the control group at day 28 [154].

Monkey Model

Despite some differences in the encircling striated muscle in the urethra [155], the outflow region of nonhuman primates is the closest to the human anatomy. Female cynomolgus monkeys share with women comorbidities common to age- and hormone-related health problems, including heart disease, osteoporosis, breast/uterine cancer, and cognitive decline [156], and also share a well-defined menarche, premenopause (characterized by a 28 day menstrual cycle), perimenopause, and postmenopause [155]. The upright posture and pelvic location of the bladder and urethra, ultrastructure of the sphincter complex, and pelvic floor support are similar to those of women [156].

Badra et al. [103] developed and characterized a nonhuman primate model of intrinsic sphincter deficiency. Injury to the sphincter complex was created by cauterizing and then transecting its pudendal innervation. Urodynamic studies were performed before and during pudendal and hypogastric nerve stimulation at baseline and 3, 6, and 12 months after injury. Sphincter function was studied in vivo by cystourethrography and ex vivo by quantitative histology and immunohistochemistry at these time points. The inflicted injury produced a 47–50% decrease in maximal urethral pressure versus baseline [103]. It also abolished the increase in maximal urethral pressure in response to pudendal and hypogastric nerve stimulation, and this effect persisted for more than 12 months after injury. Urodynamic changes were consistent with decreased skeletal and smooth muscle content [103], decreased nerve responses, and an associated decrease in somatic and adrenergic innervation in the sphincter complex.

Using this model, Badra et al. [26] assessed the long-term efficacy of autologous muscle precursor cell therapy. Maximal urethral pressure measurement and corresponding histological analysis of the structural and cellular components of the sphincter complex were performed up to 12 months after injection. It was found that cell treatment produced sustained (12 months) increases in resting somatic nerve-stimulated and adrenergic nerve-stimulated maximal urethral pressure. Treated group showed a greater percentage of sphincter area occupied by the muscle and a decreased collagen content compared to the untreated group. A translational weakness of this study was that the cells were injected only 6 weeks after muscle biopsy, which makes the approach more prophylactic than therapeutic, since a chronic, irreversible sphincter injury may not have been achieved. As pointed out previously, the animal models commonly used to simulate stress urinary incontinence in women are acute in nature and not durable and tend to recover spontaneously in a short period, while stress urinary incontinence in women most often is a chronic symptom, and spontaneous recovery is not expected. Ongoing studies in monkeys are designed to reveal whether or not this is the case.

10.3.2.2 The Adequate Cell Type Needed for Regeneration of Tissue

What is the best cell type to use for sphincter regeneration? Throughout the years different cell types have been studied and tested in animal models, and in most cases, with successful measurable results. If so, why have we failed to standardize their potential use? Maybe not all cell types have the same ability to successfully regenerate a particular target or tissue. Perhaps we need to customize the cell therapy approach to the target.

10.3.2.3 Subtyping the Stress Urinary Incontinence Population

The stress urinary continence mechanism is multifactorial and complex, and stress urinary incontinence can result from multiple etiologies and pathophysiological processes. Therefore we need actions aimed and focused at understanding the etiological and pathophysiological mechanisms of specific types of stress urinary incontinence to improve future results aimed at therapy because not all patients can be expected to respond equally to cellular therapy. This could be an underlying cause of the variation in clinical results versus results seen in preclinical studies. Personalized treatment is advocated in other urologic conditions, and characterization of the status of the sphincter and clinical investigations focused on understanding the pathophysiological mechanisms of stress urinary incontinence on an individualized basis could be expected to improve future results. As regenerative medicine approaches to treatment of pelvic floor disorders continue to be developed, attention will need to be given to these different patient cohorts and therapies possibly tailored to their needs.

10.3.2.4 Are There Alternatives to Cells?

The broad plasticity of mesenchymal stem cells was originally believed to define their therapeutic potential. However, the remarkable array of bioactive molecules produced by stem cells, the "secretome," comprises a host of diverse cytokines, chemokines, angiogenic factors, and growth factors that may mediate the diverse effects of the mesenchymal stem cells [26, 57, 157, 158]. The autocrine/paracrine role of these molecules is being increasingly recognized as key to the regulation of many physiological processes including directing endogenous and progenitor cells to sites of injury as well as mediating antiapoptosis, angiogenesis and revascularization, immune and inflammatory modulation, and wound healing and tissue repair [26]. In a rat model of simulated childbirth injury (pudendal crush + vaginal distension), Deng et al. [159] found that mesenchymal stem cells and concentrated conditioned media from cultured rat bone-marrow-derived mesenchymal stem cells provided similar protective effects on nerves and other structures. This suggested that mesenchymal stem cells act via their secretions to produce their effects. Thus, the cell and secretome approaches to therapy can be said to represent "regenerative pharmacology" defined as "the application of pharmacological sciences to accelerate, optimize, and characterize (either in vitro or in vivo) the development, maturation, and function of bioengineered and regenerating tissues" [160].

Chemokines and receptors expressed during tissue injury have been suggested to be upregulated by vaginal distension, and simulated birth trauma induced urinary incontinence [57, 157]. One of these chemokines is CXCL12, which plays a major role in cell trafficking and homing of progenitor cells to sites of injury through a receptor mechanism and enhancing cell survival once at the injury site [158, 159, 161]. Cells from the injured organ highly express CXCL12, which causes an increase of local CXCL12 levels, and peripheral and bone marrow progenitor cells follow the chemical gradient to the organ [159, 161].

References

1. Abrams P, Cardozo L, Fall M, Griffiths D, Rosier P, Ulmsten U, et al. The standardisation of terminology of lower urinary tract function: report from the Standardisation Sub-committee of the International Continence Society. Neurourol Urodyn. 2002;21(2):167–78.
2. Kalejaiye O, Vij M, Drake MJ. Classification of stress urinary incontinence. World J Urol. 2015;33(9):1215–20.
3. Norton P, Brubaker L. Urinary incontinence in women. Lancet. 2006;367(9504):57–67.
4. Hunskaar S, Lose G, Sykes D, Voss S. The prevalence of urinary incontinence in women in four European countries. BJU Int. 2004;93(3):324–30.
5. Carlson KV, Nitti VW. Prevention and management of incontinence following radical prostatectomy. Urol Clin North Am. 2001;28(3):595–612.
6. Strohbehn K, Lauria MR. Risk of urinary incontinence after childbirth: a 10-year prospective cohort study. Obstet Gynecol. 2007;109(1):202; author reply-3.
7. Cundiff GW. The pathophysiology of stress urinary incontinence: a historical perspective. Rev Urol. 2004;6(Suppl 3):S10–8.
8. Petros PEP, Ulmsten U. Urethral pressure increase on effort originates from within the urethra, and continence from musculovaginal closure. Neurourol Urodyn. 1995;14(4):337–50.
9. Kayigil O, Ahmed SI, Metin A. The coexistence of intrinsic sphincter deficiency with type II stress incontinence. J Urol. 1999;162(4):1365–6.
10. Radomski SB. Practical evaluation of post-prostatectomy incontinence. Cuaj-Can Urol Assoc. 2013;7(9-10):S186–S8.
11. Dalkin BL, Wessells H, Cui HY. A national survey of urinary and health related quality of life outcomes in men with an artificial urinary sphincter for post-radical prostatectomy incontinence. J Urol. 2003;169(1):237–9.
12. Harding CK, Thorpe AC. The surgical treatment of female stress urinary incontinence. Indian J Urol. 2010;26(2):257–62.
13. Mischinger J, Amend B, Reisenauer C, Bedke J, Naumann G, Germann M, et al. Different surgical approaches for stress urinary incontinence in women. Minerva Ginecol. 2013;65(1):21–8.
14. Hakim L, De Ridder D, Van der Aa F. Slings for urinary incontinence and the application of cell-based therapy. Adv Drug Deliv Rev. 2015;82–83:22–30.
15. Koski ME, Enemchukwu EA, Padmanabhan P, Kaufman MR, Scarpero HM, Dmochowski RR. Safety and efficacy of sling for persistent stress urinary incontinence after bulking injection. Urology. 2011;77(5):1076–80.
16. Nilsson CG, Palva K, Aarnio R, Morcos E, Falconer C. Seventeen years' follow-up of the tension-free vaginal tape procedure for female stress urinary incontinence. Int Urogynecol J. 2013;24(8):1265–9.
17. Phe V, Benadiba S, Roupret M, Granger B, Richard F, Chartier-Kastler E. Long-term functional outcomes after artificial urinary sphincter implantation in women with stress urinary incontinence. BJU Int. 2014;113(6):961–7.
18. Stanford EJ, Paraiso MF. A comprehensive review of suburethral sling procedure complications. J Minim Invasive Gynecol. 2008;15(2):132–45.
19. Osborn DJ, Dmochowski RR, Harris CJ, Danford JJ, Kaufman MR, Mock S, et al. Analysis of patient and technical factors associated with midurethral sling mesh exposure and perforation. Int J Urol. 2014;21(11):1167–70.
20. Mohr S, Siegenthaler M, Mueller MD, Kuhn A. Bulking agents: an analysis of 500 cases and review of the literature. Int Urogynecol J. 2013;24(2):241–7.
21. Martin GR. Isolation of a pluripotent cell line from early mouse embryos cultured in medium conditioned by teratocarcinoma stem cells. Proc Natl Acad Sci U S A. 1981;78(12):7634–8.
22. Thomson JA, Itskovitz-Eldor J, Shapiro SS, Waknitz MA, Swiergiel JJ, Marshall VS, et al. Embryonic stem cell lines derived from human blastocysts. Science. 1998;282(5391):1145–7.
23. Lin CS, Lue TF. Stem cell therapy for stress urinary incontinence: a critical review. Stem Cells Dev. 2012;21(6):834–43.
24. Hart ML, Izeta A, Herrera-Imbroda B, Amend B, Brinchmann JE. Cell therapy for stress urinary incontinence. Tissue Eng B Rev. 2015;21(4):365–76.
25. Tran C, Damaser MS. The potential role of stem cells in the treatment of urinary incontinence. Ther Adv Urol. 2015;7(1):22–40.
26. Badra S, Andersson KE, Dean A, Mourad S, Williams JK. Long-term structural and functional effects of autologous muscle precursor cell therapy in a nonhuman primate model of urinary sphincter deficiency. J Urol. 2013;190(5):1938–45.
27. Klauser A, Frauscher F, Strasser H, Helweg G, Kolle D, Strohmeyer D, et al. Age-related rhabdosphincter function in female urinary stress incontinence: assessment of intraurethral sonography. J Ultrasound Med. 2004;23(5):631–7; quiz 8-9.
28. Goldman HB, Sievert KD, Damaser MS. Will we ever use stem cells for the treatment of SUI?: ICI-RS 2011. Neurourol Urodyn. 2012;31(3):386–9.
29. Vaegler M, Lenis AT, Daum L, Amend B, Stenzl A, Toomey P, et al. Stem cell therapy for voiding and erectile dysfunction. Nat Rev Urol. 2012;9(8):435–47.

30. Kim JH, Lee SR, Song YS, Lee HJ. Stem cell therapy in bladder dysfunction: where are we? And where do we have to go? BioMed Res Int. 2013;2013:1–10.
31. Godfrey KJ, Mathew B, Bulman JC, Shah O, Clement S, Gallicano GI. Stem cell-based treatments for type 1 diabetes mellitus: bone marrow, embryonic, hepatic, pancreatic and induced pluripotent stem cells. Diabet Med. 2012;29(1):14–23.
32. Singla DK, Abdelli LS. Embryonic stem cells and released factors stimulate c-kit(+)/FLK-1(+) progenitor cells and promote neovascularization in doxorubicin-induced cardiomyopathy. Cell Transplant. 2015;24(6):1043–52.
33. Anker PS, Scherjon SA, Kleijburg-van der Keur C, Noort WA, Claas FH, Willemze R, et al. Amniotic fluid as a novel source of mesenchymal stem cells for therapeutic transplantation. Blood. 2003;102(4):1548–9.
34. Moorefield EC, McKee EE, Solchaga L, Orlando G, Yoo JJ, Walker S, et al. Cloned, CD117 selected human amniotic fluid stem cells are capable of modulating the immune response. PLoS One. 2011;6(10):e26535.
35. Takahashi K, Yamanaka S. Induction of pluripotent stem cells from mouse embryonic and adult fibroblast cultures by defined factors. Cell. 2006;126(4):663–76.
36. Morris SA, Daley GQ. A blueprint for engineering cell fate: current technologies to reprogram cell identity. Cell Res. 2013;23(1):33–48.
37. Wang HJ, Chuang YC, Chancellor MB. Development of cellular therapy for the treatment of stress urinary incontinence. Int Urogynecol J. 2011;22(9):1075–83.
38. Friedenstein AJ, Deriglasova UF, Kulagina NN, Panasuk AF, Rudakowa SF, Luria EA, et al. Precursors for fibroblasts in different populations of hematopoietic cells as detected by the in vitro colony assay method. Exp Hematol. 1974;2(2):83–92.
39. Ding DC, Shyu WC, Lin SZ. Mesenchymal stem cells. Cell Transplant. 2011;20(1):5–14.
40. Krause DS, Theise ND, Collector MI, Henegariu O, Hwang S, Gardner R, et al. Multi-organ, multi-lineage engraftment by a single bone marrow-derived stem cell. Cell. 2001;105(3):369–77.
41. Lavker RM, Sun TT, Oshima H, Barrandon Y, Akiyama M, Ferraris C, et al. Hair follicle stem cells. J Investig Dermatol Symp Proc. 2003;8(1):28–38.
42. Patel AN, Park E, Kuzman M, Benetti F, Silva FJ, Allickson JG. Multipotent menstrual blood stromal stem cells: isolation, characterization, and differentiation. Cell Transplant. 2008;17(3):303–11.
43. Zhang Y, McNeill E, Tian H, Soker S, Andersson KE, Yoo JJ, et al. Urine derived cells are a potential source for urological tissue reconstruction. J Urol. 2008;180(5):2226–33.
44. Wu S, Wang Z, Bharadwaj S, Hodges SJ, Atala A, Zhang Y. Implantation of autologous urine derived stem cells expressing vascular endothelial growth factor for potential use in genitourinary reconstruction. J Urol. 2011;186(2):640–7.
45. Bharadwaj S, Liu GH, Shi YG, Markert C, Andersson KE, Atala A, et al. Characterization of urine-derived stem cells obtained from upper urinary tract for use in cell-based urological tissue engineering. Tissue Eng A. 2011;17(15-16):2123–32.
46. Lang R, Liu GH, Shi YG, Bharadwaj S, Leng XY, Zhou XB, et al. Self-renewal and differentiation capacity of urine-derived stem cells after urine preservation for 24 hours. Plos One. 2013;8(1):e53980.
47. Liu GH, Pareta RA, Wu RP, Shi YG, Zhou XB, Liu H, et al. Skeletal myogenic differentiation of urine-derived stem cells and angiogenesis using microbeads loaded with growth factors. Biomaterials. 2013;34(4):1311–26.
48. Zhang Y, Atala A. Urothelial cell culture. Methods Mol Biol. 2013;1037:27–43.
49. Bharadwaj S, Liu G, Shi Y, Wu R, Yang B, He T, et al. Multipotential differentiation of human urine-derived stem cells: potential for therapeutic applications in urology. Stem Cells. 2013;31(9):1840–56.
50. Sohni A, Verfaillie CM. Mesenchymal stem cells migration homing and tracking. Stem Cells Int. 2013;2013:130763.
51. Wynn RF, Hart CA, Corradi-Perini C, O'Neill L, Evans CA, Wraith JE, et al. A small proportion of mesenchymal stem cells strongly expresses functionally active CXCR4 receptor capable of promoting migration to bone marrow. Blood. 2004;104(9):2643–5.
52. Honczarenko M, Le Y, Swierkowski M, Ghiran I, Glodek AM, Silberstein LE. Human bone marrow stromal cells express a distinct set of biologically functional chemokine receptors. Stem Cells. 2006;24(4):1030–41.
53. Docheva D, Popov C, Mutschler W, Schieker M. Human mesenchymal stem cells in contact with their environment: surface characteristics and the integrin system. J Cell Mol Med. 2007;11(1):21–38.
54. Ruster B, Gottig S, Ludwig RJ, Bistrian R, Muller S, Seifried E, et al. Mesenchymal stem cells display coordinated rolling and adhesion behavior on endothelial cells. Blood. 2006;108(12):3938–44.
55. Woo LL, Hijaz A, Kuang M, Penn MS, Damaser MS, Rackley RR. Over expression of stem cell homing cytokines in urogenital organs following vaginal distention. J Urol. 2007;177(4):1568–72.
56. Wood HM, Kuang M, Woo L, Hijaz A, Butler RS, Penn M, et al. Cytokine expression after vaginal distention of different durations in virgin Sprague-Dawley rats. J Urol. 2008;180(2):753–9.
57. Lenis AT, Kuang M, Woo LL, Hijaz A, Penn MS, Butler RS, et al. Impact of parturition on chemokine homing factor expression in the vaginal distention model of stress urinary incontinence. J Urol. 2013;189(4):1588–94.
58. Cruz M, Dissaranan C, Cotleur A, Kiedrowski M, Penn M, Damaser M. Pelvic organ distribution of mesenchymal stem cells injected intravenously after simulated childbirth injury in female rats. Obstet Gynecol Int. 2012;2012:612946.

59. Dissaranan C, Cruz MA, Kiedrowski MJ, Balog BM, Gill BC, Penn MS, et al. Rat mesenchymal stem cell secretome promotes elastogenesis and facilitates recovery from simulated childbirth injury. Cell Transplant. 2014;23(11):1395–406.
60. Woo LL, Tanaka ST, Anumanthan G, Pope JCT, Thomas JC, Adams MC, et al. Mesenchymal stem cell recruitment and improved bladder function after bladder outlet obstruction: preliminary data. J Urol. 2011;185(3):1132–8.
61. Rombouts WJC, Ploemacher RE. Primary murine MSC show highly efficient homing to the bone marrow but lose homing ability following culture. Leukemia. 2003;17(1):160–70.
62. De Becker A, Van Hummelen P, Bakkus M, Broek IV, De Wever J, De Waele M, et al. Migration of culture-expanded human mesenchymal stem cells through bone marrow endothelium is regulated by matrix metalloproteinase-2 and tissue inhibitor of metalloproteinase-3. Haematol-Hematol J. 2007;92(4):440–9.
63. Fischer UM, Harting MT, Jimenez F, Monzon-Posadas WO, Xue HS, Savitz SI, et al. Pulmonary passage is a major obstacle for intravenous stem cell delivery: the pulmonary first-pass effect. Stem Cells Dev. 2009;18(5):683–91.
64. Ghionzoli M, Cananzi M, Zani A, Rossi CA, Leon FF, Pierro A, et al. Amniotic fluid stem cell migration after intraperitoneal injection in pup rats: implication for therapy. Pediatr Surg Int. 2010;26(1):79–84.
65. Fu Q, Song XF, Liao GL, Deng CL, Cui L. Myoblasts differentiated from adipose-derived stem cells to treat stress urinary incontinence. Urology. 2010;75(3):718–23.
66. Kim SO, Na HS, Kwon D, Joo SY, Kim HS, Ahn Y. Bone-marrow-derived mesenchymal stem cell transplantation enhances closing pressure and leak point pressure in a female urinary incontinence rat model. Urol Int. 2011;86(1):110–6.
67. Lin GT, Wang GF, Banie L, Ning HX, Shindel AW, Fandel TM, et al. Treatment of stress urinary incontinence with adipose tissue-derived stem cells. Cytotherapy. 2010;12(1):88–95.
68. Wu R, Liu G, Bharadwaj S, Zhang Y. Isolation and myogenic differentiation of mesenchymal stem cells for urologic tissue engineering. Methods Mol Biol. 2013;1001:65–80.
69. Carvalho MM, Teixeira FG, Reis RL, Sousa N, Salgado AJ. Mesenchymal stem cells in the umbilical cord: phenotypic characterization, secretome and applications in central nervous system regenerative medicine. Curr Stem Cell Res Ther. 2011;6(3):221–8.
70. Caplan AI, Dennis JE. Mesenchymal stem cells as trophic mediators. J Cell Biochem. 2006;98(5):1076–84.
71. Deans RJ, Moseley AB. Mesenchymal stem cells: biology and potential clinical uses. Exp Hematol. 2000;28(8):875–84.
72. Stastna M, Van Eyk JE. Secreted proteins as a fundamental source for biomarker discovery. Proteomics. 2012;12(4–5):722–35.
73. Ryan JM, Barry FP, Murphy JM, Mahon BP. Mesenchymal stem cells avoid allogeneic rejection. J Inflamm (Lond). 2005;2:8.
74. Duijvestein M, van den Brink GR, Hommes DW. Stem cells as potential novel therapeutic strategy for inflammatory bowel disease. J Crohns Colitis. 2008;2(2):99–106.
75. Hare JM, Traverse JH, Henry TD, Dib N, Strumpf RK, Schulman SP, et al. A randomized, double-blind, placebo-controlled, dose-escalation study of intravenous adult human mesenchymal stem cells (prochymal) after acute myocardial infarction. J Am Coll Cardiol. 2009;54(24):2277–86.
76. Telukuntla KS, Suncion VY, Schulman IH, Hare JM. The advancing field of cell-based therapy: insights and lessons from clinical trials. J Am Heart Assoc. 2013;2(5):e000338.
77. Shabbir A, Zisa D, Suzuki G, Lee T. Heart failure therapy mediated by the trophic activities of bone marrow mesenchymal stem cells: a noninvasive therapeutic regimen. Am J Physiol Heart Circ Physiol. 2009;296(6):H1888–97.
78. Timmers L, Lim SK, Hoefer IE, Arslan F, Lai RC, van Oorschot AA, et al. Human mesenchymal stem cell-conditioned medium improves cardiac function following myocardial infarction. Stem Cell Res. 2011;6(3):206–14.
79. Song YS, Lee HJ, Doo SH, Lee SJ, Lim I, Chang KT, et al. Mesenchymal stem cells overexpressing hepatocyte growth factor (HGF) inhibit collagen deposit and improve bladder function in rat model of bladder outlet obstruction. Cell Transplant. 2012;21(8):1641–50.
80. Lin AS, Carrier S, Morgan DM, Lue TF. Effect of simulated birth trauma on the urinary continence mechanism in the rat. Urology. 1998;52(1):143–51.
81. Pan HQ, Kerns JM, Lin DL, Liu S, Esparza N, Damaser MS. Increased duration of simulated childbirth injuries results in increased time to recovery. Am J Physiol Regul Integr Comp Physiol. 2007;292(4):R1738–44.
82. Woo LL, Hijaz A, Pan HQ, Kuang M, Rackley RR, Damaser MS. Simulated childbirth injuries in an inbred rat strain. Neurourol Urodyn. 2009;28(4):356–61.
83. Burdzinska A, Crayton R, Dybowski B, Koperski L, Idziak M, Fabisiak M, et al. Urethral distension as a novel method to simulate sphincter insufficiency in the porcine animal model. Int J Urol. 2012;19(7):676–82.
84. Kerns JM, Damaser MS, Kane JM, Sakamoto K, Benson JT, Shott S, et al. Effects of pudendal nerve injury in the female rat. Neurourol Urodyn. 2000;19(1):53–69.
85. Pan HQ, Lin DL, Strauch C, Butler RS, Monnier VM, Daneshgari F, et al. Pudendal nerve injury

reduces urethral outlet resistance in diabetic rats. Am J Physiol Ren Physiol. 2010;299(6):F1443–50.
86. Castiglione F, Bergamini A, Bettiga A, Bivalacqua TJ, Benigni F, Strittmatter F, et al. Perioperative betamethasone treatment reduces signs of bladder dysfunction in a rat model for neurapraxia in female urogenital surgery. Eur Urol. 2012;62(6):1076–85.
87. Salcedo L, Mayorga M, Damaser M, Balog B, Butler R, Penn M, et al. Mesenchymal stem cells can improve anal pressures after anal sphincter injury. Stem Cell Res. 2013;10(1):95–102.
88. Song QX, Balog BM, Kerns J, Li Lin D, Sun YH, Damaser MS, et al. Long-term effects of simulated childbirth injury on function and innervation of the urethra. Neurourol Urodyn. 2015;34(4):381–6.
89. Kato H, Igawa Y, Khaleque MA, Nishizawa O. Bladder dysfunction after proximal urethrolysis in female dogs. Int J Urol. 1999;6(1):33–7.
90. Rodriguez LV, Chen S, Jack GS, de Almeida F, Lee KW, Zhang R. New objective measures to quantify stress urinary incontinence in a novel durable animal model of intrinsic sphincter deficiency. Am J Physiol Regul Integr Comp Physiol. 2005;288(5):R1332–R8.
91. Skaff M, Pinto E, Leite KR, Almeida FG. Development of a rabbit's urethral sphincter deficiency animal model for anatomical-functional evaluation. Int Braz J Urol. 2012;38(1):17–24.
92. Kinebuchi Y, Aizawa N, Imamura T, Ishizuka O, Igawa Y, Nishizawa O. Autologous bone-marrow-derived mesenchymal stem cell transplantation into injured rat urethral sphincter. Int J Urol. 2010;17(4):359–68.
93. Chermansky CJ, Tarin T, Kwon DD, Jankowski RJ, Cannon TW, de Groat WC, et al. Intraurethral muscle-derived cell injections increase leak point pressure in a rat model of intrinsic sphincter deficiency. Urology. 2004;63(4):780–5.
94. Lim JJ, Jang JB, Kim JY, Moon SH, Lee CN, Lee KJ. Human umbilical cord blood mononuclear cell transplantation in rats with intrinsic sphincter deficiency. J Korean Med Sci. 2010;25(5):663–70.
95. Chermansky CJ, Cannon TW, Torimoto K, Fraser MO, Yoshimura N, de Groat WC, et al. A model of intrinsic sphincteric deficiency in the rat: electrocauterization. Neurourol Urodyn. 2004;23(2):166–71.
96. Yiou R, Yoo JJ, Atala A. Restoration of functional motor units in a rat model of sphincter injury by muscle precursor cell autografts. Transplantation. 2003;76(7):1053–60.
97. Zutshi M, Salcedo LB, Zaszczurynski PJ, Hull TL, Butler RS, Damaser MS. Effects of sphincterotomy and pudendal nerve transection on the anal sphincter in a rat model. Dis Colon Rectum. 2009;52(7):1321–9.
98. Praud C, Sebe P, Bierinx AS, Sebille A. Improvement of urethral sphincter deficiency in female rats following autologous skeletal muscle myoblasts grafting. Cell Transplant. 2007;16(7):741–9.
99. Eberli D, Andersson KE, Yoo JJ, Atala A. A canine model of irreversible urethral sphincter insufficiency. BJU Int. 2009;103(2):248–53.
100. Kefer JC, Liu G, Daneshgari F. Pubo-urethral ligament injury causes long-term stress urinary incontinence in female rats: an animal model of the integral theory. J Urol. 2009;181(1):397–400.
101. Kamo I, Torimoto K, Chancellor MB, de Groat WC, Yoshimura N. Urethral closure mechanisms under sneeze-induced stress condition in rats: a new animal model for evaluation of stress urinary incontinence. Am J Physiol Regul Integr Comp Physiol. 2003;285(2):R356–65.
102. Peng CW, Chen JJ, Chang HY, de Groat WC, Cheng CL. External urethral sphincter activity in a rat model of pudendal nerve injury. Neurourol Urodyn. 2006;25(4):388–96.
103. Badra S, Andersson KE, Dean A, Mourad S, Williams JK. A nonhuman primate model of stable urinary sphincter deficiency. J Urol. 2013;189(5):1967–74.
104. Furuta A, Jankowski RJ, Pruchnic R, Egawa S, Yoshimura N, Chancellor MB. Physiological effects of human muscle-derived stem cell implantation on urethral smooth muscle function. Int Urogynecol J. 2008;19(9):1229–34.
105. Xu Y, Song YF, Lin ZX. Transplantation of muscle-derived stem cells plus biodegradable fibrin glue restores the urethral sphincter in a pudendal nerve-transected rat model. Braz J Med Biol Res. 2010;43(11):1076–83.
106. Zhao WM, Zhang C, Jin CJ, Zhang ZJ, Kong DL, Xu WH, et al. Periurethral injection of autologous adipose-derived stem cells with controlled-release nerve growth factor for the treatment of stress urinary incontinence in a rat model. Eur Urol. 2011;59(1):155–63.
107. Wu GZ, Song YF, Zheng X, Jiang ZQ. Adipose-derived stromal cell transplantation for treatment of stress urinary incontinence. Tissue Cell. 2011;43(4):246–53.
108. Corcos J, Loutochin O, Campeau L, Eliopoulos N, Bouchentouf M, Blok B, et al. Bone marrow mesenchymal stromal cell therapy for external urethral sphincter restoration in a rat model of stress urinary incontinence. Neurourol Urodyn. 2011;30(3):447–55.
109. Du XW, Wu HL, Zhu YF, Hu JB, Jin F, Lv RP, et al. Experimental study of therapy of bone marrow mesenchymal stem cells or muscle-like cells/calcium alginate composite gel for the treatment of stress urinary incontinence. Neurourol Urodyn. 2013;32(3):281–6.
110. Chun SY, Kwon JB, Chae SY, Lee JK, Bae JS, Kim BS, et al. Combined injection of three different lineages of early-differentiating human amniotic fluid-derived cells restores urethral sphincter function in urinary incontinence. BJU Int. 2014;114(5):770–83.
111. Joyce N, Annett G, Wirthlin L, Olson S, Bauer G, Nolta JA. Mesenchymal stem cells for the treatment of neurodegenerative disease. Regen Med. 2010;5(6):933–46.

112. Hardy SA, Maltman DJ, Przyborski SA. Mesenchymal stem cells as mediators of neural differentiation. Curr Stem Cell Res Ther. 2008;3(1):43–52.
113. Li GY, Zhou F, Gong YQ, Cui WS, Yuan YM, Song WD, et al. Activation of VEGF and ERK1/2 and improvement of urethral function by adipose-derived stem cells in a rat stress urinary incontinence model. Urology. 2012;80(4):953–e1.
114. Obinata D, Matsumoto T, Ikado Y, Sakuma T, Kano K, Fukuda N, et al. Transplantation of mature adipocyte-derived dedifferentiated fat (DFAT) cells improves urethral sphincter contractility in a rat model. Int J Urol. 2011;18(12):827–34.
115. Ohta Y, Takenaga M, Tokura Y, Hamaguchi A, Matsumoto T, Kano K, et al. Mature adipocyte-derived cells, dedifferentiated fat cells (DFAT), promoted functional recovery from spinal cord injury-induced motor dysfunction in rats. Cell Transplant. 2008;17(8):877–86.
116. Kwon D, Kim Y, Pruchnic R, Jankowski R, Usiene I, De Miguel F, et al. Periurethral cellular injection: comparison of muscle-derived progenitor cells and fibroblasts with regard to efficacy and tissue contractility in an animal model of stress urinary incontinence. Urology. 2006;68(2):449–54.
117. Mitterberger M, Pinggera GM, Marksteiner R, Margreiter E, Plattner R, Klima G, et al. Functional and histological changes after myoblast injections in the porcine rhabdosphincter. Eur Urol. 2007;52(6):1736–43.
118. Yiou R, Dreyfus P, Chopin DK, Abbou CC, Lefaucheur JP. Muscle precursor cell autografting in a murine model of urethral sphincter injury. BJU Int. 2002;89(3):298–302.
119. Kim YT, Kim DK, Jankowski RJ, Pruchnic R, Usiene I, de Miguel F, et al. Human muscle-derived cell injection in a rat model of stress urinary incontinence. Muscle Nerve. 2007;36(3):391–3.
120. Kim BS, Chun SY, Lee JK, Lim HJ, Bae JS, Chung HY, et al. Human amniotic fluid stem cell injection therapy for urethral sphincter regeneration in an animal model. BMC Med. 2012;10:94.
121. Bandyopadhyay B, Thakur A, Dave V, Viswanathan C, Ghosh D. A non-invasive method to evaluate the efficacy of human myoblast in botulinum-A toxin induced stress urinary incontinence model in rats. Urol J. 2013;10(4):1126–34.
122. Aref-Adib M, Lamb BW, Lee HB, Akinnawo E, Raza MM, Hughes A, et al. Stem cell therapy for stress urinary incontinence: a systematic review in human subjects. Arch Gynecol Obstet. 2013;288(6):1213–21.
123. Mitterberger M, Marksteiner R, Margreiter E, Pinggera GM, Colleselli D, Frauscher F, et al. Autologous myoblasts and fibroblasts for female stress incontinence: a 1-year follow-up in 123 patients. BJU Int. 2007;100(5):1081–5.
124. Mitterberger M, Marksteiner R, Margreiter E, Pinggera GM, Frauscher F, Ulmer H, et al. Myoblast and fibroblast therapy for post-prostatectomy urinary incontinence: 1-year followup of 63 patients. J Urol. 2008;179(1):226–31.
125. Mitterberger M, Pinggera GM, Marksteiner R, Margreiter E, Fussenegger M, Frauscher F, et al. Adult stem cell therapy of female stress urinary incontinence. Eur Urol. 2008;53(1):169–75.
126. Carr LK, Steele D, Steele S, Wagner D, Pruchnic R, Jankowski R, et al. 1-year follow-up of autologous muscle-derived stem cell injection pilot study to treat stress urinary incontinence. Int Urogynecol J Pelvic Floor Dysfunct. 2008;19(6):881–3.
127. Sebe P, Doucet C, Cornu JN, Ciofu C, Costa P, de Medina SG, et al. Intrasphincteric injections of autologous muscular cells in women with refractory stress urinary incontinence: a prospective study. Int Urogynecol J. 2011;22(2):183–9.
128. Cornu JN, Lizee D, Pinset C, Haab F. Long-term follow-up after regenerative therapy of the urethral sphincter for female stress urinary incontinence. Eur Urol. 2014;65(1):256–8.
129. Blaganje M, Lukanovic A. Ultrasound-guided autologous myoblast injections into the extrinsic urethral sphincter: tissue engineering for the treatment of stress urinary incontinence. Int Urogynecol J. 2013;24(4):533–5.
130. Gerullis H, Eimer C, Georgas E, Homburger M, El-Baz AG, Wishahi M, et al. Muscle-derived cells for treatment of iatrogenic sphincter damage and urinary incontinence in men. Sci World J. 2012;2012:1–6.
131. Stangel-Wojcikiewicz K, Jarocha D, Piwowar M, Jach R, Uhl T, Basta A, et al. Autologous muscle-derived cells for the treatment of female stress urinary incontinence: a 2-year follow-up of a polish investigation. Neurourol Urodyn. 2014;33(3):324–30.
132. Carr LK, Robert M, Kultgen PL, Herschorn S, Birch C, Murphy M, et al. Autologous muscle derived cell therapy for stress urinary incontinence: a prospective, dose ranging study. J Urol. 2013;189(2):595–601.
133. Peters KM, Dmochowski RR, Carr LK, Robert M, Kaufman MR, Sirls LT, et al. Autologous muscle derived cells for treatment of stress urinary incontinence in women. J Urol. 2014;192(2):469–76.
134. Yamamoto T, Gotoh M, Kato M, Majima T, Toriyama K, Kamei Y, et al. Periurethral injection of autologous adipose-derived regenerative cells for the treatment of male stress urinary incontinence: report of three initial cases. Int J Urol. 2012;19(7):652–9.
135. Gotoh M, Yamamoto T, Kato M, Majima T, Toriyama K, Kamei Y, et al. Regenerative treatment of male stress urinary incontinence by periurethral injection of autologous adipose-derived regenerative cells: 1-year outcomes in 11 patients. Int J Urol. 2014;21(3):294–300.
136. Lee CN, Jang JB, Kim JY, Koh C, Baek JY, Lee KJ. Human cord blood stem cell therapy for treatment of stress urinary incontinence. J Korean Med Sci. 2010;25(6):813–6.
137. Shirvan MK, Alamdari DH, Mahboub MD, Ghanadi A, Rahimi HR, Seifalian AM. A novel cell therapy

for stress urinary incontinence, short-term outcome. Neurourol Urodyn. 2013;32(4):377–82.
138. Pokrywczynska M, Adamowicz J, Czapiewska M, Balcerczyk D, Jundzill A, Nowacki M, et al. Targeted therapy for stress urinary incontinence: a systematic review based on clinical trials. Expert Opin Biol Ther. 2016;16(2):233–42.
139. Herschorn S, Radomski SB. Collagen injections for genuine stress urinary incontinence: patient selection and durability. Int Urogynecol J Pelvic Floor Dysfunct. 1997;8(1):18–24.
140. Blaganje M, Lukanovic A. Intrasphincteric autologous myoblast injections with electrical stimulation for stress urinary incontinence. Int J Gynaecol Obstet. 2012;117(2):164–7.
141. Gras S, Klarskov N, Lose G. Intraurethral injection of autologous minced skeletal muscle: a simple surgical treatment for stress urinary incontinence. J Urol. 2014;192(3):850–5.
142. Kuismanen K, Sartoneva R, Haimi S, Mannerstrom B, Tomas E, Miettinen S, et al. Autologous adipose stem cells in treatment of female stress urinary incontinence: results of a pilot study. Stem Cells Transl Med. 2014;3(8):936–41.
143. Klein G, Hart ML, Brinchmann JE, Rolauffs B, Stenzl A, Sievert KD, et al. Mesenchymal stromal cells for sphincter regeneration. Adv Drug Deliv Rev. 2015;82–83:123–36.
144. Ranganath SH, Levy O, Inamdar MS, Karp JM. Harnessing the mesenchymal stem cell secretome for the treatment of cardiovascular disease. Cell Stem Cell. 2012;10(3):244–58.
145. Lavoie JR, Rosu-Myles M. Uncovering the secretes of mesenchymal stem cells. Biochimie. 2013;95(12):2212–21.
146. Deng KL, Lin DL, Hanzlicek B, Balog B, Penn MS, Kiedrowski MJ, et al. Mesenchymal stem cells and their secretome partially restore nerve and urethral function in a dual muscle and nerve injury stress urinary incontinence model. Am J Physiol-Renal. 2015;308(2):F92–F100.
147. Jiang HH, Damaser MS. Animal models of stress urinary incontinence. Handb Exp Pharmacol. 2011;202:45–67.
148. Hong SH, Piao S, Kim IG, Lee JY, Cho HJ, Kim SW, et al. Comparison of three types of stress urinary incontinence rat models: electrocauterization, pudendal denervation, and vaginal distension. Urology. 2013;81(2):465.e1–6.
149. Koike Y, Furuta A, Suzuki Y, Honda M, Naruoka T, Asano K, et al. Pathophysiology of urinary incontinence in murine models. Int J Urol. 2013;20(1):64–71.
150. Herrera-Imbroda B, Lara MF, Izeta A, Sievert KD, Hart ML. Stress urinary incontinence animal models as a tool to study cell-based regenerative therapies targeting the urethral sphincter. Adv Drug Deliv Rev. 2015;82–83:106–16.
151. Sloff M, Simaioforidis V, de Vries R, Oosterwijk E, Feitz W. Tissue engineering of the bladder – reality or myth? A systematic review. J Urol. 2014;192(4):1035–42.
152. Eberli D, Aboushwareb T, Soker S, Yoo JJ, Atala A. Muscle precursor cells for the restoration of irreversibly damaged sphincter function. Cell Transplant. 2012;21(9):2089–98.
153. Williams JK, Eckman D, Dean A, Moradi M, Allickson J, Cline JM, et al. The dose-effect safety profile of skeletal muscle precursor cell therapy in a dog model of intrinsic urinary sphincter deficiency. Stem Cells Transl Med. 2015;4(3):286–94.
154. Burdzinska A, Crayton R, Dybowski B, Idziak M, Gala K, Radziszewski P, et al. The effect of endoscopic administration of autologous porcine muscle-derived cells into the urethral sphincter. Urology. 2013;82(3):743.e1–8.
155. Ganzer R, Kohler D, Neuhaus J, Dorschner W, Stolzenburg JU. Is the rhesus monkey (Macaca mulatta) comparable to humans? Histomorphology of the sphincteric musculature of the lower urinary tract including 3D-reconstruction. Anat Histol Embryol. 2004;33(6):355–61.
156. Kaplan JR, Manuck SB. Ovarian dysfunction, stress, and disease: a primate continuum. ILAR J. 2004;45(2):89–115.
157. Hijaz AK, Grimberg KO, Tao M, Schmotzer B, Sadeghi Z, Lin YH, et al. Stem cell homing factor, CCL7, expression in mouse models of stress urinary incontinence. Female Pelvic Med Reconstr Surg. 2013;19(6):356–61.
158. Lau TT, Wang DA. Stromal cell-derived factor-1 (SDF-1): homing factor for engineered regenerative medicine. Expert Opin Biol Ther. 2011;11(2):189–97.
159. Herberg S, Shi XM, Johnson MH, Hamrick MW, Isales CM, Hill WD. Stromal cell-derived factor-1 beta mediates cell survival through enhancing autophagy in bone marrow-derived mesenchymal stem cells. Plos One. 2013;8(3):e58207.
160. Christ GJ, Saul JM, Furth ME, Andersson KE. The pharmacology of regenerative medicine. Pharmacol Rev. 2013;65(3):1091–133.
161. Nagasawa T. CXC chemokine ligand 12 (CXCL12) and its receptor CXCR4. J Mol Med. 2014; 92(5):433–9.

11 Penis and Testis

Seok Cho and Jae Hyun Bae

11.1 Penis

11.1.1 Introduction

Reconstruction or substitution of the corpora cavernosa to restore a cosmetically acceptable phallus that would allow normal reproductive, sexual, and urinary function represents a very challenging task.

Conditions requiring penile reconstruction include penile malignancy, congenital anomaly, trauma, erectile dysfunction, Peyronie's disease, gender reassignment surgery, and post penectomy status. Penile necrosis due to infection after penile surgery such as penile prosthesis insertion is not common but not rare in diabetic patients. In those cases, partial or total penectomy has to be performed.

Penile reconstruction requires two tasks. The first one is the morphological reconstruction. In cases of post-penectomy status or congenital anomaly, the patients need to establish morphological reconstruction. The second one is the functional reconstruction. The patients with erectile dysfunction which is not response to medical treatment want to recover their erectile function first. Of course, it will be ideal to regenerate both morphological and functional recovery.

Despite the development of many surgical materials have been developed, current reconstructive techniques are limited by issues of tissue availability and compatibility. These techniques use nonurologic host tissues (skin, gastrointestinal segments, or mucosa from multiple body sites), donor tissues (cadaver fascia or cadaver or living donor kidney), heterologous tissues (bovine collagen), or artificial materials (silicone, polyurethane, or polytetrafluoroethylene) to reconstruct damaged organs. However, all of these materials can lead to significant complications resulting from immunologic rejection or functional mismatches between the native and implanted tissues such as infection, graft failure, and donor site morbidity have been reported and in these cases, the functional regeneration is impossible, therefore additional penile prosthesis insertion is needed and also, additional complications can be developed. In order to reduce or eliminate these complications, recent researchers have tried to regenerate penis to overcome the limitations of the free flap operation technique.

The functional regeneration of erectile function is now focusing on cell therapy such as stem

S. Cho (✉)
The Department of Urology, Ilsanpaik Hospital, Inje University College of Medicine, Goyang, South Korea
e-mail: envystone@hanmail.net

J.H. Bae
The Department of Urology, Korea University Ansan Hospital, Korea University College of Medicine, Ansan, South Korea
e-mail: urobae@genetherapy.or.kr

B.W. Kim (ed.), *Clinical Regenerative Medicine in Urology*, DOI 10.1007/978-981-10-2723-9_11

cell injection with or without various growth factors.

In this chapter, various trials and results will be introduced.

11.1.2 Anatomy

The penis is complex and composed of three separate cylinders: two paired cylinders called the corpora cavernosa on the dorsal side make up the majority of the penis and are the primary structure involved in erectile function. These cylinders consist of a network of large venous sinuses separated by trabeculae, dense connective tissue. The corpus spongiosum, the third cylinder is located on the ventral side and is the structure which wraps the urethra for the protection. It has the same arterial and venous relationship as the corpora cavernosa. Each of these cylinders is covered by a very tough thick sheath called the tunica albuginea. All three cylinders are encased in a thick membrane called as Buck's fascia. Finally, Collie's fascia, which is continuous with the abdominal wall, covers the entire structure and makes this whole supporting structure of the penis very tough, allowing it to take quite a bit of force and trauma.

The erectile tissue within the corpora contains arteries, nerves, muscle fibers, and venous sinuses lined with flat endothelial cells, and it fills the space of the corpora cavernosa. The cut surface of the corpora cavernosa looks like a sponge. There is a thin layer of areolar tissue that separates this tissue from the tunica albuginea.

The shaft is covered by nearly hairless skin. Under the skin lies the dense connective tissue of penile fascia. The tunica albuginea encircles all three corpora, divides the corpora proximally, but is incomplete distally.

11.1.3 Hemodynamics and Mechanism of Erection and Detumescence

The penile erectile tissue, specifically the cavernous smooth musculature and the smooth muscles of the arteriolar and arterial walls, plays a key role in the erectile process. In the flaccid state, these smooth muscles are tonically contracted, allowing only a small amount of arterial flow for nutritional purposes. The blood partial pressure of oxygen (pO2) is about 35 mmHg. The flaccid penis is in a moderate state of contraction, as evidenced by further shrinkage in cold weather and after phenylephrine injection.

Sexual stimulation triggers release of neurotransmitters from the cavernous nerve terminals. This results in relaxation of these smooth muscles and the following events: (1) dilation of the arterioles and arteries by increased blood flow in both the diastolic and systolic phases; (2) trapping of the incoming blood by the expanding sinusoids; (3) compression of the subtunical venous plexuses between the tunica albuginea and the peripheral sinusoids, reducing venous outflow; (4) stretching of the tunica to its capacity, which occludes the emissary veins between the inner circular and outer longitudinal layers and further decreases venous outflow to a minimum; (5) an increase in PO2 (to about 90 mmHg) and intracavernous pressure (around 100 mmHg), which raises the penis from the dependent position to the erect state (the full-erection phase); and (6) a further pressure increase (to several hundred millimeters of mercury) with contraction of the ischiocavernosus muscles (rigid-erection phase).

The angle of the erect penis is determined by its size and attachment to the puboischial rami (the crura) and the anterior surface of the pubic bone (the suspensory and funiform ligaments). In men with a long heavy penis or a loose suspensory ligament, the angle usually will not be greater than 90 degrees, even with full rigidity.

Three phases of detumescence have been reported in an animal study [1]. The first entails a transient intracorporeal pressure increase, indicating the beginning of smooth muscle contraction against a closed venous system. The second phase shows a slow pressure decrease, suggesting a slow reopening of the venous channels with resumption of the basal level of arterial flow. The third phase shows a fast pressure decrease with fully restored venous outflow capacity.

Erection thus involves sinusoidal relaxation, arterial dilation, and venous compression. The importance of smooth muscle relaxation has been demonstrated in animal and human studies.

11.1.4 Penile Reconstruction

Hu and colleagues suggested that patients with a penile defect are frequently extremely concerned about the extent to which their problem might affect their social status and the fidelity of their partners [2]. But, current therapies using microsurgical technology and free-flaps from autologous donor sites, combined with a penile prosthetic implant have not adequately addressed various structural disorders of the phallus. Penile implants are associated with complications including infection, erosion, and mechanical malfunction. Tissue-engineering approaches have been employed to address a number of penile disorders. Although the corporal bodies only contain two primary cell types, smooth muscle, and endothelium, engineering this tissue remains a challenge.

Only a few experimental studies have been published to date.

At first, investigators focused on morphologic reconstruction of penis using rib cartilage as a stiffener. However, this method was discouraged due to the unsatisfactory functional and cosmetic results, and then cartilage would serve as an ideal prosthesis for penile reconstruction, owing to its biomechanical properties.

Yoo et al. demonstrated that the feasibility of creating natural penile prostheses composed of cartilage [3]. Chondrocytes from articular surface of calf shoulders were seeded onto preformed cylindrical polyglycolic acid (PGA) polymer rods and then implanted in athymic mice. Rod-shaped solid cartilaginous structures with mature chondrocytes were observed in cell-seeded scaffolds. A subsequent study was performed to determine the feasibility of engineering human cartilage rods for potential use as penile prostheses [4].

Chondrocytes from the human ear were seeded on rod-shaped biodegradable-polymer scaffolds and implanted into athymic rats. After 2 months, chondrocytes formed flexible cartilaginous rods similar in size to the initial implants, and with mature chondrocytes. The mechanical properties of these cartilaginous rods were comparable with those of commercially available silicone prostheses. Autologous-engineered cartilage constructed using cells derived from rabbit ear cartilage was used for penile reconstruction in rabbits [5]. The autologous-cartilage cells were seeded on PGA and implanted into the corporal space of 10 rabbits; cartilage structures were formed within the corpora, with polymer degradation within 2 months.

Beside these efforts to develop engineering penile prosthesis, there were experiments about culturing human corporal tissues. Corporal smooth muscle is one of the major components of the phallus. In the event of pathology, and during surgical penile reconstruction for conditions such as ambiguous genitalia and exstrophy-epispadias (where there is a limited amount of native cavernosal tissue available), the creation of autologous and functional de novo corporal tissue would be beneficial.

In 1999, Park et al. investigated the possibility of developing corporal tissue by combining smooth muscle and endothelial cells. Normal human cavernosal smooth muscle cells and human endothelial cells were seeded on biodegradable polymer for implantation [6]. This is the first demonstration in tissue engineering in which capillary formation was facilitated by the addition of endothelial cells in composite tissue in vivo. This study showed that human corporal muscle and endothelial cells seeded on biodegradable polymer scaffolds are able to form vascularized cavernosal tissue when implanted in vivo.

Then initial investigation into the possible culture of human corporal SMCs combined with a polymer to generate a bioengineered corpus cavernosum has been reported [7].

Following these experiments, the same group, Kwon et al. reported that autologous corpus cavernosal smooth muscle and endothelial cells seeded on collagen matrices can form corpora cavernosa tissue structures in a rabbit model. Engineered corpora cavernosa achieved adequate structural and functional parameters [8].

The viability of engineered scaffolds was proved in vivo in rabbits in which the corpus cavernosum was replaced with an acellular collagen matrix. Chen et al. reported that normal erectile function and the ability to mate after 1 month; compared to the control animals, which were implanted with acellular matrix itself and were unable to reach erection [9]. Cavernosography examination showed that the experimental corporal bodies had structural integrity and similar pressure characteristics, when compared with the normal controls.

This work was recently taken one step further by the same group. Chen et al. demonstrated the ability to bioengineer entire pendular penile corporal bodies in a rabbit model [10]. Decellularized donor corporal bodies were seeded with autologous smooth muscle and endothelial cells in a stepwise fashion. Rabbits then underwent complete excision of the corporal bodies followed by implantation of seeded scaffolds. The group implanted with seeded scaffolds demonstrated normal intra-cavernosal pressures within 6 months of surgery, and 10 out of 12 rabbits could successfully mate with female rabbits.

A further group seeded acellular corporal collagen matrices with SMCs derived from human umbilical arteries and implanted the cellularized implants into athymic mice. Song et al. [11] reported that the scaffolds became cellularized to develop tissue with anatomical and functional properties similar to corpus cavernosum smooth muscle.

Another interesting approach has been presented by Zhang et al.: they first created a tissue-engineered corpus cavernosum seeded with muscle-derived stem cells, then described an experiment in which they transduced, via a lentivirus, the muscle-derived stem cells with vascular endothelial growth factor (VEGF) and demonstrated that after the in vivo implant, the obtained construct showed increased content of endothelial cells, smooth muscle cells, and capillaries [12, 13].

Another approach to recover erectile function by cell based therapy could be the injection of functional cells into the corpus cavernosum, e.g. in cases of severe endothelial dysfunction or decrease of smooth muscle cells. Wessells et al. suggested and experimentally investigated first for endothelial cell transplantation [14]. Fluorescently labeled autologous endothelial cells were injected into the rat corpus cavernosum. After 2 days clusters of cells were still seen within the central corporal sinusoids, whereas after one and 2 weeks remaining cells were only detected peripherally in the sinusoids beneath the tunica albuginea. Although this technique might be considered for direct repair of damaged corporal endothelium, this kind of cell based therapy for erectile dysfunction, either with endothelial or with smooth muscle cells, only seems to be promising when combined with cell manipulation by gene therapy prior to cell transfer [15–17].

Cavernous nerves can be easily damaged during penile surgery, and simply exposing the cavernous nerve can lead to ED. Therefore, methods of restoring nerve function in these patients are necessary. Different regenerative strategies used until now for repair or replacement of cavernous nerves include: use of silicone guidance tubes filled with primary Schwann cells, and nerve gap interposition with acellular nerve matrices, which showed significant nerve regeneration and ED recoverability [18]; use of sural nerve as a scaffold for cavernous nerve regeneration after radical prostatectomy, where partial improvement of ED and continence recovery were confirmed in many clinical studies [19–22]; and use of a nerve graft procedure after radical prostatectomy—a large clinical study in Japan indicated that this procedure might contribute to the recovery of urinary and sexual function [23].

11.1.4.1 Penile Enhancement

Penile enhancement has been used for both therapeutic and esthetic purposes. In a study, autologous fibroblasts from scrotal skin were harvested and seeded on poly acid-co-glycolide biodegradable (Maxpol-T) scaffolds [24]. This construct was implanted under Buck's fascia of the penile shaft of 69 patients. Mean penile girth was measured preoperatively and 6 months postoperatively in the flaccid and erect states; the mean penile girth had increased significantly after 6 months, and 94.2% of patients were satisfied with the procedure. In another study, 12 patients underwent repeated penile widening using biodegradable scaffolds enriched with expanded autologous scrotal dartos cells [25]; penile girth increased in both flaccid and erect states. Furthermore, 10–14 months after first surgery, biopsies showed highly vascularized loose tissue with collagen deposition and some inflammation, and after 22–24 months, biopsies revealed almost no inflammation and a tissue resembling normal dartos fascia. All patients were satisfied with the surgery.

11.1.4.2 Tunica Albuginea Reconstruction

Peyronie's disease is a benign, acquired wound-healing-connective-tissue disorder of the tunica albuginea that affects 3–9% of men [25]. Patients with Peyronie's disease can experience significant pain and functional problems due to plaque formation and curvature of the penis. Therapeutic options for Peyronie's disease include plication, incision and grafting of the tunica albuginea with materials such as small intestinal submucosa, fascia, and pericardium, and penile prosthesis. Graft-based therapies are associated with some complications such as recurrence of disease and graft contracture.

Schultheiss et al. isolated porcine fibroblasts from open fascia biopsies and seeded them on decellularized collagen matrices [26]. The seeded matrices were then cultivated in a bioreactor under continuous multiaxial stress for 21 days. Mechanically strained cultures of fibroblasts showed a homogeneous multilayer matrix infiltration and a regular cell alignment in the direction of strain axis after 7 days as well as increased production of extracellular matrix proteins compared to the static control.

Similarly, Joo et al. performed using mucosal graft from pigs' bladder [27]. A segment of the tunica albuginea of nine rabbits was excised, and the defect was covered with porcine bladder acellular matrix. Two months after implantation, the graft sites exhibited excellent healing without contracture, and the fusion between the graft and the neighboring normal TA appeared to be well established. There were no significant histological differences between the implanted tunica and the normal control tunica at 6 months after implantation.

Recently, Imbeault et al. used human fibroblasts to form a fibroblast sheet—human umbilical vein endothelial cells (HUVECs) were seeded on fibroblasts sheets and wrapped around a tubular support, and HUVECs were seeded into the lumen of the fibroblast sheets [25]. This endothelialized tubular graft was structurally similar to normal tunica albuginea with adequate mechanical resistance. In an attempt to engineer tunica albuginea, autologous fibroblasts were seeded onto a PGA scaffold and sutured into the same-sized defect in rat corpora. The seeded graft had features similar to the sham group, in terms of erectile response to cavernous nerve stimulation and retraction of graft 4 months after surgery, while the unseeded-graft control group had poor outcomes for these factors. Studies on the reconstruction of the penile tissue are shown in Table 11.1.

Table 11.1 List of study about penile tissue for which regenerative medicine therapies are being reported

Tissue	Material	Target	Authors	Years
Penile prosthesis				
Cartilage	Chondrocytes	Athymic mice	Yoo JJ et al.	1998
Cartilage	Autologous cartilage cells	Rabbit	Yoo JJ et al.	1999
Cartilage	Chondrocytes from human ear	Athymic rat	Kim BS et al.	2002
Corporal tissues				
Smooth muscle and endothelial cells	Human cavernosal smooth muscle cells and endothelial cells	Athymic mice	Park HJ et al.	1999
Smooth muscle	Human corporal smooth cells	Athymic mice	Kershen RT et al.	2002
Smooth muscle and endothelial cells	Rabbit cavernosal smooth muscle and endothelial cells	Rabbit	Kwon TG et al.	2002
Corpus Cavernosum	Autologous cavernosal tissue on cellular collagen matrix	Rabbit	Yoo JJ et al.	2006
Smooth muscle cells	Human umbilical artery smooth muscle cells	Athymic mice	Song LJ et al.	2009
Penile corporal body	Autologous cavernosal tissue	Rabbit	Chen KL et al.	2010
Corpus cavernosum	Muscle derived stem cells	Rabbit	Ji C et al.	2011
Corpus cavernosum	Muscle derived stem cells with human vascular endothelial growth factor	Rabbit	An G et al.	2013
Cavernous nerve				
Cavernous nerve	Acellular nerve graft	Rat	Connolly SS et al.	2008
Penile enhancement				
	Fibroblast	Human	Djinovic R et al.	2011
	Scrotal dartos cell	Human	Imbeault A et al.	2011
Tunica albuginea				
Tunica albuginea	Porcine fibroblast	In vitro	Shultheiss D et al.	2004
Tunica albuginea	Porcine bladder acellular matrix	Rabbit	Joo KW et al.	2006
Tunica albuginea	Human fibroblast	In vitro	Imbeault A et al.	2011

11.2 Testis

11.2.1 Introduction

In males, androgens, in particular testosterone, are known to have many important physiological actions, including effects on muscle, bone, central nervous system, prostate, bone marrow, and sexual function. There are many causes of a bilateral loss or congenital absence of the testicles, which compromises endocrine, reproductive, and psychological function in men. The testes can be congenitally absent in conditions such as unresolved cryptorchidism due to bilateral maldescent or atrophy. Acquired causes of anorchia include trauma, torsion, toxic damage from alcohol or chemotherapy and infections, such as mumps, orchitis or bilateral orchiectomy, which may be performed as a treatment for primary testicular cancer. There are many sequelae for patients without testes. Patients are sterile, have increased risk of osteoporosis and pathologic fractures, and are subject to further complications, such as sexual dysfunction, mood changes, severe psychological trauma, altered fat distribution, reduced stamina, and muscle wasting [28].

Current strategies of treatment in anorchic patients focus on the medical management of androgenic hormone replacement therapy to avoid the complications associated with hypogonadism. And such therapy may increase muscle strength, stabilize bone density, improve osteoporosis, and restore secondary sexual characteristics, including libido and erectile function. Surgically, the main intervention is the implantation of a prosthetic testis, a procedure which is well accepted by patients and has been shown to subjectively improve body image [29].

11.2.2 Anatomy

The testes are two glandular organs, which secrete the semen, suspended in the scrotum by the spermatic cords. In mammals, the testes are located outside the body because spermatogenesis in mammals is more efficient at a temperature somewhat less than the core body. When the temperature needs to be lowered, the cremasteric muscle relaxes and the testicles are lowered away from the warm body and are able to cool.

The tunica albuginea has smooth muscle cells that course through predominantly collagenous tissue. These smooth muscle cells may impart contractile capability to the capsule, may affect blood flow into the testis, and promote the flow of seminiferous tubule fluid out of the testis.

Under the tunica albuginea, a tough fibrous shell, the testis contains very fine coiled tubes called seminiferous tubules. The tubes are lined with a layer of cells that, from puberty into old age, performed spermatogenesis.

From the cellular point of view the human testis is a complex organ comprising germ cells and a variety of somatic cells such as Sertoli, Leydig, endothelial, fibroblast, macrophage, and peritubular myoid cells.

Testicles are components of both the reproductive system (being gonads) and the endocrine system (being endocrine glands). The testis has two functions: The first is spermatogenesis, which occurs in the seminiferous tubules. The seminiferous tubules consist of germ cells and supporting cells and are a unique environment for gamete production. Support cells include the Sertoli cells and fibrocyte and myoid cells of the basement membrane. The germ cells include a slowly dividing stem cell population, more rapidly proliferating spermatogonia and spermatocytes, and metamorphosing spermatids. Sertoli cells have irregularly shaped nuclei, prominent nucleoli, low mitotic index, and exhibit unique tight junctional complexes between adjacent Sertoli cells. These tight junctions are the strongest intercellular barriers in the body. They divide the seminiferous tubule space into basal (basement membrane) and adluminal (lumen) compartments. This anatomic arrangement forms the basis for the "blood-testis barrier" and allows spermatogenesis to occur in an immunologically privileged site. Sertoli cells serve as "nurse" cells for spermatogenesis, nourishing developing germ cells that are arranged between Sertoli cell cytoplasmic projections.

The second functions of testis are secretion of steroid hormones (androgens) by Leydig cells in the interstitial tissue. Testosterone blood levels change dramatically during human fetal, neonatal, and adult life. A testosterone peak occurs in the human fetus between 12 and 18 weeks of gestation. Another testosterone peak occurs at approximately 2 months of age. A third testosterone peak occurs during the 2nd or 3rd decade of life. After this, there is a plateau, and then a slow decline with age. The testosterone peaks correspond temporally to the following developmental events: (1) the differentiation and development of the fetal reproductive tract, (2) the neonatal organization or imprinting of androgen-dependent target tissues, (3) the masculinization of the male at puberty, and (4) the maintenance of growth and function of androgen-dependent organs in the adult.

11.2.3 Transplantation of Testicular Tissue

Testicular transplantation had its foundation not as a treatment modality for the recognized complications of anorchism, but for arguably less ethical reasons. The origins emerged from the quest to find the fountain of eternal youth; in this pursuit, the science of endocrinology and experimentation with testicular transplants began in the late nineteenth century. The French physiologist Charles-Édouard Brown-Séquard [30] presented the results of an experiment about injecting himself with crushed testicular extracts from young dogs and guinea pigs. He reported this technique successfully made him feel younger, along with increasing his sexual potency and heightening his intellect. This early work reinforced the belief that youth rejuvenation could be achieved through the replacement of endogenous substances, which then led to an increased interest in sex gland transplantation.

Voronoff in 1923 was the first to use chimpanzee and baboon organs for treating patients. The success described in this case quickly resulted in the procedure gathering interest. But that times none of surgeons used microsurgery to join blood vessels of the graft to the host's circulation, resulting in ischemic necrosis preceded by organ rejection [31].

With the advent of synthetic testosterone and other steroid hormones in 1935, the clinical indications of testicular transplantation as a form of androgen therapy gradually became obsolete.

In 1978, Silber [32] published a case report claiming the first entire human testis transplant. He described the case of two genetically identical male twins, one born with two normally functioning testicles and the other born with none. Using microvascular techniques, Silber successfully transplanted a testicle from the healthy twin into the anorchic brother, and, within 2 h of the operation, normal levels of testosterone were measured in both the donor and the recipient. FSH and LH levels also gradually normalized, and, over several months, the sperm count and motility also reached normal levels in the recipient.

The problems arising from the size of the testis and its fibrous capsule led some transplanters to use sliced or minced organs. Kearns, who reimplanted testicular tissue subcutaneously in a victim of accidental castration, reported the most plausible case [33].

According to this report, testosterone was being produced by the autograft, but without the normal architecture of the seminiferous epithelium, it is hard to understand how germ cell transfer could have restored spermatogenesis. Therefore, efforts to develop tissue grafting for the purpose of improving testosterone levels in hypogonadal men are more likely to succeed than are attempts at restoring fertility. The former goal appears to be simple, requiring the transfer of interstitial cells (Leydig cells), which are readily isolated from the donor testes by means of collagenase. Interstitial cells grafted in castrated rodents resulted in partial restoration of body weight, and testosterone levels above those of controls [34–36]. A number of vehicles and several implantation sites for interstitial cells have been tried, but none fully replaced testicular androgen production.

However, in 2009, Sun reported the successful restoration of androgen production in prepubertal rats undergoing Leydig cell transplantation. Serum testosterone levels were reported as normal 12 weeks post-transplantation with a high rate of survival and functionality of transplanted cells [37].

In the twenty-first century, attention turned to testicular transplantation mainly as a therapeutic intervention for infertility. This scenario arises in oncology patients who face courses of chemotherapy that jeopardize their fertility. Cryopreservation of sperm leaves these patients with a finite resource of fertility and is associated with only modest rates of successful future conception. Cryopreservation is also not a suitable option for prepubertal patients who have not yet started to produce sperm [38].

Complex immunologic principles provide a further unique challenge in testicular transplantation. The testes are known to be a site of immunologic privilege, which means they can tolerate the introduction of antigens without the development of an immune response.

Spermatogonial transplantation and reconstitution of donor cell spermatogenesis was shown in recipient mice by Nagano et al. [39]. The same group was able to show the restoration of fertility in mice with the transplantation of male germ-line stem cells [40]. Subsequently Lo et al. reported that spermatogonial stem cells can be isolated and enriched with specific markers [41]. In 2003, Shinohara et al. presented that Sertoli cells, the main component of the testicular germ-cell niche, were recovered and dissociated from testes of donor male mice and microinjected into recipient testes, forming mature seminiferous tubule structures, and supporting spermatogenesis [42]. Kanatsu-Shinohara et al. showed the germ-line niche transplantation was also able to restore fertility in mice [43].

According to Fijak and Meinhardt, further research into the immune system of the testis may open new avenues for future successful testicular transplantation. It is apparent, however, that despite more than 100 years of effort, evidence of successful entire testis transplantation between humans remains scarce [44].

More recently, Raya et al. reported that engineered testicular prosthesis using chondrocytes seeded onto testicular-shaped polymer scaffolds can be implanted in vivo, and can release testosterone for a prolonged period. Furthermore, the levels of testosterone release can be maintained within the physiologic range. Periodic reinjection may potentially provide permanent physiologic hormonal replacement. This novel technology may be beneficial for patients who require testicular prostheses and chronic hormone supplementation [45]. This technique can be used in prepubertal boys who undergo chemotherapy to save fertility after puberty.

Testosterone and sperm production are the main functions of the testis, and can be disrupted by congenital or acquired problems. However, sperm production is still beyond our ability for now.

11.2.4 Testosterone Delivery System

Patients with testicular dysfunction require androgen replacement for somatic development. Conventional treatment for testicular dysfunction consists of periodic intramuscular injections of chemically modified testosterone or, more recently, skin patch applications. However, long-term nonpulsatile testosterone therapy is not optimal and can cause multiple problems, including fluid and nitrogen retention, excessive erythropoiesis, hypertension, and bone density changes. In addition, fluctuating serum testosterone levels may occur, and frequent treatments may be required. Due to these problems, alternate treatment modalities, involving more physiological and longer-acting systems for androgen delivery, have been pursued. To address the problem, a system was designed in which Leydig cells, which produce most of the testosterone in the male, were microencapsulated in an alginate-poly-L-lysine solution and injected into castrated animals. Serum testosterone was measured serially; the animals were able to maintain testosterone levels in the long term [46].

Other studies have shown that testicular prostheses created with chondrocytes in bioreactors could be loaded with testosterone, and that these prostheses could provide controlled testosterone release into the bloodstream when implanted. The prostheses were implanted in athymic mice with bilateral anorchia, and testosterone was released long term, maintaining the androgen level at a physiologic range [45]. One could envision combining the Leydig cell technology described above with engineered prostheses for the long-term functional replacement of androgen levels.

References

1. Bosch RJ, Bénard F, Aboseif SR, Stief CG, Lue TF, Tanagho EA. Penile detumescence. Urol. 1991: 867–71.
2. Hu W, Lu J, Zhang L, Wu W, Nie H, Zhu Y, et al. A preliminary report of penile transplantation: part 2. Eur Urol. 2006;50:1115–6.
3. Yoo JJ, Lee I, Atala A. Cartilage rods as a potential material for penile reconstruction. J Urol. 1998;160:1164–8.
4. Kim B-S, Yoo JJ, Atala A. Engineering of human cartilage rods: potential application for penile prostheses. J Urol. 2002;168:1794–7.
5. Yoo JJ, Park HJ, Lee I, Atala A. Autologous engineered cartilage rods for penile reconstruction. J Urol. 1999;162:1119–21.
6. Park HJ, Yoo JJ, Kershen RT, Moreland R, Atala A. Reconstitution of human corporal smooth muscle and endothelial cells in vivo. J Urol. 1999;162:1106–9.

7. Kershen RT, Yoo JJ, Moreland RB, Krane RJ, Atala A. Reconstitution of human corpus cavernosum smooth muscle in vitro and in vivo. Tissue Eng. 2002;8:515–24.
8. Kwon TG, Yoo JJ, Atala A. Autologous penile corpora cavernosa replacement using tissue engineering techniques. J Urol. 2002;168:1754–8.
9. Yoo JJ, Chen K-L, Atala A. Total penile corpora cavernosa replacement using tissue engineering techniques. The FASEB Journal. 2006;20(5):A885–A.
10. Chen K-L, Eberli D, Yoo JJ, Atala A. Bioengineered corporal tissue for structural and functional restoration of the penis. Proc Natl Acad Sci U S A. 2010;107:3346–50.
11. Song L-J, Xu Y-M, Li C, Fu Q, Cui L, Hu X-Y. Construction of cavernosum smooth muscle using umbilical artery smooth muscle cells seeded on acellular corporal collagen matrices. Int J Androl. 2009;32:514–23.
12. Ji C, Min F, Liang W, Chen Y, Pan S, Bin L, et al. Construction of tissue-engineered corpus cavernosum with muscle-derived stem cells and transplantation in vivo. BJU Int. 2011;107:1638–46.
13. An G, Ji C, Wei Z, Chen H, Zhang J. Engineering of corpus Cavernosum using vascular endothelial growth factor-expressing muscle-derived stem cells seeded on acellular corporal collagen matrices. Urology. 2013;81:424–31.
14. Wessells H, Williams SK. Endothelial cell transplantation into the corpus cavernosum: moving towards cell-based gene therapy. J Urol. 1999;162:2162–4.
15. Chancellor MB, Tirney S, Mattes CE, Tzeng E, Birder LA, Kanai AJ, et al. Nitric oxide synthase gene transfer for erectile dysfunction in a rat model. BJU Int. 2003;91:691–6.
16. Deng WW, Bivalacqua TJ, Chattergoon NN, Hyman AL, Jeter JR, Kadowitz PJ. Adenoviral gene transfer of eNOS: high-level expression in ex vivo expanded marrow stromal cells. Am J Phys Cell Phys. 2003;285:C1322–C9.
17. Tirney S, Mattes CE, Yoshimura N, Yokayama T, Ozawa H, Tzeng E, et al. Nitric oxide synthase gene therapy for erectile dysfunction: comparison of plasmid, adenovirus, and adenovirus-transduced myoblast vectors. Mol Urol. 2001;5:37–43.
18. Connolly SS, Yoo JJ, Abouheba M, Soker S, McDougal WS, Atala A. Cavernous nerve regeneration using acellular nerve grafts. World J Urol. 2008;26:333–9.
19. Kim ED, Scardino PT, Hampel O, MILLS NL, Wheeler TM, Nath RK. Interposition of sural nerve restores function of cavernous nerves resected during radical prostatectomy. J Urol. 1999;161:188–92.
20. Kim ED, Nath R, Slawin KM, Kadmon D, Miles BJ, Scardino PT. Bilateral nerve grafting during radical retropubic prostatectomy: extended follow-up. Urology. 2001;58:983–7.
21. Singh H, Karakiewicz P, Shariat SF, Canto EI, Nath RK, Kattan MW, et al. Impact of unilateral interposition sural nerve grafting on recovery of urinary function after radical prostatectomy. Urology. 2004;63:1122–7.
22. Sim HG, Kliot M, Lange PH, Ellis WJ, Takayama TK, Yang CC. Two-year outcome of unilateral sural nerve interposition graft after radical prostatectomy. Urology. 2006;68:1290–4.
23. Namiki S, Saito S, Nakagawa H, Sanada T, Yamada A, Arai Y. Impact of unilateral sural nerve graft on recovery of potency and continence following radical prostatectomy: 3-year longitudinal study. J Urol. 2007;178:212–6.
24. Djinovic R. Penile corporoplasty in Peyronie's disease: which technique, which graft? Curr Opin Urol. 2011;21:470–7.
25. Imbeault A, Bernard G, Ouellet G, Bouhout S, Carrier S, Bolduc S. Surgical option for the correction of Peyronie's disease: an autologous tissue-engineered endothelialized graft. J Sex Med. 2011;8:3227–35.
26. Schultheiss D, Lorenz RR, Meister R, Westphal M, Gabouev AI, Mertsching H, et al. Functional tissue engineering of autologous tunica albuginea: a possible graft for Peyronie's disease surgery. Eur Urol. 2004;45:781–6.
27. Joo K-J, Kim B-S, Han J-H, Kim C-J, Kwon C-H, Park H-J. Porcine vesical acellular matrix graft of tunica albuginea for penile reconstruction. Asian J Androl. 2006;8:543–8.
28. Kumar P, Kumar N, Thakur DS, Patidar A. Male hypogonadism: symptoms and treatment. J Adv Pharm Technol Res. 2010;1:297–301.
29. Incrocci L, Bosch JL, Slob AK. Testicular prostheses: body image and sexual functioning. BJU Int. 1999;84:1043–5.
30. Schultheiss D, Denil J, Jonas U. Rejuvenation in the early 20th century. Andrologia. 1997;29:351–5.
31. Deschamps JY, Roux FA, Saï P, Gouin E. History of xenotransplantation. Xenotransplantation. 2005;12:91–109.
32. Silber SJ. Transplantation of a human testis for anorchia. Fertil Steril. 1978;30:181–7.
33. Kearns WM. Testicular transplantation: successful autoplastic graft following accidental castration. Ann Surg. 1941;114:886.
34. FOX M, BOYLE PF, HAMMONDS JC. Transplantation of interstitial cells of the testis. Br J Urol. 1973;45:696–701.
35. BOYLE PF, Fox M, Slater D. Transplantation of interstitial cells of the testis: effect of implant site, graft mass and ischaemia. Br J Urol. 1975;47:891–8.
36. Tai I, Sun A. Microencapsulation of recombinant cells: a new delivery system for gene therapy. FASEB J. 1993;7:1061–9.
37. Sun J, Xi Y-B, Zhang Z-D, Shen P, Li H-Y, Yin M-Z, et al. Leydig cell transplantation restores androgen production in surgically castrated prepubertal rats. Asian J Androl. 2009;11:405.
38. Blackhall FH, Atkinson A, Maaya M, Ryder WDJ, Horne G, Brison DR, et al. Semen cryopreservation, utilisation and reproductive outcome in men treated for Hodgkin's disease. Br J Cancer. 2002;87:381–4.
39. Nagano M, Brinster RL. Spermatogonial transplantation and reconstitution of donor cell spermatogenesis in recipient mice. APMIS. 1998;106:47–57.

40. Ogawa T, Dobrinski I, Avarbock MR, Brinster RL. Transplantation of male germ line stem cells restores fertility in infertile mice. Nat Med. 2000;6:29–34.
41. Lo KC, Brugh VM, Parker M, Lamb DJ. Isolation and enrichment of murine spermatogonial stem cells using rhodamine 123 mitochondrial dye. Biol Reprod. 2005;72:767–71.
42. Shinohara T, Orwig KE, Avarbock MR, Brinster RL. Restoration of spermatogenesis in infertile mice by Sertoli cell transplantation. Biol Reprod. 2003;68:1064–71.
43. Kanatsu-Shinohara M, Miki H, Inoue K, Ogonuki N, Toyokuni S, Ogura A, et al. Long-term culture of mouse male germline stem cells under serum-or feeder-free conditions. Biol Reprod. 2005;72:985–91.
44. Fijak M, Meinhardt A. The testis in immune privilege. Immunol Rev. 2006;213:66–81.
45. Raya-Rivera AM, Baez C, Atala A, Yoo JJ. Tissue engineered testicular prostheses with prolonged testosterone release. World J Urol. 2008;26:351–8.
46. Machluf M, Atala A. Emerging concepts for tissue and organ transplantation. Graft. 1998;1:31–7.